# The Reviewer's Guide to Quantitative Methods in the Social Sciences

*The Reviewer's Guide to Quantitative Methods in the Social Sciences* is designed for evaluators of research manuscripts and proposals in the social and behavioral sciences, and beyond. Its 31 uniquely structured chapters cover both traditional and emerging methods of quantitative data analysis, which neither junior nor veteran reviewers can be expected to know in detail. The book updates readers on each technique's key principles, appropriate usage, underlying assumptions, and limitations. It thereby assists reviewers to offer constructive commentary on works they evaluate, and also serves as an indispensable author's reference for preparing sound research manuscripts and proposals. Key features include:

**Comprehensive Coverage**—Thirty-one chapters cover virtually all of the classic and emerging quantitative techniques, thus helping reviewers to evaluate a manuscript's methodological approach to data analysis. In addition, the volume serves as an indispensable reference tool for those designing their own research.

**Unique Chapter Format**—For ease of use, all chapters follow the same structure.

- The opening page of each chapter defines and explains the purpose of that statistical method.
- The next one or two pages provide a table listing various criteria that should be considered when evaluating and applying that methodological approach to data analysis.
- The remainder of each chapter contains numbered sections corresponding to the criteria listed in the opening table. Each section explains the role and importance of that particular criterion.

**Expert Chapter Authors**—Chapters are written by methodological and applied scholars who are expert in the particular quantitative method being reviewed.

**Gregory R. Hancock** is Professor and Chair of the Department of Measurement, Statistics, and Evaluation at the University of Maryland and is Director of their Center for Integrated Latent Variable Research.

**Ralph O. Mueller** is Professor and Dean of the College of Education, Nursing, and Health Professions at the University of Hartford, Connecticut.

# The Reviewer's Guide to Quantitative Methods in the Social Sciences

**Editors**
Gregory R. Hancock
Ralph O. Mueller

Routledge
Taylor & Francis Group

NEW YORK AND LONDON

First published 2010
by Routledge
270 Madison Avenue, New York, NY 10016

Simultaneously published in the UK
by Routledge
2 Park Square, Milton Park, Abingdon, Oxon OX14 4RN

*Routledge is an imprint of the Taylor & Francis Group, an informa business*

© 2010 Taylor and Francis

Typeset in Minion by Swales & Willis Ltd, Exeter, Devon
Printed and bound in the United States of America
on acid-free paper by Edwards Brothers, Inc.

*Library of Congress Cataloging in Publication Data*
The reviewer's guide to quantitative methods in the social sciences /
Gregory R. Hancock, Ralph O. Mueller.
p. cm.
Includes bibliographical references and index.
1. Social sciences—Research—Methodology. 2. Social sciences—Statistical
methods. I. Hancock, Gregory R. II. Mueller, Ralph O.
H62.R466 2010
300.72—dc22
2009030780

ISBN 10: 0–415–96507–1 (hbk)
ISBN 10: 0–415–96508–X (pbk)
ISBN 10: 0–203–86155–8 (ebk)

ISBN 13: 978–0–415–96507–1 (hbk)
ISBN 13: 978–0–415–96508–8 (pbk)
ISBN 13: 978–0–203–86155–4 (ebk)

Dedicated to the unheralded reviewers whose words frustrate and challenge us—and make our work stronger for it.

# Contents

# Figures

# Tables

# Preface

## Volume Background

A cornerstone of research institutions and agencies around the world is the creation of new knowledge that often is generated through utilizing quantitative research methods. The dissemination of such knowledge through specialized journals, application-oriented outlets, and technical reports is typically filtered through a rigorous peer review process to ensure work of the highest possible quality. Reviewers, and the editors they serve, thus operate in the critical role of gatekeeper and must, collectively, be held accountable for the integrity of the research enterprise.

The quantitative skills that reviewers bring to this role tend to fall into two categories—expertise in methods they use fairly regularly and competently in their own research, and knowledge from academic or professional training that has largely laid dormant since that initial exposure. This limiting state of affairs, which is exacerbated as cohorts of new researchers are trained in increasingly advanced data analysis techniques, can force quantitatively uninitiated reviewers to confine their critical commentary primarily to the content-area portions of a manuscript. In the end, these reviewers are operating by assuming that someone else is tending the methodological gate, and editors are left in the difficult position of burdening those few reviewers who are quantitatively current while constantly having to update their stable of adjunct reviewers with highly specific areas of methodological expertise.

In all fairness, reviewers, whether novice or veteran, should not be expected to have a command of all data analysis methods used in modern research. We believe that they should, however, maintain some broad and evolving awareness of methodological developments, and make an honest attempt to evaluate the analytical methods that are at the core of a manuscript's intellectual contribution. The current volume, *The Reviewer's Guide to Quantitative Methods in the Social Sciences*, was born out of a desire to assist reviewers in meeting this professional responsibility. In particular, this volume is designed as a reference tool specifically for reviewers of manuscripts and proposals in the social sciences and beyond, addressing a broad range of traditional and emerging quantitative techniques and providing easy access to critical information that authors should have considered and addressed in their submissions. How this volume's structure is specifically geared toward reviewers is explained in the next section.

## Volume and Chapter Structure

This volume has 31 chapters arranged alphabetically by title, with each chapter addressing a particular quantitative method or area. These include the sound practice of research (e.g., research design, survey sampling, power analysis), the general linear model (e.g., analysis of variance, multiple regression, hierarchical linear modeling, discriminant analysis), the generalized linear model (e.g., logistic regression, log-linear analysis), measurement (e.g., item response theory, generalizability theory, multidimensional scaling), and latent structure methods (e.g., latent class analysis, latent growth curve models, structural equation modeling). The structure within each chapter is the same, consisting of the following three sections:

- Method overview
- Table of desiderata
- Explications of desiderata.

*Method overview.* Each chapter starts with a brief introduction and overview of the method(s) being addressed in the chapter. This is not meant to teach the reader the topic area per se, but rather to provide an orientation and possible refresher. This section concludes with useful references for the reader who wants more introductory and/or advanced information on that chapter's topic.

*Table of desiderata.* On the second page of each chapter begins a numbered list of key elements (desiderata) that should be addressed in any study using that chapter's method(s). This table serves to provide essential evaluation criteria that a reviewer should consider when judging a manuscript's methodological approach to data analysis, including the technique's key principles, appropriate usage, underlying assumptions, and limitations. For each desideratum the section(s) of a manuscript in which the specific issue should most likely be addressed is denoted with (I) Introduction, (M) Methods, (R) Results, and (D) Discussion.

Typically the desiderata appear in a single table; in some cases this table is partitioned by special applications of a certain method (e.g., Chapter 20, Multidimensional Scaling). In a couple of cases the desiderata are presented in two separate tables due to the bifurcated nature of the topic (e.g., Chapter 8, Factor Analysis, with tables for exploratory and confirmatory methods). The user of this volume will also note that there are many desiderata in common across chapters, including such elements as making connections between research questions and the analytic method at hand, explicitly addressing how missing data and outliers were handled, reporting the software and version used, and so forth. Although these could have been culled for a separate table at the beginning of the volume, we believe that having them contained within each chapter as they pertain to the specific method is in keeping with the reference nature of this guide. In this manner, a chapter's table(s) of desiderata may be used by a reviewer as a checklist to evaluate a manuscript's methodological soundness. Explications of each desideratum then follow.

*Explications of desiderata.* Following each chapter's table(s) of desiderata are corresponding explications for each numbered desideratum. For a reader already thoroughly familiar with a particular desideratum, a given explication may be unnecessary; we expect, however, that most readers will benefit from the supporting explanation, elaboration, and any additional references specific to that desideratum. For example, if a desideratum calls for a manuscript to explicitly examine the assumptions underlying a particular analytical technique, the explication might provide a treatment of the inferential consequences of failing to examine those assumptions as well as the preferred methods for conducting and presenting the results of such an examination. The explications of the desiderata constitute the main body of each chapter, justifying each desideratum in light of accepted practice and the supporting methodological literature.

## The Future of This Volume

Any time someone offers recommendations as to proper practice or conduct, reasonable people will disagree. The current volume, methods, and desiderata are no different. In fact, we are quite certain that some knowledgeable readers will take issue with particular chapter authors' portrayal of a given method's best practice. That said, in the preparation of this volume we have encouraged chapter authors to try to convey what each technique's methodological literature considers to be the most accepted practices associated with a given method. Further, we understand that even if there is currently agreement on a technique's best practices, the methodological literature might determine otherwise in the future.

We therefore view this volume as a living resource. As quantitative methodologies evolve, not just in preferred practice of existing techniques but also in the development of new techniques, we expect each subsequent edition of this guide to adapt as well. Toward this end we welcome readers' correspondence with chapter authors and editors in an on-going dialog about specific methods and their desiderata, thus helping to keep this volume continually responsive to changes in appropriate practice. Our hope is that this will create a dynamic resource for reviewers of manuscripts and proposals, as well as for the authors themselves (e.g., applied researchers, university faculty, and graduate students) when designing their own research projects.

Gregory R. Hancock and Ralph O. Mueller
June 2009

# Acknowledgements

We acknowledge the many individuals at the University of Maryland and The George Washington University who supported us throughout this project. We offer our deepest thanks to Lane Akers at Taylor and Francis, whose encouragement and friendship saw us through this effort from initial idea until final realization. And above all, we convey our immeasurable gratitude to those who have seen us through all of our endeavors, successful and otherwise, with unwavering love and support—our parents Bob and Marta, and Axel and Nora.

# 1

# Analysis of Variance
## *Between-Groups Designs*

### Alan J. Klockars

Between-groups analysis of variance (ANOVA) is one of the most commonly used techniques to analyze data when the intent of the research is to determine whether an independent variable (IV), typically in the form of qualitatively different treatment groups, causes variation in an outcome measure, the dependent variable (DV). Causal statements rest on the random assignment of subjects to the various treatment groups defining the levels of the IV. The broad area of ANOVA consists of significance tests to determine if a non-random relationship exists between IVs and DVs, multiple comparison procedures to investigate more thoroughly the nature of such a relationship, and measures of the strength of the relationship.

The following is an overview of some of the types of designs and experiments that can be analyzed by *between-groups ANOVA*. The number of IVs will determine the type of experiment. For example, a single IV is called a *one-way experiment*. The levels of the IV in the one-way ANOVA typically have fixed treatment groups, reflecting specific differences in treatments of interest to the experimenter. There are also ANOVA experiments that have a number of different IVs. If, for example, there were three IVs that were completely crossed with one another, it would be called *three-way factorial* experiments. The term *design* is related to the nature of the observations collected in the experiment. The two most common designs covered here are *randomized group* and *randomized block* designs. Repeated measures designs (see Chapter 2, this volume) are also common.

Crossing additional fixed IVs with the first IV in a *factorial experiment* typically enriches the major theory being investigated, providing for additional main effects as well as interactions between the factors. A *random factor* is one that has the levels of the treatment groups randomly chosen from some universe of possible levels, rather than specific treatments of interest. It may be crossed with the fixed IVs yielding a *mixed model*, or it may be nested within the fixed IV in a *hierarchical experiment*. These experiments relate to the generalizability of the fixed effect over the universe of levels of the random factor. Individual difference variables may also be incorporated into ANOVA as another factor of the experiment, as in a *randomized block* design, or as a continuous control variable with an *analysis of covariance* (ANCOVA). These IVs can be used to reduce random variability and explore the interaction between the individual differences and the fixed treatment effects. There is a rich literature in ANOVA including texts written for experimental design, such as Kirk (1995), Maxwell and Delaney (2004), and Keppel and Wickens (2004).

Table 1.1 Desiderata for Analysis of Variance: Between-Groups Designs

| Desideratum | Manuscript Section(s)* |
|---|---|
| 1. Each independent variable is defined so that discrete groups (treatment or otherwise) exist that are related to the research questions, hypotheses, theory, and the literature reviewed. | I |
| 2. A rationale is given for the simultaneous inclusion of two or more independent variables within the research design. | I |
| 3. The presence of any covariate is justified relative to the research question(s) being addressed. | I |
| 4. The validity of the outcome measure is justified relative to the research question(s) being addressed. | I |
| 5. A rationale is given for the number of participants, the source of the participants, and any selection/exclusion criteria used. | M |
| 6. The research design is explained in detail, including the nature of the IVs (fixed or random), the significance level used, and the method of assignment of participants to groups. | M |
| 7. Assumptions underlying the tests of significance, as well as psychometric properties of the outcome measure and any covariate, are discussed. | M, R |
| 8. Multiple comparison procedures to be used are indicated and related to the research question(s) being addressed. | M, R |
| 9. Any error term used reflects the unique characteristics of the design. | M, R |
| 10. In randomized block designs the number of blocks, treatment of the blocks as fixed or random, and method of block creation is discussed. | M, R |
| 11. In mixed models the use of an IV as a random factor is explained and all analyses are appropriate considering the type of IVs involved. | M, R |
| 12. In hierarchical designs the rationale for nesting is explained, the analysis clearly acknowledges the dependence in the data, and the data warrant using ANOVA rather than multi-level modeling. | M, R |
| 13. In incomplete designs, or complex variants of other designs, sufficient references are provided to understand the design. | M, R |
| 14. Adequate statistical information (e.g., means, effect sizes, confidence intervals) is provided to facilitate interpretation of the results. | R |
| 15. Sufficient information is presented about the results of the test(s) of significance, including error terms and degrees of freedom. | R |
| 16. Statistically significant interactions are analyzed and interpreted to clearly indicate the nature and strength of the interaction. | R |
| 17. Appropriate language relative to the meaning and generalizability of the finding is used. | D |

* *Note*: I = Introduction, M = Methods, R = Results, D = Discussion

## 1. Independent Variables

A between-groups analysis of variance (ANOVA) requires that unique, mutually exclusive groups be created. Typically, the groups reflect (1) differences on fixed treatment variables, (2) levels of an individual difference variable, or (3) levels of a random variable. Fixed treatment groups are created that differ in some aspect of the way the participants are treated. The differences between treatments capture specific differences of interest and thus are treated as a fixed factor with either qualitatively different treatments or treatments having different levels of intensity of some ordered variable. In a randomized block design, besides treatment differences, the groups differ with regard to levels of an individual difference variable (see Desideratum 10). When the intent of the experiment is to generalize the treatment variable over levels of a random factor, the groups differ with regard to the levels of that random factor. For example, in an experiment to test for the effectiveness of differences types of reading prompts (a fixed treatment factor) the experimenter may group subjects into high, moderate, and low reading ability (an individual difference variable) and/or may vary a random factor, such as the specific book studied or the teacher presenting the material, to address the generalizability of the prompt effect over the levels of this additional, random factor (see Desideratum 11).

The treatment groups are the operational definitions of the theoretical independent variable of interest. In the report of the research the relations described are generally in terms of the theoretical construct the groups were created to capture (e.g., stress) not in terms of the operations involved (e.g., group was told the assignment counted for 50% of their grade). The way the treatments are defined must clearly capture the essence of the theoretical variable. The authors must defend the unique characteristics of treatment groups as clearly indicating the crucial differences desired to address the theoretical variable of interest. The treatments must have face validity in that the reader sees the obvious linkage between the theoretical variable of interest and the operations used to create that variable.

Where the treatments are levels selected from along a continuum of potential levels (e.g., dosages of a medication, hours of instruction), the authors must defend the choices made with particular attention to the extreme levels. The number of levels chosen should be defended, based, in part, on the expected functional relation between level and outcome. In some instances where the relation is of greater interest than group mean differences, the authors should be encouraged to explore a regression-based analysis (see Chapter 21, this volume).

There are a number of common shortcomings in the construction of treatment groups that can result in low power or confounded interpretations. For example, differences in the wording of reading prompts may have a subtle effect too small to be detected by an experiment unless an extremely large sample size is used. Low power may be the result of treatments that were implemented for too short a period or with insufficient intensity to be detected. The power of tests of significance to detect differences will also be reduced if an experiment includes a large number of levels of the treatment variable. This is particularly true when many of the levels have similar and moderate treatment effects. Confounded effects can happen when groups differ in multiple ways, any one of which may produce observed differences. This may happen inadvertently such as if one type of prompt required the study period to be longer than any of the other types of prompts. Any difference found might be either the differences in prompts or the differences in study time. In other experiments the intent is to compare treatment groups that differ on complex combinations of many differences resulting in uncertainty regarding the "active ingredient" that actually produced any group differences found.

## 2. Inclusion of Two or More Independent Variables

Many settings use factorial experiments where several different IVs are crossed to facilitate the exploration of both main effects and interactions. Typically one IV is central to an investigator's research program with the other IVs included for control and/or to expand upon on the theory being explored. Authors should defend the inclusion of all IVs relative to their relation to the central theory they are testing. IVs should not be included if they are irrelevant to that theory. The explanation of an IV's role should include both the main and interaction effects anticipated.

Fixed IVs generally enrich theory. The inclusion of an IV that is a random factor is meant to show the generalizability of the central findings across levels of the random factor. The choice of the random factor should be justified relative to the need for greater generalizability concerning the main effect (see Desideratum 10).

Factorial experiments provide more information than a single factor experiment but the authors should recognize the costs involved in terms of complexity of interpretation. In factorial experiments the authors must continually interpret findings that are summed over many different conditions. The more complex the experiment, the more difficult it is to understand what, if anything, really differs. Interactions alter the interpretation of the main effects, but when an interaction is not statistically significant authors often over-generalize main effects ignoring the possibility that the statistically non-significant interaction may be due to lack of power (i.e., a Type II error).

Factorial experiments with unequal numbers of observations in treatment groups require special attention as the main effects and interactions are not orthogonal. The strategy to deal with the nonorthogonality should be indicated. The most commonly used is to report results based on *Type III Sums of Squares*.

## 3. Covariates

Authors should justify the inclusion of any covariate(s) relative to the theory being tested. Covariates can increase the power of the statistical tests within the ANOVA but can also increase the Type I error rate if their selection allows for the capitalization on chance. Multiple covariates and those that have been selected a posteriori capitalize on random variability and should not be used. Covariate scores must not be a function of the treatment group and should be available before random assignment takes place. Experiments in which the covariate is obtained at the same time as the outcome have the likelihood that the covariate scores will be altered due to treatment. A primary function of the covariate is to provide a way to statistically equate groups to which the treatments are applied; this function would be invalidated if the treatment changed the covariate score used.

The use of ANCOVA adds several assumptions to the analysis that are not germane to an ANOVA. Some, such as the covariate being a fixed factor and measured without error, are likely to be violated but that is true of almost every use of ANCOVA. Thus, there is little value in requiring every research paper using a covariate to include a discussion of these. However, authors should have considered and discuss any likely violation of the assumptions of linear relation between the covariate and the outcome measure, homogeneous residual variances within treatments, and homogeneity of regression coefficients within treatments.

A violation of the assumption of homogeneous regression coefficients is itself a major finding analogous to an interaction between the covariate and the treatments. In most computer software the testing of homogeneity of regression coefficients requires more steps than the test of the main effects for treatment and is often not reported. This is unfortunate as the finding of differences in slopes should be considered an important finding and related to the literature of *Aptitude-Treatment Interactions*. In a similar manner the finding that there are differences in the variability of outcome scores about the regression lines for different treatments may be a finding worthy of reporting and discussion.

## 4. Validity of the Outcome Measure

As the treatment groups are the embodiment of the theoretical independent variable, the outcome measure is the embodiment of the theoretical dependent variable. The authors must defend the measure used through, for example, establishing sufficient construct validity within the population being studied. The reader should be able to see the obvious fit between the measure and the construct. The authors should cite relevant literature to support the use of the measure.

For ANOVA the outcome measure must provide values on which arithmetic computations can be meaningfully performed (see Desideratum 7). All analyses and decisions made in tests of significance are about the characteristics of the outcome measure, not directly on the theoretical dependent variable. The outcome variable is the set of numbers that we have collected. If the means for groups differ statistically significantly on the outcome variable it strongly suggests that treatment had an effect. The researcher, however, does not typically want to discuss differences on the unique measure used but rather about differences on the theoretical construct the test is suppose to tap. Only to the extent that the outcome measure provides a valid measurement of the theoretical construct will the conclusions drawn be relevant to the underling theory. The theoretical variable must be sufficiently characterized so that the appropriate link to the observed variable can be established.

The overall mean and variance of the outcome measure must be such that potential differences among groups can be detected. Outcomes where subjects have scores clustered near the minimum or maximum possible on the instrument provide little opportunity to observe difference. This will reduce power and obscure differences that might really be present. Authors often employ measures used in previously published literature. Of concern is the appropriateness of the measure in the new setting, such as with different age or ability levels. In factorial experiments, particularly with randomized blocks designs where one of the factors is an individual difference measure on which subjects have been stratified, the inappropriateness of the measure at some levels of the blocking variable may incorrectly appear as an interaction of the blocks with the treatments. For example, consider a learning experiment where the impact of the treatment was measured by an achievement test and the subjects were blocked on ability. If it were true that one treatment was generally superior to the others across all levels of ability the appropriate finding would be a main effect for treatment. However, if the outcome measure were such a difficult test that subjects in all but the highest ability group obtained essentially chance scores (which would not show treatment effects) while those in the highest ability group showed the true treatment effect, the effect would be declared an aptitude-treatment interaction.

## 5. Study Participants

The authors must indicate the source of their participants in such a way that readers will be able to evaluate the generalizability of their findings. The authors should have a case study as their model for their description of the participants and setting. To determine if the findings speak to a unique reader's situation the reader has to know as much about the participants and the setting as possible. A statistically significant difference among the mean scores of different treatment groups indicates that, within some specific population, the additive effects of the different treatments were unlikely to have all been equal. The concern of the readers is whether those differences apply in a setting of concern to them.

Among the conditions that should be described are the ages of the participants, how they were recruited, whether inducements were included to obtain participation, and any selection criteria that were used to exclude some participants. Sample specific demographic information should be included to better understand the implications of the study. Information about the number of individuals declining to participate and/or dropping out during the study should be provided along with a description of any quality check that may have been conducted to detect data deemed unresponsive.

The rationale for the number of participants used should be explained through a priori sample size determination (see also Chapter 24, this volume). Sample size should reflect the importance of the question, the expected magnitude of difference, and the cost and availability of subjects. Because of the direct relation between power and sample size, the number of participants should be considered in discussing the findings. Where differences that were expected to be statistically significant are non-significant a post hoc power analysis should be conducted to indicate the sample size that would have been required given the observed results. On the other hand, authors using extremely large sample sizes should explain their reasoning as well.

## 6. Research Design

The Methods section should clearly describe the characteristics of the experiment so that reader can easily evaluate the appropriateness of the Results and Discussion sections relative to the actual experiment conducted. The Methods section should include a summary of the research design that allows the reader to review all of the critical elements of the experiment. The information should be sufficiently complete so that the reader can anticipate the sources of variation in the dependent variable(s) and the degrees of freedom for each source.

The characteristics summarized should include the following. The authors should indicate that the research was a true experimental design with random assignment of subjects. It should be clearly indicated in the summary if one group of participants were measured at multiple times (see Chapter 2, this volume) or if aspects of the research involve a quasi-experimental or non-experimental design where intact groups were used (see Chapter 26, this volume).

The number of factors and the number of levels of each factor should be stated as well as the resulting total number of treatment groups. The summary should provide the abbreviated names of the factors that will be used in other sections of the paper. If appropriate, the authors should state that the factors are completely crossed although the readers will assume a completely crossed experiment unless specifically informed that a factor was either incompletely crossed (as in a lattice design) or was nested. In some multifactor experiments there are treatment groups that do not completely fit within the factorial structure. If a reading study varied the types of prompts given to students, a second factor might be whether the prompts were given immediately prior to reading the passage or immediately after. A control group having no prompts could not incorporate this placement factor since the factor presumes the presence of some sort of prompt. The presence of this type of treatment group should be carefully acknowledged.

Each factor should be identified as either a fixed or random factor within the experiment. If the experiment includes random factors the author should clearly indicate the appropriate error term for the treatment effects and interactions that are impacted by the presence of the random factor. The author should also provide any technical term that might help summarize the experimental design (e.g., Latin square).

The number of participants should be indicated, clearly specifying if the cited number refers to those in a unique cell or in some higher level of aggregation. The authors should additionally indicate the number of participants on which each mean is based in complex experiments where the number of observations may be unclear. In experiments involving repeated measures on some factors (see Chapter 2, this volume) the need for clarity regarding number of participants and number of observations is particularly important.

The significance level to be used should be clearly indicated. In a factorial experiment where several main effects and interactions will be tested the authors should indicate if each main effect and interaction will be considered a family, with Type I error rate controlled separately at alpha ($\alpha$) for each family, or if a more conservative level of control will be used. It should also be stated if the authors are not reporting omnibus tests for the various factors but rather proceeding directly to multiple comparison procedures (see Desideratum 8) that provide the required control over Type I error. While this approach is perfectly legitimate for controlling Type I error, it is sufficiently unusual at present so that readers should be informed. The authors should defend any tests, including multiple comparison procedures, that are to be applied as one-tailed tests. For any one-tailed applications a convincing argument must be provided for the triviality of any potential findings that are counter to the expected direction that justifies ignoring their potential existence.

## 7. Assumptions and Psychometric Properties of Outcome Measure and Covariate(s)

Parametric statistical procedures such as ANOVA have a number of assumptions on which their mathematical bases are built. In many cases assumptions are difficult if not impossible to completely satisfy. However, ANOVA is also relatively robust with regard to violations of assumptions, particularly with equal sample sizes across groups. Authors should scrutinize their studies relative to the assumptions and discuss any that they believe will be seriously violated along with whatever justification they have for continuing with ANOVA. In ANOVA the following assumptions are made: outcome variables are measured on at least an interval scale and are normally distributed, individual errors are random and independent, and variances within treatment populations are homogeneous.

The first assumption of interval scale has attracted considerable debate over the years, with some maintaining that equal intervals are necessary in order to use ANOVA while others argue that the numbers themselves do not necessarily have any properties until one assumes those specific properties. The first position places the burden on the researcher to prove an interval scale before proceeding. The alternative asserts that all analyses are done on the observed scores and those scores do not know what kind of numbers they are. This second position requires the researcher to be cautious in making the leap from talking about differences between the numbers to talking about differences on an associated theoretical dependent variable. I side with the latter, more liberal, position meaning that the analysis can be conducted with a scale of relatively unknown properties but caution must be exercised in making the inferential leap to the dependent variable. Thus, if researchers have an ordered, multi-step scale, they may proceed with ANOVA but should not imply they have an interval scale but rather that they are treating the numbers as if they were an interval scale, realizing that the lack of fit affects the degree to which the research answers the questions that were raised. Additionally, I do not advocate a long statement defending this position as all papers would say essentially the same thing, wasting journal space.

The second assumption, regarding a normal distribution, is routinely violated as the normal curve a continuous, two-parameter mathematical model while scores are discrete points along a scale. Even in a more liberal and approximate form—a symmetric, unimodal distribution—it is still routinely violated. However, simulation studies have demonstrated that violations seldom have any serious impact on the actual Type I error rate. The critical features needed to allow researchers to ignore this assumption are equal sample sizes and similar, though potentially non-normal, distributions. Any author who anticipates that the scores may be seriously non-normal should defend the decision to use ANOVA based on information regarding the similarity of the shapes and the distribution of subjects to the groups.

The independence and randomness of errors is crucial for controlling Type I error at the stated $\alpha$ level. Independence and randomness of errors are typically assumed to have been met when random assignment to groups is used. In quasi-experimental designs with intact groups the errors may be correlated due to factors such as self-selection of group members and groups having a common history. These factors can result in an underestimate of the random variability and an inflated Type I error rate.

Homogeneity of treatment population variances is an assumption that has minimal impact when violated if equal sample sizes are used. When unequal samples are present, the issue has been studied extensively (the so-called *Behrens-Fisher problem*). Several modified tests of significance can be used when heterogeneity of variance might be present, including tests by Brown and Forsythe, James, and Welch. The authors should discuss their approach to handling this problem if there is a likelihood that it may exist. Unfortunately, heterogeneity of variance is seldom viewed as a research finding in and of itself, with a test of significance to determine if the differences are sufficient to reject the null hypothesis of equal population variances. Treatments that systematically compress or expand the distribution of scores can be interesting apart from whether there are also mean-level treatment effects.

## 8. Multiple Comparison Procedures

A multiple comparison procedure (MCP) that controls Type I error rate experimentwise may be conducted without having to conduct an omnibus *F* test. Of the commonly used methods all but Least

Significant Difference (which is not recommended) control error-rate experimentwise. Using an omnibus test as a gateway to conducting multiple comparisons will result in an unnecessarily conservative experimentwise Type I error rate.

The choice of major strategies to conduct MCPs should be based on the stated research hypotheses. The MCPs should have a one-to-one correspondence to the hypotheses. There are four general scenarios that commonly occur: (1) tests of planned, theory-derived contrasts; (2) tests of all possible pairs of means; (3) tests of each experimental group against a control group; and (4) tests that were not planned but rather suggested by the pattern found in the observed means.

The first three scenarios each have several methods that may be used. The authors should clearly specify which method they chose for the analysis. In general, Bonferroni's inequality (often called *Dunn's test*) is a potential method for controlling Type I error in any of the first three cases. The advantage of Bonferroni's approach is that it is conceptually simple and easy to calculate; however, methods with greater power exist for each strategy. For the second scenario, Tukey's Honestly Significant Difference test controls the experimentwise error rate adequately for any number of treatment groups. For the third scenario, Dunnett's procedure is most commonly used. And as for the fourth scenario, involving comparisons/contrasts suggested by the observed data, Scheffé's test is required to adequately control Type I error. This is, however, the only scenario for which Scheffé's method is appropriate.

Note that each of the first three strategies also has sequential testing methods available. Sequential methods redefine the critical value as a result of having rejected previously tested hypotheses. They control Type I error experimentwise but are subject to a slightly greater number of Type I errors per experiment. Thus, when a Type I error occurs within an experiment there is a slightly greater likelihood that a second error might occur due to the change in the critical value used. Sequential methods also are problematic with regard to constructing confidence intervals. Authors especially interested in creating confidence intervals should avoid using a sequential method. A fairly comprehensive and critical review of traditional multiple comparison procedures, as well as more modern alternatives, may be found in Hancock and Klockars (1996).

## 9. Error Terms

Each test of significance within an ANOVA depends on having a denominator (error term) of the $F$ ratio whose expected value includes all of the terms in the expectation of the numerator except the one term whose existence is denied by the null hypothesis. While the correct error term is generally available in an experimental design text, there are some relatively common scenarios in which researchers use inappropriate error terms. The primary concern is related to the presence of random factors within an experiment. If all factors are fixed the variability of subjects within treatment groups ($MS_{s/a}$) is the appropriate denominator in all but the most complex experiments. The degrees of freedom for the $MS_{s/a}$ is large compared to any other source of variation within the experiment. When a random variable is introduced to test the generalizability of the fixed factors the random variability of levels of the random factor becomes the appropriate error term. If the random factor is crossed with the fixed factors the interaction of the factors will be the error. If the random factor is nested within the levels of the fixed factor the variability of the levels of the random factor within each level of the fixed factor will be appropriate. In either case the number of degrees of freedom will be considerably smaller than the degrees of freedom associated with $MS_{s/a}$. Ignoring the proper error term in experiments with random factors in favor of $MS_{s/a}$ to capitalize on its greater number of degrees of freedom can seriously inflate the Type I error rate reported.

Reports of ANOVAs often make it difficult to determine if the proper error term has been used because the denominator is labeled as the generic "Error MS" or "$MS_{error}$" without any indication of what it measures. This problem is partly caused by some statistical software packages such as SPSS using the "Error MS" label in the output. While the degrees of freedom will often allow for the proper identification, the "Error MS" should be defined in specific terms. Also, the degrees of freedom for the Error MS should always be reported. With the exception of sources of variation within incomplete designs, all Error MS terms are either measures of variability between elements of some random factor (e.g., subjects, or randomly selected book to read) or the interaction of the elements of that random factor with a fixed factor. These sources are then often pooled over random groups. The definition of an Error MS should include both the source or interacting sources and the levels over which it was pooled. The Error MS that is appropriate for the $F$ test of the main effect or interaction should also be the basis of the standard error of the contrasts used in multiple comparisons.

## 10. Randomized Block Designs

The term *randomized block* will be used to describe the case where an individual difference variable is used to create levels (blocks) of a factor to be included as part of a mixed factorial experiment (see Desideratum 11), and where participants are then randomly assigned to treatments from within the blocks. An alternative usage of the term is found when the blocks have only enough subjects to have a single subject per block/treatment combination. Our usage of the term is a special type of factorial experiment and those desiderata appropriate for factorial experiments are also appropriate here (see, e.g., Desiderata 1, 2, and 6). The rationale for including the individual difference variable should be clearly stated. This rationale should distinguish between a motive of including the individual difference variable to remove variability in scores that would otherwise be considered random, and thus increasing the power of the main effect, and a motive of exploring the relation between the individual difference variable and the treatment variable via a test of the interaction.

The particular measure of the individual difference variable used to create blocks should also be justified as a valid measure in the experimental setting. When the motive for including the individual difference variable is to explore the way that the treatments provide differing effects depending on the level of the blocks, the purpose is to address a relationship that should be theoretically grounded. The measure of the individual difference variable must be a valid measure of the theoretical construct in order to make meaningful interpretations of the interaction. This is less of a concern when reduction of the error variance is the sole motive for including the individual difference measure as there are no theoretically grounded tests that depend on the meaning of the individual difference variable.

The number of blocks used should be justified relative to the motive for including the individual difference variable, the strength of the relation between the individual difference variable and the outcome measure, and the quality of the individual difference measure. The method of creating blocks should be explained with particular attention to whether the blocks were independently defined followed by the recruitment of participants who fit into those blocks, or were blocks created post hoc out of a set of $n$ available participants. If independently defined levels of the blocking measure were used the authors should explain how the points used to define the levels were determined, with particular concern for the blocks on both ends of the continuum.

The blocking variable can be considered either random or fixed, depending on how it was created and the extent to which the results are to be generalized. This distinction is important in determining whether the subjects-within-treatments (for fixed) or the block-by-treatment interaction (for random) should be used as the appropriate error term. If subjects-within-treatments is used as the error term, the generalizations should be carefully limited to the specific levels of the blocking variable created or observed. However, the block-by-treatment interaction as an error term allows for greater generalization across the universe of possible blocks.

## 11. Mixed Models

The term *mixed model* is commonly used to describe a factorial experiment in which at least one of the factors is a random variable and at least one of the others is fixed. The primary purpose of a random factor is to argue for the generalizability of the treatment effect averaged across all levels of the random factor rather than just over the specific levels sampled. The authors must explain the rationale for the specific random variable chosen.

The authors should discuss the universe of levels from which the sample of levels of the random factor was selected. Some random factors such as "book chosen" in a reading experiment may have a wide universe and come close to represent a random selection. For other random factors, such as the clinic within which the experiment is conducted, may be very limited, including only those in a narrow geographical region. The degree to which we feel we can generalize is dependent on the breadth of the universe. The authors should be careful not to interpret the absence of an interaction between the random and fixed factor as proving that the treatment works equally well in all levels of the random factor. All generalizations should be averaged across the universe of levels.

As with the randomized block design discussed in Desideratum 10, the choice of appropriate error term depends on the assumptions made about the type of variables in the experiment. In mixed models one of the factors is assumed to be a random variable and thus the error term for a fixed main effect is the interaction between the fixed and random variable. If there are few levels of the random factor, this error term will have few degrees of freedom and the associated $F$ test will have little power.

In factorial experiments involving more than one random factor there is generally no mean square that is a valid error term for the treatment effects of greatest interest. So-called *quasi-F* tests exist that can provide an approximate $F$ test but, unless crucial to the question at hand, an experiment should involve no more than one random factor.

In some situations the levels of the random variable may be viewed as replications of the fixed portion of the experiment in a different setting. This leads to separate analyses for each level of the random factor and the inability to attach a probability level to the possible presence of an interaction between the treatment and the random factor. When viewed as replications rather than parts of a single, large experiment with many levels of the random factor, the results may be summarized along with other experiments in a meta-analysis (see Chapter 19, this volume)

An ANOVA of a mixed experiment relies on relatively strict assumptions concerning the variability of subjects within levels and works best with balanced experiments where there are an equal number of participants in each cell. In situations where there may be very different variances and systematically different sample sizes, authors should consider pursuing an analysis method with less stringent assumptions (e.g., multi-level modeling).

## 12. Hierarchical Experiments

Hierarchical experiments are similar to mixed models except that the random variable is typically nested within the levels of the fixed factor rather than crossed with them (see Desideratum 11). For an experiment that could be conducted either way, the mixed model is preferable in that the levels of the treatment variable would be repeated within each of the levels of the random variable. However, it is often the case that a level of the random variable can only be administered within one of the treatments, or that the levels of the random variable are unique to a specific treatment. Consider a study to determine if listening to jazz, classical, and rock music differentially impacts studying. The random variable would be particular musical pieces that are unique to a specific genre. Note that a control group with no music would probably be outside the hierarchical structure (except that subjects would still probably be nested in cells). Research involving subjects working together in teams are typically hierarchical design where the group is the random factor.

The authors should explain the rationale for selecting the nested variable, including the importance of being able to generalize the treatment effect averaged over all levels of the random variable. The breadth of the universe of levels from which the levels were chosen should be described including the procedures for selecting the specific levels.

Authors must acknowledge the dependence in the nested data by using an error term that includes this dependence. Main effects and interactions of fixed factors typically have error terms that involve the variability of the levels of the random factor rather than the variability of the subjects with the cells. In complex designs special care should be taken by authors to clearly indicate the mean square that was used as the error term for each test along with its number of degrees of freedom.

The analysis of hierarchical experiments with ANOVA is problematic under a number of commonly encountered situations. There is an assumption of homogeneity of residual variances within cells both within levels of the random and fixed factors that, when violated, can lead to inflation of Type I error rates. Equal sample size lessens the impact. If samples are systematically different in the number of subjects and the nature of the treatments leads one to believe the assumptions are likely to be violated, then the researchers should consider reanalysis using a multi-level modeling approach (see, Chapter 10, this volume).

### 13. Incomplete and Other Complex Designs

Authors using Latin squares or other incomplete designs for the creation of treatment groups should provide a rationale to explain the advantage of the reduced number of treatment groups given the reduced amount of information provided. The rationale should include the authors' defense of no interactions between the factors that are incompletely crossed. In the Discussion section, the authors should acknowledge the confounding of the main effects with higher-order interactions inherent in the design.

The error term should be identified sufficiently to allow readers to understand the analysis. For complex experimental designs, where a complete description may take extensive space, references should be provided.

### 14. Necessary Statistical Information

A clear distinction should be made between the existence of a relation between the IV and the DV and the strength/direction of that relation. Tests of significance along with the probability levels might lead one to infer that a non-chance relation exists, but this does not satisfy our responsibility to describe both the strength and the direction of the relation. Multiple comparisons provide a description of the direction or nature of the relation, leaving only a need to present evidence as to the strength of the relation. In the last 20 years there has been a clear movement by professional organizations such as the American Psychological Association to require specific information about the strength of hypothesized relations. Effect sizes are commonly used to describe the difference between means scaled using the standard deviation of the scores (see Chapter 7, this volume). Effect size measures of one type or another are particularly important in making mean differences comparable in meta-analyses (see Chapter 19, this volume).

A number of measures of association between the IV and DV have been developed that provide a way of indicating the proportion of the variance present in the set of DV scores that is related to the differences introduced by treatment. The more common measures are $\varepsilon^2$, $\omega^2$, and $\eta^2$. Each is like a squared correlation in that the interpretation is related to the proportion of explained variance. At present no one index is universally accepted as the standard. Individual fields of study should agree on a measure so that understanding can be developed through usage.

## 15. Test(s) of Statistical Significance

For any test of significance reported, the authors should provide the test statistic (e.g., $F$, $t$, $q$), the source used as the error term and its degrees of freedom, and either a statement about the probability level associated with the observed test statistic or an indication whether the finding was statistically significant at the chosen $\alpha$ level. ANOVA summary tables are space consuming and may not be needed for simple designs. More complex designs should have more complete reporting of the analysis so that the reader can understand a specific finding from within the context of the total experiment. Because main effects are averaged over the fixed levels of the other IVs in an experiment, it is important to keep all of the IVs in mind. Interactions require analyses that are described in Desideratum 16.

The significance level of a finding, often communicated by asterisks (e.g., * $p < .05$, ** $p < .01$), should not be confused with the magnitude of the finding. Magnitude of effect should be indicated by effect size (Desideratum 14). If more information is desired specifically about the probability level of the test statistic under the null hypothesis, the exact probably can be given.

## 16. Statistically Significant Interactions

Interactions in factorial experiments often contain the most interesting information in the experiment and should receive comparable analyses to those applied to the main effects. The analysis of interactions should reflect the multiple comparison strategy that was (or would have been) used to analyze the main effects. Assume for simplicity that factor A has 3 levels and factor B has 2 levels. If the A main effect was analyzed by comparing all possible pairs of levels of A (A1 vs. A2, A1 vs. A3, and A2 vs. A3), then the A $\times$ B interaction should analyze the 2 $\times$ 2 tables created by all possible pairs of the levels of A crossed with B (A1A2 $\times$ B, A1A3 $\times$ B, A2A3 $\times$ B). The test can be conducted with Tukey's test or any of the other methods that control Type I error for a family of comparisons. Graphical as well as analytic presentations can be used to indicate the source of the interaction.

Main effects found along with statistically significant interactions should be carefully described to indicate that, while the differences among the means of the main effect (averaged across all levels of the other variable) are statistically significantly, those differences are altered by the level of the other variable. Main effects may still be able to be meaningfully interpreted even when their factor is involved in an interaction, but the interpretation still must acknowledge the complexity in the findings.

Higher-order interactions should not be immediately dismissed as Type I errors simply because a follow-up analysis would require considerable effort. If the finding is not central to the study, the authors might not be inclined to do an extensive investigation into the nature of the interaction; however, the authors must provide sufficient descriptive information (i.e., cell means and appropriate error terms) so that the interested reader could conduct further inquiry.

Simple main effects provide an alternative approach to analysis following a statistically significant interaction. In the A $\times$ B example above the authors might decide simply to test the difference between the two levels of B for each of the three A levels, resulting in three tests of significance. Authors taking this approach forgo the opportunity to talk about the relative difference between the B levels across the levels of A. Any implied difference about the relative size of the B effect for different levels of A based on the B difference being statistically significant in one instance but not in the other must not be made.

## 17. Language Relative to Meaning and Generalizability

The logic of hypothesis testing presents some challenges in the use of language that authors should carefully consider. In particular, the nature of an indirect proof causes problems in correctly stating what the results of a study indicate. All statements should acknowledge the probabilistic nature of research in the social and behavioral sciences. In other words, we cannot prove anything; we infer there is a difference among populations because it is highly unlikely that the data we obtained would have happened if the null hypothesis were true.

A retained null hypothesis should never be taken as proving the null hypothesis is true. While many researchers would like to show that one method is as good as another, we cannot prove that the treatments have no difference. The researchers can calculate confidence intervals about the observed difference that can be used to argue that there is no "practical" difference between groups, but that is quite difference from proving the null hypothesis. In the literature that argues against the use of significance tests one of the common complaints is the incorrect usage of language regarding the meaning of a *statistically significant* difference.

The term *significant* should never be used in referring to the outcome of a test of significance without the additional word *statistical*. Statistical significance provides an unambiguous description of the results of a test of significance, whereas without the *statistical* qualifier we are left unsure whether the author is describing the outcome of the test of significance or arguing for the difference having real world importance.

Generalizations should be made with great caution. Statistically justified generalizations are made back to the theoretical population from which random sampling occurred. The problem is that almost all experimentation is done on samples of convenience. We use volunteers from our university or students in schools who have agreed to work with us. In the strictest sense we have no physical set of subjects from which we sampled. Rather, we have available participants whom we have randomly assigned to treatments. All generalizing is then an extension beyond the actual data. Authors should make all generalizations as speculative. In the Methods section they should provide the reader with enough information about all of the conditions under which the experiment was conducted so that the reader is able to evaluate the similarity of the situation to one of concern to the reader. Similar considerations are germane with mixed models where there is a random selection of levels of a treatment variable. Here the idea of a random sample may be more defensible, but the reader must be told both the process used and the extent of the universe of treatment levels from which those used is a sample in order to be able to evaluate the extent to which generalizations are warranted.

## References

Hancock, G. R., & Klockars, A. J. (1996). The quest for $\alpha$: Developments in multiple comparison procedures in the quarter century since Games (1971). *Review of Educational Research, 66,* 269–306.

Keppel, G., & Wickens, T. D. (2004). *Design and analysis: A researcher's handbook* (4th ed.). Upper Saddle River, NJ: Pearson Education.

Kirk, R. E. (1995). *Experimental design: Procedures for the behavioral sciences* (3rd ed.). Monterey, CA: Brooks-Cole.

Maxwell, S. E., & Delaney, H. D. (2004). *Designing experiments and analyzing data: A model comparison approach* (2nd ed.). Mahwah, NJ: Erlbaum.

# 2

# Analysis of Variance
## *Repeated Measures Designs*

### Lisa M. Lix and H. J. Keselman

A repeated measures design, also known as a *within-subjects design*, in which study participants are measured $K$ times on the same dependent variable, is one of the most common research designs in the social, behavioral, and health sciences. The design arises in both experimental and observational settings. Repeated measurements arise when a study participant is exposed to two or more experimental conditions (i.e., factor levels) such as different dosage levels of the same drug, or when a participant is observed at multiple points in time. A key characteristic of the data is correlation among the measurements for each study participant.

One advantage of adopting a repeated measures design is that, for a fixed sample size, it will generally result in greater precision of parameter estimates and more efficient inferential analyses than a between-subjects design. In addition, research questions about individual growth or maturation can only be effectively investigated in repeated measures designs.

This chapter focuses on procedures to analyze repeated measures data that are continuous; a brief discussion of procedures for discrete data is provided in Desideratum 3. Procedures for continuous data range from the simple to the complex. A simple procedure is to conduct dependent-sample $t$ tests for pairs of measurement occasions. The limitation of this approach is that it does not provide information about the overall, or *omnibus*, within-subjects effect(s). The repeated measures analysis of variance (ANOVA) $F$ test is the conventional procedure for testing hypotheses about omnibus within-subjects effects. This procedure makes stringent assumptions about the structure of the covariance matrix of the repeated measurements. Alternatives to the repeated measures ANOVA $F$ test may be more suitable for many of the data-analytic conditions encountered by researchers in the social, behavioral, and health sciences. Alternative procedures include: (a) an adjusted degrees of freedom (df) procedure, which modifies the df of the repeated measures ANOVA critical value using information about the covariance matrix, (b) repeated measures multivariate analysis of variance (MANOVA), which makes no assumptions about the structure of the covariance matrix of the repeated measurements (except across any between-subjects factors in the design), (c) the multiple regression model (MRM), which allows the researcher to characterize the variances and covariances of the repeated measurements using a small number of parameters, (d) the random-effects (e.g., mixed-effects) model, which allows the researcher to describe and test subject-specific variation in repeated measures data, and (e) approximate df procedures, which do not assume that the data follow

Table 2.1 Desiderata for Analysis of Variance: Repeated Measures Designs

| Desideratum | Manuscript Section(s)* |
|---|---|
| 1. The type of repeated measures design is specified (i.e., number of within- and between-subjects factors and number of levels of each). | I, M |
| 2. Issues of statistical power have been considered and the sample size is reported. | M |
| 3. The number and type of dependent (i.e., response) variable(s) is specified. | M |
| 4. Assumptions about the distribution of the dependent variable(s) are evaluated and an appropriate test procedure is selected. | M |
| 5. Assumptions about the covariance structure of the repeated measurements are evaluated and used to guide the selection of a test procedure. | M |
| 6. The pattern and rate of missing observations is considered. The method adopted to handle missing observations is identified. | M, R |
| 7. The method used to conduct a priori or post hoc multiple comparisons of within-subjects factor levels is specified. The method adopted for testing multiple dependent variables, if present, is specified. | M |
| 8. The name and version of the selected software package is reported. | M, R |
| 9. Exploratory analyses of the repeated measures data are summarized. | R |
| 10. The results for omnibus tests of within-subjects effects and multiple comparisons are reported. The criterion used to assess statistical significance is specified. | R |
| 11. Consideration is given to reporting effect sizes and confidence intervals. | R |
| 12. The strengths and limitations of a repeated measures design are considered and threats to the validity of study findings are discussed. | D |

* *Note*: I = Introduction, M = Methods, R = Results, D = Discussion

a normal distribution or that covariances are equal (i.e., homogeneous) across any between-subjects factors in the design.

Comprehensive resources that discuss procedures for the analysis of continuous repeated measures data include Hedeker and Gibbons (2006), Fitzmaurice, Laird, and Ware (2004), Littell, Pendergast, and Natarajan (2000), and Singer and Willett (2003). Several of these sources also discuss procedures for the analysis of discrete repeated measures data.

Desiderata for the analysis of repeated measures data are given in Table 2.1. A detailed discussion of each item is provided below.

## 1. Types of Repeated Measures Designs

Information about the characteristics of a repeated measures design is used to assess the overall appropriateness and validity of the hypothesis-testing strategy. The simplest repeated measures design is one in which a single group of study participants is measured on one dependent variable at two or more occasions or for two or more experimental conditions. Consider an example in which a study cohort is observed repeatedly after being introduced to a new therapeutic treatment. Suppose the researcher is interested in investigating the treatment's effect on the quality of life of study participants. An appropriate null hypothesis for such a design is that there is no change in average quality of life ratings over time. If this omnibus hypothesis is rejected, multiple comparisons might be conducted to test for a mean difference between pairs of measurement occasions. A priori contrasts among the measurement occasions could be conducted instead of a test of the omnibus hypothesis (see Desideratum 7).

Factorial repeated measures designs contain two or more repeated measures factors. The simplest factorial design is one in which a single group of study participants is measured on a single dependent variable for each possible combination of the levels of two factors. For example, suppose that a researcher investigates psychological well-being of a cohort exposed to experimental stimuli that represent all combinations of sex (male, female) and facial expression (positive, neutral, negative). The dependent variable in this example is psychological well-being. The null hypothesis for the within-subjects interaction effect is that the effect of sex of the stimuli on psychological well-being is constant at each level of the facial expression factor. If the null hypothesis is rejected, multiple interaction contrasts might be conducted to identify the combination of factor levels that contribute to rejection of the omnibus hypothesis. If the interaction effect is not significant, the researcher can choose to test the main effects (i.e., sex, facial expression; see Desideratum 10).

A mixed design, also referred to as a *split-plot* repeated measures design, contains both between-subjects and within-subjects factors. The simplest mixed design contains a single within-subjects factor, a single between-subjects factor, and a single dependent variable. For example, study participants might be randomly assigned to control and intervention groups prior to being measured at successive points in time on their reading comprehension. In a mixed design, the researcher is primarily interested in testing whether there is a group-by-time interaction, that is, whether the change over time on the dependent variable (e.g., reading comprehension) is the same for control and intervention groups, although main effects will be of interest if the interaction is not significant (see Desideratum 10).

## 2. Sample Size and Statistical Power

Statistical power and sample size ($N$) must be considered early on in the design of a study. Statistical power is the probability that an effect will be detected when it exists in the population. In a repeated measures design, calculating the sample size to achieve a desired level of statistical power requires information about the pattern and magnitude of the within- and between-subjects effect(s), the variances and covariances over the measurement occasions, the number of measurement occasions, the level of significance ($\alpha$), and the choice of analysis procedures (Overall & Doyle, 1994). Information about the magnitude of effects, as well as the pattern of variances and covariances of the repeated measurements, can often be obtained from previous research.

Existing statistical software packages can often be used to calculate sample size requirements for simple repeated measures designs when the repeated measures ANOVA $F$ test or MANOVA procedure are used to analyze the data. Calculating sample size for more complex designs, including those with multiple dependent variables or for analysis procedures such as the MRM or random-effects models, is less straightforward and the researcher is advised to consult with a statistician having expertise in this area.

Sample size may also dictate the choice of analysis procedures. For example, if $N$ is less than the number of measurement occasions ($K$), the repeated measures MANOVA procedure cannot be used to test within-subjects effects. Furthermore, if the ratio $N/K$ is small, the covariance parameter estimates may be unstable. In mixed designs, the ratio of the group sizes is also an important consideration in the choice of procedures; if group sizes are unequal and equality (i.e., homogeneity) of the group covariances is not a tenable assumption, then the repeated measures ANOVA and MANOVA procedures may result in invalid inferences (e.g., too many false rejections of null hypotheses).

## 3. Number and Type of Dependent Variables

Repeated measures designs may be either univariate or multivariate in nature. A multivariate repeated measures design is one in which measurements are obtained from study participants on $P$ dependent variables at each occasion. In multivariate data there are two sources of correlation: (a) within-individual within-variable correlation, and (b) within-individual between-variable correlation. The latter arises because the measurements obtained on the dependent variables at a single occasion are almost always related.

One approach to analyze multivariate repeated measures data is to conduct $P$ tests of within-subjects effects, one for each dependent variable. This method can be substantially less powerful than a multivariate analysis, which simultaneously tests within-subjects effects for the set of $P$ outcomes. Several procedures have been proposed to test multivariate within-subjects main and interaction effects (Vallejo, Fidalgo, & Fernandez, 2001). Two conventional procedures are the *doubly multivariate model* (DMM) and *multivariate mixed model* (MMM) procedures, which are extensions of repeated measures MANOVA and ANOVA, respectively, to the case of two or more dependent variables. The choice between these two procedures is a function of sample size and one's assumptions about the data. The DMM cannot be applied to datasets in which $N/(P \times K)$ is less than one. Moreover, when this ratio is small, covariance parameter estimates may be unstable. The MMM procedure makes stringent assumptions about the covariance structure of the repeated measurements and dependent variables. When these assumptions are not satisfied, the MMM will result in invalid inferences.

The MRM procedure is one alternative to these conventional procedures. It allows the researcher to define the covariance matrix of the repeated measurements and dependent variables using a small number of parameters. A parsimonious (i.e., simple) structure for the MRM is a separable structure, in which the covariance matrix of the repeated measurements is assumed to be the same for each dependent variable. It is advantageous to assume a separable covariance structure when sample size is small because it requires estimation of fewer covariance parameters than when an unstructured covariance is assumed and therefore the estimates will be more stable.

In either univariate or multivariate repeated measures data, the dependent variables may have a continuous or discrete scale. For the latter, the outcome might be the presence (or absence) of a response or a count of the number of times a response occurs. Generalized linear models are a unified class of models for the analysis of discrete data. They have been extended to the case of correlated observations. Binary repeated measurements can be analyzed using an extension of logistic regression, repeated counts of rare events can be analyzed using an extension of Poisson regression, and repeated ordinal measurements can be analyzed using an extension of multinomial regression for repeated measures data. Two different types of generalized linear models for repeated measurements are marginal models and random-effects models. The choice between these two approaches is largely a function of the research purpose; a marginal model is used to make inferences about the average response in the population while the random-effects model is used to make inferences about the response of the average individual in the population.

## 4. Distributional Assumptions

Repeated measures ANOVA, MANOVA, and MRM procedures rest on the assumption of multivariate normality. If the repeated measurements are distributed as multivariate normal, then the data for each measurement occasion is normally distributed and the joint distribution of the data for all measurement occasions is normally distributed. However, even when the data for each measurement occasion are normally distributed, the set of measurements might not follow a multivariate normal distribution (Keselman, 2005).

Assessing potential departures from a multivariate normal distribution is critical to selecting a valid analysis procedure. The researcher can compute measures of skewness (symmetry) and kurtosis (tail weight) for the marginal distributions (i.e., for each measurement occasion), as well as measures of multivariate skewness and kurtosis. Values near zero (assuming an adjusted measure of kurtosis) are indicative of a normal distribution. Tests of univariate and multivariate normality, such as the Shapiro-Wilk test for univariate normality and Mardia's test for multivariate normality, are sensitive to even slight departures from a normal distribution. Data exploration tools, such as normal probability plots, might be useful for assessing departures from a multivariate normal distribution; these are discussed in further detail in Desideratum 9. Alternatively, one may simply bypass these assessments of the data distribution in favor of a test procedure that is robust (i.e., insensitive) to departures from multivariate normality.

Procedures such as the repeated measures ANOVA and MANOVA are sensitive to the presence of outliers in the data distribution. Outliers inflate the standard error of the mean resulting in reduced sensitivity to detect within-subjects effects. One approach to overcome the biasing effects of nonnormality is to adopt robust measures of central tendency and variability. Trimmed means and Winsorized variances have been extensively studied as alternatives to conventional least-squares means and variances in the analysis of repeated measures data (Keselman, Wilcox, & Lix, 2003). A trimmed mean is obtained by removing an a priori determined percentage of the largest and smallest observations at each measurement occasion and computing the mean of the remaining observations. A commonly recommended trimming percentage is 20% in each tail of the distribution. The trimmed mean will have a smaller standard error than the least-squares mean when the data are sampled from a heavy-tailed distribution (i.e., a distribution containing outliers or extreme scores). To compute the Winsorized variance, the smallest non-trimmed score replaces the scores trimmed from the lower tail of the distribution and the largest non-trimmed score replaces the scores removed from the upper tail. These non-trimmed and replaced scores are called Winsorized scores. A Winsorized mean is calculated by applying the usual formula for the mean to the Winsorized scores; a Winsorized variance is calculated using the sum of squared deviations of Winsorized scores from the Winsorized mean. The Winsorized variance is used instead of the trimmed variance, because the standard error of a trimmed mean is a function of the Winsorized variance. Test procedures based on robust estimators have demonstrated good performance (i.e., accurate Type I rates and acceptable levels of statistical power to detect non-null treatment effects) for analyzing repeated measures data. Computer programs that use the trimmed mean and Winsorized variance to test within-subjects effects are discussed in Desideratum 8.

Another approach to deal with the biasing effects of nonnormality is to transform the data prior to analyzing it. Rank transform procedures, in which observations are ranked prior to applying an existing procedure for analyzing repeated measurements, are appealing because they can be easily implemented using existing statistical software packages (Conover & Iman, 1981). One limitation is that they cannot be applied to tests of within-case interaction effects, because the ranks are not a linear function of the original observations. Therefore, ranking may introduce additional effects into the statistical model that were not present in the original data. Ranking may also alter the pattern of correlations among the repeated measurements, which can be particularly problematic for the repeated measures ANOVA procedure, which makes specific assumptions about the correlation structure of the data. Thus, rank transform procedures, while insensitive to departures from a normal distribution of responses, must be used with caution.

A nonparametric bootstrap resampling method has also been proposed for the analysis of non-normal repeated measures data (Berkovits, Hancock, & Nevitt, 2000). Under this methodology, a normal-theory procedure is used to test within-subjects effects; however, the critical value for evaluating statistical significance is based on the empirical sampling distribution of the test statistic rather than a theoretical critical value (e.g., a critical value from an *F* distribution). The method proceeds as follows: A bootstrap dataset is obtained by randomly sampling with replacement from the original data. The data are centred using the mean of the within-subjects effects, to approximate the sampling distribution of the null hypothesis. A test of the within-subjects effect is computed from the bootstrap dataset. This process is repeated $B$ times. The $B$ test statistics are ranked in ascending order. For a test at the $\alpha$ level of significance, the $B \times (1-\alpha)$th observation in the ranked set of observations, which corresponds to the $100 \times (1-\alpha)$ percentile, is used to approximate the critical value. The choice of $B$ depends on the goals of the analysis; $B \geq 1000$ is recommended for constructing confidence intervals around parameter estimates, while $300 \leq B \leq 1000$ is sufficient for conducting inferential analyses.

## 5. Covariance Structure of the Repeated Measurements

Procedures for analyzing repeated measures data vary widely in their assumptions about the structure of the covariance matrix; evaluation of the covariance structure is therefore critical to the selection of a valid method of analysis. The repeated measures ANOVA procedure assumes that the covariance matrix of the repeated measurements has a spherical structure. For *sphericity* to be satisfied, the population variances of the differences between pairs of repeated measures factor levels must be equal. Furthermore, for mixed designs, the more stringent assumption of *multisample sphericity* must be satisfied; this requires equality of the common variance of the pairwise repeated measures differences across levels of the between-subjects factor. The sphericity assumption is not likely to be satisfied in data arising in social, behavioral, and health sciences research. For example, when measurements are obtained at multiple points in time, it is often the case that the variance increases over time. Moreover, a test of the sphericity assumption is sensitive to departures from a multivariate normal distribution. Therefore, the repeated measures ANOVA procedure cannot routinely be recommended in practice.

The *approximate df ANOVA* procedure is one alternative when sphericity is not a tenable assumption. The *repeated measure MANOVA* procedure is another alternative; it does not make any assumptions about the structure of the common covariance matrix of the repeated measurements. However, both the adjusted df ANOVA and repeated measures MANOVA procedures do assume homogeneity of the group covariance matrices across any between-subjects factor levels in the design. These procedures are not robust under departures from covariance homogeneity, particularly when group sizes are unequal. If the group with the smallest sample size exhibits a larger degree of variability among the covariances than the group with the largest sample size, then tests of within-subjects effects will tend to produce liberal Type I error rates, above the nominal $\alpha$ level (i.e., too often inferring there are within-subjects effects when none are present). Conversely, if the group with the smallest sample size exhibits the smallest degree of variability of the covariances, then tests of within-subjects effects will tend to produce conservative error rates, below the nominal $\alpha$ level (i.e., too often failing to detect real population effects). Unfortunately, a likelihood ratio procedure to test the null hypothesis of covariance homogeneity is sensitive to departures from a multivariate normal distribution, as well as to small sample sizes.

When it is not reasonable to assume that covariances are homogeneous, the researcher is recommended to bypass these analysis procedures in favor of an approximate df procedure (Keselman, Algina, Lix, Wilcox, & Deering, 2008). The approximate df procedure, which is a multivariate and multi-group generalization of the non-pooled two-group $t$ test, has been extensively studied for both univariate and multivariate repeated measures designs when covariances are heterogeneous. It will result in valid inferences about within-subjects effects provided that sample size is not too small. However, the approximate df procedure does assume that the repeated measures data follows a multivariate normal distribution. When multivariate normality is not a tenable assumption, then the approximate df procedure should be implemented by substituting trimmed means and Winsorized variances for the usual least-squares means and variances. A computer program for this procedure is described in Desideratum 8.

The MRM procedure allows the researcher to model the covariance matrix of the repeated measurements in terms a small number of parameters. Heterogeneous covariance structures can also be accommodated for mixed designs if homogeneity of group covariances is not a tenable assumption. There are several different covariance structures that can be fit to one's data. Autoregressive and Toeplitz structures assume that the correlation among repeated measurements is a function of the lag, or interval, between two measurement occasions. Some covariance structures assume that the variances of the measurement occasions are constant, while other structures allow for heterogeneous variances. For example, the random coefficients structure is a flexible structure that models subject-specific variation characterized by non-constant variances and non-constant correlations.

When a parsimonious covariance structure is specified for the repeated measurements, the MRM procedure will result in a more powerful test of within-subjects effects than the repeated measures MANOVA procedure. However if the covariance structure is incorrectly specified, tests of within-subjects effects may be biased, resulting in erroneous inferences.

Graphic techniques and summary statistics to aid in selecting an initial model(s) for the covariance structure are described in Desideratum 9. Measures of model fit and/or inferential analyses are used to select a model for the covariance structure. If the candidate covariance structures are nested, then a likelihood ratio test can be used to select one of these structures as the final model. Two covariance structures are nested if one is a special case of another. For example, a *compound symmetric* covariance structure, which assumes that all variances are equal and all covariances are equal, is a special case of an unstructured covariance model, which does not assume that either variances or covariances are equal. Caution is advised when adopting the likelihood ratio test because it is sensitive to multivariate non-normality and small sample size. Aikake's Information Criterion (AIC) and Schwarz's Bayesian Information Criterion (BIC) are two well-known criteria for assessing model fit for non-nested covariance structures. The BIC penalizes models more severely for the number of parameter estimates than does the AIC, therefore the latter is favored over the former. These information criteria should only be used for comparing the covariance structures of models that contain the same regression parameters.

## 6. Missing Observations and Loss to Follow-Up

Missing data are a concern in repeated measures designs because of the potential loss of statistical efficiency and/or bias in parameter estimates due to differences between observed and unobserved data. Although researchers may devote substantial effort to reduce the amount of missing data, some loss of data is inevitable, particularly in medical research. Therefore an assessment of the amount and type of missing data is essential in a study involving repeated measurements.

Missing data can be either monotone or intermittent. Both patterns can appear in the same dataset. A *monotone*, or drop-out, pattern arises if a participant is observed on a particular occasion but not on subsequent occasions. Study drop-out may arise for a number of reasons, including death or illness, or lack of interest in continuing a study. An *intermittent* pattern is one in which there are "holes" in the data because a study participant will have at least one observed value following a missing observation.

The rate of missingness is the proportion of missing observations to the total number of observations in the dataset. As the rate of missingness increases, statistical efficiency decreases and the probability of biased inference increases.

Repeated measures ANOVA, adjusted df ANOVA, and repeated measures MANOVA procedures assume a complete set of measurements for each study participant. If data are incomplete, the researcher is faced with the following choices. First, one could simply exclude from the analysis all study participants with at least one missing observation (so called *casewise* or *listwise* deletion). The final sample size available for analysis can be extremely small if the rate of missing observations is large. A second option is to choose an analysis procedure based on maximum-likelihood estimation (e.g., *full information* maximum likelihood estimation), which does not result in deletion of participants with missing observations. Another choice is to impute missing values, using single or multiple imputation methods, in order to obtain a complete data set for subsequent analysis.

Examples of single imputation methods include mean substitution or last observation carried forward. These methods are not widely recommended, particularly when the rate of missing observations is large, because they do not account for random variation in the missing observations. Single imputation will therefore result in model parameters with underestimated error variances. Multiple imputation, the preferred approach (Little & Rubin, 2002), generates $M$ plausible values for each missing observation, yielding $m$ pseudo-complete datasets. The $M$ datasets are analyzed using complete-case methods. The results from the $M$ analyses are combined using simple arithmetic formulae. There is no single value of $M$ that is recommended in practice, although Schafer (1999) suggests that between three and ten imputations will likely be sufficient for the majority of missing data problems. The value of $M$ depends on the rate of missing observations. One strategy for choosing $M$ is to conduct several sets of $M$ imputations, starting with a small value of $M$ and evaluating whether parameter estimates are relatively stable across these independent sets of imputations. If the estimates demonstrate wide variability, then $M$ should be increased and the stability of the estimates re-evaluated.

Imputation-based analyses will result in unbiased tests of within-subjects effects only if the missing data are ignorable. There are three mechanisms by which data may be incomplete (Little & Rubin, 2002): (a) missing completely at random (MCAR), (b) missing at random (MAR), and (c) missing not at random (MNAR). MCAR means the probability that an observation is missing is independent of either observed or unobserved responses. MAR means the probability that an observation is missing depends only on the pattern of observed responses. All other missing data mechanisms are MNAR, or non-ignorable. Unfortunately there are no formal tests of the null hypothesis that the missing data follow a MAR pattern instead of a MNAR pattern. Pattern selection models or pattern mixture models are recommended to reduce biases when the data are assumed to be MNAR. Pattern selection models rely on multiple imputations, under a variety of assumptions about the missing data mechanism, to test effects and/or estimate model parameters. Pattern mixture models develop a categorical predictor variable for the different patterns of missing data and this predictor variable is included in the statistical model for testing within-subjects effects. No single missing data model can be uniformly recommended to reduce bias in parameter estimates; the choice depends on the type and rate of missing observations, the number of measurement occasions, and the magnitude of the within-subjects effects. Shen, Beunckens, Mallinckrodt, and Molenberghs (2006) proposed using sensitivity analyses when the missingness is assumed to be non-ignorable, to assess whether the findings for different missing data models produce consistent results. Sensitivity analysis techniques can also be used to identify influential observations in analyses of missing data.

## 7. Multiple Comparison Procedures and Multiple Testing Strategies

When conducting post hoc or a priori multiple comparisons, such as pairwise contrasts among within-subjects factor levels, the researcher will typically wish to adopt a procedure to control the familywise error rate (FWER), the probability of committing at least one Type I error, for the set of tests. The well-known Dunn-Bonferroni procedure conducts each of $C$ comparisons at the $\alpha/C$ level of significance. The Bonferroni method is simple to implement, but may not be as powerful as other procedures, such as Hochberg's (1988) step-up procedure. This procedure orders the $p$-values from smallest to largest so that $p_{(1)} \leq p_{(2)} \leq \ldots \leq p_{(C)}$, to test the corresponding hypotheses $H_{(1)}, \ldots, H_{(C)}$. The sequence of testing begins with the largest $p$-value, $p_{(C)}$, which is compared to $\alpha$. Once a hypothesis is rejected, then all hypotheses with smaller $p$-values are also rejected by implication. For example, if $p_{(C)} \leq \alpha$, then all $C$ hypotheses are rejected. If the null hypothesis corresponding to the largest $p$-value, $H_{(C)}$, is accepted, the next $p$-value, $p_{(C-1)}$ is evaluated using the $\alpha/2$ criterion. More generally, the decision rule is to reject $H_{(w')}$ ($w' \leq w$; $w = C, \ldots, 1$) if $p_{(w)} \leq \alpha/(C - w + 1)$. An assumption underlying Hochberg's procedure is that the tests are independent, which is unlikely to be satisfied in a repeated measures design. However, Hochberg's procedure will control the FWER for several situations of dependent tests, making this procedure applicable to most multiple comparison situations that social scientists might encounter (Sarkar & Chang, 1997). Multiple comparison procedures for correlated data are discussed later in this Desideratum.

In factorial and mixed designs, multiple comparisons to probe interaction effects should be conducted using interaction contrasts (Lix & Keselman, 1996). Tests of simple main effects, which involve examining the effects of one factor at a particular level of the second factor, may also assist researchers in examining the interaction. A significant interaction implies that at least one contrast among the levels of one factor is different at two or more levels of the second factor. Tetrad contrasts are one type of interaction contrasts that are a direct extension of pairwise contrasts for probing marginal (i.e., main) effects. In a two-way design, a tetrad contrast involves testing for the presence of an interaction between rows and columns in a $2 \times 2$ sub matrix of the data matrix, or in other words, of testing for a difference between two pairwise differences. Control of the FWER for a set of tetrad contrasts can be achieved using an appropriate multiple comparison procedure as described previously.

In multivariate repeated measures data, one strategy to conduct multiple comparisons is to follow a significant omnibus multivariate effect with post hoc multivariate multiple comparisons. For example, a significant multivariate within-subjects interaction indicates that the profiles of the repeated measurements are not parallel for two or more levels of the between-subjects factor for some linear combination of the dependent variables. Multivariate interaction contrasts are an appropriate choice for probing this effect.

There are many kinds of comparisons that might be tested in a multivariate design. Bird and Hadzi-Pavlovic (1983) distinguish among *strongly restricted contrasts*, which are defined for between- and/or within-subjects factor levels on a single dependent variable, and *moderately restricted contrasts*, which are defined for between-subjects and/or within-subjects factor levels for two or more dependent variables. A third type, the *unrestricted contrast*, is defined as the maximum contrast for the first linear discriminant function, that is, the linear combination of coefficients that maximizes the distance between the means of the dependent variables. Unrestricted contrasts can be difficult to interpret because the coefficients are usually fractional, while strongly restricted contrasts are the easiest to interpret because they focus on only a single dependent variable. At the same time, a simultaneous test procedure (e.g., Bonferroni) to control the FWER for all possible strongly restricted contrasts will have very low power to detect significant effects because it uses a stringent criterion to evaluate each test statistic. A more powerful approach is to conduct a small set of a priori multivariate contrasts on the between-subjects or within-subjects factor levels for the set of dependent variables, using a stepwise multiple comparison procedure, such as Hochberg's (1988) procedure, to control the FWER for the set of tests.

Another approach to probe multivariate repeated measures data is to conduct tests of within-subjects effects for each of the $P$ dependent variables, adopting a significance criterion to control the FWER that is also adjusted for the correlation among the dependent variables. The Bonferroni method and its stepwise counterparts assume that the dependent variables are independent and will therefore result in conservative tests of within-subjects effects on the $P$ dependent variables, particularly when $P$ is large. Alternate approaches that adjust for correlation include Roy's (1958) step-down analysis and resampling-based methods.

In a step-down analysis, the researcher rank orders the dependent variables in descending order of importance and then conducts tests of within-subjects effects using an analysis of covariance (ANCOVA) approach in which higher-ranked dependent variables serve as covariates for tests on lower-ranked variables. Under the null hypothesis and assuming that the data are normally distributed, the step-down test statistics, $F_l$ ($l = 1, \ldots, P$) and $p$-values, $p_p$, are conditionally independent. The FWER for the set of step-down tests is controlled to $\alpha$ using a multiple comparison procedure such as Hochberg's (1988) method. A step-down analysis is an appropriate method if the researcher is able to specify an a priori ordering of the dependent variables; this is often the case when some outcomes have a greater theoretical importance to the researcher than others.

Westfall and Young (1993) described a step-down resampling-based multiple testing procedure that also adjusts for the correlation among multiple dependent variables. Their procedure uses a permutation method, in which the observations are reshuffled or rerandomized. The permutation procedure is implemented as follows: A permuted dataset is obtained by reshuffling the original observations. A test of the within-subjects effect is computed for each of the $P$ dependent variables. The test statistic in the permutated dataset that corresponds to the maximum test statistic in the original dataset is used to evaluate statistical significance for each of the $P$ dependent variables. This process is repeated $B$ times. The $p$-value for the $m$th dependent variable ($m = 1, \ldots, P$) is the proportion of permutations in which the maximal criterion exceeds the value of the $m$th test statistic in the original dataset. A critical issue in implementing this multiple testing procedure is ensuring that the data are re-randomized correctly. For example, to test the within-subjects interaction effects in a mixed design the data must be doubly randomized, that is, reshuffled among rows as well as among columns of the original data matrix.

## 8. Software Choices

Procedures for the analysis of repeated measures data are available in software packages commonly used by researchers in the social, behavioral, and health sciences including SPSS, SAS, Stata, and R. Reporting the name and version of statistical software is recommended because not all packages will rely on the same default options for estimating model parameters or testing within-subjects effects. Options available for imputing missing values and conducting computationally intensive re-sampling techniques may not be the same in all software packages. As well, potential biases in analytic results can be more easily detected when the researcher gives full disclosure of the computational details. For example, Keselman, Algina, Kowalchuk, and Wolfinger (1999) found that the default test statistics implemented in the MIXED procedure in version 6.1 of SAS could result in liberal or conservative rates of Type I error even when the covariance structure of the repeated measurements was correctly specified; this problem has been rectified in more recent versions of the software.

Syntax to implement repeated measures ANOVA, MANOVA, and MRM procedures are described in a number of sources. For example, Littell, Pendergast, and Natarajan (2000) provided SAS code to implement the MRM using PROC MIXED. Singer and Willett (2003) offer downloadable programs to analyze within-subjects effects in multiple software packages. The approximate df procedure for testing within-subjects effects in the presence of covariance heterogeneity is not currently available in commercial statistical software packages. A program written in the SAS/IML language to implement this solution is available at the first author's website: http://www. usask.ca/sph/faculty_staff/our_faculty/Lisa-Lix.html. Numeric examples that demonstrate this software for a variety of research designs are provided, along with documentation about its implementation. Tests can be conducted using least-squares means and variances or trimmed means and Winsorized variances. The program will evaluate statistical significance of tests of within-subjects effects using either a critical value from an $F$ distribution or a bootstrap critical value. As well, it will compute robust effect size estimates and robust confidence intervals; these are described in more detail in Desideratum 11.

## 9. Exploratory Analysis Techniques

Graphic techniques and summary statistics are used to evaluate the tenability of derivational assumptions that underlie different methods for the analysis of repeated measures data and to aid in the selection of an appropriate model for the covariance structure of the repeated measurements under the multiple regression model (MRM). The results of exploratory analyses should be summarized in a manuscript; they provide an assurance that the choice of analysis procedures is justified.

Profile plots of the data for individual study participants are used to assess the magnitude of subject-specific variation in the data and whether that variation is increasing or decreasing across measurement occasions, which could result in violations of the assumption of sphericity. Scatter plots for pairs of measurement occasions can aid in the identification of potential outliers or influential observations. A *normal probability plot*, or *normal quantile plot*, is a scatter plot of the percentiles of the data versus the percentiles of a population from a normal distribution. If the data do come from a normally distributed population, the resulting points should fall closely along a straight line.

Summary statistics such as correlation coefficients and variances can aid in the selection of a model for the covariance structure of the repeated measurements. The *correlogram*, which plots the average correlation among the measurement occasions against the number of lags ($h$) between the occasions ($h = 1, \ldots, K - 1$), can also be used for this purpose. Finally, change scores between pairs of measurement occasions might be useful for identifying informative post hoc contrasts for probing within-subjects effects.

## 10. Reporting Test Statistic Results

For completeness, the test procedure(s) used to conduct all analyses should be specified. For example, when the repeated measures MANOVA procedure is used to analyze data arising from a multi-group mixed design, there are four different test statistics that can be used to test the within-subjects interaction: the Pillai-Bartlett trace, Roy's largest root criterion, Wilks's lambda, and the Hotelling-Lawley trace. These statistics represent different ways of summarizing multivariate data. When the design contains two groups, all of these tests reduce to Hotelling's $T^2$, a multivariate extension of the two-sample $t$ statistic. All four multivariate criteria rest on the assumption of a normal distribution of responses and homogeneity of group covariances. Olson (1976) found the Pillai-Bartlett trace to be the most robust of the four tests when the multivariate normality assumption is not tenable, and is sometimes preferred for this reason.

For the MRM, several statistics are available to test hypotheses about covariance structures and within-subjects effects. Tests about covariance parameters can be made using a Wald $z$ statistic, which is constructed as the parameter estimate divided by its asymptotic standard error. However, this test statistic can produce erroneous results when sample size is small. The likelihood ratio test for comparing nested covariance structures, which asymptotically follows a $\chi^2$ distribution when the data are normally distributed, is sensitive to small sample sizes. Specifically Type I error rates may exceed the nominal $\alpha$. For testing hypotheses about within-subjects main and interaction effects, a Wald $F$ statistic can be used; it has good performance properties in large samples (Gomez, Schaalje, & Fellingham, 2005). Improved performance in small sample sizes can be obtained either by adjusting the df of the test statistic or modifying the test statistic value (Kenward & Roger, 1997); this option is available in SAS software.

## 11. Effect Sizes and Confidence Intervals

An effect size describes the magnitude of a treatment effect (see Chapter 7, this volume). Reporting effect sizes in addition to hypothesis testing results is required in some journal editorial policies and is supported by the American Psychological Association's Task Force on Statistical Inference. One commonly reported measure of effect size is Cohen's $d$. In a repeated measures design, this measure is computed as the standardized difference of the means for two within-subjects factor levels, taking into account the correlation between the measurement occasions. A confidence interval should also be reported for an effect size to provide information about the precision of the estimate. The noncentral $t$ distribution is used to construct a confidence interval when the data are normally distributed. Effect size measures need not be limited to the case of only two within-subjects factor levels; Keselman et al. (2008) discussed this issue in detail.

When the data are not normally distributed, the coverage probability of the confidence interval is poor (Algina, Keselman, & Penfield, 2005), and may become worse as the correlation among the measurement occasions increases. One option is to use an empirical method, such as the bootstrap, to construct a confidence interval. A bootstrap dataset is obtained by randomly sampling with replacement from the original data. An effect size measure is computed from the bootstrapped dataset. This process is repeated $B$ times. The $B$ effect sizes are ranked in ascending order. The $B \times (\alpha/2)$ and $B \times (1 - \alpha/2)$ observations of the empirical distribution represent the upper and lower limits of the $100 \times (1 - \alpha)\%$ confidence interval, respectively. A measure of effect size that is insensitive to departures from a multivariate normal distribution can be obtained by using the trimmed mean and Winsorized variance in place of the usual least-squares mean and variance.

Cohen's effect size assumes homogeneity of group covariances in mixed designs, because the denominator, or "standardizer," of the effect size is based on an estimate of error variance that averages across levels of the between-subjects factor. As a result, when covariances are heterogeneous and group sizes are unequal this measure will be systematically affected by the sample sizes used in the study. An alternate approach is to adopt a standardizer for computing Cohen's effect size that is not based on a pooled estimate of error variance. The SAS/IML program for the approximate df procedure that was described in Desideratum 8 can be used to compute an effect size that is insensitive to covariance heterogeneity. As well, it will compute a confidence interval for an effect size that does not rest on the assumption of a normal distribution of responses; this is accomplished using a bootstrap method.

## 12. Strengths and Limitations of Repeated Measures Designs

There are several potential threats to the validity of research findings in a repeated measures design that should be evaluated in a manuscript. The single-group design lacks a control group for comparison, therefore maturation effects may be impossible to distinguish from time effects. Adopting a *cohort sequential design*, in which the time of entry into the study is staggered, is one approach to estimate these two separate effects.

Within-subject effects might be a result of respondent fatigue, practice effects, or response shift (i.e., a change in the meaning of one's evaluation of the target construct), rather than true change in the dependent variable. An active area of research in the quality of life literature is around the use of statistical methods, such as structural equation modeling, to detect response shift in repeated measures designs (Schwartz & Sprangers, 1999). A *cross-over design*, in which the provision of treatments is counter-balanced among study participants, can also be used to test for carry-over effects due to fatigue or maturation. Another approach is to externally validate the study results in a different population than the one from which the study sample was selected.

High rates of participant attrition can also threaten the validity of study findings. As noted in Desideratum 6, sensitivity analysis is one approach to assess potential bias in study parameters as a result of missing data.

Other threats to validity are not unique to repeated measures designs. Some examples include selection bias due to lack of random assignment to treatment and control groups and experimenter bias, when the individuals who are conducting an experiment have an inadvertent effect on the outcome.

## References

Algina, J., Keselman, H. J., & Penfield, R. D. (2005). Effect sizes and their intervals: The two repeated measures case. *Educational and Psychological Measurement, 65*, 241–258.

Berkovits, I., Hancock, G. R., & Nevitt, J. (2000). Bootstrap resampling approaches for repeated measures designs: Relative robustness to sphericity and nonnormality violations. *Educational and Psychological Measurement, 60*, 877–892.

Bird, K. D., & Hadzi-Pavlovic D. (1983). Simultaneous test procedures and the choice of a test statistic in MANOVA. *Psychological Bulletin, 93*, 167–178.

Conover, W. J., & Iman, R. L. (1981). Rank transformation as a bridge between parametric and nonparametric statistics. *The American Statistician, 35*, 124–129.

Fitzmaurice, G. M., Laird, N. M., & Ware, J. H. (2004). *Applied longitudinal analysis*. Hoboken, NJ: Wiley.

Gomez, E. V., Schaalje G. B., & Fellingham, G. W. (2005). Performance of the Kenward-Roger method when the covariance structure is selected using AIC and BIC. *Communications in Statistics—Simulation and Computation, 34*, 377–392.

Hedeker, D., & Gibbons, R. D. (2006). *Longitudinal data analysis*. Hoboken, NJ: Wiley.

Hochberg, Y. (1988). A sharper Bonferroni procedure for multiple tests of significance. *Biometrika, 75*, 800–802.

Kenward, M. G., & Roger, J. H. (1997). Small sample inference for fixed effects from restricted maximum likelihood. *Biometrics, 53*, 983–997.

Keselman, H. J. (2005). Multivariate normality tests. In B. S. Everitt & D. C. Howell (Eds.). *Encyclopedia of statistics in behavioral science* (Vol. 3, pp. 1373–1379). Chichester, England: Wiley.

Keselman, H. J., Algina, J., Kowalchuk, R. K., & Wolfinger, R. D. (1999). A comparison of recent approaches to the analysis of repeated measurements. *British Journal of Mathematical and Statistical Psychology, 52*, 63–78.

Keselman, H. J., Algina, J., Lix, L. M., Wilcox, R. R., & Deering, K. N. (2008). A generally robust approach for testing hypotheses and setting confidence intervals for effect sizes. *Psychological Methods, 13*, 110–129.

Keselman, H. J., Wilcox, R. R., & Lix, L. M. (2003). A generally robust approach to hypothesis testing in independent and correlated groups designs. *Psychophysiology, 40*, 586–596.

Littell, R. C., Pendergast, J., & Natarajan, R. (2000). Tutorial in biostatistics: Modelling covariance structure in the analysis of repeated measures data. *Statistics in Medicine, 19*, 1793–1819.

Little, R. J. A., & Rubin, D. B. (2002). *Statistical analysis with missing data* (2nd ed.). New York: Wiley.

Lix, L. M., & Keselman, H. J. (1996). Interaction contrasts in repeated measures designs. *British Journal of Mathematical and Statistical Psychology, 49*, 147–162.

Olson, C. L. (1976). On choosing a test statistic in multivariate analyses of variance. *Psychological Bulletin, 83*, 579–586.

Overall, J. E., & Doyle, S. R. (1994). Estimating sample sizes for repeated measures designs. *Controlled Clinical Trials, 15*, 100–123.

Roy, S. N. (1958). Step down procedure in multivariate analysis. *Annals of Mathematical Statistics, 29*, 1177–1187.

Sarkar, S. K., & Chang, C-K. (1997). The Simes method for multiple hypothesis testing with positively dependent test statistics. *Journal of the American Statistical Association, 92*, 1601–1608.

Schafer, J. L. (1999). Multiple imputation: a primer. *Statistical Methods in Medical Research, 8*, 3–15.

Schwartz, C.E., & Sprangers, M.A. (1999). Methodological approaches for assessing response shift in longitudinal health-related quality of life research. *Social Science and Medicine, 48*, 1531–1548.

Shen, S., Beunckens, C., Mallinckrodt, C., & Molenberghs, G. (2006). A local influence sensitivity analysis for incomplete longitudinal depression data. *Journal of Biopharmaceutical Statistics, 16*, 365–384.

Singer, J. D., & Willett, J. B. (2003). *Applied longitudinal data analysis: Modeling change and event occurrence*. New York: Oxford University Press.

Vallejo, G., Fidalgo, A., & Fernandez, P. (2001). Effects of covariance heterogeneity on three procedures for analyzing multivariate repeated measures designs. *Multivariate Behavioral Research, 36*, 1–27.

Westfall, P. H., & Young, S. S. (1993). *Resampling based multiple testing*. New York: Wiley.

# 3

# Canonical Correlation Analysis

**Xitao Fan and Timothy R. Konold**

Pioneered by Hotelling (1935), canonical correlation analysis (CCA) focuses on the relation between two sets of variables, each consisting of two or more variables. In some applications, the two sets may be described in terms of independent and dependent variables, although such designations are not necessary. There are a variety of ways to study relations among groups of variables. The general goal of CCA is to uncover the relational pattern(s) between two sets of variables by investigating how the measured variables in two distinct variable sets combine to form pairs of *canonical variates*, and to understand the nature of the relation(s) between the two sets of variables. CCA has often been conceptualized as a unified approach to many univariate and multivariate parametric statistical testing procedures (Knapp, 1978; Thompson, 1991), and even a unified approach to some nonparametric procedures (Fan, 1996; Knapp, 1978). The close linkage between CCA and other statistical procedures suggests that the association between two sets of variables often needs to be understood in our statistical analyses: "most of the practical problems arising in statistics can be translated, in some form or the other, as the problem of measurement of association between two vector variates **X** and **Y**" (Kshirsagar, 1972). From this perspective, CCA has been considered as a general representation of the *general linear model* (Knapp, 1978; Thompson, 1984), unless we consider *structural equation modeling* (see Chapter 28, this volume) as the most general form of the general linear model that takes measurement error into account (Thompson, 2000). Interested readers are encouraged to consult additional sources for more technical treatments of CCA (Johnson & Wichern, 2002; Chapter 10), for more readable explanations and discussions of CCA (Thompson, 1984, 1991), and for understanding the linkages between CCA and other statistical techniques (Bagozzi, Fornell, & Larcker, 1981; Fan, 1997; Knapp, 1978). Recommended desiderata for studies involving CCA are presented in Table 3.1 and are discussed in the subsequent sections.

## 1. Appropriateness of Canonical Correlation Analysis

Canonical correlation analysis (CCA) is an analytic technique for examining the multivariate relation(s) between two sets of constructs/variables, with each set consisting of two or more variables. Through CCA, it is hoped that the multivariate relational pattern between the two sets of variables can be more parsimoniously understood and described. Early in a manuscript, the link between the substantive research issue(s) and CCA as the analytic approach for investigating the substantive

Table 3.1 Desiderata for Canonical Correlation Analysis

| Desideratum | Manuscript Section(s)* |
|---|---|
| 1. Substantive research issues are presented, and associated reasons why CCA is an appropriate and rational analytic choice are discussed. | I, M |
| 2. Two natural/logical variable sets, each consisting two or more variables, are explicitly justified within the context of the substantive research issues. | I, M |
| 3. Path diagrams, if presented, aid readers' understanding of the conceptual canonical model and the various interpretive facets of the canonical analysis. | M |
| 4. Summary statistics for the two sets of measured variables are presented, including sample size and within-set and between-set correlations. | R |
| 5. Canonical correlations and statistical testing of these canonical correlations are presented and discussed. | R |
| 6. Canonical function coefficients (standardized and/or unstandardized) are presented and discussed. | R, D |
| 7. Canonical structure coefficients are presented and discussed. | R, D |
| 8. Based on canonical function and structure coefficients, reasonable interpretations of canonical functions (canonical variates) are offered. | R, D |
| 9. Canonical adequacy and redundancy coefficients are presented and discussed, in light of some known limitations. | R, D (Optional) |
| 10. Canonical functions (variates) are related back to the substantive research issues. | R, D |
| 11. Clear presentation of CCA results to facilitate readers' understanding of CCA findings | R, D |

* *Note*: I = Introduction, M = Methods, R = Results, D = Discussion

issue(s) should be explicated. The early discussion related to the substantive issue(s) should lay the foundation for the later introduction of CCA as a logical/rational analytic choice for investigating the substantive issue(s). In the Methods section of the manuscript, links between the substantive issue(s) and CCA should be more carefully articulated, and the case for CCA as an appropriate analytical choice for the research issue(s) should be made explicit. Oftentimes, the link between the substantive research issue(s) and CCA as the analytic choice is presented or established through discussion of how the two sets of constructs/variables are involved in the substantive research, and how the relational pattern between the two sets of constructs/variables is the focus of the research. For example, Lewis (2007) laid the foundation for CCA as an appropriate analytic choice through discussion in the Introduction of an interest in examining the relation between perception of risk and social norms (first set of variables) and alcohol involvement measures (second set of variables) in a college population.

Canonical correlation analysis may be used to address a wide range of substantive issues in education, psychology, and the social sciences in general. For example, McDermott (1995) examined canonical relations between children's demographic characteristics (age, gender, ethnicity, social class, region, community size, and their interactions) and measures of cognitive ability, academic achievement, and social adjustment. McIntosh, Mulkins, Pardue-Vaughn, Barnes, and Gridley (1992) examined canonical relations between a set of verbal and a set of nonverbal measures of ability; and Dunn, Dunn, and Gotwals (2006) employed CCA in a multivariate validity study for establishing the construct validity of a new measure on sport perfectionism by relating the subscales of the new measure to those of an established measure.

## 2. Two Logical Sets of Variables

In CCA, two sets of variables are examined with the goal of understanding the multivariate relational pattern between the two sets as more parsimoniously operationalized by the *canonical correlation*. Conceptually, this relation can be described as a bivariate correlation between two "synthetic" variables, each of which is based on a linear combination of one set of variables involved in the analysis. In CCA, each of two variable sets consists of two or more variables, and the variables within each set should form a natural/logical group. In addition, there should be a reasonable expectation that the two sets of variables are substantively related, and that the relation between the two sets of variables is of potential research interest.

Early in the manuscript the two sets of variables should be discussed in terms of why the relation between them is of research interest. Furthermore, there should be some indication as to why the variables within each set are included. For example, an industrial psychologist may be interested in understanding how a set of employee satisfaction variables (e.g., career satisfaction, supervisor satisfaction, and financial satisfaction; based on employees' responses to a survey) relates to a set of employees' job characteristics (e.g., variety of tasks required by the position, position responsibility, and position autonomy; based on supervisors' responses for the positions held by each employee). In this situation, the investigator may reason that employees' satisfaction variables and their position characteristics form two logical groups of variables, and that there is a reasonable expectation that the two sets of variables are related in one or more ways. The two sets of variables might have a complicated relational pattern that would not be obvious through simple inspection of the bivariate correlations. Here, CCA may help to uncover the relational pattern via a more parsimonious representation of the association between the two sets.

A possible example of a poorly conceptualized match between two variable sets might involve the pairing of either of the two sets of variables described above, with a set of employees' physical measurements (e.g., measurements of height, waist, and pulse rate). Here, it would be a formidable task to justify why employees' satisfaction variables and their physical measurements would be naturally and logically grouped into two inter-related variable sets. Further, it would be harder to justify the expectation that job satisfaction variables are somehow related to the physical measurements. The author(s) of the manuscript should provide a reasonable justification for the two groups of variables used in CCA in making the case that CCA is an appropriate analytic choice for the issue(s) at hand.

## 3. Path Diagrams

Depending on the nature of the manuscript, path diagrams may be considered to help readers understand CCA and its major interpretive facets. In practitioner-oriented substantive journals that have little focus on quantitative methods, such path diagrams typically are not needed. However, in more quantitatively oriented substantive journals, such diagrams can be helpful in aiding readers' understanding of the CCA analytic model and its various interpretive facets.

As an illustrative example, Figure 3.1 presents a model in which one set of observed variables ($\mathbf{X}$) consists of three variables ($x_1, x_2, x_3$), and the other set of observed variables ($\mathbf{Y}$) consists of two variables ($y_1, y_2$). The single-headed arrows from the observed variables to the unobserved *canonical variates* ($X_1^*$ and $Y_1^*$, and $X_2^*$ and $Y_2^*$) denote the presumed direction of influence, and the canonical variates are derived by using *canonical weights* (also called *function coefficients*) to linearly combine the observed variables ($x_1, x_2, x_3$, using the *a* weights, and $y_1, y_2$ using the *b* weights). In a manuscript, the *standardized weights* (see Desideratum 6 below about standardized vs. unstandardized function coefficients) may be inserted into the figure so that readers can more easily see the contribution of each observed variable to its canonical variate. The curved double-headed arrow linking the pair of

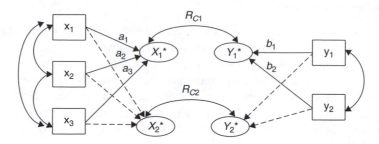

**Figure 3.1** Graphic Representation of Canonical Correlation Design.

canonical variates represents the canonical correlation. The total number of canonical correlations (i.e., the number of pairs of canonical variates) possible is equal to the number of the observed variables in the smaller of the two sets, though not all of the canonical correlations may be statistically/practically meaningful. In the example illustrated in Figure 3.1, two canonical correlations ($R_{C1}$ and $R_{C2}$) are possible. Here again, actual canonical correlations values can be placed in the figure. Last, the double-headed arrows, linking the observed variables in each of the two sets, reflect that the correlations among them are taken into account in the derivation of the canonical function coefficients.

Figure 1 depicts the *first* pair of canonical variates:

$$X_1^* = a_1 (x_1) + a_2 (x_2) + a_3 (x_3)$$
$$Y_1^* = b_1 (y_1) + b_2 (y_2)$$

The Pearson product-moment correlation coefficient between these two canonical variates is the first canonical correlation coefficient $R_{C1}$, which is the maximum of all possible canonical correlation coefficients that can be extracted from the variables. In a similar vein, the second pair of canonical variates ($X_2^*$ and $Y_2^*$) can be constructed and a second canonical correlation coefficient ($R_{C2}$) can be obtained, as shown in Figure 3.1. The construction of the second pair of canonical variates is subject to the orthogonality condition: canonical variates in the second pair ($X_2^*$, $Y_2^*$) are subject to the constraint that they are not correlated with either of the canonical variates in the previous pair ($X_1^*$, $Y_1^*$). In other words, all correlations across pairs (i.e., $r_{X1^*X2^*}$, $r_{Y1^*Y2^*}$, $r_{X1^*Y2^*}$, $r_{X2^*Y1^*}$) are zero. If additional pairs of canonical variates can be extracted, as is the case when more observed variables are included in the design, all subsequent pairs of canonical variates are subject to this orthogonality condition relative to all previously extracted canonical pairs.

## 4. Summary Statistics of Measured Variables

As previously indicated, two sets of logically grouped variables are involved in CCA. It is important that the summary statistics of the two sets of variables are presented in the manuscript. Both within-set and between-set correlations of the two sets of variables should be presented in a form that is easy for readers to see the within-set and between-set relational patterns. Providing such summary statistics serves the general purposes of allowing readers to run secondary data analyses, or allowing readers to replicate the CCA results presented in the manuscript if they have questions regarding the results presented.

As an example of providing such summary statistics, an industrial psychologist (same researcher previously mentioned) is using CCA to investigate how a set of job satisfaction variables (e.g., career satisfaction, supervisor satisfaction, and financial satisfaction; based on employees' responses to a

Table 3.2  Summary Statistics of Two Sets of Variables ($N = 200$)

| | | | | | | |
|---|---|---|---|---|---|---|
| Job Satisfaction | | | | | | |
| Career | 1.00 | | | | | |
| Supervisor | 0.55 | 1.00 | | | | |
| Finance | 0.22 | 0.21 | 1.00 | | | |
| Job Characteristics | | | | | | |
| Task Variety | 0.31 | 0.42 | 0.39 | 1.00 | | |
| Responsibility | 0.32 | 0.42 | 0.01 | 0.26 | 1.00 | |
| Autonomy | 0.32 | 0.56 | 0.37 | 0.53 | 0.12 | 1.00 |
| Std | 20.41 | 31.10 | 2.20 | 25.77 | 27.44 | 15.82 |

Note: Variable means have no relevance in CCA, and thus are not presented here.

survey) is related to a set of employees' job characteristics (e.g., variety of tasks required by the position, position responsibility, and position autonomy; based on supervisors' responses for each employee on the three aspects of a position). Table 3.2 illustrates how a table of summary statistics for the two sets of variables could be presented. These summary statistics are informative for the readers from the perspective that the correlation pattern is organized into within-set correlation matrices (correlations among variables within each set) and between-set correlation matrix (correlations among variables from different sets: *job satisfaction* variables with *job characteristics* variables).

## 5. Canonical Correlation Coefficients and Statistical Testing

In CCA, the bivariate Pearson product-moment correlation between two canonical variates within a pair is the canonical correlation coefficient. Unlike the Pearson correlation between two observed variables, however, canonical correlation coefficients do not take on negative values. This is because the direction of a canonical variate [e.g., $Y_1^* = b_1 (y_1) + b_2 (y_2)$] can be reversed by multiplying all canonical function coefficients (e.g., $b_1$ and $b_2$) by $-1$. In other words, in the multidimensional space in which the set of multiple variables reside, directionality is arbitrary. For this reason, all canonical correlation coefficients are defined as positive, ranging from 0 to 1.

Like many other statistical techniques (e.g., regression analysis), CCA is a statistical maximization procedure through which the relation between two canonical variates is maximized. The maximization property of the procedure ensures that the first canonical correlation coefficient based on the first pair of canonical variates is the largest among all possible canonical correlation coefficients, and the second canonical correlation coefficient from the second pair of canonical variates is smaller than the previous one, but larger than all *remaining* canonical correlation coefficients, and so on.

In CCA, the first question a researcher typically asks is, "Do I have anything?" (Thompson, 2000, p. 301). In other words, the researcher asks if the results appear to indicate that there is some "true" association between the two sets of variables. This question can be addressed from two supplemental perspectives, one being statistical, and the other being practical/substantive.

CCA not only maximizes whatever true population relation may exist between two sets of variables, but it also maximizes any random relation introduced by sampling error. As a result, canonical correlation coefficients can vary in magnitude as a result of sampling error. For this reason, it is necessary to statistically test the canonical correlation coefficients to help ensure that the obtained canonical correlations represent real population relations between the two sets of variables, rather than simply chance relations due to sampling error. Statistically, the question of "Do I have anything?" can be answered by conducting tests of statistical significance for canonical correlation coefficients to determine the probability that the canonical correlation of this or greater magnitude could arise when no "true" relation exists between the **X** and **Y** variable sets in the population).

Similar to some other multivariate techniques (e.g., discriminant analysis; see Chapter 6, this volume), as explained elsewhere (e.g., Fan, 1997), because of the complexity of sampling distribution theory of canonical correlation coefficients (Johnson & Wichern, 2002; Kshirsagar, 1972), the likelihood ratio test in CCA is a sequential testing procedure, instead of testing each individual canonical function. For example, consider a CCA that yields three canonical functions, with their respective canonical correlation coefficients. Here, there will be three sequential likelihood ratio tests. The first tests all three canonical functions combined, the second tests the second and third canonical functions combined, and the last tests the third canonical function by itself. Assuming that only the first test is statistically significant, and the latter two are not, this pattern of results leads to the conclusion that the first canonical function is statistically significant, but the latter two are not. Here, our conclusion sounds as though we have conducted significance tests for each individual canonical function, when in reality we have not. Strictly speaking, only the last test in this sequence is a true test for an individual canonical function.

Practically and substantively, results based on statistical significance testing should be augmented by measure(s) of effect size (Wilkinson & APA Task Force on Statistical Inference, 1999). In CCA, canonical correlation coefficients and their squared values can be taken as a gauge of effect size. In a manuscript, these canonical correlation coefficients, and their squared values, should be clearly presented and discussed. This is particularly true for the statistically significant canonical functions. In general, CCA is a large-sample analytic method, and the validity of the likelihood ratio tests for canonical functions depends on a reasonably large sample size. Large sample size leads to high statistical power, and as a result, some trivial canonical association(s) may be statistically significant. Although there is no "rule of thumb" here, there are some general recommendations. For example, Stevens (2002) recommended the lower limit of sample size of 20 times as many cases as the number of variables for CCA for the purpose of interpreting the first canonical correlation. Barcikowski and Stevens (1975) recommended 40 to 60 times as many cases as the number of variables for two canonical correlations.

Measures of effect size help temper over-interpreting statistically significant results that may be of low practical importance. As discussed in Cohen (1988), while squared bivariate correlation coefficient ($r^2_{xy}$) values of 0.02, 0.13, and 0.25 are considered as small, medium, and large effect sizes, for canonical correlation coefficient, the benchmarks for small, medium, and large effect sizes are dependent on the number of variables in the two sets (**X** and **Y**). For example, for a CCA with two variables in each of **X** and **Y** sets, squared canonical correlation values of 0.04, 0.24, and 0.45 are considered as small, medium, and large effect sizes. But for a CCA with four variables in each of the two sets, we need squared canonical correlation coefficient values of 0.06, 0.34, and 0.59 for small, medium, and large effect sizes. In other words, what is a small or large effect size for canonical correlation coefficient depends on the set sizes used in the CCA. For details concerning effect size benchmarks for canonical correlation, readers may consult Cohen (1988, Chapter 10).

## 6. Canonical Function Coefficients

Once we ascertain that there are meaningful canonical correlations between the two sets of variables (see Desideratum 5), we proceed to examine the nature of the canonical correlations ("Where do what I have originate?" Thompson, 2000). For this purpose, function coefficients in CCA will help us to understand the nature of the meaningful canonical correlations.

Similar to other multivariate techniques (e.g., discriminant analysis; see Chapter 6, this volume), CCA produces multiple sets of coefficients. It is important that author(s) provide correct interpretations of these coefficients, including canonical function coefficients. For each of the two sets of variables (e.g., $X$ or $Y$ in the diagram for Desideratum 3), there is a unique set of function coefficients for each canonical function (i.e., canonical variate pair). As a result, there are generally multiple sets of function coefficients, assuming we have more than one canonical function. These canonical function coefficients (i.e., canonical weights) are used to linearly combine the observed variables in each set to obtain a canonical variate for that set, and they are derived to optimize the correlation between the pair of canonical variates. They are not necessarily derived to extract the maximum variance from the observed variables. In CCA, the function coefficients serve the primary purpose of determining the canonical variates.

Canonical function coefficients are available in both unstandardized and standardized forms. Both forms represent the partial and unique contribution of an observed variable to its canonical variate, after controlling for the variable's relation with others in the set. In addition to serving as weights for deriving a canonical variate, these coefficients are often used to gauge individual variables' relative importance to the resulting canonical variate. Because the variables in a set are often on different measurement scales that will affect the values of unstandardized function coefficients, unstandardized coefficients are generally not useful for assessing the relative importance of the variables. In comparison, standardized coefficients and their associated standardized forms of the variables are placed on the same scale so that they can be more easily compared in terms of their relative contributions. A larger coefficient is interpreted to indicate that a given variable contributes more unique variance to a canonical variate than another variable with a smaller coefficient. In a manuscript, it should be clear that discussions of unique and relative contributions of a variable be based on standardized function coefficients.

Because standardized function coefficients are based on a variable's partial relation with the canonical variate after the variable's association with other variables in the set has been removed, the overall variable/variate association might be under- or overestimated through interpretation of the standardized function coefficients. For example, low standardized function coefficients might underestimate a variable's association with the canonical variate when the variable under consideration is strongly related to both the canonical variate and other variables in the set. In addition, suppression effects can result in sign changes. Because of these issues, function coefficients alone themselves could lead to ambiguity in our understanding about canonical functions.

## 7. Canonical Structure Coefficients

Canonical structure coefficients provide another mechanism through which linkages between observed variables and their canonical variates can be examined. Unlike CCA function coefficients, which are affected by inter-variable correlations (similar to regression coefficients in regression analysis), a canonical structure coefficient measures the zero-order correlation between a given variable in a set and a given canonical variate of that set. As such, it reflects the overall degree of association between each variable and the resulting canonical variate that is at the center of the canonical correlation.

There is a general consensus that it is essential that canonical structure coefficients be considered in order to develop good understanding about the canonical functions (e.g., Pedhazur, 1997; Thompson, 2000). Structure and function coefficients of a variable with a canonical variate may be similar, or they may be quite different. When the function and structure coefficients are consistent, either both being low (the variable has little to do with the canonical variate) or both being high (the variable contributes a lot to the canonical variate), it typically does not present any difficulty in interpretation of the results. However, when there is a divergence between function and structure coefficients, some caution is needed in our interpretation. For example, when the function coefficient is low while the structure coefficient is high, the low function coefficient should not lead to the conclusion

| Function Coefficients | | | |
|---|---|---|---|
| | | Low | High |
| Structure Coefficients | Low | **A**<br>The variable shares little with the canonical variate. | **B**<br>A suppression effect is very likely the reason for the moderate/high function coefficient. The variable may actually share little with the canonical variate. |
| | High | **C**<br>The variable shares a lot with the canonical variate, but its contribution to the canonical variate overlaps with other variables in its set due to the collinearity between this variable and some other variable(s) in the set. | **D**<br>The variable shares a lot with the canonical variate. There is little collinearity between the variable and other variable(s) in the set. |

**Figure 3.2** Relational Patterns of Function and Structure Coefficients.

that the variable shares little with the canonical variate; on the contrary, the variable shares a lot with the canonical variate, but its contribution to the canonical variate overlaps with another or other variables in its own set. In another situation where the function coefficient is moderate or high in absolute value and the structure coefficient is very low, the high function coefficient should not lead to the conclusion that the variable shares a lot with the canonical variate. In this situation, it is very likely that the variable shares little with the canonical variate, and the high function coefficient is the result of a suppression effect, similar to the phenomenon in regression analysis (e.g., Horst, 1941; Lancaster, 1999). These situations, and their implications, are summarized in Figure 3.2.

For structure coefficients, values greater than or equal to .32 are often interpreted as practically meaningful. This comes from the fact that squared structure coefficients represent the amount of shared variance between the observed variable and its canonical variate, and that $.32^2$ represents approximately 10% common variance. More importantly, however, a coefficient is usually considered relative to other coefficients in the set, and the relative magnitudes of the coefficients often play an important role in defining a canonical variate. In a manuscript, CCA structure coefficients should be routinely reported and interpreted to help derive an understanding of the nature of the canonical variates.

## 8. Interpretation of Canonical Functions (Canonical Variates)

Understanding and interpreting the canonical correlation (function) is largely dependent on a meaningful explanation of what the canonical variates represent. The meaning of the canonical variate is usually inferred based on the pattern of coefficients (function and structure coefficients) associated with each of the variables. Similar to loadings in factor analysis (see Chapter 8, this volume), subsets of variables within a set with high and low weights help us to understand the nature of the resulting variate by revealing which variables are most closely associated with it and which variables are not. Here the researcher is faced with the substantive challenge of understanding the essence of different coefficients with the goal of providing a more parsimonious description of the canonical variate in terms of a construct it is attempting to capture based on the observed variables. For this purpose, variable function coefficients with relatively higher absolute values are given greater emphasis, while lower values are marginalized in the interpretation. The signs of these function coefficients should also be taken into consideration in relation to the scaling of the measured variables. In other words, negative relations between the variables and the canonical variate should be considered and discussed in relation to the substantive labeling of the resulting canonical variate.

As indicated above (see Desideratum 6), when function coefficients are considered, only the variable's unique shared variance with the canonical variate is taken into account. By contrast, structure coefficients consider how a given variable relates to its canonical variate without the interference of how the variable relates to other variables in the set. In this sense, structure coefficients are not confounded by a given variable's relation with other variables in the set. For the purpose of interpreting and labeling the canonical variates, both function and structure coefficients should be considered. In general, when we consider a variable's function and structure coefficients for the purpose of interpreting its canonical variates, the patterns described in both Quadrants A and B in Figure 3.2 would suggest that the variable contributes little to the substantive meaning of the canonical variate. On the other hand, the patterns described in both Quadrants C and D in Figure 3.2 would suggest that the variable has considerable contribution to the substantive meaning of the canonical variate.

Misinterpretation can easily occur in CCA. For example, unseasoned CCA users may misinterpret the Quadrant C pattern (low function coefficient and high structure coefficient) to mean that the variable contributes little to the substantive meaning of the canonical variate, because of the low function coefficient. Similarly, the Quadrant B pattern (high function coefficient and very low structure coefficient) can also be misinterpreted to mean that the variable contributes a lot to the substantive meaning of the canonical variate, when in reality the high function coefficient may be the result of a suppression effect. In a manuscript, it should be clear that the interpretation of canonical functions (canonical variates) is based on the joint consideration of both function coefficients and structure coefficients. As Levine (1977) emphasized, if one wants to understand the nature of canonical association beyond the computation of the canonical variate scores (such computation relies solely on function coefficients), one has to interpret structure coefficients.

## 9. Canonical Adequacy and Redundancy Coefficients

CCA is designed to maximize the canonical correlation (i.e., the correlation between two canonical variates in a pair). It may be of interest to know how much variance a given variate (e.g., $X_1^*$ in Figure 3.1) can extract from the variables of its own set (i.e., $x_1$, $x_2$, and $x_3$ in Figure 3.1), or how much variance a given variate (e.g., $X_1^*$ in Figure 3.1) can extract from the variables of the other set (i.e., $y_1$, $y_2$, and $y_3$ in Figure 3.1). Canonical adequacy and redundancy coefficients, respectively, are designed for such purposes.

*Canonical Adequacy Coefficients.* Structure coefficient measure the correlation between a variable and its canonical variate, and the squared structure coefficient represents the proportion of variance in the variable that is shared with its canonical variate. *Canonical adequacy coefficients* are associated with each canonical variate, and measure the average of all the squared structure coefficients for one set of variables as related to a given canonical variate formed from this set. Canonical adequacy coefficients describe how well a given canonical variate represents the original variables in its set, and quantitatively, it is the proportion of variance in the set of the variables (e.g., $x_1$, $x_2$, and $x_3$ in Figure 3.1) that can be reproduced by a canonical variate of its own set (e.g., $X_1^*$).

Related to this adequacy concept described above, we may also be interested in knowing what proportion of variance in a measured variable is associated with the extracted canonical functions. This percentage of variance in a variable associated with the extracted canonical functions is called *communality* ($h^2$), which is defined in the same way as in factor analysis (see Chapter 8, this volume). If a measured variable has very low communality relative to other variables in a CCA analysis, it suggests that this variable is behaving differently from other variables in the set, and probably does not belong to this set of variables.

*Canonical Redundancy Coefficients.* Between a pair of canonical variates, the redundancy index (Stewart & Love, 1968) measures the percentage of variance of the original variables of one set (e.g., $x_1$, $x_2$, and $x_3$ in Figure 3.1) that may be predicted from the canonical variate derived from the other set (e.g., $Y_1^*$). In other words, when we want to describe how much variance the $Y_1^*$ variate shares with the set of $X$ variables, or how much variance the $X_1^*$ variate shares with the set of $Y$ variables (i.e., $y_1$, $y_2$, and $y_3$ in Figure 3.1), we use canonical redundancy coefficients. A canonical redundancy coefficient for a canonical variate (e.g., $Y_1^*$) can be computed as the product of the canonical adequacy coefficient of its counterpart (i.e., adequacy coefficient for $X_1^*$) multiplied by the squared canonical correlation.

There is some controversy surrounding redundancy coefficients in terms of whether they should in fact be interpreted in CCA (Roberts, 1999). The primary concern for redundancy coefficients is that CCA is designed to maximize the canonical correlation, and it does *not* attempt to maximize the redundancy coefficient. As a result, "it is contradictory to routinely employ an analysis that uses functions coefficients to optimize [the canonical correlation], and then to interpret results (Rd) not optimized as part of the analysis" (Thompson, 1991, p. 89). Because of this and some other concerns, we caution that redundancy coefficient is not always meaningful in CCA. Researchers who use this statistic should present appropriate interpretations within their research contexts (e.g., in a multivariate concurrent validity study where there is theoretical expectation for high redundancy coefficients).

## 10. Substantive Research Issues

CCA provides a variety of interpretive frameworks that should be tied into the substantive question(s) that were outlined early in the manuscript. In general, license for these interpretations comes by way of at least one reliable (i.e., statistically significant) canonical correlation; failure to achieve this suggests that the two sets of variables cannot combine in a way to produce a variate pair correlation that is statistically greater than 0. Assuming the data have passed this threshold, researchers need to relate the resulting canonical functions to the initially proposed research questions. For each statistically significant canonical correlation, the relation between the two variates should be described both in terms of the resulting effect size and in terms of substantive interpretation of the nature of the variate pair in terms of what they represent and why this does or does not align with expectations. As described in previous sections, interpreting and labeling of the variates is best accomplished through the joint use of function and structure coefficients. Although in practice there is a tendency to focus the discussion on which variable(s) contribute most to the variates' definition and how this relates to the substantive problem, it is also important to consider which variables do not factor into the relations as indicated by very low coefficients.

## 11. Presentation of CCA Results

As discussed in, for example, Desiderata 6 and 7, CCA typically produces multiple sets of different coefficients, which can be quite a challenge for readers to grasp. In a manuscript, it is expected that the major CCA results are adequately presented in some tabular form such that the readers can easily find the major outcomes of the analysis. For each statistically significant canonical function, it is typical to present (1) canonical correlation and/or its squared value, (2) canonical function coefficients, and (3) canonical structure coefficients. In addition, canonical adequacy, canonical redundancy coefficients, and communality for each measured variable may also be presented. For the illustrative data previously presented in Table 3.2, there are three possible canonical functions, with only the first two being

Table 3.3  Example of Presentation of CCA Analysis Results

| | Function I | | | Function II | | | |
|---|---|---|---|---|---|---|---|
| | $B^a$ | $r^b$ | $r^{2\,c}$ | $B$ | $r$ | $r^2$ | $h^{2\,d}$ |
| **Job Satisfaction** | | | | | | | |
| $x_1$ – Career | 0.08 | 0.61 | 0.37 | −.25 | −.24 | 0.06 | 0.43 |
| $x_2$ – Supervisor | 0.83 | 0.95 | 0.90 | −.36 | −.30 | 0.09 | 0.99 |
| $x_3$ – Financial | 0.32 | 0.51 | 0.26 | 0.98 | 0.85 | 0.71 | 0.97 |
| *Adequacy* | | | 0.51 | | | 0.29 | |
| $Rd\,(R_{x \cdot ys})$ | | | 0.25 | | | 0.03 | |
| $R_C$ | | *0.70* | *0.49* | | *0.31* | *0.10* | |
| **Job Characteristics** | | | | | | | |
| $Rd\,(R_{Y \cdot xs})$ | | | 0.25 | | | 0.03 | |
| *Adequacy* | | | 0.52 | | | 0.27 | |
| $y_1$ – Task Variety | 0.24 | 0.72 | 0.52 | 0.74 | 0.48 | 0.22 | 0.74 |
| $y_2$ – Responsibility | 0.40 | 0.54 | 0.29 | −.91 | −.72 | 0.52 | 0.81 |
| $y_3$ – Autonomy | 0.69 | 0.87 | 0.76 | −.04 | 0.24 | 0.06 | 0.82 |

*Notes*

a  standardized function coefficients.

b  structure coefficients (when associated with the variables), or canonical correlation coefficients (when not associated with original variables).

c  squared structure coefficients (when associated with the variables); Adequacy and Redundancy coefficients, and squared canonical correlation coefficients.

d  communality: % variance in a variable jointly extracted by the two canonical functions.

statistically significant ($p < .05$). Based on the recommendation of Thompson (2000), Table 3.3 illustrates the major CCA results and how they might be presented in tabular form.

From this presentation, interested readers can more easily find all the relevant CCA statistics, such as the two canonical correlation coefficients (0.70 and 0.31, respectively), function and structure coefficients for the two functions, adequacy and redundancy coefficients associated with each canonical variate in each of canonical functions, and so forth. Based on the example data presented in Table 3.3, one may tentatively infer that the first canonical function probably represents a general positive relation between job characteristics and employee satisfaction, although Position Autonomy and Satisfaction With the Ssupervisor appear to play more important role in defining this first canonical function. The second canonical correlation is primarily defined by the *negative* relation between Financial Satisfaction and Job Responsibility, suggesting that there might be a perception that financial compensation is not in agreement with a position's responsibility.

The presentation in Table 3.3 is succinct and reasonably complete. In a manuscript involving CCA as the major analytic technique, information similar to that presented in Table 3.3 should be expected.

## References

Bagozzi, R. P., Fornell, C., & Larcker, D. F. (1981). Canonical correlation analysis as a special case of a structural relations model. *Multivariate Behavioral Research, 16,* 437–454.

Barcikowski, R., & Stevens, J. P. (1975). A Monte Carlo study of the stability of canonical correlations, canonical weights, and canonical variate-variable correlations. *Multivariate Behavioral Research, 10,* 353–364.

Cohen, J. (1988). *Statistical power analysis for the behavioral sciences* (2nd ed.). Hillsdale, NJ: Erlbaum.

Dunn, J. G. H., Dunn, J. C., & Gotwals, J. K. (2006). Establishing construct validity evidence for the Sport Multidimensional Perfectionism Scale. *Psychology of Sport and Exercise, 7,* 57–79.

Fan, X. (1996). Canonical correlation analysis as a general analytical model. In B. Thompson (Ed.), *Advances in social science methodology* (Vol. 4, pp. 71–94). Greenwich, CT: JAI Press.

Fan, X. (1997). Structural equation modeling and canonical correlation analysis: What do they have in common? *Structural Equation Modeling: A Multidisciplinary Journal, 4,* 65–79.

Horst, P. (1941). The role of predictor variables which are independent of the criterion. *Social Science Research Bulletin, 48,* 431–436.

Hotelling, H. (1935). The most predictable criterion. *Journal of Educational Psychology, 26,* 139–142.

Johnson, R. A., & Wichern, D. W. (2002). *Applied multivariate statistical analysis* (5th ed.). Upper Saddle River, NJ: Prentice-Hall, Inc.

Knapp, T. R. (1978). Canonical correlation analysis: A general parametric significance testing system. *Psychological Bulletin, 85,* 410–416.

Kshirsagar, A. M. (1972). *Multivariate analysis.* New York: Marcel Dekker.

Lancaster, B. P. (1999). Defining and interpreting suppressor effects: Advantages and limitations. In B. Thompson (Ed.), *Advances in social science methodology* (Vol. 5, pp. 139–148). Stamford, CT: JAI Press.

Lewis, T. F. (2007). Perceptions of risk and sex-specific social norms in explaining alcohol consumption among college students: Implications for campus interventions. *Journal of College Student Development, 48,* 297–310.

McDermott, P. A. (1995). Sex, race, class, and other demographics as explanations for children's ability and adjustment: A national Appraisal. *Journal of School Psychology, 33,* 75–91.

McIntosh, D. E., Mulkins, R., Pardue-Vaughn, L., Barnes, L. L., & Gridley, B. E. (1992). The canonical relationship between the Differential Ability Scales upper preschool verbal and nonverbal clusters. *Journal of School Psychology, 30,* 355–361.

Pedhazur, E. J. (1997). *Multiple regression in behavioral research: Explanation and prediction* (3rd ed.). Fort Worth, TX: Harcourt Brace College Publishers.

Roberts, J. K. (1999). Canonical redundancy (Rd) coefficients: They should (almost never) be computed and interpreted. In B. Thompson (Ed.), *Advanced in social science methodology* (Vol. 5, pp. 333–341). Stamford, CT: JAI Press.

Stevens, J. (2002). *Applied multivariate statistics for the social sciences* (4th ed.). Hillsdale, NJ: Erlbaum.

Stewart, D. K., & Love, W. A. (1968). A general canonical correlation index. *Psychological Bulletin, 70,* 160–163.

Thompson, B. (1984). *Canonical correlation analysis: Uses and interpretation.* Thousand Oaks, CA: Sage.

Thompson, B. (1991). A primer on the logic and use of canonical correlation analysis. *Measurement and Evaluation in Counseling and Development, 24,* 80–95.

Thompson, B. (2000). Canonical correlation analysis. In L. G. Grimm & P. R. Yarnold (Eds.), *Reading and understanding more multivariate statistics* (pp. 285–316). Washington DC: American Psychological Association.

Wilkinson, L., & APA Task Force on Statistical Inference. (1999). Statistical methods in psychology journals: Guidelines and explanations. *American Psychologist, 54,* 594–604.

# 4
# Cluster Analysis

Dena A. Pastor

The term *cluster analysis* is generally used to describe a set of numerical techniques for classifying objects into groups based on their values on a set of variables. The intent is to classify objects into groups such that objects within the same group have similar values on the set of variables and objects in different groups have dissimilar values. The objects classified into groups are most typically persons and the variables used to classify objects can either be categorical or continuous. Cluster analysis can be used as a data reduction technique to reduce a large number of observations into a smaller number of groups. It can also be used to generate a classification system for objects or to explore the validity of an existing classification scheme. Unlike other multivariate techniques, such as discriminant analysis (see Chapter 6, this volume), group membership is not known but instead is imposed on the data as a result of applying the technique. Because objects are classified into groups even if no grouping structure truly exists, analyses beyond the simple classification of objects into groups are essential. It is through these analyses that researchers can provide support for the replicability and validity of their particular cluster solution.

Cluster analytic methods can be classified as being either *model based* or *non-model based*. This chapter focuses primarily on non-model-based cluster analytic methods as these are the most commonly used. Model-based clustering methods, such as finite mixture modeling, utilize probability models, whereas non-model-based cluster analytic methods do not utilize such statistical models. Because a statistical model is not utilized, non-model-based methods are not formally considered to be inferential statistics and are more appropriately classified as numerical algorithms. Readers interested in model-based methods are referred to, for example, Everitt and Hand (1981).

Popular statistical software packages such as SAS and SPSS can be used to perform non-model-based cluster analysis as well as the replicability and validity analyses. Although a readable overview of cluster analysis can be found in Aldenderfer and Blashfield (1984), this chapter will more heavily rely on the more thorough and current treatment of the topic provided by Everitt, Landau, and Leese (2001). Table 4.1 displays the specific desiderata for applied cluster analytic studies, each of which will be explained further below.[1]

## 1. Target Population and Reasons for Classification

Because the purpose of a cluster analytic study is to classify objects into groups, an important question to answer at the forefront of a cluster analytic study is: "What objects are being classified?" Although

Table 4.1 Desiderata for Cluster Analysis

| Desideratum | Manuscript Section(s)* |
|---|---|
| 1. The target population of objects to be clustered is described and the reasons for wanting to classify these objects into clusters are stated. | I |
| 2. Justification is provided for the specific variables that were chosen as the basis for classifying objects into groups; if applicable, reliability and validity evidence for variables is provided. | I, M |
| 3. Any transformation of variables is described and justification provided. | M |
| 4. Any weighting of variables is described and justification provided. | I, M |
| 5. The characteristics of the sample are described and justification for use of the sample is provided. | M |
| 6. Outlying cases and missing data are addressed. | M |
| 7. If utilized, the proximity measure for capturing the similarity or dissimilarity between objects on the set of variables is explicated. | M |
| 8. The specific cluster analytic method used to classify objects into clusters is described in sufficient detail for replication (e.g., hierarchical: single linkage, Ward's method; non-hierarchical: $k$-means). | M |
| 9. The procedures used for choosing the final cluster solution are explained (e.g., dendrogram, amalgamation coefficients, statistical indices). | M |
| 10. The methods and the external variables used to assess the replicability and validity of the final cluster solution are described. | M |
| 11. The software and specific procedures utilized in the software are reported. | M |
| 12. Descriptive statistics for all variables are reported. | R |
| 13. Indices or figures used in deciding upon the final cluster solution are provided. | R |
| 14. The final cluster solution is presented, including a description of the characteristics of each cluster. | R |
| 15. Results are provided from the assessment of the solution's replicability and validity. | R |
| 16. The theoretical and/or practical utility of the final cluster solution is addressed and the results are interpreted in light of the research questions and target population. | D |
| 17. Suggestions for future analyses that could be used to provide support for the replicability and validity of the cluster solution are provided. | D |

* *Note*: I = Introduction, M = Methods, R = Results, D = Discussion

the objects or entities to be clustered in the social sciences typically are people, cluster analysis can be used to classify a wide variety of objects, including schools, countries, or words. The particular objects to be classified and the target population should be made explicit in the manuscript. For instance, the objects to be classified may be college students and the target population may be college students at four-year liberal arts institutions in America.

After defining the objects and target population, the manuscript should provide an answer to the question of: "What goals are to be accomplished by classifying objects into groups?" Cluster analysis may be pursued for the purposes of data reduction, to generate a classification scheme, to validate an existing classification scheme, or to explore the relations among variables. Readers are directed to Romesburg (1984) for a more thorough list of possible goals in a cluster analytic study. Answering the question of why objects need to be classified is an important piece of a cluster analytic manuscript since it justifies the use of the technique and also provides a framework for which the utility of the results can be gauged.

## 2. Variables Used to Classify Objects into Groups

In cluster analysis, objects are assigned to clusters based on their values on a set of variables. Because it is the particular variables chosen by the researcher that drive the resulting classification scheme, readers should be provided with answers to the following questions: "Why were the particular variables selected for the study?" and "What particular manifestations of the variables were used?"

The authors need to address why the particular variables selected for use in the cluster analysis are thought to be essential for separating objects into groups. The justification for the use of certain variables over others should be linked to goals that the authors hope to attain by classifying objects into groups. This is an important component of a cluster analytic study since it is well known that cluster solutions are highly dependent upon the variables used; the inclusion of even a single irrelevant variable can complicate the discovery of true clusters (Milligan, 1980).

Given the dependency of the cluster solution on the variables, it is imperative that the variables are described in detail in the Methods section. Different classification schemes may result if a different manifestation of a variable is utilized (e.g., test anxiety measured using self-report vs. galvanic skin response). When tests or scales are used to measure variables, supporting reliability and validity evidence should be provided (see Chapter 25, this volume).

If it is not obvious, the level of measurement (categorical, continuous) of the variables should also be reported. Although reporting the level of measurement is not typical, it is important in this context because the choice of proximity measure (see Desideratum 7) depends on whether variables are categorical, continuous, or a mix of both metrics.

## 3. Transformation

Many sources advise that variables be transformed prior to their use in a cluster analysis (e.g., to $z$-scores). Because there is a wide variety of transformations available and because the cluster solutions resulting from the use of untransformed versus transformed variables can differ, answers should be supplied to the following questions: "Were variables transformed and if so, why and how were they transformed?"

When the variables used to classify objects are on different scales (e.g., age, SAT scores, grade point average, gender), it is commonly recommended that the variables be transformed in some manner in order to avoid the overpowering of variables with larger scales on the classification of objects into groups. Even if all variables are measured on the same scale (e.g., all variables are 7-point Likert items), transformation is still recommended if the variables differ widely in their variability since variables with larger variances more heavily influence the cluster solution. Any transformation of the variables should be reported and justified. Because a wide variety of transformations exist, authors need to report the particular transformation method employed. For instance, transformation can be achieved by standardizing the variable or dividing the variable by its range. Other transformation methods are described by Milligan and Cooper (1988).

Although the cluster analysis literature often advocates for the transformation of variables on different scales or with disparate variances, authors need to carefully consider first whether transformation is even necessary, and if so, the ramifications of transformation. Whether transformation is necessary or not depends on whether the clustering method (see Desideratum 8) is invariant to transformations of the data. For instance, some clustering methods will yield the same solution regardless of whether variables are utilized in their untransformed or transformed forms (e.g., non-hierarchical methods minimizing the determinant of the within-cluster dispersion matrix, **W**). If the clustering method is not invariant to variable transformations, the manuscript would be strengthened by exploring the replicability of the solution using untransformed and transformed variables (see Desideratum 10).

There is another undesirable side effect associated with variable transformation. To explain this side effect, it is important to understand that the total variance of a variable consists of both between-group and within-group variance, with the groups here being the yet to be identified clusters. If a considerable amount of a variable's variance is between-groups, then the untransformed variable will have a strong influence on the cluster solution. Unfortunately, when a variable is transformed by standardization using the total variance, the influence and discriminating power of the variable is weakened (Fleiss & Zubin, 1969). For this reason, authors may want to consider the use of other transformation methods that utilize a variable's within-group variance (with some approximation for the yet to be identified groups) as opposed to its total variance (see section 3.6 of Everitt et al., 2001). Authors may also want to consult simulation studies examining the performance of different transformation methods (e.g., Milligan & Cooper, 1988; Steinley, 2004). Regardless of which transformation is adopted, an examination into the extent to which solutions replicate across different transformation methods is advised.

## 4. Weighting

When variables are equally weighted, each variable influences the resulting classification scheme to the same degree. When variables are differentially weighted, some variables have a stronger influence on the resulting classification scheme than others. Sometimes differential weighting of variables is explicit and intended. Other times, differential weighting occurs implicitly through the use of variables on different scales, heterogeneous variances, strong variable relations or the transformation method employed. The manuscript should thus address the question: "Were the variables differentially weighted?"

Sometimes the weighting of the variables is done explicitly, to heighten the influence certain variables have on the resulting classification scheme. When variables are weighted explicitly, the particular weighting method should be described in detail and justification provided for the differential weighting of variables. The differential weighting of variables is related to the discussion in Desideratum 3 pertaining to the selection of variables. Just as justification needs to be provided for excluding variables (which essentially is giving such variables a weight of zero), it also needs to be provided in the Introduction section for included variables that are weighted more heavily than others. In other words, a case needs to be made as to why the more heavily weighted variables should be allowed to have a stronger influence on the classification of objects into clusters.

The differential weighting of objects that occurs implicitly, as opposed to explicitly, should also be addressed by the authors. In Desideratum 3, it was mentioned that untransformed variables with larger scales (SAT vs. GPA) or with larger variances will more heavily influence the cluster solution. If untransformed variables with heterogeneous scales or variances are utilized, the differential weighting of variables that may result should be addressed. If used, authors should also acknowledge any effects the transformation method has on variable weighting (see Desideratum 3).

Implicit weighting of variables can also occur when variables that are highly related to one another are used in the clustering procedure. Related variables are more heavily weighted because the characteristics they share have more of an influence on the resulting classification scheme. If there are strong relations among variables, the authors should explain how they identified such relations (e.g., bivariate correlations) and how they chose to handle the multicollinearity. Some options include the elimination of redundant variables or the representation of related variables through the creation of composites or factors using principal components analysis or factor analysis, respectively (see Chapter 8, this volume). If the latter approach is taken, it should be noted that solutions using the composites or factors may differ in undesirable ways from those using the original variables (Chang, 1983; Green & Krieger, 1995; Schaffer & Green, 1998).

Another option for handling variables that are highly related is to use Mahalanobis distance as the proximity measure (see Desideratum 10). A pooled within-cluster dispersion matrix, **W**, is utilized in the Mahalanobis distance proximity measure to adjust for the relations among variables. Because Mahalanobis distance is calculated before the clusters are formed, various approximations of **W** have been proposed. When there is little to no relation among variables, Mahalanobis distance equals the squared Euclidean distance proximity measure computed using standardized variables. Mahalanobis distance should only be used as a proximity measure if justification can be provided for pooling **W** across the unidentified clusters and if both standardization (see Desideratum 3) and an equal weighting of variables is desired.

## 5. Sample

The information provided in this desideratum is intended to assist the authors in answering the following questions: "What particular sample of objects was used in the study and why?" and "How were data collected from that sample?" Even though non-model-based cluster analytic techniques do not qualify as being inferential statistical procedures per se, care still needs to be placed on sample selection if the researcher desires to generalize their findings to a larger population. To facilitate the generalizability of the cluster solution, a large sample representative of the target population should be acquired. Multiple samples should always be sought so that the replicability of cluster solutions across samples can be explored (see Desideratum 10). As is typical in most manuscripts, the Methods section should include a detailed description of the sample, including the sample selection method, sample size and sample characteristics. Giving readers access to this information allows them to judge the adequacy of the sample and the generalizability of the results.

## 6. Outlying Cases and Missing Data

Other questions that need to be answered about the data include: "Were there any outlying cases or missing data and if so, what was their nature and how were they handled?" As with most numerical techniques, outlying cases (objects) can overly influence the results of a cluster analysis (Milligan, 1980). It is recommended that the Methods section contain an explanation of the methods by which outliers were identified, including any univariate (e.g., histograms), bivariate (e.g., scatterplots), or multivariate (e.g., Mahalanobis distance) techniques. The number and nature of the identified outlying cases should be stated and the treatment of outlying objects described. Similarly, the nature and extent of missing data should be described and methods used to treat missing data, such as listwise deletion or multiple imputation, should be explicitly stated and justified.

## 7. Proximity Measures

Proximity measures, which may or may not be used by a clustering method, capture the similarity or dissimilarity between two objects on the set of variables. Because a wide variety of different proximity measures exist and because cluster solutions resulting from the use different proximity measures can differ, answers should be supplied to the following questions: "Was a proximity measure used, and if so, which proximity measure was used?"

Proximity measures can either be obtained directly, perhaps by acquiring ratings on a Likert scale as to the extent to which two objects are similar, or indirectly through calculation of measures that utilize the variables' values for the two objects. Regardless of whether the proximity measures were obtained directly or indirectly, the result is an $n \times n$ matrix of such measures, where $n$ equals the number of objects. A popular dissimilarity measure for continuous variables ($x$) is the Euclidean distance measure. For $J$ variables, the Euclidean distance for objects $g$ and $h$ is calculated as

$$d_{gh} = \sqrt{\sum_{j=1}^{J} (x_{gj} - x_{hj})^2},$$

(1)

with larger values of $d_{gh}$ being indicative of less similarity between the two objects on the set of variables. The Euclidean distances are calculated for all possible pairs of $n$ objects and stored in an $n \times n$ matrix. [Note that sometimes the square root is omitted in Equation 1, the resulting measure being the *squared* Euclidean distance.]

The majority of clustering methods (see Desideratum 8) operate on the $n \times n$ matrix of proximity measures. If a proximity measure is used, it should be described in sufficient detail and justification should be provided for the particular proximity measure chosen. Proximity measures are commonly classified by whether they are appropriate for only continuous variables, only categorical variables, or a mix of both.

If proximity measures are calculated for binary variables, the authors should state how the proximity measure handles joint absences. For instance, consider a binary variable that is coded 0 to represent the absence of a characteristic and 1 to represent the presence of a characteristic. Objects that share an absence of the characteristic are considered to be joint absences and objects that share a presence of the characteristic are considered joint presences. With binary variables of this type it is important to consider whether joint absences should be weighted the same as joint presences. In other words, is a shared absence of the characteristic just as telling about the similarity between objects as a shared presence of the characteristic? Because proximity measures for binary variables differ in not only how they weight joint absences and presences, but also in how they weight matches (0/0, 1/1) and non-matches (0/1), it is important to describe the proximity measure and justify its weighting scheme.

Explanation should also be provided for how proximity measures were calculated for any ordinal or nominal variables. If variables of mixed levels of measurement are utilized, the authors need to explicate how the different metrics were handled when creating the proximity matrix.

Proximity measures cannot only be classified by the kind of variables for which they are best suited, but also by the kind of clusters they are most inclined to capture. For example, correlations are well suited for identifying clusters that differ in their shape, which is the cluster's pattern of values across variables. Distance measures, on the other hand, are better suited for identifying clusters that differ in their elevation (the average value of the cluster across variables). A proximity measure should then be chosen that corresponds well with the unknown clusters' characteristics. A guess as to what these characteristics are could be provided by theory, previous research, or preliminary visualizations of the data (see Chapter 2 of Everitt et al., 2001). A manuscript would be strengthened by including a discussion of the methods used to anticipate the clusters' characteristics and the ability of the proximity measure to capture such characteristics.

## 8. Cluster Analytic Method

One of the most important decisions made by the researcher executing a cluster analytic study is the method chosen to separate objects into clusters. Most of the methods available are classified as being either hierarchical or non-hierarchical. The information necessary to report in a manuscript is first described for hierarchical methods and then for non-hierarchical methods.

*Hierarchical.* With hierarchical methods, no set number of clusters is specified a priori. Instead, $n$ partitions are made of the data, either beginning with all objects in their own cluster and progressing to a single cluster (*agglomerative*) or beginning with all objects in a single cluster and progressing until each object is in its own cluster (*divisive*). Whether the author is using an agglomerative or divisive partitioning should be made explicit. All hierarchical methods result in nested clusters, where the clusters from solutions with more clusters are nested within those with less clusters. If hierarchical methods are chosen, the authors need to provide a rationale for choosing a method that yields a nested classification of the data.

Hierarchical clustering methods differ in the rules that are used to classify objects into clusters. The different rules that are used are often reflected in the method's name. Popular hierarchical methods include: single linkage, complete linkage, average linkage, centroid linkage, median linkage, and Ward's method. The name of the hierarchical method should be provided along with a description of the method and justification for why a particular hierarchical method was chosen. Any alternative names for the method (e.g., single linkage is also known as nearest neighbor), should also be provided. Further information about hierarchical clustering methods can be found in Chapter 4 of Everitt et al. (2001).

*Non-hierarchical.* The majority of non-hierarchical methods require the user to specify the number of clusters before implementing the method. It is typical for this number to remain fixed during implementation of the procedure, although some non-hierarchical procedures allow the number of clusters to change (algorithmic methods; see Steinley, 2006). It is therefore important for users to report if their particular non-hierarchical method required the number of clusters to be specified a priori and if this number was fixed during the implementation of the procedure.

Non-hierarchical methods use an optimization algorithm to iteratively reallocate objects across clusters for the purposes of maximizing or minimizing some numerical clustering criterion. There are various different optimization algorithms and clustering criteria that can be used in non-hierarchical cluster analysis. An overview of the clustering criteria and optimization methods used in non-hierarchical methods is provided below.

Clustering criteria differ in the type of values that are desirable and in the type of matrix upon which they operate. If high values of the clustering criterion are associated with desirable cluster properties (such as high separation among clusters), then the goal is to allocate objects to clusters in order to maximize the criterion. Conversely, if low values of the clustering criterion are associated with desirable cluster properties (such as low heterogeneity within clusters), then the goal is to allocate objects to clusters in order to minimize the criterion. Some clustering criteria operate on a proximity matrix, while others, particularly those used with continuous variables, operate on the dispersion matrix of the variables.

The dispersion matrix is often represented in matrix form as $\mathbf{T}$, the total multivariate dispersion matrix, which can be decomposed into within-cluster ($\mathbf{W}$) and between-cluster matrices ($\mathbf{B}$). Objects are often reallocated to clusters for the purpose of minimizing the sum of the within-cluster or error sums of squares across all variables, which is equivalent to minimizing the trace of $\mathbf{W}$ (or maximizing the trace of $\mathbf{B}$). This criterion can also operate on a proximity matrix since it happens to be equivalent to minimizing the sum of squared Euclidean distances between objects and their cluster means.

Although the criterion of minimizing the trace of $\mathbf{W}$ is popular, there are several disadvantages associated with its use. First, different results are obtained when one uses raw versus standardized variables. Second, the criterion tends to produce spherically shaped clusters (which is desirable if clusters are truly spherical in shape, undesirable otherwise). To avoid the dependency of the solution on the scale of the variables (raw or standardized), Friedman and Rubin (1967) suggested the use of two other cluster criteria based on the dispersion matrix: minimizing the determinant of $\mathbf{W}$ and maximizing the trace of $\mathbf{BW}^{-1}$. The former criterion is the most popular of their suggestions and is desirable in that it does not create spherical clusters. Unfortunately, minimizing the determinant of $\mathbf{W}$ does share an undesirable property with minimizing the trace of $\mathbf{W}$: both methods impose clusters on the data of roughly the same size and shape. To overcome this disadvantage, cluster criterion utilizing $\mathbf{W}$ within each cluster were developed by Scott and Symons (1971), Maronna and Jacovkis (1974), and Symons (1981). Given the large variety of clustering criterion and the implications associated with their use, the criterion utilized needs to be described and justified.

Optimization algorithms are the different processes that are used to find the $k$ cluster solution that maximizes or minimizes the clustering criterion. Optimization algorithms are necessary because it is computationally prohibitive to calculate the clustering criterion for all possible partitions of $n$ objects into $k$ clusters. The process used by optimization algorithms known as hill-climbing algorithms entails four steps:

1. Initially partitioning the objects into $k$ groups.
2. For each object, computing the change in the clustering criterion that would result if the object were reallocated to a different cluster.
3. Reallocating those objects whose move results in the largest change in the clustering criterion.
4. Repeating steps 2 and 3 until no object can be moved to improve the clustering criterion.

In the manuscript, each of these four steps needs to be explicated in detail. Describing the first step is essential due to the vast number of different ways in which objects can be initially assigned to clusters. For instance, the partitioning from a hierarchical clustering method can be used as the initial classification or objects can be randomly assigned to partitions. Alternatively, cluster centroids (cluster seeds), which are the multivariate mean of the variables, can be randomly generated or supplied by the user. Another option is to choose $k$ observations whose cluster centroids serve as cluster seeds. Once seeds are chosen, objects are initially allocated to the cluster whose seeds they are nearest to in distance or objects are initially assigned on the basis of optimizing the clustering criterion. Because different initialization methods are known to lead to different solutions (Blashfield, 1976; Friedman & Rubin, 1967), it is important to describe the initialization method utilized.

Simulation studies have indicated that several of the initialization methods are prone to producing locally optimal solutions (Steinley, 2003). Locally optimal solutions are those that prematurely terminate and thus fail to truly maximize or minimize the clustering criterion. Researchers are encouraged to examine simulation studies such as Steinley's to investigate if their initialization method has a tendency to terminate at a local optimum. If local optima are commonly encountered with the chosen initialization method, the authors need to address how that problem was investigated and overcome. One means by which the problem can be investigated is by examining the extent to which the solution replicates when different initialization methods are utilized (see Desideratum 10).

In the second and third steps of the hill-climbing algorithm, the change in the clustering criterion is computed and objects whose movement results in the most desirable change in the criterion are reallocated. A popular non-hierarchical method, known as the $k$-means algorithm, alters these steps somewhat by computing the distance between an object's set of variable values and each cluster's centroids and allocating the object to the cluster that they are closest to in distance. Under certain conditions, the $k$-means approach is equivalent to using the four step hill-climbing procedure with the criterion of minimizing the trace of $\mathbf{W}$. Many of the aforementioned disadvantages associated with minimizing the trace of $\mathbf{W}$ also then apply to the $k$-means procedure. This relationship between methods was noted to emphasize the possibility of a correspondence between the different methods.

A detailed description of the second and third steps is necessary since implementations of these steps vary across non-hierarchical methods. For instance, updates to the clustering criterion (or to cluster centroids in $k$-means) may happen after the reallocation of each object, after a fixed number of object reallocations, or after all objects have been reallocated. This further emphasizes the need for the steps of a non-hierarchical clustering method to be described in detail. Simply supplying the name of the procedure (e.g., $k$-means, hill-climbing) does not provide readers with enough detail to replicate the study or judge the adequacy of the clustering methods used.

The most popular clustering method is therefore not always the best. Authors need to demonstrate that their choice of clustering method was not based on convenience or popularity, but on careful review of research investigating the properties of the various methods and consideration of their cluster's anticipated characteristics. In the absence of any knowledge regarding the cluster characteristics, it is recommended that authors choose a method (and, if applicable, proximity measures) well suited to uncover clusters with a variety of characteristics.

## 9. Choosing Final Cluster Solution

As mentioned previously, the ultimate goal of a cluster analysis is to classify objects into $k$ homogeneous groups. Cluster analysis is often criticized because of the subjectivity associated with the number of clusters to retain; therefore, it is essential that an answer is provided to the following question: "What methods were used to decide upon the final cluster solution?"

It is critical that the procedures used to arrive at the particular solution of $k$ clusters are described. It is recommended that a variety of different methods be used in deciding upon the final solution, with agreement across methods strengthening the retention of a particular solution. It is strongly suggested that the interpretability of the solution along with its correspondence to theory and past research also be considered. Several of the plots and indices that can be used in deciding on the final cluster solution are described below.

*Hierarchical.* With hierarchical clustering methods, there are $n$ solutions from which to choose, meaning there are as many solutions as there are objects. Commonly used procedures to select the final solution from $n$ solutions include inspection of the *dendrogram* (a tree-like plot showing the progression of objects being merged into clusters) or inspection of a graph plotting the number of clusters by some form of the average within cluster variance (often referred to as the fusion or amalgamation coefficient). The point at which the function flattens in these graphs directs the number of clusters to retain. Rather than looking at graphs of such coefficients, the coefficients can be listed with a large increase in the value from one solution to the next indicating that clusters were merged that were quite distinct from one another and that the solution just prior to the increase should be retained.

Because use of the dendrogram or procedures based on the average within cluster variance are highly subjective, other statistical rules for use with hierarchical clustering methods have been created and investigated. For instance, a simulation study by Milligan and Cooper (1988) examined the ability of 30 different statistical indices to detect the correct number of clusters. Some of the better performing indices included those of Calinski and Harabasz (1974), Duda and Hart (1973), and Beale (1969). In 1996, Milligan suggested that two or three of these better performing indices be utilized in hierarchical cluster analysis, with agreement among the indices providing stronger evidence for retention of a particular solution.

Rather than focusing on all $n$ solutions, a researcher typically chooses a smaller, more manageable number of solutions upon which to focus. For instance, only solutions ranging from 2 to 7 clusters may be seriously considered. It is important that researchers provide the range of clusters they considered plausible final solutions, along with justification for focusing on this range.

*Non-hierarchical.* Although the number of clusters is specified beforehand in non-hierarchical clustering methods, it is strongly encouraged that a variety of different solutions specifying a different $k$ be examined. Thus, one should avoid examining only a single solution of $k$ clusters and instead explore a variety of different non-hierarchical solutions. The exact values of $k$ that were examined should be made explicit in the manuscript. When more than one solution is examined, a flattening of the function of the final clustering criterion plotted against $k$ can be used to choose among solutions. As with the plots described for use with the hierarchical methods, the subjectivity associated with the interpretation of such a plot is considered problematic.

Although not extensively used or studied, the indices examined by Milligan and Cooper (1988) can be adapted for use with non-hierarchical methods, so long as the index is not limited to use with only nested clusters or hierarchical clustering methods (Steinley, 2006). Consult Steinley (2006) for a description of how these indices can be applied for determining $k$ in $k$-means clustering.

Steinley also describes algorithmic methods for arriving at the number of clusters in $k$-means clustering. With algorithmic methods, the number of clusters specified by the user is allowed to change during the implementation of the algorithm. The number of clusters may change through the use of user-supplied criteria, which may dictate the merging of similar clusters, the splitting of heterogeneous clusters, or the elimination of small clusters. Other algorithmic methods he describes use criteria whose values are not supplied by the user, but dictated by the data. Any use of algorithmic methods to generate a final cluster solution warrants their detailed description and justification in the manuscript.

While determining the number of clusters, the researcher may also question whether any clustering structure whatsoever exists in the data. As noted by Everitt et al. (2001), such tests for the absence of structure are not usually used in practice. The authors point to a review by Gordon (1998) for readers interested in these tests from a theoretical standpoint.

In summary, the Methods section should include a description of the plots, indices or algorithms used to decide upon the final cluster solution and how these procedures were used. Any additional criteria used in determining the number of clusters should also be described, including the interpretability of the solution or its correspondence with previous research or theory.

## 10. Replicability and Validity

Even if no structure exists in the data, clusters will be created as a result of implementing the clustering method. In fact, clustering methods are often criticized because they may create structure, regardless of whether structure truly exists. This is similar to exploratory factor analysis, where a solution is provided even if no true factor structure exists. To address this criticism, it is essential that the classification scheme resulting from a cluster analysis is shown to be consistent and meaningful. Thus, a cluster analytic study is not complete until it answers the following questions: "Is the final cluster solution replicable across conditions (e.g., samples, clustering methods, transformation methods)?" and "Is the final cluster solution meaningful?"

*Replicability.* The replicability of the cluster solution refers to the consistency or the stability of the solution across different conditions. Throughout this chapter it has been suggested that authors explore the extent to which their solution replicates across different transformation methods, weighting methods, proximity measures, and clustering methods. Evidence that a particular solution is robust to changes in such specifics of the cluster analysis procedure should be provided.

More importantly, the extent to which the final cluster solution replicates across different samples should be presented. Ideally two samples would be collected for this purpose, although in practice it may be more feasible to randomly split a single sample into two halves. If sample size permits, replicability could be examined by independently implementing the clustering method in each sample and then exploring the extent to which the same cluster solution was found in both samples. More formal procedures for replication are described by McIntyre and Blashfield (1980) and Breckenridge (1989, 2000).

*Validity.* If a solution replicates across samples, more faith can be placed in the generalizability and meaningfulness of the results. Although replicability is necessary for a solution's validity, evidence beyond replicability is needed for the solution to be considered meaningful, authentic, and valid. Validity evidence for the cluster solution can be acquired by showing that the clusters relate to external variables in ways anticipated by theory, logic and previous research. Validity thus involves embedding the solution in a program of construct validity.

The variables used to provide validity evidence for the cluster solution are called "external" because they are not the same variables used to create the clusters. Although it may seem appealing to illustrate the meaningfulness of the solution by showing that clusters significantly differ from one another in their "internal" variable values, it is not useful or surprising information since the clusters were created to be maximally different from one another on these variables.

Because the validity of a solution is essential to a cluster analytic study, the external variables need to be carefully selected. A description of the external variables and a justification for their use in assessing the validity of the cluster solution should be provided early in the manuscript, ideally in the introduction. The Methods section should contain a detailed description of the external variables as well as the analyses used to assess the relationship between the clusters and external variables. Possible analyses might include a discriminant analysis (see Chapter 6, this volume) or a multinomial logistic regression (see Chapter 17, this volume), with the clusters regressed on the external variables. Alternatively, a series of ANOVAs or chi-squares may be used to examine the bivariate relationship between clusters and continuous or categorical external variables, respectively. For reporting purposes, readers are encouraged to refer to the chapters in this book corresponding to the statistical methods chosen for their validity analyses.

## 11. Software

Blashfield (1977) found that when different software packages are used, different solutions may be obtained even if the same clustering method is applied to the same data. Although this finding may not hold true today, it still underscores the need to provide answers to the following questions: "What is the name and version of the software program used?" and "If applicable, which specific facility was used within that software program?"

A variety of different software programs can be used for cluster analysis purposes (see the Appendix of Everitt et al., 2001). Some programs are limited to cluster analysis (e.g., Clustan), whereas others are more general statistical software programs (e.g., SAS, SPSS). In the Methods section the author should report the name and version of software program used. If applicable, the specific facility used within the software program should also be mentioned. For instance, if SAS is utilized the specific procedure within SAS (e.g., proc cluster, proc fastclus, proc modeclus) should also be provided.

## 12. Descriptive Statistics

A common requirement of almost all quantitative research is the reporting of descriptive statistics. Descriptive statistics (e.g., means, standard deviations; frequencies for categorical variables) for all internal and external variables should be reported or made available to the reader. In the context of cluster analysis, descriptive statistics are important because they can provide information as to the level of measurement (categorical, continuous) of the internal and external variables. This information enables readers to judge the appropriateness of proximity measures (if used) and the analyses used for validation purposes. The descriptive statistics for the internal variables are particularly useful because they convey information about the variables for the overall sample, which may be of interest when the same information is reported by cluster (see Desideratum 14). Measures of association between all internal variables should also be provided for the purposes of multicollinearity assessment (see Desideratum 4).

## 13. Indices and Figures

In Desideratum 8, authors were instructed to describe the figures and/or indices that were to be used in deciding upon the final cluster solution. Such figures or indices should be displayed in the Results section or made available to the reader upon request. It is recommended to report indices for the range of cluster solutions considered (not just the final solution) so that readers can judge for themselves the solution to be favored. The bottom line is that all information used by the researchers to decide on the final solution should be shared with the reader. In addition to figures or indices, this may include the interpretability of the solution and its correspondence with previous research and theory.

The extent to which indices or figures were in agreement should be discussed. If multiple solutions were considered on the basis of indices or figures, the process used to arrive at the final solution should be described. For example, suppose indices favored equally a 2-cluster and 3-cluster solution. The 2-cluster may be selected over the 3-cluster solution if the extra cluster in the latter solution did not appear to be sufficiently distinct from another cluster or if the extra cluster was not interpretable in light of theory and previous research. Whenever multiple solutions are favored, each of the solutions should be described and the process used in selecting the final solution explained.

This is also a sensible place in the manuscript for the authors to mention if any small or outlying clusters in the favored solutions were found. Small clusters are those that capture a very minor proportion of the sample and outlying clusters are those capturing objects with extreme values on the variables. Small or outlying clusters should be described and the authors need to convey whether they consider these clusters to be theoretically important or irrelevant. Commonly, the presence of a small or outlying cluster directs which cluster solution is retained. For instance, a small cluster formed when specifying $k$ clusters may point to retention of the $k-1$ cluster solution.

## 14. Final Cluster Solution

The results section should clearly state which cluster solution was chosen and a thorough description of the resulting clusters. The clusters can be described by providing a table with the proportion of the sample in each cluster and relevant descriptive statistics (e.g., means, standard deviations, bivariate correlations) for the variables by cluster. It may also be useful to provide a figure to illustrate the cluster characteristics. For instance, a graph of the variable means by cluster may nicely convey the cluster profiles. Other graphs are described in section 8.6 of Everitt et al. (2001).

It is recommended that clusters be referred to by a letter or number (Cluster 1, Cluster 2 or Cluster A, Cluster B) rather than by a name. Although this recommendation may result in seemingly uninformative names for the clusters, it is beneficial in that it allows for a more objective interpretation of the clusters and helps to avoid the creation of misleading names. For instance, Meece and Holt (1993) used cluster analysis to classify 5th and 6th grade students on measures of achievement goal orientation. They labeled their third cluster "low mastery-ego," because this cluster was lower than other clusters in their average levels of mastery and ego goals. Although they explained that "low" in the label meant "low" in relation to the other clusters, "low" could also be interpreted in relation to the scale of the mastery and ego goals. The label for this cluster could easily be misinterpreted as meaning that the cluster is characterized by students with low mastery and ego goals, when, in fact, the averages for this cluster fell at the midpoint of the scale.

## 15. Replicability and Validity Results

In Desideratum 9 it was recommended that the methods used to assess the replicability and validity of the final cluster solution be provided. It is in the Results section that the author needs to present the results of the replicability and validity analyses in enough detail to allow the reader to judge for themselves the consistency and meaningfulness of the final solution.

With regard to replicability, the extent to which the same solution was championed across different conditions (e.g., samples, transformation methods, weighting methods, proximity measures, and clustering methods) should be reported, and for each condition the characteristics of the clusters should be described (see Desideratum 14). When different conditions are examined using the same sample, it is important to present the extent to which objects were classified in the same cluster.

A wide variety of statistical analyses can be pursued to examine the validity of a cluster solution. Any statistical method appropriate for examining the relation of a nominal variable (e.g., cluster membership) with one or more categorical or continuous variables is appropriate for this purpose. Therefore, discriminant analysis, multinomial logistic regression, MANOVA, ANOVA, chi-square or log-linear modeling could be used for validation purposes. When reporting the results of these analyses, readers should consult the chapters in this book that correspond to the particular statistical analyses being utilized.

## 16. Utility of Final Cluster Solution

A thoughtful examination of the study's results should be provided in the discussion section. This examination should provide answers to the following questions: "Did the final cluster solution correspond to the clusters anticipated based on previous research or theory?" and "Do the authors consider the results of the replicability and validity analyses to be supportive?" Possible explanations should be provided for any unanticipated results, including any results that do not support the replicability and validity of the cluster solution. If a meaningful cluster solution was obtained with evidence supporting its replicability and validity, then a discussion should ensue as to whether the goals sought in classifying objects into groups were met. This discussion ties the results of the study back to the purposes stated in the introduction for classifying objects into groups (see Desideratum 1). The utility of the cluster solution needs also to be addressed. In other words, the ways in which the classification scheme can be used in future research or in applied settings should be explicated.

## 17. Future Analyses

It is a well-known fact in measurement that the collection of reliability and validity evidence for an instrument or scale is a never-ending process. The same can be said about cluster analysis. Once a solution is obtained, continuous evidence needs to be gathered to support the consistency and authenticity of the solution. If a seemingly stable and meaningful cluster solution is found, the authors should provide next steps in establishing the replicability and validity of the cluster solution. Suggestions may include the conditions under which the replicability should be further assessed or other external variables that should be used in future validity studies.

## Note

1  Although this chapter focuses on non-model based clustering methods, many of the desiderata in Table 4.1 also apply to model-based clustering methods. Desiderata *generally* pertaining to model-based approaches include: 1–2, 5–6, 8–17. Because the specifics of these desiderata will differ for model-based approaches, the literature on model-based classification should be consulted.

## References

Aldenderfer, M. S., & Blashfield, R. K. (1984). *Cluster analysis.* Newbury Park, CA: Sage.

Beale, E. M. L. (1969). Euclidean cluster analysis. In Bulletin of the International Statistical Institute: Proceedings of the 37th Session (London), Book 2 (pp. 92–94). Voorbug, Netherlands: ISI.

Blashfield, R. K. (1976). Mixture model tests of cluster analysis. Accuracy of four agglomerative methods. *Psychological Bulletin, 83,* 377–385.

Blashfield, R. K. (1977). The equivalence of three statistical packages for performing hierarchical cluster analysis. *Psychometrika, 42,* 429–431.

Breckenridge, J. N. (1989). Replicating cluster analysis: Method, consistency, and validity. *Multivariate Behavioral Research, 24*, 147–162.

Breckenridge, J. N. (2000). Validating cluster analysis: Consistent replication and symmetry. *Multivariate Behavioral Research, 35*, 261–285.

Calinski, R. B., & Harabasz, J. (1974). A dendrite method for cluster analysis. *Communications in Statistics, 3*, 1–27.

Chang, W. C. (1983). On using principal components before separating a mixture of two multivariate normal distributions. *Applied Statistics, 32*, 267–275.

Duda, R. O., & Hart, P. E. (1973). *Pattern cassification and scene analysis.* New York: Wiley.

Everitt, B. S., & Hand, D. J. (1981). *Finite mixture distributions.* London: Chapman & Hall.

Everitt, B. S., Landau, S., & Leese, M. (2001). *Cluster analysis* (4th ed.). London: Arnold Publishers.

Fleiss, J. L., & Zubin, J. (1969). On the methods and theory of clustering. *Multivariate Behavioral Research, 4*, 235–250.

Friedman, J. H., & Rubin, J. (1967). On some invariant criteria for grouping data. *Journal of the American Statistical Association, 62*, 1159–1178.

Gordon, A. D. (1998). Cluster validation. In C. Hayashi, N. Ohsumi, K. Yajima, H.-H. Bock, & Y. Baba (Eds.), *Data science, classification and related methods* (pp. 22–39). Tokyo: Springer-Verlag.

Green, P. E., & Krieger, A. M. (1995). Alternative approaches to cluster-based market segmentation. *Journal of the Market Research Society, 37*, 221–239.

McIntyre, R. M., & Blashfield, R. K. (1980). A nearest-centroid technique for evaluating the minimum-variance clustering procedure. *Multivariate Behavioral Research, 2*, 225–238.

Maronna, R., & Jacovkis, P. M. (1974). Multivariate clustering procedures with variable metrics. *Biometrics, 30*, 499–505.

Meece, J., & Holt, K. (1993). A pattern analysis of students' achievement goals. *Journal of Educational Psychology, 85*, 582–590.

Milligan, G. W. (1980). An examination of the effect of six types of error perturbation on fifteen clustering algorithms. *Psychometrika, 45*, 325–342.

Milligan, G. W. (1996). Clustering validation: Results and implications for applied analyses. In P. Arabie, L. J. Hubert, & G. De Soete (Eds.), *Clustering and classification* (pp. 341–375). River Edge, NJ: World Scientific Publication.

Milligan, G. W., & Cooper, M. C. (1988). A study of standardization of variables in cluster analysis. *Journal of Classification, 5*, 181–204.

Romesburg, C. H. (1984). *Cluster analysis for researchers.* Belmont, CA: Lifetime Learning Publications.

Schaffer, C. M., & Green, P. E. (1998). Cluster-based market segmentation: Some further comparisons of alternative approaches. *Journal of the Market Research Society, 40*, 155–163.

Scott, A. J., & Symons, M. J. (1971). Clustering methods based on likelihood ratio criteria. *Biometrics, 27*, 387–398.

Steinley, D. (2003). K-means clustering: What you don't know may hurt you. *Psychological Methods, 8*, 294–304.

Steinley, D. (2004). Standardizing variables in K-means clustering. In D. Banks, L. House, F. R. McMorris, P. Arabie, & W. Gaul (Eds.), *Classification, clustering and data mining applications* (pp. 53–60). New York: Springer.

Steinley, D. (2006). K-means clustering: A half-century synthesis. *British Journal of Mathematical and Statistical Psychology, 59*, 1–34.

Symons, M. J. (1981). Clustering criteria and multivariate normal mixtures. *Biometrics, 39*, 35–43.

# 5

# Correlation and Other Measures of Association

Jason W. Osborne

Correlation and, more generally, association is a basic staple of inferential statistics, often easily computed with handheld calculators, simple spreadsheet software, and even by hand. Most modern researchers compute correlations and other indices of relatedness without thinking deeply about their choices. Despite being basic in the current pantheon of analytic options, measures of association are important, almost ubiquitous, and can easily produce errors of inference if not utilized thoughtfully and with attention to detail.

The goal of correlational analyses is to assess whether two variables of interest covary or are related, and ultimately, to draw conclusions that allow the researcher to speak to some issue in the "real world." In the best of instances, the researcher has: (a) a theoretical rationale for exploring this issue, (b) high quality measurement of variables of interest, (c) an appropriate analytical approach, (d) attention to detail in terms of ensuring assumptions of the analytic approach are met, and (e) followed best practices in performing the analysis and interpreting the result(s).

The primary focus of this chapter will be the most common index of association, Pearson's Product Moment Correlation Coefficient, $r$, but will also include brief discussions of other measures of relation, such as alternative correlation coefficients and odds ratios. Related but more advanced correlational procedures such as multiple regression, logistic regression, multilevel modeling, and structural equation modeling are beyond the scope of this chapter (but see Chapters 21, 17, 10, and 28, this volume). Of course, as modern statistics incorporates these and all ANOVA-type analyses into a single general linear model (GLM), many of my comments will apply across a broad range of techniques. Contemporary treatments of issues in this chapter are included in classic textbooks such as Cohen, Cohen, West, and Aiken (2002), Pedhazur (1997), Aiken and West (1991), Tabachnick and Fidell (2001) and Osborne (2008b), among others. Specific desiderata for studies using correlation/relational methodologies are presented in Table 5.1 and explicated subsequently.

## 1. Substantive Theories and Measures of Association

Conducting "fishing expeditions," such as examining large correlation matrices to look for ideas to explore, is really a sub-optimal and limiting way to go about trying to understand the world. Rather,

Table 5.1 Desiderata for Correlation and Other Measures of Association

| Desideratum | Manuscript Section(s)* |
|---|---|
| 1. A succinct literature review clearly situates the current study in the field's context. The substantive theory or rationale that led to the investigated relation(s) is explained. | I |
| 2. The goals and the correlational nature of the research question(s) are clearly stated. | I |
| 3. The variables of interest are explicitly identified and operationalized. | I/M |
| 4. The sampling framework and sampling method(s) are clearly defined and justified. | M |
| 5. Results from (preferably a priori) power analyses that are in line with the chosen sampling strategy are reported. | M |
| 6. If data to be analyzed are nested, multi-level in nature, or otherwise more appropriate for multi-level modeling, those methods are used. | M/R |
| 7. Relevant psychometric characteristics (preferably of the current data but possibly from previous, similar studies) are presented and discussed. At minimum this should include reliability and factor structure (if applicable). Variables with unacceptable reliability should not be included in analyses. | M/R |
| 8. Fundamental descriptive statistics of the variables are presented and discussed (e.g., measurement scale, mean, variance/standard deviation, skewness and kurtosis). | M/R |
| 9. If preliminary analyses suggest that data on variables of interest are not reasonably normally distributed, appropriate actions are taken to normalize the data or subsequent analytic strategies that accommodate significant deviations from normality are chosen (and justified as appropriate). | M/R |
| 10. Missing data, if present, are appropriately dealt with. | M/R |
| 11. Authors report how outliers/fringeliers were defined, identified, and, if any were present, how they were dealt with. | M/R |
| 12. The testing of assumptions underlying the analyses are presented. | M/R |
| 13. Where variables violate distributional assumptions of Pearson $r$, alternative correlational coefficients are used. | |
| 14. Multiple zero-order analyses are not reported unless defensible corrections for increased Type I error rates are employed. Multiple analyses should be combined into a single multivariate analysis where possible. | R |
| 15. Authors use semipartial and partial correlations where appropriate, and interpret them correctly | R/D |
| 16. Where appropriate, an interpretation of results takes variable transformations into account. | R/D |
| 17. Appropriate effect size measures are reported and interpreted. | R/D |
| 18. $p$ values are interpreted correctly. Effect sizes, not $p$ values, should guide the narrative and discussion. | R/D |
| 19. Curvilinear relationships or interactions, when found, should be presented graphically. | R/D |
| 20. Discussion of correlational analyses refrains from making causal inferences. | R/D |

* *Note*: I = Introduction, M = Methods, R = Results, D = Discussion

good research flows from good theory. Thus, I believe that reviewers should evaluate any type of quantitative analysis from within the knowledge base of the discipline that anchors the research.

Of course, data can inform theory just as easily as theory can drive research. Yet, unless we clearly ground the current research being reported in the context of what has come before and is currently being discussed (where appropriate), we risk merely reinventing the wheel. It is my position that no research should occur without clearly articulating how previous knowledge has led to the specific questions addressed within the study at hand.

Thus, reviewers should see a thorough review of the literature that has led to the current study, and if citations only go back a decade or two, reviewers should make sure authors are not missing seminal research in the field. Few lines of research have roots less than 10–20 years old.

## 2. Goals and Correlational Nature of the Research Question(s)

Reviewers should demand clearly stated, operationalizable goals and objectives that clearly lend themselves to correlational/relational analyses. All too often one sees goals for group differences, growth, or change inappropriately explored through correlational methods (as well as correlational hypotheses are tested via ANOVA-type designs). Researchers should be encouraged to align their analytic strategies to their goals/data as closely as possible and reviewers need to enforce best practices. In the case of correlational analyses, the goals and objectives should clearly be focused on the relational nature of two or more variables. Words such as "group difference" or "causal" should not be included in correlational hypotheses.

## 3. Operationalizing the Variables of Interest

Researchers interested in the social sciences often intuitively understand the need to explicitly identify and describe the variables of interest while at the same time being challenged to operationalize important, but almost unmeasurable constructs. One can have the most thorough and impressive literature review, sound theory and rationale, and important goals, but without being able to *validly and reliably* measure the constructs at hand the research is little more than a thought experiment. Operationalization (defining specifically how one did or will measure the constructs at hand) is the *conditio sine qua non* of science, but this is often the place in the logic of the research project where the scientific quality of the project breaks down. Reviewers must demand high quality operationalization in order to get high quality research outcomes.

Because we are dealing with variable relations in this chapter, we must know (a) exactly what variables the researchers were examining for a potential association, (b) how the authors defined each considered construct, (c) what *specific* operations the researchers utilized to measure each variable, and (d) how successfully the researchers measured what they thought they measured. For example, some researchers have studied whether students' use of instructional technology is related to student achievement. But what, exactly, do they mean by "instructional technology" or "student achievement?" In the modern context, researchers have referred to instructional technology as a catch-all phrase for computers, the internet, personal data assistants (PDAs), calculators, smartboards, student response systems, on-line assessment systems, word processors, multimedia learning systems, and so forth. And, of course, student achievement can refer to many different things, from high-stakes end-of-grade tests mandated in many states to ephemeral concepts such as change in knowledge or acquisition of skills. One could imagine many different studies purporting to test whether instructional technology is related to student achievement, all using radically different operationalizations of each variable, each potentially coming to divergent conclusions at least partly because of a different operational definition rather than a flaw in the basic premise. Thus, the reader should have a very precise picture of exactly what was meant by instructional technology, and how and how well student use of instructional technology was assessed. The last point refers to the quality of measurement, which will be discussed below.

Finally, it is at this point that the reviewer makes an assessment as to whether the basic measurement was adequate or not. For example, if I am assessing student use of instructional technology, and I merely ask the student to indicate on a survey whether they used the internet:

a everyday,

b twice a week or more,

c at least once a month, or

d less than once a month

have I measured this construct with high precision and quality? Not at all. Leaving aside the inherent issues with self-report data, the question—and potential response—is imprecise (we do not know *why* students were using the internet or what they did with it). Yet I could easily correlate resulting data with scores on a high-stakes test and publish conclusions that may be erroneous. It is important that reviewers disallow poorly operationalized research from becoming part of the knowledge base of the field, as it can skew results dramatically.

## 4. Justification of Sampling Method(s)

There are many studies with interesting conceptual frameworks but flawed sampling methodology. For example, the social science literature contains studies purporting to study working adults but using adolescent psychology students; or investigations intending to generalize results to school age children but utilizing college students; or studies of the U.S. population as a whole (with all the racial/ethnic, chronological, religious, and developmental diversity) but using samples that are quite homogeneous in nature.

Strong inference requires good sampling. The sample must be representative of the population of interest in order for the results to generalize. Studies should clearly describe the intended population and the methods used to recruit and retain participants for the study that are representative of that population, along with response rates, and so forth. The results should confirm that the sampling strategy was successful by, where possible, showing that sample characteristics are not meaningfully different from known population characteristics.

It is not always the case that we want a sample that is purely representative of the population as a whole. Sometimes a researcher will want to compare different subgroups, and where subgroups are not equally common within a population, a stratified sampling framework should be used,[1] the rationale for how the strata were chosen and sampled should be defended and the analyses and discussion of the results should take it into account. For example, if a researcher wanted to explore the relation between income and life satisfaction across the lifespan, that researcher should be careful to use stratified sampling to gather equal numbers of people at various ages because random sampling from the population will not yield equal numbers of individuals of each age range (see also Chapter 30, this volume).

## 5. Power Analysis

Statistical power is the probability of rejecting a null hypothesis when indeed it is false given a particular effect size, alpha level, sample size, and analytic strategy.[2] Jacob Cohen (e.g., Cohen, 1962, 1988) spent many years encouraging the use of power analysis in planning research, reporting research, and in interpreting results (particularly where null hypotheses are not rejected). Authors were discussing the issue of power more than 60 years ago (e.g., Deemer, 1947), yet few authors in the social sciences today (only about 2% in educational psychology) reported having tested or used power (Osborne, 2008d). The concept of power is complementary to significance testing and effect size and, Cohen and others argued, a necessary piece of information in interpreting research results (see Chapters 7 and 24, this volume).

Null Hypothesis Statistical Testing (NHST) has been reviled in the literature by many as counterproductive and misunderstood (e.g., Fidler & Cumming, 2008; Thompson, 2002). The ubiquitous

$p < .05$ criterion is probably not disappearing from scholarly social science any time soon despite the fact that almost all null hypotheses are ultimately false in an absolute sense (e.g., rarely is a correlation coefficient exactly 0.00; rarely are two means exactly equal to all decimal places), and thus, given sufficient power, even the most minuscule effect can produce a $p < .05$ (see also Tukey, 1991). Thus, the reporting of effect sizes (in the case of simple correlation, $r$ and $r^2$ are the effect sizes to report), which tells us generally how important an effect is, is crucial.

Power is critical in two different aspects of research. First, Cohen and others argued that no prudent researcher should conduct research without first making a priori analyses to determine the probability of correctly rejecting the null hypothesis. Researchers who fail to do this risk committing Type II errors (failing to reject a false null hypothesis), thus wasting the time and effort conducting underpowered research and worse, risk causing confusion in the field by publishing conflicting results that may not be conflicting at all, merely a collection of studies that would have all been in accord had all been sufficiently powered. Additionally, researchers who fail to do a priori power analyses risk gathering too much data to test their hypotheses—if a power analysis indicates that data from $n = 100$ participants would be sufficient to detect a particular effect, gathering a sample of $n = 400$ is a waste of resources.

Second, power analyses are useful in order to shed light on null results. For example, when a researcher fails to reject a null hypothesis with an a priori power of .90 to detect anticipated effects, s/he can be fairly confident of having made a correct decision. However, when a null hypothesis is not rejected, but there is low power, it is unclear as to whether a Type II error has occurred.

Further, something not generally discussed in the social science literature is that low power has implications for Type I error rates in bodies of literature. In the ideal case of strong power (i.e., almost all real effects are detected) and the almost ubiquitous alpha of .05 (i.e., few false conclusions of effects where there are none), a relatively small proportion of studies achieving statistical significance in a field would result from Type I errors. However, maintaining an alpha of .05 but considering a sub-optimal situation where a large group of studies have low power (e.g., .20), in fact a much larger relative proportion of published studies would contain Type I errors because so few true effects are being detected (Rossi, 1990). Thus, ironically, poor power in studies can inflate rates of Type I error across research domains, meaning many real effects are being missed (and possibly not published as a result), while at the same time a large number of Type I errors are being published relative to those studies where true effects are being detected. This may, in turn, lead to seemingly conflicting results in a line of research, and give rise to apparent controversies in fields that traditionally have poor power to detect effects which are, in the end, more likely a result of poor power than conflict of a substantive nature.

In sum, statistical power is an important concept, but authors and reviewers tend to neglect this piece of the empirical puzzle. Authors should report a priori (preferably) or a posteriori power to detect effects, and reviewers should insist on seeing it.

## 6. Nested Data

People tend to exist within organizational structures, such as families, schools, business organizations, churches, towns, states, and countries. In education, students exist within hierarchical social structures that can include families, peer groups, classrooms, grade levels, schools, school districts, states, and countries. Workers exist within production or skill units, businesses, and sectors of the economy, as well as geographic regions. Health care workers and patients exist within households and families, medical practices and facilities (e.g., a doctor's practice, or hospital), counties, states, and countries. Many other communities exhibit hierarchical data structures as well.

In addition, Raudenbush and Bryk (2002) discussed two other types of data hierarchies that are less obvious: repeated-measures data and meta-analytic data. Once one begins looking for hierarchies in data, it becomes obvious that data repeatedly gathered on an individual are hierarchical, as all the observations are nested within individuals (who are often nested within other organizing structures). While there are other adequate procedures for dealing with repeated measures data, the assumptions relating to them being rigorous (see Chapter 2, this volume), the procedures relating to hierarchical modeling require fewer assumptions (see Chapter 10, this volume). Also, when researchers are engaged in the task of meta-analysis—the analysis of a large number of existing studies (see Chapter 19, this volume)—it should become clear that participants, results, procedures, and experimenters are nested within experiment or study.

Hierarchical, or nested, data present several problems for analysis. First, individuals that exist within hierarchies tend to be more similar to each other than people randomly sampled from the entire population. For example, students in a particular third-grade classroom are more similar to each other than to students randomly sampled from the population of third-graders. This is because students are not randomly assigned to classrooms from the population, but rather are assigned to schools based on geographic factors. Thus, students within a particular classroom tend to come from a community or community segment that is more homogeneous in terms of morals and values, family background, socio-economic status, race or ethnicity, religion, and even educational preparation than the population as a whole. Further, students within a particular classroom share the experience of being in the same environment—the same teacher, physical environment, and similar experiences—which may lead to increased homogeneity over time.

*The problem of independence of observations.* The previous discussion indicated that, often, study participants tend to share certain characteristics (e.g., environment, background, experience, demographics) and hence their data are not fully independent. However, most analytic techniques require independence of observations as a primary assumption. Because this assumption is violated in the presence of hierarchical data, ordinary least squares regression produces standard errors that are too small (unless these so-called *design effects* are incorporated into the analysis[3]). When assumptions of independence are not met, an undesirable outcome is that the smaller standard errors bias significance testing toward inappropriate rejection of null hypotheses.

*The problem of how to deal with cross-level data.* Going back to the example of a third-grade classroom, it is often the case that a researcher is interested in understanding how environmental variables (e.g., teaching style, teacher behaviors, class size, class composition, district policies or funding, or state or national variables) affect individual outcomes (e.g., achievement, attitudes, retention). But given that outcomes are gathered at the individual level whereas other variables are assessed at the classroom, school, district, state, or nation levels, the question arises as to what the unit of analysis should be and how one should deal with the cross-level nature of the data.

One way researchers traditionally attempt to deal with such data is to assign classroom or teacher (or school or district) characteristics to all students, bringing the higher-level variables down to the student level. The problem with this approach is non-independence of observations as all students within a particular classroom assume identical scores on a variable.

Another way researchers have attempted to deal with such data has been to aggregate data up to the level of the classroom, school, or district levels. Thus, one could assess the effect of teacher or classroom characteristics on *average* classroom achievement. However, this traditional approach is problematic in that: (a) much (up to 80–90%) of the individual variability on the outcome variable is lost, potentially leading to dramatic under- or overestimation of observed relations between variables (Raudenbush & Bryk, 2002), and (b) the outcome variable changes significantly and substantively from individual achievement to average classroom achievement.

Neither of these conventional and common strategies for dealing with multilevel data can be considered best practice. Neither allows the researcher to examine truly important and interesting questions, such as the effect of a particular teacher characteristic on student learning. The only methodologically sound way to deal with nested data is through multilevel modeling. Both aggregation and disaggregation can lead to wildly mis-estimated effects, with some effects (particularly in aggregated analyses) overestimated by 100% or more (see Osborne, 2008c). Thus, nested data not analyzed with multilevel modeling should be treated as suspect and might be rejected by reviewers.

Further, performing multilevel analyses such as hierarchical linear modeling has no drawbacks. If observations are not independent the results are correct estimates of effects accounting for non-independence, protecting the researcher from an error of inference or substantially mis-estimated effects. If observations are truly independent, multilevel analyses will exactly reproduce the results of simple correlation or Ordinary Least Squares (OLS) regression analyses.

## 7. Psychometric Characteristics

In a recent survey of the educational psychology literature in top-tier journals, only about one in four studies (26%) reported any type of basic psychometric information on the measures utilized in the research being reported (Osborne, 2008d). Because reliable and valid measurement is a hallmark of good science and a necessary condition for *any* statistical analysis such as correlation or regression, authors should report basic psychometric information on their measures, such as internal consistency and/or other forms of reliability and validity (see Chapter 25, this volume). Further, reviewers should ensure that reliability of measurement is acceptable in magnitude.

Acceptable standards for reliability have been debated for decades, and often depend on the type of reliability being reported. Many authors assume that internal consistency estimates (Cronbach's alpha) of .70 and above are acceptable (e.g., Nunnally, 1978), and despite the fact that social science journals are filled with alphas of .70 and below, this should not be considered quality measurement. In fact, an alpha of .70 represents measurement that is approximately 50% error variance, and even when reliability is .80, effect sizes are attenuated dramatically (when alphas are in the .80 to .85 range, correlations can be attenuated approximately 33%; Osborne, 2003). Of course, internal consistency estimates are not always appropriate or ideal (e.g., reliability of behavioral observation data), but reviewers should have *some* reasonable indicator of reliability and validity reported in every paper, and the reviewers should be convinced that the data are sufficiently reliable to be taken seriously.

Where scales are used to measure constructs, authors should be encouraged to present evidence that the scale structure they are endorsing for further analysis is the best, most parsimonious representation of that scale via confirmatory factor analysis (see Chapter 8, this volume) or other modern measurement analysis (e.g., Rasch measurement or IRT, see Chapter 12, this volume).[4] Factor structure matters because if authors are creating composites of items that are not measuring a homogeneous construct, they are introducing error into their measurement, which significantly attenuates effect size estimates and power to detect effects.

In sum, authors should research their instruments and, when possible, choose those that tend to produce highly reliable data. If authors are dealing with data that have less than optimal reliability, simple analyses such as correlation and regression can be manually disattenuated via common formulae (Cohen et al., 2002; Pedhazur, 1997) or through other, more sophisticated means such as latent variable modeling (see Chapter 28, this volume).

## 8. Descriptive Statistics

The presentation and discussion of descriptive statistics is a desideratum because readers should have basic information about the variables at hand, because it facilitates meta-analyses of studies—a best practice for a variety of reasons (see Chapter 19, this volume)—and re-analyses of data can be easily accomplished if basic statistics are reported. Further, descriptive information helps readers (and authors) understand the data on which the main analyses are based.

## 9. Normality of Data

Normally distributed data are desirable in most cases, even when utilizing nonparametric techniques. Few authors report normality statistics for their data, and fewer still report correcting for non-normality despite the fact that it is often relatively easy to do, involving simple mathematical manipulation or removal of outliers. It should be noted that mis-applications of these transformations can make matters worse, rather than better, so authors and reviewers should not only assess whether data are acceptably normal (skew between −1.0 and 1.0, closer to 0 is better), but whether authors dealt with non-normality appropriately. For example, it is often the case that data transformations are most effective if the minimum value of a distribution is anchored at a value of 1.0 exactly (e.g., Osborne, 2008a). Further, for most simple, common transformations to work, original data need to be positively skewed. For distributions with a negative skew, the distribution needs to be *reflected* or reversed by multiplying each data point by −1.0 and adding a constant to bring the minimum value to 1.0 and the distribution needs to be reflected again after the transformation is completed to return the data to their original order. Correlations and most other common statistical analyses benefit significantly from these practices, making the results more generalizable.

## 10. Missing Data

Missing data are common in the social sciences, yet perusal of the literature shows that few authors report whether there were any missing data, how missing data issues were addressed, and perhaps equally important, whether participants with missing data were significantly different in some way than those with complete data (i.e., whether the missing data were randomly distributed). Many authors use listwise deletion, simple removal of participants' data with any missing score on any variable in the analysis. However, this approach is sub-optimal and can lead to significant changes in the nature of the data and decreases in generalizability. Another popular approach, mean substitution, can have unintended consequences (such as artificially decreasing the standard deviation of the variable) and should be avoided. Multiple imputation is a better alternative than either of the aforementioned options (e.g., Cole, 2008). Irrespective, reviewers should expect to see a discussion and justification of exactly how authors dealt with missing data, if present.

## 11. Outliers and Fringeliers

While there has been (and continues to be) great discussion in the methodology literature regarding the definition of outliers and whether or not they have significant effects on effect estimates (summarized in Osborne & Overbay, 2004), empirical examples of their effects are common. Outliers, and fringeliers (i.e., extreme scores near $z = \pm 3.00$), can have significant effects on the accuracy of effect estimates. Osborne and Overbay (2004) demonstrated that even in reasonably large samples of up to 400 or more subjects, a handful of outliers with scores that are only slightly outside the distribution ($z$ scores of 3.00 to 3.50) can cause substantial errors in parameter estimates and in inference. Across hundreds of simulations, summarized below in Table 5.2, 70–100% of correlation estimates were statistically significantly more accurate after identification and removal of fringeliers, and errors of inference were reduced dramatically (note also that errors of inference were disturbingly prevalent with just a few outliers present).

Thus, reviewers should expect authors to report checking for outliers and fringeliers, and report how they were handled. There are several ways authors can handle extreme scores such as these, including deleting them, recoding them to some more reasonable value, selectively weighting cases to reduce their influence on outcomes, or using data transformations to reduce their influence.

Table 5.2 Effects of Outliers on Correlations

| Population r: | N: | Average initial r | Average cleaned r | t | % more accurate | % errors before cleaning | % errors after cleaning | t |
|---|---|---|---|---|---|---|---|---|
| $r = -.06$ | 52 | .01 | −.08 | 2.5** | 95% | 78% | 8% | 13.40*** |
| | 104 | −.54 | −.06 | 75.44*** | 100% | 100% | 6% | 39.38*** |
| | 416 | 0 | −.06 | 16.09*** | 70% | 0% | 21% | 5.13*** |
| $r = .46$ | 52 | .27 | .52 | 8.1*** | 89% | 53% | 0% | 10.57*** |
| | 104 | .15 | .50 | 26.78*** | 90% | 73% | 0% | 16.36*** |
| | 416 | .30 | .50 | 54.77*** | 95% | 0% | 0% | — |

Source: Osborne & Overbay, 2004.

*Note*: 100 samples were drawn for each row. Outliers were actual members of the population who scored at least $z = 3$ on the relevant variable.
With $N = 52$, a correlation of .274 is significant at $p < .05$. With $N = 104$, a correlation of .196 is significant at $p < .05$. With $N = 416$, a correlation of .098 is significant at $p < .05$, twotailed.
** $p < .01$, *** $p < .001$.

## 12. Underlying Assumptions

Many authors are under the erroneous impression that most statistical procedures are "robust" to violations of most assumptions. Writers such as Zimmerman (1998) have pointed out that violations of assumptions can lead to serious consequences, and when violations of more than one assumption are present it is not safe to assume the validity of the analysis results.

Authors routinely fail to report testing assumptions (only 8.3% of top-tier educational psychology articles reported testing assumptions in 1998–99; Osborne, 2008d), which could indicate that authors fail to test assumptions or authors fail to report associated results. In either case, this is a serious issue as the quality of the corpus of knowledge depends on high quality research analyses, and failing to test assumptions can lead to serious errors of inference and mis-estimation of effects.

For example, failing to use multilevel modeling to test relations when data are nested (independence of error terms or independence of observations assumption) can cause substantial mis-estimation of effects (see Desideratum 6). Failure to measure variables reliably (see Desideratum 7) can lead to serious underestimation of effects in simple (zero-order) relations and either under- or overestimation of effects when variables are controlled for depending on the relative reliability of the covariates and variables of interest (e.g., Osborne & Waters, 2002). As Osborne and Waters (2002) pointed out, failure to test for and account for other issues such as curvilinearity (assumption of linear relationship) can lead to serious errors of inference and underestimation of effects.

In sum, while authors are ultimately responsible for reporting results from testing underlying assumptions, reviewers and editors are equally culpable when such results do not appear in published research reports. Reviewers should demand that authors report results from testing each assumption of a utilized statistical method.

## 13 Pearson r Alternatives

In many textbooks authors prominently discuss alternative correlation coefficients (e.g., Cohen et al., 2002, pp. 29–31) such as the *point-biserial* correlation (a simplified Pearson r appropriate for when one variable is continuous and one is dichotomous), *phi* (a simplified Pearson r when both variables are dichotomous), and *Spearman rank* correlation $(r_s)$, a simplified formula for when data are sets of rank ordered data. Reviewers should *not* demand authors use these alternative computations of the Pearson product-moment correlation, as they are archaic. Essentially, before massive computing power was available in every office, these formulae provided computational shortcuts for people computing correlations by hand or via hand calculator. Researchers using statistical software have no true use for them.

However, there are other correlation coefficients that might be more desirable than $r$. For example, the *tetrachoric* correlation (Pearson, 1901) is particularly good for examining the relationship between raters where the rating is dichotomous (presence or absence of some trait or behavior), while the *polychoric* correlation is designed to examine correlations for ordered-category data (e.g., quality of a lesson plan), another way to measure rater agreement. Importantly, one assumption of these two measures is that while the ratings are dichotomous or categorical, they assume that the latent or underlying variable is continuous (e.g., one can have the presence of particular characteristics of autism to a greater or lesser extent although we might rate them as present/absent). They estimate what the correlation between raters would be if ratings were made on a continuous scale.

Robust measures of association, such as Kendall's *tau*, Spearman's *rho*, and winsorized correlations have been developed to attempt to measure association in the presence of violation of assumptions (e.g., presence of outliers), but Wilcox (2008) concluded that none of these measures of association are truly robust in the face of moderate violations of assumptions. The recommended technique currently appears to be the *skipped correlation coefficient* (Wilcox, 2003).

In the case of a truly continuous and dichotomous or polytomous variable (as opposed to a dichotomous or categorical variable that has an underlying continuous distribution), logistic regression may be considered a best practice in determining association between two variables (see Chapter 17, this volume). While logistic regression has been slow to be adopted in the social sciences, it does carry many benefits, including appropriate mathematical handling of discrete categorical data. One challenge to authors using logistic regression involves the correct interpretation of the measure of association, the *odds ratio* (OR) or index of *relative risk* (RR), the preferred, but more difficult to obtain statistic. A lengthy treatise on these statistics, best practices, and correct interpretation is beyond the scope of this chapter, but see Osborne (2006).

Reviewers should be vigilant about authors who have substantial violations of assumptions and proceed to use these techniques to ameliorate these violations. The best way to deal with violations of assumptions is to deal explicitly with the violation, such as removal of outliers, transformation to improve normality, and so forth. Where this is not possible, robust methods such as the skipped correlation coefficient is a good alternative currently. Reviewers seeing authors use archaic correlation coefficients (e.g., point-biserial, phi, or the Spearman rank order coefficient) should be skeptical as to the authors' quantitative prowess, as this seems to be an indicator of outdated training. Finally, in the specific case of examining inter-rater agreement, alternative nonparametric estimators such as polychoric and tetrachoric correlations are good practices and should be recommended to authors if they are not present (see also Chapter 11, this volume).

## 14. Univariate vs. Multivariate Analyses

Large zero-order correlation tables are often reported in journals. Since the introduction of significance levels ($p < .05$),[5] largely attributed to Fisher (1925),[6] one issue has been maintaining a low rate of Type I errors across an entire set of analyses rather than for each individual analysis. For example, if one reports a correlation table for five variables, the lower triangular matrix contains 10 separate correlations. Assuming researchers are using the traditional $\alpha = .05$ criterion to test the correlations, this means that the Type I error rate across the family of correlations can *greatly* exceed .05. This seems unacceptable, but is routinely done. To combat this issue, early statisticians developed corrections for this expanded Type I error rate.

Many different types of corrections for this issue are available (e.g., Bonferroni adjustments), although these types of corrections tend to reduce power and increase the probabilities of a Type II error. Reviewers should expect authors to address inflated Type I error rates in some way, either through a type of correction such as Bonferroni or, preferably, through reducing the number of correlations reported by focusing on the truly important correlations (Curtin & Schulz, 1998) or through the use of multivariate methods. To the extent that $p < .05$ remains an important criterion (and there is a good deal of discussion attempting to reduce reliance on this criterion in the methodology literature, e.g., Killeen, 2008), authors should seek to keep a $= .05$ for each major, substantive *hypothesis* (e.g., familywise or experimentwise error rate) rather than for each statistical test.

## 15. Semipartial and Partial Correlations

A *semipartial* correlation is a correlation between two variables (variables 1 and 2), controlling for a third variable (that is correlated in some significant way to both of the other variables), with the common notation $sr_{(12.3)}$. Similarly, *partial* correlations are correlations between two variables with the effect of a third removed, and use the common notation $pr_{(12.3)}$. However, there is an important conceptual (and mathematical) difference between the two. For example, consider calculating a correlation between an independent variable (IV) such as motivation and a dependent variable (DV) such as intentions to attend college, controlling for a covariate such as student grades. The semipartial correlation removes the effect of grades from the IV (motivation) but *not* from the DV (intentions). Thus, a semipartial correlation, when squared, communicates the unique variance accounted for in the DV by the IV. This is useful when researchers want to examine which variables make unique contributions to predicting outcomes (e.g., Does motivation predict intentions to attend college once student grades are accounted for?), or which variables account for the most unique variance once others are controlled for (e.g., Does motivation or student grades have a stronger unique effect on intention to attend college?). The semipartial correlation does not, technically, remove the effect of the other variable from the analysis, as it is still in the DV, and authors and researchers should carefully examine interpretations of semipartials so that they are not mis-interpreted. Partial correlations, on the other hand, remove the effect of the covariate (e.g., student grades) from both the DV and IV, giving you a more "pure" measure of that relationship without the effect of the covariate in either. This is most commonly used when researchers want to remove the effect of confounding or extraneous variables to get a measure of the relationship if one could hold other variables constant (e.g., Holding grades constant, does motivation have any association with intentions to attend college?).

Finally, when the underlying assumptions of correlation and regression are not met, partialing or controlling for other variables, as mentioned above, can yield unpredictable results. For example, if student grades are not reliably measured, and we try to control for it, only a fraction of the variable can actually be removed. The rest of the variance attributable to grades is still in the analysis, and thus if grades and motivation are highly correlated, the effect of motivation may be substantially mis-estimated. As another example, if the effect of grades is curvilinear, and only the linear effect is covaried, then again part of the effect of grades remains in the analyses, causing mis-estimation of other effects (curvilinearity is discussed in more detail in Desideratum 19).

Reviewers should pay careful attention that authors use semipartial and partial correlation when appropriate and interpret them accurately. If variables to be used in these types of analyses are not reliable, reviewers might insist on discussion of this issue, use of alternative analyses that can correct for low reliability (e.g., structural equation modeling; see Chapter 28, this volume), or reject the analysis as untenable as it is not testing what the authors believe it is testing.

## 16. Data Cleaning and Data Transformation

Data cleaning is often necessary and needs to be considered when interpreting the analysis results. For example, transformations are tricky to get right computationally (Osborne, 2008a), and also complicate interpretation. First, it is easy to create missing data when doing a transformation (e.g., some scores might be converted to missing data as it is impossible to take the square root or natural log of a negative number). Second, some transformations to reduce skew require the data to be *reflected* (the order of the scores is reversed with the highest scores becoming the lowest and the lowest scores becoming the highest) prior to a transformation being applied. If the data are not reflected after the transformation, the correlation computed afterward will be exactly opposite of what it should be (e.g., 0.50 rather than –0.50) which could lead to misinterpretation of the results.

Even assuming a transformation is applied successfully, using best practices, the researcher is still left with one or more transformed variables, an altered version of the original construct. Is it straightforward to say that the correlation with log of a variable means the same as the correlation with the original variable? Not technically, and probably not in practice, either. So authors need to be clear with readers in interpreting results.

Reviewers need to see evidence that where data transformations are necessary (e.g., highly skewed variable), authors use best practices (see Osborne, 2008a for guidelines on best practices in utilizing some common data transformations) and are mindful that analyses used transformed data when discussing the implications of their results.

## 17. Interpretation of $p$ Values

For many decades, $p < .05$ has been the primary indicator to a researcher that an effect is "important." In fact, many authors have been known to place more importance on effects that produce smaller $p$ values. Terms like "more significant" and "highly significant" *are common in published research, as is the use of different indicators* for smaller $p$ values (e.g., using asterisks to indicate significance level: * $p < .05$, ** $p < .01$, *** $p < .001$). There are two issues that reviewers need to attend to: misuse of $p$ as a proxy for importance of an effect and misinterpretation of $p$.

It may not be surprising that some researchers use $p$ as a proxy for importance of an effect or for effect size itself (discussed in more detail in Desideratum 18). One of the primary factors in the magnitude of $p$ is effect size. But other factors also help determine $p$, making it not usable for this purpose: sample size is a primary determiner of $p$. Reviewers need to ensure that authors disentangle effect size/importance with significance level.

Once statistically significant effects are identified, effect sizes and confidence intervals around effect sizes should guide discussion as to importance of the findings (see Chapter 7, this volume). Authors should focus on effect sizes in interpreting results, and reviewers should closely monitor manuscripts for author hyperbole.

Finally, several authors have proposed modern alternatives to $p$, given that $p$ usually does not provide the information researchers seek. One of the most promising alternatives is Killeen's (2008) $p_{(rep)}$, the *probability of replication*, a statistic that is directly related to what most researchers are interested in, and easily calculated from information most statistical software provides.

## 18. Effect Size Measures

Effect sizes are simple to calculate in the context of correlation and regression analyses. Unlike other types of analyses such as ANOVA and $t$ tests, correlation coefficients are themselves effect sizes, as are $\beta$s and multiple $R$s in regression (for more information on this point, see, e.g., Cohen, 1988; Cohen et al., 2002). Effect sizes tend to come in two general categories: a standardized index of strength and a percentage of variance accounted for. $R$, $\beta$, $sr$, $pr$, and $r$ are all indices of strength (as are analogous effect size indexes for ANOVA type designs, such as $d$, $f$, and $\omega^2$). Because indices of strength tend to be on different scales, it is good practice for authors to report effect sizes that represent the percentage of variance accounted for. In the case of association, $r^2$, $sr^2$, $pr^2$, and $R^2$ are appropriate. Some authors (e.g., Thompson, 2002) have argued that effect sizes should be accompanied by confidence intervals for effect sizes. Since this last technique is not commonly available via statistical packages at this time (it is available through the R statistical package, freely available at http://www.r-project.org/), its use should be seen as a suggestion rather than a requirement.

Reviewers should "reality check" authors' claims about strength of association. Authors might over-exaggerate the strength of the effects in their research in hopes of making the projects more publishable. For example, correlations of $r = .30$ only represent 9% variance accounted for, and as such, do not generally qualify as "strong" effects or "good" support for consistency. Rules of thumb for what constitutes a strong, moderate, or weak effect for various effect size indices are widely disseminated (e.g., Cohen, 1962, 1988; see also Chapter 24, this volume) and are also available on the internet.

## 19. Curvilinear Relations and Interactions

Few researchers in the social sciences examine or discuss curvilinear effects. Is this because there are so few curvilinear effects or because researchers fail to look for them? Until recently, it was difficult to test for curvilinear effects explicitly within statistical software frameworks. However, most popular modern statistical packages include simple diagnostic options to identify possible curvilinearity (e.g., residual plots) and contain curvilinear regression options, both of which allow researchers to test for curvilinear effects. Further, as human nature is complex and interesting, it is likely that many associations are curvilinear in nature. Note the well-known association between arousal (or less accurately, anxiety) and performance, such that increased arousal leads to increased performance. However, researchers also noted that there is an optimal level of arousal, and beyond that, higher levels tend to be associated with diminishing performance until, at some theoretical level of extreme arousal, performance would be as low as extremely low arousal. In fact, the classic arousal-performance curve discussed here will produce correlations close to $r = 0.00$ if curvilinearity is not accounted for, leaving a robust and important effect undetected or mis-estimated. Thus, it is in the interest of the researcher, as well as the scientific community, to examine associations for curvilinearity. When found, these effects are usually best conveyed to readers as graphs that show the nature of the curvilinearity; thus, reviewers should expect a graph to accompany the effect description/interpretation, and best practices in graphing should be followed.

## 20. Causal Inferences

The "cardinal sin" of correlational analysis is to make causal statements or inferences. For example, one might read, "Because availability of instructional technology is associated with stronger growth in student achievement, schools should make stronger efforts to include instructional technology in the daily life of students." Anyone having taken a basic research methods course should remember that any relation could have at least three possible causal bases: (a) variable A could cause B, (b) variable B could cause A, and (c) the relationship between A and B could be caused by a third variable, C, that is not measured. Authors should be careful to respect the limits of correlational methodologies and take care not to make causal inferences or statements. Reviewers should carefully police these same issues. Too often the lay public and policymakers (who might not be trained to understand these distinctions) are misled into causal thinking by correlational research.[7]

Consider the following example. There has long been an assumption that, since individuals with higher blood cholesterol are more likely to have adverse medical outcomes such as heart attacks, reducing cholesterol should reduce risk for adverse medical outcomes. There was even some research suggesting that people who, through change in diet and/or exercise, reduce their cholesterol levels also reduce their risk for heart attack and other adverse outcomes. Thus, when drug companies developed drugs that reduce cholesterol without change in diet and exercise, the assumption was that this reduction would similarly *cause* decreases in risk of adverse outcomes. Unfortunately, recent trials of cholesterol lowering drugs have failed to show medical benefits in the same way as reducing cholesterol through other means. This is an example of correlation-based causal inferences (e.g., decreased serum cholesterol and decreasing risk of adverse outcomes) failing to be supported. This is not merely an esoteric methodological issue. Research can have profound implications on real outcomes for real people. Imagine a patient taking one of these drugs for years expecting a medical benefit when in fact they should have been making lifestyle changes instead. How many students around the world are laboring through educational interventions that are based solely on misapplied correlational data? Reviewers need to be vigilant that authors correctly avoid causal inferences when reporting results from correlational research.

## Notes

1 Stratified sampling refers to the process of grouping members of the population into relatively homogeneous subgroups before sampling. The strata should be mutually exclusive: every element in the population must be assigned to only one stratum, and ideally, every element of the population should be represented by a stratum.

2 Many authors have acknowledged the significant issues with null hypothesis statistical testing (NHST) and some (see particularly Killeen, 2008) have proposed alternatives such as the probability of replication as a more interesting/useful replacement. Interested readers should examine Thompson (1993) and Fidler (2005) as an introduction to the issues.

3 Design effects are often calculated and disseminated in large, standardized surveys with complex sampling designs, such as government surveys or samples. They are methodologically challenging to calculate, but if known, can be incorporated into analyses as an adjustment to effect sizes and significance tests so as to account for the artificially small standard errors present in certain types of samples, such as those with nested data. Multilevel modeling is generally a more elegant and effective way to deal with this issue.

4 Exploratory factor analysis (see Chapter 8, this volume) is often difficult to defend because of low replicability and lack of inferential statistics.

5 Those interested in the history leading to the origins of $p < .05$ should read Cowles and Davis's (1982) excellent history of significance testing.

6 Although the concept of formal, norm-guided significance testing itself is traceable to Gosset (1908) and Wood & Stratton (1910).

7 Because this chapter is focusing on relatively simple associational analyses, simple associational analyses should not be couched in causal terms. However, more sophisticated analytic techniques, such as structural equation models based on strong theory can use correlational data to support or refute hypotheses. However, at some point causal statements need to be assessed via methodologies (e.g., double blind experimental studies) that are designed to test causal inference more effectively.

## References

Aiken, L. S., & West, S. (1991). *Multiple regression: Testing and interpreting interactions.* Thousand Oaks, CA: Sage.

Cohen, J. (1962). The statistical power of abnormal-social psychological research: A review. *Journal of Abnormal and Social Psychology, 65,* 145–153.

Cohen, J. (1988). *Statistical power analysis for the behavioral sciences* (2nd ed.). Hillsdale, NJ: Lawrence Erlbaum.

Cohen, J., Cohen, P., West, S., & Aiken, L. S. (2002). *Applied multiple regression/correlation analysis for the behavioral sciences.* Mahwah, NJ: Erlbaum.

Cole, J. C. (2008). How to deal with missing data. In J. W. Osborne (Ed.), *Best practices in quantitative methods.* Thousand Oaks, CA: Sage.

Cowles, M. D., & Davis, C. (1982). On the origins of the .05 level of statistical significance. *American Psychologist, 37,* 553–558.

Curtin, F., & Schulz, P. (1998). Multiple correlations and Bonferroni's correction. *Biological Psychiatry, 44,* 775–777.

Deemer, W. L. (1947). The power of the *t* test and the estimation of required sample size. *Journal of Educational Psychology, 38,* 329–342.

Fidler, F. (2005). *From statistical significance to effect estimation: Statistical reform in psychology, medicine, and ecology.* Doctoral Dissertation, University of Melbourne.

Fidler, F., & Cumming, G. (2008). The new stats: Attitudes for the 21st century. In J. W. Osborne (Ed.), *Best practices in quantitative methods* (pp. 1–14). Thousand Oaks, CA: Sage.

Fisher, R. A. (1925). *Statistical methods for research workers.* Edinburgh: Oliver & Boyd.

Gosset, S. W. S. (1908). The probable error of a mean. *Biometrika, 6,* 1–25.

Killeen, P. R. (2008). Replication statistics. In J. W. Osborne (Ed.), *Best practices in quantitative methods* (pp. 103–124). Thousand Oaks CA: Sage.

Nunnally, J. C. (1978). *Psychometric theory* (2nd ed.). New York: McGraw Hill.

Osborne, J. W. (2003). Effect sizes and the disattenuation of correlation and regression coefficients: Lessons from educational psychology. *Practical Assessment, Research & Evaluation, 8*(99).

Osborne, J. W. (2006). Bringing balance and technical accuracy to reporting odds ratios and the results of logistic regression analyses. *Practical Assessment, Research & Evaluation, 11*(7).

Osborne, J. W. (2008a). Best practices in data transformation: The overlooked effect of minimum values. In J. W. Osborne (Ed.), *Best practices in quantitative methods* (pp. 197–204). Thousand Oaks, CA: Sage.

Osborne, J. W. (2008b). *Best practices in quantitative methods.* Thousand Oaks, CA: Sage.

Osborne, J. W. (2008c). A brief introduction to hierarchical linear modeling. In J. W. Osborne (Ed.), *Best practices in quantitative methods* (pp. 445–450). Thousand Oaks, CA: Sage.

Osborne, J. W. (2008d). Sweating the small stuff in educational psychology: How effect size and power reporting failed to change from 1969 to 1999, and what that means for the future of changing practices. *Educational Psychology, 28,* 1–10.

Osborne, J. W., & Overbay, A. (2004). The power of outliers (and why researchers should ALWAYS check for them). *Practical Assessment, Research & Evaluation, 9*(6).

Osborne, J. W., & Waters, E. (2002). Four assumptions of multiple regression that researchers should always test. *Practical Assessment, Research & Evaluation, 8*(2).

Pearson, K. (1901). Mathematical contribution to the theory of evolution. VII: On the correlation of characters not quantitatively measurable. *Philosophical Transactions of the Royal Society of London, A 195,* 1–47.

Pedhazur, E. J. (1997). *Multiple regression in behavioral research: Explanation and prediction.* Fort Worth, TX: Harcourt Brace College Publishers.

Raudenbush, S. W., & Bryk, A. S. (2002). *Hierarchical linear models: Applications and data analysis methods.* (Vol. 1). Thousand Oaks, CA: Sage.

Rossi, J. S. (1990). Statistical power of psychological research: What have we gained in 20 years? *Journal of Counseling and Clinical Psychology, 58,* 646–656.

Tabachnick, B. G., & Fidell, L. S. (2001). *Using multivariate statistics* (4th ed.). New York: Harper Collins.

Thompson, B. (1993). The use of statistical significance tests in research: Bootstrap and other alternatives. *Journal of Experimental Education, 61,* 361–377.

Thompson, B. (2002). What future quantitative social science research could look like: Confidence intervals for effect sizes. *Educational Researcher, 31,* 24–31.

Tukey, J. W. (1991). The philosophy of multiple comparisons. *Statistical Science, 6,* 100–116.

Wilcox, R. (2003). *Applying contemporary statistical techniques.* San Diego: Academic Press.

Wilcox, R. (2008). Robust methods for detecting and describing associations. In J. W. Osborne (Ed.), *Best practices in quantitative methods* (pp. 263–279). Thousand Oaks, CA: Sage.

Wood, T. B., & Stratton, F. J. M. (1910). The interpretation of experimental results. *Journal of Agricultural Science, 3,* 417–440.

Zimmerman, D. W. (1998). Invalidation of parametric and nonparamteric statistical tests by concurrent violation of two assumptions. *Journal of Experimental Education, 67,* 55–68.

# 6

# Discriminant Analysis

Carl J Huberty

In the 1930s, Sir Ronald A. Fisher (1890–1962) initiated, in print, *discriminant analysis* in the context of classifying a plant into different species using flower measurements as predictor scores. What was conducted is currently termed a *predictive discriminant analysis* (PDA). The design of an empirical study in which a PDA would be conducted involves $N$ *analysis units*, $J$ groups of units, and $p$ *predictor* variables. The basic purpose of such a study is to determine a prediction rule that consists of $J$ composites of the $p$ predictors used to assign each of the $N$ units to one of the $J$ groups. For example, a PDA might be used to predict whether or not a college student drops out of school using predictors such as high school and college academic performance, intelligence, family income, age, and anxiety.

A second aspect of discriminant analysis involves describing separation among $J$ groups based on $J - 1$ *linear* composites of the $p$ *outcome* variables. This is a *descriptive discriminant analysis* (DDA) and is a follow-up to a multivariate analysis of variance (MANOVA; see Chapter 23, this volume). A DDA might, for example, be used to study the differences among job types (e.g., machinery maintenance, operator, and mechanic) using several abilities (e.g., ability to prepare written reports, ability to train others, ability to work with minimum supervision, or ability to learn new skills).

Details about PDA and DDA can be found in Huberty (2002, 2005), Huberty and Olejnik (2006), McLachlan (2004), and Rencher (2002). Software packages most often used for PDA and for DDA are SAS and SPSS; for a discussion on the use of these packages for PDA and DDA, see Huberty and Lowman (1997). For specific desiderata to help evaluate manuscripts that involve PDA or DDA applications, see Table 6.1a (PDA) and Table 6.1b (DDA) and the supporting explanations to follow.

## P1. Analysis Purpose

A PDA is conducted for the purpose of predicting membership of $N$ analysis units into one of $J$ groups of units using measures on $p$ predictor variables for each unit. This analysis is done basically for a *practical* purpose. For example, a researcher might be interested in predicting $J = 3$ types of child behavior disorders using the $p = 8$ predictor variables of aggression, hyperactivity, anxiety, depression, attention problems, atypicality, adaptability, and social skills. Here, the purpose of using PDA is to assess the accuracy of the collection of the eight predictors at predicting group membership (behavior disorder). More specifically, here it is of interest to determine the most, and least, important predictors of the three types of child behavior disorders.

Table 6.1a  Desiderata for Predictive Discriminant Analysis

| Desideratum | Manuscript Section(s)* |
|---|---|
| P1. The selection of PDA as an appropriate analysis tool for predicting group membership based on a set of predictor variables is justified based on related literature and the study's overall practical and predictive purpose. | I |
| P2. The overall research design is described, including the grouping variable, the set of predictor variables, and the sampling plan. | I, M |
| P3. Preliminary analyses are conducted and their results reported. Specifically, the assumptions of multivariate normality and homogeneity of dispersion are checked, and missing data and outliers are addressed. Also, the name and version of the software package used are reported. | M, R |
| P4. The PDA is conducted by determining appropriate weights to form composites of predictor variables. These composites—the linear or quadratic classification functions (LCFs of QCFs, respectively)—allow for the classification of analysis units into groups. | M, R |
| P5. In addition to descriptive statistics (e.g., means and standard deviations) for each predictor variable in each group, results from preliminary data-screening analyses should be provided. A table of weights for the LCFs or QCFs and a classification table should be presented and discussed based on the specific predictive purpose of the study. | R, D |

\* *Note*: I = Introduction, M = Methods, R = Results, D = Discussion

Table 6.1b  Desiderata for Descriptive Discriminant Analysis

| Desideratum | Manuscript Section(s)* |
|---|---|
| D1. The selection of DDA is justified as an appropriate follow-up analysis to statistically significant MANOVA results for determining linear composites of outcome variables that produced group differences. Justifications are based on related literature and the study's overall theoretical and descriptive purpose. | I |
| D2. The overall research design is described, including the grouping and outcome variables, and the sampling plan. | I, M |
| D3. Preliminary analyses are conducted and their results reported. Specifically, the assumptions of multivariate normality and homogeneity of dispersion are checked, statistically significant MANOVA results are reported, and missing data and outliers are addressed. Also, the name and version of the software package used are reported. | M, R |
| D4. The DDA is conducted by determining appropriate weights to form composites of outcome variables after statistically significant MANOVA results. These composites—the linear discriminant functions (LDFs)—assist in the interpretation of the substantive or theoretical dimensions on which the groups differ. | M, R |
| D5. In addition to descriptive statistics (e.g., means and standard deviations) for each outcome variable in each group, results from the MANOVA and other preliminary data-screening analyses should be provided. Tables including weights for the LDFs and structure $r$ values should be presented and discussed based on the specific descriptive purpose of the study. In particular, structure $r$ values should be used to help define the LDFs from a substantive/theoretical perspective as dimensions that are responsible for group differences. | R, D |

\* *Note*: I = Introduction, M = Methods, R = Results, D = Discussion

## P2. Study Design

With the study's purpose in mind, a researcher's next step is to determine a grouping variable (with two or more levels) and a collection of predictor variables. Of course, these two variable types would be considered in connection with the sampling plan used. This plan might involve a random sample, a convenience sample, or a random subset of analysis units from some data base. Irrespective of sampling method, the sampling plan must be explicitly described and the sample size must be justified. Two sample size charts for a PDA study are given in Huberty and Olejnik (2006, pp. 310, 356). The selection of predictors may be based on previous studies and/or a less theoretical and more practical rationale, especially when an existing data set is used.

## P3. Preliminary Analysis

When considering a PDA, applied researchers should have in mind a *data matrix*. Such a matrix has $N$ rows and $1 + p$ columns, with the first column typically indicating membership in one of the $J$ groups. Although a data matrix would not be reported in a PDA study, it should be mentioned in the manuscript that it was examined for missing data and obvious outliers, and how those latter issues were addressed.

Another preliminary data investigation is the checking of PDA assumptions. First, are the predictors continuous? If so, is there approximate multivariate normality in the $J$ populations? If categorical predictors are involved, see Huberty and Olejnik (2006, pp. 366–369). Second, are the $J$ matrix determinants approximately equal? A statistical test of determinant equality is available in the SAS and SPSS programs; authors should clarify which package and which version was used for data analysis.

If one has approximately equal covariance matrices (as assessed, for example, using the Box M test), then a *linear* prediction rule is used. This rule involves $J$ linear composites of the $p$ predictors; these are linear classification functions (LCFs). If, on the other hand, the $J$ matrices are judged *not* to be approximately equal, then one uses $J$ *quadratic* composites, or quadratic classification functions (QCFs).

Also, consideration of reliability and validity information of the predictor scores enhances the quality of a PDA study. As in most empirical research, investigators that utilize PDA are attempting to make inferences to larger populations, and score reliability and validity are essential for such generalizations (see Chapter 25, this volume).

Finally, another decision must be made after the data set to be analyzed is established. This decision pertains to the *prior probabilities* of group memberships, reflecting the relative sizes of the $J$ populations. For example, in a two-group context, if one corresponding population is estimated to be one-half the size of the other, then the two prior probabilities would be .33 and .67. These probabilities might be based on relative group sizes. Group sizes, however, may not correctly reflect the priors. Previous research might serve as a determiner. More appropriately, the researcher might refer to some experts in the connected field of study to set the prior probabilities.

## P4. Final Analysis

With the prior probabilities set and judgment made regarding the equality of covariance matrices, a prediction rule is determined. Using the resulting classification functions (LCFs or QCFs), a $(J \times J)$ classification table would result. This table gives the number of *hits* (correct classifications) in the main diagonal cells, and the number of *misses* (incorrect classifications) in the off-diagonal cells. Based on this information, a judgment of classification accuracy of the obtained rule—linear or quadratic—can be made, across all $J$ groups or for each group.

It is strongly urged that an external classification rule, rather than an internal rule, be used. An *internal* rule is one that is based on the available data, and is then used to classify the same data set. An *external* rule is built on one data set and then applied to another data set. A popular external rule is the *leave-one-out* (L-O-O) rule. With this approach, a classification rule is based on $N-1$ units, and used to classify the single left-out unit. As such, $N$ rules and classifications are carried out. This external rule is easily conducted with the SAS DISCRIM software routine, either linear or quadratic. Note that the SPSS DISCRIMINANT routine can also be used for a linear L-O-O rule; but, the SPSS quadratic L-O-O results are not correct.

It might be of interest to assess the relative predictive value of the $p$ predictor variables. This can be easily accomplished by conducting $p$ L-O-O PDAs, each with $p-1$ predictors. Thus, one would obtain $p$ classification results. The hit rate of interest—for all $J$ groups, or for a particular group—that is lowest when a given predictor is deleted indicates that predictor is most important. Thus, a predictor ordering may be established. Also, a predictor may be deleted for the final analysis if, when it is deleted, the hit rate of interest is greater than when that predictor is included in the classification rule. This method of variable deletion is strongly preferred over the stepwise strategies available in standard software packages.

Finally, there are two types of estimated probabilities, each associated with an analysis unit. One is a posterior probability, the estimated probability of a unit belonging to a particular group. This probability is used to assign a unit to a particular group and is equivalent to using a unit-to-group distance measure, a Mahalanobis distance from a unit score vector to a group mean vector (i.e., a centroid). The other type of estimated probability is a *typicality* probability. This probability reflects the proportion of units in a group that have predictor scores near the mean scores for the group; they are reported in the SPSS DISCRIMINANT output but not in the SAS DISCRIM output.

## P5. Reporting and Discussing PDA Results

As mentioned earlier, the mean and standard deviation (or variance) for each predictor in all groups should be reported in a table. This table can also include the $p \times p$ error (or pooled) correlation matrix (see, for example, Huberty & Olejnik, 2006, p. 119). Also, the results for the test of equality of the $J$ covariance matrices should be reported; supplementally, the (natural) logarithms of the determinants of the $J$ covariance matrices might be reported as well (see Huberty, 2002, pp. 587–588). Finally, authors should report if one or more of the original $p$ predictors were deleted, based on indications from the all-but-one variable L-O-O analyses that the hit rate of interest increased when a predictor is left out of the analysis.

Once the final data set is determined, a PDA is conducted; an abundance of information is available, much of which is important in reporting results. To start, it should be reported how outliers were dealt with, if present. The posterior probabilities should also be examined to see if there are some "in-doubt" analysis units; that is, are there some units that are as close to one group center as to another group center? The score vectors of such units might lead to the discovery of some particular unit "types," that is, the score vectors of a collection of in-doubt units should be examined to determine if some unique characteristic(s) of these analysis units exist. Also, so that in-doubt units are excluded from affecting the hit-rates, a THRESHOLD command can be used with the SAS DISCRIM routine (for more detail, see Huberty & Olejnik, 2006, pp. 307–309). Having addressed any in-doubt units, a classification table is reported and discussed. The discussion should pertain to hit rates on the main diagonal, and include a comparison of these values to hit rates expected just by chance (see Huberty & Olejnik, 2006, Ch. 16). An example of a linear L-O-O classification table is given in Table 6.2. From this table, the total-group hit rate is $(16 + 99 + 13)/264 \approx .485$. Only for Group 2 is the hit-rate respectable (.811).

A reasonable question is whether or not an obtained hit rate, individual-group or total-group, is better than chance. For details of assessing the effectiveness of classification rules, see Huberty and

Table 6.2  A Linear L-O-O Classification Table

| | | Predicted Group | | | |
|---|---|---|---|---|---|
| | | 1 | 2 | 3 | $n_j$ |
| | 1 | 16 (.211) | 55 | 5 | 76 |
| Actual Group | 2 | 12 | 99 (.811) | 11 | 122 |
| | 3 | 6 | 47 | 13 (.197) | 66 |
| Total | | 34 | 201 | 29 | 264 = N |

Note: Group hit rates are in parentheses

Olejnik (2006, Chapter 16). In general, it is advisable to consider chance assessment in every PDA study.

The weights for each of $J$ LCFs or QCFs should be also reported, or at least be made available through the author. The reason for this suggestion is that these weights might be used for prediction/classification by researchers who have data on the same set of $p$ predictor variables. This allows for potential cross-validation of the PDA results.

Finally, any manuscript in which a PDA is reported should include a restatement of the predictive and practical purpose of the study as part of the relevant interpretive discussion. The classification results should also be summarized and related to previous relevant prediction studies. A review of some occurring problems in reporting the use of PDA is given in Huberty and Hussein (2003).

## D1.  Analysis Purpose

Suppose a researcher has a total of $N$ analysis units contained in $J$ groups drawn respectively from $J$ populations, and the intent is to infer if there are any differences among the populations with respect to a collection of $p$ outcome variables. The basic analysis for this context is a multivariate analysis of variance (MANOVA; see Chapter 23, this volume). Suppose, also, that the MANOVA results indicated that there are, indeed, significant group differences; that is, a small $p$-value and a large effect-size are associated with the MANOVA analysis. It should be noted that for $J > 2$, group differences may be assessed across all $J$ groups, for pairs of groups, or for some multiple-group contrast(s). It is for any of these three contexts that a *descriptive discriminant analysis* (DDA) would be appropriate. The purpose of this MANOVA/contrast follow-up analysis is to provide a *description* of the resulting group differences. Authors should use existing literature to explicate the largely *theoretical* purpose of DDA, as opposed to the largely *practical* purpose of PDA (see Desideratum P.1).

## D2.  Study Design

The design of a study that involves a DDA is basically the same as that involving a PDA (see Desideratum P.2), but with three important differences. First, the role of predictors and outcomes is reversed relative to PDA, such that the $p$ variables involved are *outcome* variables while the group membership information serves in a predictive role. Second, in a DDA context, there may be more than one grouping variable as in, for example, a two-factor design (see Chapter 1, this volume). Third, the collection of $p$ outcome variables in a DDA context, in some way, form a substantive or theoretical construct, whereas in a PDA context that is not the case. That is, one purpose of DDA is to identify some outcome variable construct or constructs to which group differences may be attributed.

## D3. Preliminary Analysis

To start, the data matrix is to be examined for missing and obvious outlying measures on the $p$ outcome variables. A basic DDA analysis is typically a follow-up to statistically significant MANOVA results. The DDA techniques discussed here pertain to one-factor MANOVA designs (but see Huberty & Olejnik, 2006, Chapter 8, for multiple-factor designs) with $N$ analysis units, $J$ groups, and $p$ outcome variables.

If there is some interest in outcome variable deletion, stepwise analysis methods should not be used; see Huberty and Olejnik (2006, pp. 103–106) for a discussion of an appropriate deletion method. For the original or resulting set of $J$ outcome variables, there are $c = \min(p, J-1)$ linear composites of the $p$ outcome variables, the *linear discriminant functions* (LDFs). For example, with $p = 8$ and $J = 3$, there are $c = 2$ LDFs to consider in describing group differences that were judged to be statistically significant based on MANOVA results.

## D4. Final Analysis

What is statistically determined to this point is a matrix with $p$ rows and $c$ columns. Entries in this matrix are the $p$ outcome variables and $c$ sets of LDF weights. [There are statistical tests to determine the number of LDFs to retain for interpretation purposes (Huberty & Olejnik, 2006, pp. 89–91).] With $p = 8$ and $J = 3$, suppose the $c = 2$ LDFs are to be retained for interpretation purposes. What is then examined is a two-dimensional LDF plot of the three group LDF centroids; see Figure 6.1. In this plot, $LDF_1$ separates $G_1$ and $G_2$ from $G_3$, while $LDF_2$ separates $G_1$ and $G_3$ from $G_2$.

An important aspect of the final analysis pertains to the definitions of the LDFs that are retained based on the statistical tests performed. To define or label the retained LDFs, *structure r* values are obtained (via SAS DISCRIM or SPSS DISCRIMINANT). These structure $r$ values are the $p$ correlations for each of the LDFs with the outcome variables (see Huberty & Olejnik, 2006, p. 8, for more detail); they should be reported in manuscripts utilizing DDA (as in Table 6.3).

## D5. Reporting DDA Results

For descriptive purposes, the mean and standard deviation (or variance) for each outcome variable in all groups are to be reported in a table. If some outcome variables were deleted, a rationale should be given. This information should be supplemented with a $p \times p$ error correlation matrix. The computer

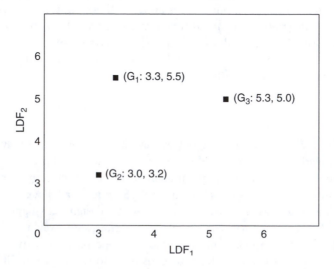

**Figure 6.1** LDF Plot.

Table 6.3  Structure $r$ values for $J = 3$ groups

| Variable | $LDF_1$ | $LDF_2$ |
|---|---|---|
| $X_1$ | −.03 | **.51** |
| $X_2$ | **.58** | .03 |
| $X_3$ | .19 | **−.46** |
| $X_4$ | **−.61** | .12 |
| $X_5$ | **.48** | .23 |
| $X_6$ | **.63** | .18 |
| $X_7$ | .29 | .09 |
| $X_8$ | .07 | **.55** |

results for the test of equality of the $J$ covariance matrices should also be reported; this may include the determinants or the (natural) logarithm of these $J$ matrices.

A DDA would not be conducted unless statistically significant MANOVA results were found. So, the MANOVA $p$-value and effect-size value need to be reported based on the $N$ analysis units and $p$ outcome variables (unless a rationale is provided for deleting some variables).

It might be of interest to assess certain group contrast effects, such as pairwise group contrasts or more complex contrasts. The resulting pairwise test statistic and effect-size estimate should be reported. For a pairwise group comparison there is a single LDF. With $J > 2$ groups there is a maximum of $J − 1$ LDFs. In the latter situation, a test of dimensionality would be conducted and reported. This determines the number of LDFs to consider in the reporting of DDA results. Suppose two LDFs are to be considered. Then, an *LDF plot* should be reported. An example of such a plot is given in Figure 6.1. For this hypothetical example it seems obvious that $LDF_1$ accounts for the difference of Group 3 from Groups 1 and 2. With respect to $LDF_2$, Group 2 appears to be different from Groups 1 and 3.

From a substantive or theoretical standpoint, it is of interest to identify the underlying constructs represented by the two LDFs. Because this identification is based on structure $r$ values, authors should report and interpret a $p \times c$ table of these correlations, as illustrated in Table 6.3. The interpretation is made by examining the structure $r$ values for each LDF (see the bold values in Table 6.3). For example, $LDF_1$ is defined by variables $X_2$, $X_4$, $X_5$, and $X_6$, while $LDF_2$ is defined by variables $X_1$, $X_3$, and $X_8$. The labeling of $LDF_1$ is based on what underlying construct is identified by the combination of variables $X_2$, $X_4$, $X_5$, and $X_6$. Likewise, the label for $LDF_2$ is based on variables $X_1$, $X_3$, and $X_8$.

Finally, any manuscript in which a DDA is reported should include a restatement of the theoretical and descriptive purpose of the study as part of the relevant interpretive discussion. The associated MANOVA results should be summarized, and the resulting LDF structure should be related to previous theoretical research.

## References

Huberty, C. J. (2002). Discriminant analysis. In J. Meij (Ed.), *Dealing with the data flood* (pp. 585–600). The Hague, Netherlands: Study Centre for Technology Trends.

Huberty, C. J. (2005). Discriminant analysis. In B. S. Everitt & D. C. Howell (Eds.), *Encyclopedia of statistics in behavioral science* (pp. 499–505). Chichester, England: Wiley.

Huberty, C. J., & Hussein, M. H. (2003). Some problems in reporting use of discriminant analysis. *Journal of Experimental Education, 71*, 177–191.

Huberty, C. J., & Lowman, L. L. (1997). Discriminant analysis via statistical packages. *Educational and Psychological Measurement, 57*, 759–784.

Huberty, C. J., & Olejnik, S. (2006). *Applied MANOVA and discriminant analysis* (2nd ed.). Hoboken, NJ: Wiley.

McLachlan, G. J. (2004). *Discriminant analysis and pattern recognition* (2nd ed.). New York: Wiley.

Rencher, A. C. (2002). *Methods of multivariate analysis* (2nd ed.). New York: Wiley.

# 7

# Effect Sizes and Confidence Intervals[1]

Geoff Cumming and Fiona Fidler

An *effect size* (ES) is simply an amount of something of interest. It can be as simple as a mean, a percentage increase, or a correlation; or it may be a standardized measure of a difference, a regression weight, or the percentage of variance accounted for. Most research questions in the social sciences are best answered by finding estimated ESs, meaning *point estimates* of the true ESs in the population. Grissom and Kim (2005) provided a comprehensive discussion of ESs, and ways to calculate ES estimates.

A *confidence interval* (CI), most commonly a 95% CI, is an *interval estimate* of a population ES. It is an interval that extends above and below the point ES estimate; it indicates the *precision* of the point estimate. The *margin of error* (MOE) is the length of one arm of a CI. The most common CIs are symmetric, and for these the MOE is half the total width of the CI. The MOE is our measure of precision. Cumming (2007a) provided a brief overview of CIs and their meaning, and Cumming and Finch (2005)[2] provided an introduction, explained the advantages of CIs, and described a number of *rules of eye* to assist the understanding and interpretation of CIs. Smithson (2002) provided a more detailed account of CIs.

In the social sciences, statistical analysis is still dominated by null hypothesis significance testing (NHST). However, there is extensive evidence that NHST is poorly understood, frequently misused, and often leads to incorrect conclusions. It is an urgent research priority that social scientists shift from relying mainly on NHST to using better techniques, particularly those including ESs, CIs, and meta-analysis (MA). The best reference on statistics reform is the excellent book by Kline (2004). Wilkinson and the Task Force on Statistical Inference (TFSI) (1999) is a further source of good advice. Our aim in this chapter is to assist authors and manuscript reviewers to make the vital transition from over-reliance on NHST to more informative methods, including ESs, CIs, and MA.

## 1. Formulation of Main Questions as Estimation

An astronomer wishes to know the age of the Earth; a chemist measures the boiling point of an interesting new substance. These are the typical questions of science. Correspondingly, in the social sciences we wish to estimate how seriously divorce disrupts adolescent development, or the effect of a type of psychotherapy on depression in the elderly. The chemist reports her result as, for example, $27.35 \pm 0.02°C$, which signals that 27.35 is the point estimate of the boiling point, and 0.02 is the

Table 7.1 Desiderata for Effect Sizes and Confidence Intervals

| Desideratum | Manuscript Section(s)* |
|---|---|
| 1. The main questions to be addressed are formulated in terms of estimation and not simply null hypothesis significance testing. | I |
| 2. Previous research literature is discussed in terms of effect sizes, confidence intervals, and from a meta-analytic perspective. | I |
| 3. The rationale for the experimental design and procedure is explained and justified in terms of appropriateness for obtaining precise estimates of the target effect sizes. | I, M |
| 4. The dependent variables are described and operationalized with the aim that they should lead to good estimates of the target effect sizes. | M |
| 5. Results are presented and analyzed in terms of point estimates of the effect sizes. | R |
| 6. The precision of effect sizes is presented and analyzed in terms of confidence intervals. | R |
| 7. Wherever possible, results are presented in figures, with confidence intervals. | R, D |
| 8. Effect sizes are given substantive interpretation. | D |
| 9. Confidence intervals are given substantive interpretation. | D |
| 10. Meta-analytic thinking is used to interpret and discuss the findings. | D |

\* *Note*: I = Introduction, M = Methods, R = Results, D = Discussion

precision of that estimate. Correspondingly, it is most informative if the psychologist reports the effect of the psychotherapy as an ES—the best estimate of the amount of change the therapy brings about—and a CI (e.g., 95%) to indicate the precision of that estimate. This approach can be contrasted with the less informative dichotomous thinking (there is, or is not, an effect) that results from NHST.

In expressing their aims, authors should use language such as:

- We estimate the extent of …
- We will assess the influence of … on …
- Our aim is to find how large an effect … has on …
- We investigate the nature of the relation between … and …
- We will estimate how well our model fits these data …

Expressions like these naturally lead to answers that are ES estimates. Contrast these with statements like, "We investigate whether the new treatment is better or worse than the old"; "We examine whether there is a relation between … and …." These statements suggest that a mere dichotomous yes-or-no answer would suffice. Almost certainly the new treatment has *some* effect; our real concern is whether that effect is tiny, or even negative, or is positive and usefully large. It is an estimate of ES that answers these latter questions.

Examine the wording used to express the aims and main questions of the manuscript, especially in the abstract and introduction, but also in the title. Replace any words that convey dichotomous thinking with words that ask for a quantitative answer.

## 2. Previous Literature

Traditionally, reviews of past research in the social sciences have focused on whether previously published studies have, or have not, found an effect. A review, or the Introduction section to a manuscript, may reduce to a mere list of studies that found a statistically significant effect, versus those whose results failed to reach statistical significance. That is an impoverished and even misleading approach, which ignores the sizes of effects observed, and the fact that many negative results are likely to have been Type II errors attributable to low statistical power.

The American Psychological Association (APA) *Publication Manual* stated, "It is almost always necessary to include some measure of effect size in the Results section" (APA, 2010, p. 34). If an educational intervention increased mean reading age by 6.5 months, then 6.5 is our point estimate of the ES for that intervention. The *Manual* further stated, "Confidence intervals ... can be an extremely effective way of reporting results. Because confidence intervals combine information on location and precision ..., they are, in general, the best reporting strategy" (APA, p. 34). The above educational study may have found the increase to be 6.5 months of reading age, with a 95% CI of [6.5 ± 3.0], or [3.5, 9.5]. Narrow CIs indicate more precise estimates, and so the shorter a CI the better.

Past research should, wherever possible, be discussed in terms of the point and interval estimates obtained for the effects of interest. To researchers in many disciplines, that would not need stating. In the social sciences, however, many articles rely heavily on NHST, omit vital information about ES estimates, and conclude only that some intervention did or did not make a statistically significant difference (e.g., $p < .05$). Penetrating criticisms of NHST have been published by leading social scientists over more than half a century (e.g., Meehl, 1978), and advocacy of alternative techniques has been growing in volume and detail, especially in recent years (e.g., Fidler & Cumming, 2007). Other disciplines, notably medicine, have traveled at least part way along the road of statistical reform (Fidler, Cumming, Burgman, & Thomason, 2004). Kline (2004) identified 13 erroneous beliefs about $p$ values and their use, and explained why NHST causes so much damage. He summarized the statistical reform debate, and proposed a well-informed and balanced approach to improving statistical practice by shifting from reliance on NHST to widespread use of ESs and CIs, together with MA. We recommend Kline's book; its position is similar to the position we take in this chapter.

Point estimates of ESs encompass a diversity of types of measures. Most simply, an ES is a mean or other measurement in the original measurement units: The average extent of masked priming was 27 ms; the mean improvement after therapy was 8.5 points on the Beck Depression Inventory; the regression of annual income against time spent in education was 3,700 dollars/year. Alternatively, an ES measure may be unit-free: After therapy, 48% of patients no longer met the criteria for the initial clinical diagnosis; the correlation between hours of study and final grade was .52; the odds ratio for risk of unemployment in young adults not in college is 1.4, for males compared with females. Some ES measures indicate percentage of variance accounted for, such as $R^2$, as often reported in multiple regression (Howell, 2002, Ch. 15), and $\eta^2$ or $\omega^2$, as often reported with analysis of variance (Howell, Ch. 11). An important class of ES measures are *standardized* ESs, including Cohen's $d$ and Hedges' $g$, which are differences—typically between an experimental and a control group—measured in units of some relevant standard deviation (SD), for example the pooled SD of the two groups. Cumming and Finch (2001) explained Cohen's $d$ and how to calculate CIs for $d$. Grissom and Kim (2005) is an excellent source of assistance with the calculation and presentation of a wide variety of ES measures.

The introduction to the manuscript should focus on the ES estimates reported in past research, to provide a setting for the results to be reported. It is often helpful to combine the past estimates, and meta-analysis (MA) allows that to be done quantitatively. Hunt (1997) gave a general introduction to MA, and explanation of its importance. Lipsey and Wilson (2001) provided an overview of how a MA should be conducted, and Chapter 19 of this volume discusses MA in more detail.

Figure 7.1 is a *forest plot*, which presents the hypothetical results of six past studies, and their combination by MA. The result of each study is shown as a point estimate, with its CI. The result of the MA is a weighted combination of the separate point estimates, also shown with its CI. This CI on the result is usually much shorter, indicating greater precision, as we would expect given that results are being combined over multiple studies. Some medical journals now routinely require the introduction to

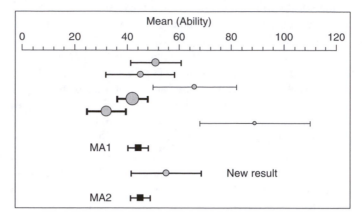

Note: A forest plot, showing the results of six fictitious studies, each as a mean ability score, with 95% CI. These are the upper six dots, whose sizes and line thicknesses indicate sample size and study variance: Large dots and heavy lines signal large sample size and small variance and, therefore, a high weighting in the meta-analysis. MA1 is the weighted combination mean for those six studies, with its 95% CI. The hypothetical new result is similarly displayed as a mean and CI, with size again indicating weight in the meta-analysis. MA2 is the weighted combination mean for all seven studies, with its 95% CI.

**Figure 7.1**  A Forest Plot.

each article to cite a MA—or wherever possible to carry out and report a new MA if none is available—as part of the justification for undertaking new research. That is a commendable requirement. Forest plots summarize a body of research in a compact and clear way; they are becoming common and familiar in medicine, and should be used more widely. Considering a number of estimates in combination, eventually including new results, is *meta-analytic thinking*, which we discuss below.

## 3. Experimental Design and the Precision of Estimates

Traditionally, statistical power estimates have been used to guide selection of the sample size $N$ required if a planned study is to have a reasonable chance of identifying effects of a specified size (assuming they exist). This approach requires an estimate of error variability, preferably based on past research or a pilot study, and specification of what size of effect is likely, or is of theoretical interest. The TFSI stated that, "Because power computations are most meaningful when done before data are collected ..., it is important to show how effect-size estimates [to be used in power calculations] have been derived from previous research and theory" (Wilkinson et al., 1999, p. 596).

The power approach was advocated by Jacob Cohen, and his book (Cohen, 1988) provided tables and advice (see also Chapter 24, this volume). An Internet search readily identifies freely available software to carry out power calculations, including G*Power (http://www.psycho.uni-duesseldorf. de/aap/ projects/gpower/). The power approach can be useful, but statistical power is defined in the context of NHST, and has meaning only in relation to a specified null hypothesis. Null hypotheses are almost always statements of zero effect, zero difference, or zero change. Rarely is such a null hypothesis at all plausible, and so it a great advantage of CIs that no null hypothesis need be formulated. In addition, CIs offer an improved way to do the job of statistical power.

The TFSI recognized that CI width, or the MOE, is the appropriate measure of precision, or of the sensitivity of the experiment: "Once the study is analyzed, confidence intervals replace calculated power in describing results" (Wilkinson et al., 1999, p. 596). An important advance in statistical practice is routine use of precision, meaning the MOE, in planning a study, as well as in discussion and interpretation of results. Di Stefano, Fidler, and Cumming (2005) described such an alternative approach that avoids NHST and the need to choose a null hypothesis. It is based on calculation of what sample size is needed to give a CI a chosen expected width: How large must $N$ be for the expected 95% CI to be no wider than, for example, 60 ms? Given a chosen experimental design, what sample size is needed for the expected MOE to be 0.2 units of Cohen's $d$? As with power, an estimate of variability is usually required, but no null hypothesis need be stated.

Justification of the experimental design and chosen sample size should appear as part of the rationale at the end of the Introduction section, or in the Methods section. It is often omitted from journal articles, having been overlooked by authors and reviewers, or squeezed out by strict word limits. Providing such justification is, however, especially important in cases where using too small a sample is likely to give estimates so imprecise that the research is scarcely worth doing, and may give misleading results. It is ethically problematic to carry out studies likely to give such inaccurate results. The converse—studies with such a large sample of participants that effects are estimated with greater precision than is necessary—are also ethically problematic, although these tend to be less common. The best way to justify a proposed design and sample size is in terms of the precision of estimates—the expected MOE—likely to be given by the results.

## 4. Dependent Variables

Specifying the experimental questions in terms of estimation of ESs leads naturally to choice of the dependent variables (DVs), or measures, that are most appropriate for those questions. Choose the operationalization of each DV that is most appropriate for expressing the ESs to be estimated, and that has adequate measurement properties, including reliability and validity. The aim is to choose measures that (1) relate substantively most closely to the experimental questions, and therefore will give results that are meaningful and interpretable, and (2) are most likely to give precise estimates of the targeted population effects.

In the Introduction section there may be discussion of methods used in past research, and this may help guide the choice of measures. In the Methods section there may be reference to published articles that provide information about the development of particular measures, and their psychometric properties. One important consideration is that the results to be reported should be as comparable as possible with previous research, and likely future research, so that meta-analytic combination over studies is as easy as possible. It can of course be a notable contribution to develop and validate an improved measure, but other things being equal it is advantageous to use measures already established in a field of research.

Choice of measures is partly a technical issue, with guidance provided by psychometric evidence of reliability and validity in the context of the planned experiment. It is also, and most importantly, a substantive issue that requires expert judgment by the researchers: The measures must tap the target concepts, and must give estimates of effects that can be given substantive and useful interpretation in the research context.

## 5. Results: Effect Sizes

The main role of the Results section is to report the estimated ESs that are the primary outcomes of the research. We mentioned in Desideratum 2 the wide range of possible ES measures, and emphasized that many of these are as simple and familiar as means, percentages, and correlations. In many cases it is possible to transform one ES measure into a number of others; Kirk (1996, 2003) provided formulas for this purpose. A correlation, for example, can be transformed into a value of Cohen's $d$. It is a routine part of MA to have to transform ES estimates reported in a variety of ways into some common measure, as the basis for conducting the MA. In medicine, odds ratio or log odds ratio are frequently used as the common ES measure, but in the social sciences Cohen's *d*, or some other standardized measure of difference (such as Hedges' *g*) is more frequently chosen as the basis for MA.

The authors of a manuscript need to consider which ES measures to report, bearing in mind ease of substantive interpretation, and the needs of future researchers wishing to include the results in some future MA. Often it will be best to present results in the original measurement scale of a DV, for simplicity and ease of interpretation, and also in some standardized form to assist both the comparison of results over different studies and the conduct of future MA. For example, an improvement in depression scores might be reported as mean change in score on the Beck Depression Inventory (BDI) because such scores are well known and easily interpreted by researchers and practitioners in the field. However, if the improvement is also reported as a standardized score the result is easily compared with, or combined with, the results of other studies of therapy, even where they have used other measures of depression. Similarly, a regression coefficient could be reported both in raw form, to assist understanding and interpretation, and as a standardized value, to assist comparison across different measures and different studies. In any case, it is vital to report SDs, and mean square error values, so that later meta-analysts have sufficient information to calculate whichever standardized ES measures they require.

A standardized measure of difference, such as Cohen's *d*, can be considered simply as a number of standard deviations. It is in effect a *z* score. It is important to consider which SD is most appropriate to use as the basis for standardization. Scores on the BDI, and changes in BDI scores, could be standardized against a published SD for the BDI. The SD unit would then be the extent of variation in some BDI reference population. That SD would have the advantage of being a stable and widely available value. Similarly, many IQ measures are already standardized to have a SD of 15. Alternatively, a change in BDI score could be expressed in units of the pre-test SD in a sample of participants. That would be a unit idiosyncratic to a specific study, and containing sampling error, but it might be chosen because it applies to the particular patient population being studied, rather than the BDI reference population. As so often is the case in research, informed judgment is needed to guide the choice of SD for standardization. When a manuscript reports a Cohen's *d* value, or any other standardized measure, it is essential that it makes clear what basis was chosen for standardization; when any reader interprets a standardized measure it is critical to have clearly in mind what SD units are being used.

It may be objected that much research has the aim not of estimating how large an effect some intervention has, but of testing a theory. However, theory testing is most informative if considered as a question of estimating goodness of fit, rather than of rejecting or not rejecting a hypothesis derived from the theory. A goodness of fit index, which may be a percentage of variance, or some other measure of distance between theoretical predictions and data, is an ES measure, and point and interval estimates of goodness of fit provide the best basis for evaluating how well the theory accounts for the data (Velicer et al., 2008).

## 6. Results: Confidence Intervals

Wilkinson et al. (1999) advised that: "Interval estimates should be given for any effect sizes involving principal outcomes. Provide intervals for correlations and other coefficients of association or variation whenever possible" (p. 599). The *Publication Manual* (APA, 2010) also recommended CIs: "Whenever possible, provide a confidence interval for each effect size reported to indicate the precision of estimation of the effect size" (p. 34) It specified (p. 117) the following style for reporting CIs in text.

At the first occurrence in a paragraph write: "The mean decrease was 34.5 m, 95% CI: [12.0, 57.0], and so. ..." On later occasions in the paragraph, if the meaning is clear write simply: "The mean was 4.7 cm [−0.8, 10.2], which implies that ...," or, "The means were 84% [73, 92] and 65% [53, 76], respectively ...," or, "The correlation was .41 [.16, .61]. ..." The units should not be repeated inside the square brackets. Note that in the last example, which gives the 95% CI on Pearson's *r* = .41, for *N* = 50, the interval is not symmetric about the point estimate; asymmetric intervals are the norm when the variable has a restricted range, as in the cases of correlations and proportions.

If an author elects to use the conventional confidence level of 95%, this should be stated the first time a CI appears in a paragraph, and the simple bracket format used thereafter to signal that the values in the brackets are the lower and upper limits respectively of the 95% CI. We recommend general use of 95% CIs, for consistency and to assist interpretation by readers, but particular traditions or special circumstances may justify choice of 99%, 90%, or some other CIs. If an author elects to use CIs with a different level of confidence, then that should be stated in every case throughout the manuscript, for example: "The mean improvement was 1.20 scale points, 90% CI [–0.40, 2.80]."

In a table, 95% CIs may similarly be reported as two values in square brackets immediately following the point estimate. Alternatively, the lower and upper limits of the CIs may be shown in separate labeled columns.

Altman, Machin, Bryant, and Gardner (2000) explained how to calculate CIs for a range of variables widely used in medicine, and provided software to assist. The variables covered include correlations, proportions, odds ratios, and regression coefficients. Cumming and Finch (2001) explained how to calculate CIs for Cohen's *d*. Grissom and Kim (2005) also provided advice on how to calculate CIs for many measures of ES.

## 7. Figures with CI Error Bars

Wilkinson et al. (1999) advised that authors should: "In all figures, include graphical representations of interval estimates whenever possible" (p. 601). We agree, and Cumming and Finch (2005) discussed the presentation and interpretation of error bars in figures. A serious problem is that the familiar graphic used to display error bars in a figure, as shown in Figure 7.2, can have a number of meanings. The extent of the bars could indicate SD, standard error (SE), a 95% CI, a CI with some other level of confidence, or even some other measure of some variability. Cumming, Fidler, and Vaux (2007) described and discussed several of these possibilities. The most basic requirement is that any figure with error bars must include a clear statement of what the error bars represent. A reader can make no sense of error bars without being fully confident of what they show, for example 95% CIs, rather than SDs or SEs.

CIs are interval estimates and thus provide inferential information about the ES of interest. CIs are therefore almost always the intervals of choice. In medicine it is CIs that are recommended and routinely reported. In some research fields, however, including behavioral neuroscience, SE bars (error bars that extend one SE below and one SE above a mean) are often shown in figures. Unless sample size is less than about 10, SE bars are about half the width of the 95% CI, so it is easy to translate visually between the two. But SE bars are not accurately and directly inferential intervals, so CIs should almost always be preferred.

Cumming et al. (2007) found that, in leading psychology journals, from 1998 to 2006 there was an increase from 11% to 38% of articles that included at least one figure with error bars. That is a dramatic and welcome increase. However, even recently, 47% of those articles showed SE bars, and 34% did not make clear what the error bars represented. It is encouraging that many more authors are now including error bars in figures, but it remains a major problem that they often do not appreciate the desirability of using CIs, and the critical importance of making clear in every case what the error bars represent.

Figure 7.2 shows means with CIs for a hypothetical two-group experiment with a repeated measure. A treatment group was compared with a control group, and three applications of an anxiety scale provided pre-test, post-test, and follow-up measures. The figure illustrates several important issues. First, a knowledgeable practitioner might feel that the CIs are surprisingly and discouragingly wide, despite the reasonable group sizes ($N = 23$ and 26). It is an unfortunate reality across the social sciences that error variation is usually large. CI width represents accurately the uncertainty inherent in a set of data, and we should not shoot the messenger by being critical of CIs themselves for being too wide. The

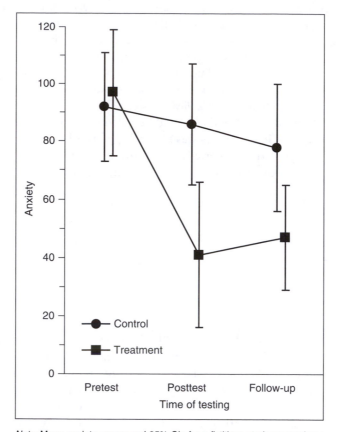

Note: Mean anxiety scores and 95% CIs for a fictitious study comparing a Treatment ($n = 23$) and a Control ($n = 26$) group, at each of three time points: pre-test, post-test, and follow-up. Means have been displaced slightly so all CIs can be clearly seen.

**Figure 7.2** Error Bar Display.

problem is NHST, with its simplistic reject or don't reject outcome, which may delude us into a false sense of certainty, when in fact much uncertainty remains. Cohen (1994) said, "I suspect that the main reason they [CIs] are not reported is that they are so embarrassingly large!" (p. 1002). We should respond to the message of large error variation by making every effort to improve experimental design and use larger samples, but must acknowledge the true extent of uncertainty by reporting CIs wherever possible.

Cumming and Finch (2005) provided rules of eye to assist interpretation of figures such as Figure 7.2. For means of two independent groups, the extent of overlap of the two 95% CIs gives a quick visual indication of the approximate $p$ value for a comparison of the means. If the intervals overlap by no more than about half the average of the two MOEs, then $p < .05$. If the intervals have zero overlap—the intervals touch end-to-end—or there is a gap between the intervals, then $p < .01$. In Figure 7.2, the control and treatment means at pre-test, for example, overlap extensively, and so $p$ is considerably greater than .05. At post-test, however, the intervals have only a tiny overlap, so at this time point $p$ for the treatment vs. control comparison is approximately .01. At follow-up, overlap is about half the length of the average of the two overlapping arms (the two MOEs), and so $p$ is approximately .05.

It is legitimate to consider overlap when the CIs are on independent means, but when two means are paired or matched, or represent a repeated measure, overlap of intervals is *irrelevant* to the comparison of means, and may be misleading. Further information is required, namely the correlation between the two measures, or the SD of the *differences*. For this reason it is not possible to assess in Figure 7.2 the *p* value for any within-group comparison, such as the pre-test to post-test change for the treatment group. Belia, Fidler, Williams, and Cumming (2005) reported evidence that few researchers appreciate the importance of the distinction between independent and dependent means when interpreting error bars. If CIs in figures are to be used to inform the interpretation of data—as we advocate—it is vital that figures make very clear the status of each independent variable. For between-subject variation, or independent means, intervals can be directly compared. For within-subject variation, a repeated measure, or dependent means, intervals may not be compared.

This important issue was illustrated further by Cumming and Finch (2005). If it seems puzzling, think of it in terms of the two familiar *t* tests: For two independent means, the independent *t* test is based on variation within the two groups; this variation determines the two CIs, and so those CIs are informative about the comparison of the two means. For paired data, by contrast, the paired *t* test is based on the variation in the paired differences, which is often in practice much smaller than the variation within either group. The variation in the differences is not represented by the CIs on the two means, and so these intervals are irrelevant to assessment of the mean difference—which can, however, be assessed if we have the CI on the difference, as shown in Figure 7.3 for the difference in each group between pre-test and post-test.

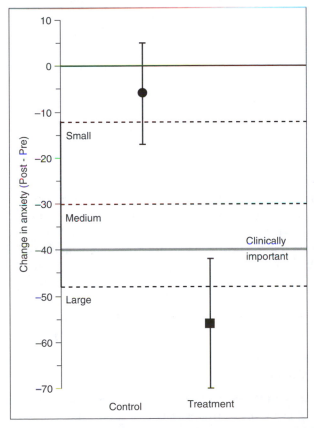

*Note*: Mean change in anxiety score from pre-test to post-test, for Treatment and Control groups, with 95% CIs, for the data shown in Figure 2. Dotted lines indicate reference values for changes considered small, medium, and large, and the grey line indicates the change considered to be clinically important.

**Figure 7.3** Confidence Intervals for Mean Differences.

It is a problem that many current software packages do not sufficiently support the preparation of figures with error bars. In Figure 7.2, for example, the means are slightly offset horizontally so that all CIs can be seen clearly, but few packages make it easy to do this. One solution is to use Microsoft Excel. Figure 7.2 was prepared as an Excel scatterplot, which requires the horizontal and vertical coordinates for each point to be specified, so means can readily be displayed with a small horizontal offset.

In summary, the Results section should report point and interval estimates for the ESs of interest. Figures, with 95% CIs shown as error bars, should be presented wherever that would be informative. Every figure must make clear what error bars represent, and must describe the experimental design so a reader can understand whether each independent variable varies between or within subjects.

## 8. Interpretation of Effect Sizes

A primary purpose of the Discussion section is to present a substantive interpretation of the main ES estimates, and to draw out the implications. One unfortunate aspect of NHST is that the term *significant* is used with a technical meaning—a small *p* value was obtained—whereas in common language the word means "important." Kline (2004) recommended the word simply be dropped, so that if a null hypothesis is rejected we would say "a statistical difference was obtained." The common practice of saying "a significant difference was obtained" almost insists that a reader regard the difference as important, whereas it may easily be small and of trivial importance, despite yielding a small *p* value. Judging whether an ES is large or important is a key aspect of substantive interpretation, and requires specialist knowledge of the measure and the research context. We recommend that, if reporting NHST, either avoid the term "significant," as Kline recommended, or make its technical meaning clear by saying "statistically significant." When discussing the importance of a result, use words other than "significant," perhaps including "clinically important," "educationally important," or "practically important."

Cohen (1988, pp. 12–14) discussed the need for reference standards for the interpretation of common standardized and unit-free ES measures. He suggested standards that have become well known and widely used. For example, for Pearson correlation he suggested that values of .1, .3, and .5 can be regarded as small, medium, and large, respectively; and for Cohen's *d* he suggested similar use of .2, .5, and .8. An advantage of such standards is that ESs can be compared across different measures. He argued that the values and labels he chose are likely to be judged reasonable in many situations in the social sciences, but he stated clearly that they were arbitrary, and "were set forth throughout with much diffidence, qualifications, and invitations not to employ them if possible" (p. 532). Sometimes numerically tiny differences may have enormous theoretical importance, or indicate a life-saving treatment of great practical value. Conversely, a numerically large effect may be unsurprising and of little interest or use. Knowledgeable judgment is needed to interpret ESs (How large? How important?), and a Discussion section should give reasons to support the interpretations offered, and sufficient contextual information for a reader to come to an independent judgment.

## 9. Interpretation of Confidence Intervals

The correct way to understand the level of confidence, usually 95%, is in relation to indefinitely many replications of an experiment, all identical except that a new random sample is taken each time. If the 95% CI is calculated for each experiment, in the long run 95% of these intervals will include the population mean $\mu$, or other parameter being estimated. For our sample, or any particular sample, the interval either does or does not include $\mu$, so the probability that this particular interval includes $\mu$ is 0 or 1, although we will never know which. It is misleading to speak of a probability of .95, because that suggests the population parameter is a variable, whereas it is actually a fixed but unknown value.

Here follow some ways to think about and interpret a 95% CI (see also Cumming & Finch, 2005):

- The interval is one from an infinite set of intervals, 95% of which include $\mu$. If an interval does not contain $\mu$, it probably only just misses.
- The interval is a set of values that are *plausible* for $\mu$. Values outside the interval are relatively implausible—but not impossible—for $\mu$. (This interpretation may be the most practically useful.)
- We can be 95% confident that our interval contains $\mu$. If in a lifetime of research you calculate numerous 95% CIs in a wide variety of situations, overall, 95% of these intervals will include the parameters they estimate, and 5% will miss.
- Values around the center of the interval are the best bets for $\mu$, values towards the ends (the lower and upper limits) are less good bets, and values just outside the interval are even less good bets for $\mu$ (Cumming, 2007b).
- The lower limit is a likely lower bound of values for $\mu$, and the upper limit a likely upper bound.
- If the experiment is replicated, there is on average an 83.4% chance that the sample mean (the point estimate) from the replication experiment will fall within the 95% CI from the first experiment (Cumming, Williams, & Fidler, 2004). In other words, a 95% CI is approximately an 83% *prediction interval* for the next sample mean.
- The MOE is a measure of precision of the point estimate, and is the likely largest error of estimation, although larger errors are possible.
- If a null hypothesized value lies outside the interval, it can be rejected with a two-tailed test at the .05 level. If it lies within the interval, the corresponding null hypothesis cannot be rejected at the .05 level. The further outside the interval the null hypothesized value lies, the lower is the $p$ value (Cumming, 2007b).

The last interpretation describes the link between CIs and NHST: Given a CI it is easy to note whether any null hypothesized value of interest would be rejected, given the data. Note, however, the number and variety of interpretations of a CI that make no reference to NHST. We hope these will become the predominant ways researchers think of CIs, as CIs replace NHST in many situations.

Authors may choose from the options above to guide their use of CIs to interpret their results. They may, for example, give substantive interpretation of the point estimate, and of the lower and upper limits of the CI, and thus cover the likely full range of plausible values for the parameter being estimated. Cumming et al. (2007) found that reporting of CIs in leading psychology journals increased from 4% to 11% of articles, between 1998 and 2005–2006. However, in only 24% of cases where CIs were reported were the intervals interpreted, or used explicitly to support data interpretation. As the *Publication Manual* recommends, "wherever possible, base discussion and interpretation of results on point and interval estimates" (APA, 2010, p. 34).

Figure 7.3 shows for the two groups the mean differences between pre-test and post-test, for the data presented in Figure 7.2. The figure includes 95% CIs on those differences, and there are reference lines that indicate the amounts of improvement judged by the researchers to be small, medium and large, and of clinical importance. These lines were created by adding further data series, with labels, to the Excel scatterplot used to present the mean differences with CIs. The CIs in Figure 7.3 allow us to conclude that, for the control group, the change from pre-test to post-test is around zero, or at most small; for the treatment group the change is of clinical importance, and likely to be large or even very large.

## 10. Meta-analytic Thinking

Figure 7.1 shows, as we mentioned earlier, the meta-analytic combination of hypothetical results from previous research. It also shows the point and interval estimates found in the current experiment, and a second meta-analysis that combines these with the earlier results. Considering that particular effect, our experiment has advanced the state of knowledge from the first to the second of the combined ES estimates, marked with square symbols in Figure 7.1. The Introduction and Discussion sections of the manuscript should both consider current research in the context of past results and likely future studies. This is meta-analytic thinking (Cumming & Finch, 2001), and it guides choice of what statistics are most valuable to report, and how results are interpreted and placed in context. Wilkinson et al. (1999) made several statements that outline what is needed:

> Reporting and interpreting effect sizes in the context of previously reported effects is essential to good research. ... Reporting effect sizes also informs ... meta-analyses needed in future research. ... Collecting intervals across studies also helps in constructing plausible regions for population parameters (p. 599).
>
> ...
>
> Do not interpret a single study's results as having importance independent of the effects reported elsewhere in the relevant literature. ... The results in a single study are important primarily as one contribution to a mosaic of study effects (p. 602).

A forest plot (Figure 7.1) can display point and interval ES estimates expressed in any way—as original units, or in standardized form, or as some unit-free measure. For many types of research a forest plot can conveniently summarize current and past research in terms of estimation, and thus bring together all the ES, CI, and MA components that we have discussed in this chapter. It is important that authors, reviewers, and editors work together to help advance the social sciences as much as possible from the blinkered, dichotomous thinking of NHST to the richer and more informative research communication described in this chapter.

### Notes

1 This research was supported by the Australian Research Council.
2 Cumming and Finch (2001, 2005), Cumming, Williams, and Fidler (2004), and Cumming (2007b) are accompanied by components of ESCI ("ess-key"; Exploratory Software for Confidence Intervals), which runs under Microsoft Excel. Components of ESCI are available from www.latrobe.edu.au/psy/esci.

### References

Altman, D. G., Machin, D., Bryant, T. N., & Gardner, M. J. (2000). *Statistics with confidence* (2nd ed.). London: British Medical Journal.
American Psychological Association. (2010). *Publication manual of the American Psychological Association* (6th ed.). Washington, DC: Author.
Belia, S., Fidler, F., Williams, J., & Cumming, G. (2005). Researchers misunderstand confidence intervals and standard error bars. *Psychological Methods, 10*, 389–396.
Cohen, J. (1988). *Statistical power analysis for the behavioral sciences* (2nd ed.). Hillsdale, NJ: Erlbaum.
Cohen, J. (1994). The earth is round (*p* < .05). *American Psychologist, 49*, 997–1003.
Cumming, G. (2007a). Confidence intervals. In G. Ritzer (Ed.) *The Blackwell encyclopedia of sociology* (Vol. II, pp. 656–659). Oxford, UK: Blackwell.
Cumming, G. (2007b). Inference by eye: Pictures of confidence intervals and thinking about levels of confidence. *Teaching Statistics, 29*, 89–93.
Cumming, G., Fidler, F., Leonard, M., Kalinowski, P., Christiansen, A., Kleinig, A., Lo, J., McMenamin, N., & Wilson, S. (2007). Statistical reform in psychology: Is anything changing? *Psychological Science, 18*, 230–232.

Cumming, G., Fidler, F., & Vaux, D. L. (2007). Error bars in experimental biology. *Journal of Cell Biology, 177*, 7–11.

Cumming, G., & Finch, S. (2001). A primer on the understanding, use, and calculation of confidence intervals that are based on central and noncentral distributions. *Educational and Psychological Measurement, 61*, 530–572.

Cumming, G., & Finch, S. (2005). Inference by eye: Confidence intervals, and how to read pictures of data. *American Psychologist, 60*, 170–180.

Cumming, G., Williams, J., & Fidler, F. (2004). Replication, and researchers' understanding of confidence intervals and standard error bars. *Understanding Statistics, 3*, 299–311.

Di Stefano, J., Fidler, F., & Cumming, G. (2005). Effect size estimates and confidence intervals: An alternative focus for the presentation and interpretation of ecological data. In A. R. Burk (Ed.), *New trends in ecology research* (pp. 71–102). Hauppauge, NY: Nova Science.

Fidler, F. & Cumming, G. (2007). The new stats: Attitudes for the twenty-first century. In J. W. Osborne (Ed.), *Best practice in quantitative methods* (pp. 1–12). Thousand Oaks, CA: Sage.

Fidler, F., Cumming, G., Burgman, M., & Thomason, N. (2004). Statistical reform in medicine, psychology and ecology. *Journal of Socio-Economics, 33*, 615–630.

Grissom, R. J., & Kim, J. J. (2005). *Effect sizes for research: A broad practical approach.* Mahwah, NJ: Erlbaum.

Howell, D. C. (2002). *Statistical methods for psychology* (5th ed.). Pacific Grove, CA: Duxbury.

Hunt, M. (1997). *How science takes stock: The story of meta-analysis.* New York: Russell Sage.

Kirk, R. E. (1996). Practical significance: A concept whose time has come. *Educational and Psychological Measurement, 56*, 746–759.

Kirk, R. E. (2003). The importance of effect magnitude. In S. F. Davis (Ed.), *Handbook of research methods in experimental psychology* (pp. 83–105). Malden, MA: Blackwell.

Kline, R. B. (2004). *Beyond significance testing: Reforming data analysis methods in behavioral research.* Washington, DC: American Psychological Association.

Lipsey, M. W., & Wilson, D. B. (2001). *Practical meta-analysis.* Thousand Oaks, CA: Sage.

Meehl, P. E. (1978). Theoretical risks and tabular asterisks: Sir Karl, Sir Ronald, and the slow progress of soft psychology. *Journal of Consulting and Clinical Psychology, 46*, 806–834.

Smithson, M. (2002). *Confidence intervals.* Thousand Oaks, CA: Sage.

Velicer, W. F., Cumming, G., Fava, J. L., Rossi, J. S., Prochaska, J. O., & Johnson, J. (2008). Theory testing using quantitative predictions of effect size. *Applied Psychology: An International Review, 57*, 589–608.

Wilkinson, L., & Task Force on Statistical Inference (1999). Statistical methods in psychology journals. Guidelines and explanations. *American Psychologist, 54*, 594–604.

# 8

# Factor Analysis

## *Exploratory and Confirmatory*

**Deborah L. Bandalos and Sara J. Finney**

Factor analysis is a method of modeling the covariation among a set of observed variables as a function of one or more latent constructs. Here, we use the term *construct* to refer to an unobservable but theoretically defensible entity, such as intelligence, self-efficacy, or creativity. Such constructs are typically considered to be latent in the sense that they are not directly observable (see Bollen, 2002, for a more detailed discussion of latent constructs). The purpose of factor analysis is to assist researchers in identifying and/or understanding the nature of the latent constructs underlying the variables of interest. Technically, these descriptions exclude *component analysis*, which is a method for reducing the dimensionality of a set of observed variables through the creation of an optimum number of weighted composites. A major difference between factor and component analysis is that in the latter all of the variance is analyzed, whereas in factor analysis, only the shared variance is analyzed. For this reason, factor analysis is sometimes referred to as *common factor analysis.* In many ways, however, component analysis is very similar to common factor analysis, and many of the desiderata for exploratory factor analysis presented here apply equally to component analysis. Given that the goal of component analysis is to explain as much observed variance as possible via the weighted composites and not, as in common factor analysis, to model the relations among variables as functions of underlying latent variables, those desiderata relating to the importance of theory for factor analysis do not apply to component analysis (see Widaman, 2007, for a detailed explanation of the conceptual and mathematical distinction between exploratory factor analysis and principal components analysis).

Two broad classes of factor analytic methods are described in this chapter: exploratory factor analysis (EFA) and confirmatory factor analysis (CFA). Although these two methods both model the observed covariation among variables as a function of latent constructs, in EFA the purpose of such models is typically to identify the latent constructs or to generate hypotheses about their possible structures, whereas the purpose of CFA is to evaluate hypothesized structures of the latent constructs and/or to develop a better understanding of such structures. Thus, CFA is a specific form of structural equation modeling (SEM; see Chapter 28, this volume). Whereas EFA can be carried out using conventional statistics software such as SPSS and SAS, CFA requires the use of specialized software such as AMOS (Arbuckle, 2007), EQS (Bentler, 2006), LISREL (Jöreskog & Sörbom, 2007), or Mplus (Muthén & Muthén, 2007).

Table 8.1a Desiderata for Exploratory Factor Analysis

| Desideratum | Manuscript Section(s)* |
|---|---|
| E1. Theory and/or previous research supporting the construct(s) under investigation are synthesized. | I |
| E2. Exploratory vs. confirmatory analysis is justified. | I, M |
| E3. Measured variables that operationalize the construct are presented and thoroughly justified in terms of both quantity and content. (Here, information on reliability and validity, if appropriate, is included.) | I, M |
| E4. Sampling method(s) and sample size(s) are discussed and justified. | M |
| E5. Data are screened for outliers and the method of handling missing data is discussed. The correlation matrix is presented and/or measures of amenability of data to factoring are presented anddiscussed. Summary statistics of measured variables are presented. If raw data were analyzed, information on obtaining access to the data is provided. | M, R |
| E6. Name and version of software package is reported. | M, R |
| E7. Method of extraction is discussed and justified. | M, R |
| E8. Method(s) used to determine the number of factors are discussed and decision is justified. A number of factor solutions are explored and this is clearly presented. If a model is championed over others, there is clear justification as to why this model is superior. | M, R |
| E9. Method of rotation is stated and justified. | M, R |
| E10. Justification is provided for any variables that are eliminated; model is reanalyzed after variables are eliminated. | R |
| E11. Parameter estimates (e.g., pattern/structure coefficients, factor correlations) are presented and discussed for final model. | R |
| E12. Percentages of variance accounted for (both total and by each factor) are provided and discussed. | R |
| E13. Appropriate interpretation of factors is provided. | R, D |
| E14. Factor score determinacy is evaluated. | R |
| E15. Factor quality is evaluated using reliability and, if possible, validity evidence. | R |
| E16. Appropriate caveats are provided; importance of replication and limitations of the study are discussed. | D |

* *Note:* I = Introduction, M = Methods, R = Results, D = Discussion

For more in depth treatment of exploratory factor analysis we recommend texts by Gorsuch (1983), Comrey and Lee (1992), and McDonald (1985). For a more contemporary treatment, we recommend Pett, Lackey, and Sullivan (2003). More in-depth treatments of CFA can be found in SEM textbooks such as those by Byrne (1998, 2001), Bollen (1989), Kline (2005), and Loehlin (2004). Brown (2006) devotes a complete text to the use of confirmatory factor analysis for applied research. Specific desiderata are provided for EFA in Table 8.1a and for CFA in Table 8.1b, and are elucidated in the subsequent sections of this chapter.

## E1. Theory and/or Previous Research

Many of the decisions made in EFA are, at least to some degree, subjective. Although statistically and/or mathematically based guidelines exist for some decisions (e.g., determining the number of factors), many important decisions are made on the basis of congruence with theory and/or previous research. Familiarity with the theory and research findings regarding the construct to be studied is therefore essential in EFA studies. Although some may argue that this reliance on prior theory can be self-serving, it should be kept in mind that EFA is, as its name implies, more an exploratory than an

Table 8.1b  Desiderata for Confirmatory Factor Analysis

| Desideratum | Manuscript Section(s)* |
|---|---|
| C1. Theory and/or previous research leading to the model(s) under investigation are synthesized; a set of a priori specified competing models, represented by path diagrams, is preferred. | I |
| C2. Confirmatory versus exploratory analysis is justified. | I, M |
| C3. Measured variables that operationalize the construct are thoroughly presented and their relationship with the factor is explained. (Here, information on reliability and validity, if available, is included.) | I, M |
| C4. Sampling method(s) and sample size(s) are discussed and justified. | M |
| C5. Data are screened for outliers and nonnormality, and the method of handling missing data is discussed. Summary statistics of measured variables are presented. If raw data were analyzed, information on obtaining access to the data is given. | M, R |
| C6. Method of estimation is discussed and justified. | M |
| C7. Name and version of software package is reported. | M, R |
| C8. Problems with model convergence, improper estimates, and/or model identification are reported and discussed. | R |
| C9. Recommended data-model fit indices, including standardized residuals, are presented and discussed. | R |
| C10. For competing models, comparisons are made using statistical tests (for nested models) or information criteria (for non-nested models). | R |
| C11. Post hoc model modifications, if made, are justified on the basis of both theoretical and statistical criteria. If cross-validation is not possible, issues with post hoc model modification are discussed. | R, D |
| C12. If data-model fit is adequate, parameter estimates, standard errors, and measures of variance accounted for are presented and discussed. Method used to determine saliency of loading coefficients is provided and justified, and appropriate interpretation of factors is provided. | R, D |
| C13. Factor quality is evaluated using reliability and, if possible, validity evidence. | R, D |
| C14. Appropriate caveats are provided; importance of replication and limitations of the study, including equivalent models, are discussed. | D |

* *Note*: I = Introduction, M = Methods, R = Results, D = Discussion

inferential methodology. As such, its role is to generate hypotheses rather than to provide strict inferential tests of a priori hypotheses. Given these purposes, reliance on theory and prior research is not only appropriate, it is fundamental to the hypothesis-generating process.

In some cases, researchers might argue that the latent construct or scale being analyzed is new and that no underlying theory is available. Such an argument is implausible unless one is willing to assume that the variables being analyzed were selected at random. If this is not the case then some theory, however rudimentary, must have guided the selection of the variables and this theory should be explicated to the extent possible.

## E2. Exploratory vs. Confirmatory

Although exploratory and confirmatory analyses are often referred to as though they represent a dichotomy, the distinction between the two is really more a matter of degree. Certainly, it is possible to use EFA in a confirmatory manner or to use CFA in an exploratory manner. However, because the CFA model is much more restrictive than that of EFA, most experts recommend that EFA be used for situations in which minimal research has been conducted regarding the structure of the construct or measure of interest. Use of CFA in such situations often results in gross misfit of the model to the data. In situations such as these, there are innumerable ways in which the model can be respecified to improve fit, and a researcher can easily be overwhelmed and led astray by allowing the estimation of parameters that do not lead to the generating structure. Alternative EFA models, on the other hand, typically represent a more finite set of models that differ on criteria such as the number of factors, method of extraction, and type of rotation. Thus, the choice of models to compare tends to be much clearer in EFA. In addition, exploring many different models via EFA may uncover interesting and conceptually plausible structures that may have gone unstudied if CFA were employed.

A common situation in which one must choose between EFA and CFA is the investigation of items that have been written to measure a construct hypothesized to have several dimensions. Researchers will often claim a priori knowledge of the underlying structure based on the fact that the items have been written to measure specific aspects of the construct. However, in our experience, items are rarely aware of the scale for which they have been written and often fail to behave as they should. So unless there is empirical evidence to support such a claim, it is probably best to begin by conducting an EFA in such situations. If a clear, interpretable structure emerges, CFA can be employed, using an *independent* sample, to further test the structure that was championed from the exploratory analyses.

As a general guideline, EFA should be used for situations in which the variables to be analyzed are either newly developed or have not previously been analyzed together, or when the theoretical basis for the factor analysis model (i.e., number of factors, level of correlation among factors) is weak. In such situations, it is not possible to specify the model a priori in sufficient detail to conduct a CFA. Therefore, in our view, CFA should only be used if the structure of the variables has been previously studied using EFA with an independent source of data.

### E3. Measured Variables

The number and nature of the factors are dictated by the observed variables that are analyzed. When conducting an EFA, the researcher should have an in-depth understanding of the construct under study (see Desideratum E1). This understanding should, in turn, inform the researcher's justification of how the observed variables cover the breadth of the construct. The number of variables to be factored should also be informed by the complexity of the construct. For example, EFA is often used to assess the dimensionality of newly created measures that represent latent constructs (e.g., motivation, anxiety). If the latent construct is believed to consist of multiple dimensions, it is imperative that several variables that represent these dimensions are included on the measure, and ultimately in the factor analysis. Again, the author should be clear that the EFA solution is completely dependent on the variables being factored; an "expected" factor will not emerge if the variables do not capture the construct adequately.

The variables can take many forms: items from a scale, subscale scores, or direct measures of subject characteristics (e.g., number of occurrences of behaviors, heart rate). Although some methodologists recommend against factoring items due to their generally low reliability and lack of a continuous scale, such analyses should not be problematic if the items have at least five scale points and are reasonably intercorrelated (see Desideratum E5 for more information on this point). If analyzing items or subscales that are the composite of a set of items, the author should present the actual items or at least example items whenever possible (if there are no copyright or item security issues). If the specific items or subscales have been studied previously, any reliability or validity evidence should be presented. Likewise, if direct measurements are being factored, the manner in which these variables were gathered and any prior work supporting the manner in which they were gathered should be presented. Without a clear presentation of each observed variable, it is difficult for a reader to interpret the factor solution (i.e., what do the factors represent?). Therefore, the number, type, and examples of the variables to be factored must be presented.

## E4. Sampling Method(s) and Sample Size(s)

Unlike the vast majority of statistical methods with which quantitative researchers are familiar, EFA is generally used in a descriptive and exploratory, rather than inferential, manner. This is one reason that standard errors and statistical tests of fit are generally not available for these methods.[1] (The exception to this is maximum likelihood factor analysis, discussed in Desideratum E7.) Although sampling theory need not be invoked to obtain factor analytic solutions, it is still the case that results can only be generalized to samples similar to that on which the analyses have been conducted, unless previous empirical evidence exists for broader generalizations. Therefore, it is important that researchers describe the makeup of their sample in sufficient detail that readers can determine the degree to which the results might generalize to populations in which they are interested.

Although many rules of thumb have been suggested for determining an adequate sample size for EFA studies, recent studies have found that the sample size necessary to obtain accurate parameter estimates depends on characteristics of the data. The primary parameter of interest in factor analytic studies is the factor loading. There are two types of loadings, known as *structure coefficients* and *pattern coefficients*. Structure coefficients represent zero-order correlations between variables and factors. Pattern coefficients represent the unique effect of a factor on a variable, with the effects of all other factors partialed out. For situations in which there is only one factor, or in which factors are uncorrelated, structure and pattern coefficients are equivalent. In other cases, however, both need to be taken into account in interpreting factors. The sample size needed to obtain accurate estimates of pattern and structure coefficients depends on the level of *communality* of the variables, which represents the amount of variance in the variables that is accounted for by the factor solution, the number of variables per factor, and the interactions of these two conditions. Specifically, samples of 100 may be sufficient to obtain accurate estimates with three factors measured by three to four variables each if communalities are at least .7, but if communalities are lower than .5, sample sizes of at least 300 would be needed. With more factors, larger samples are needed. For example, in the latter situation (communalities < .5 and three to four variables per factor) a sample size of 500 would be needed if the number of factors were increased to seven (Hogarty, Hines, Kromrey, Ferron, & Mumford, 2005; MacCallum, Widaman, Zhang, & Hong, 1999; MacCallum, Widaman, Preacher, & Hong, 2001; Velicer & Fava, 1998).

## E5. Data Screening

Most EFA estimation methods are not based on strict assumptions about the nature of the sample and how it was obtained. However, this does not mean that these aspects of the study can be safely ignored. Similar points can be made with regard to the nature of the variables to be analyzed and their distributions. Although use of factor analysis does not assume normality or continuousness of the variables to analyzed, variables with nonnormal distributions and/or few scale points, as well as outlying observations, can have substantial effects on the results of EFAs.

It is fairly well known in the factor analytic literature that variables with similar levels of skew and/or kurtosis can form artifactual factors. This occurs because variables that are similarly distributed are more highly correlated with each other than variables with different distributions, given that all else is equal. In the educational literature, such factors are termed "difficulty factors" and occur because easy (negatively skewed) and difficult (positively skewed) achievement and aptitude items tend to form factors irrespective of item content. This can result in solutions that are challenging to interpret at best, and misleading at worst. The severity of such problems increases with the level of nonnormality. In general, if absolute skewness and kurtosis values are no greater than 2.0 (some researchers suggest a more liberal standard of 7.0 for kurtosis), little or no distortion should occur. However, it is important that researchers provide information on the distributions of the variables to be analyzed so that readers are alerted to the potential for such problems. Our preference is for a table in which the mean and standard deviation of each variable, along with values of skewness and kurtosis, are reported. If the number of variables being analyzed is so large as to preclude this option, the range of values or a description of the levels of skewness and kurtosis should be provided.

On a related note, the level of measurement associated with the variables should be clearly presented to readers. Because EFA estimation is typically based on analysis of Pearson Product-Moment (PPM) correlation matrices, violations of the assumptions underlying PPM correlations can result in bias of EFA parameters. More specifically, continuousness of the variables and a linear relationship between the variables and factors are necessary to obtain accurate results. These assumptions are violated when data are dichotomous or ordinal in nature. Although there should be little bias with five or more ordered categories, for variables with fewer categories the analysis of the PPM correlation matrix can produce biased results. Researchers should note how they addressed this issue. A common solution is to use a correlation matrix that takes the categorical nature of the variables into account, such as a matrix of tetrachoric (for dichotomous scales) or polychoric (for ordinal scales) correlations (see Finney & DiStefano, 2006, for a review of this literature), coupled with a special estimator (see Desideratum C6). However, other methods (full-information; Bolt, 2005; Jöreskog & Moustaki, 2001; McLeod, Swygert, & Thissen, 2001; Moustaki, 2007; Swygert, McLeod, & Thissen, 2001) can be used as well.

Outliers can affect EFA results, just as they do other analyses. For this reason, the data should be screened for multivariate outliers and the results of these analyses reported. If outliers are found, the researcher may or may not decide to delete these. Although many researchers delete outliers as a matter of course, our view is that a thoughtful analysis of the nature and possible causes of the outlying cases should precede any such decision. Such an analysis may provide information on subpopulations for which the factor model does not hold, or other potential avenues for further investigation. If the researcher ultimately decides to delete the outlying cases from the analyses, our preference is that the analyses be conducted with and without the outliers and the results of both sets of analyses reported (space permitting).

## E6. Software

Commonly used software packages differ in terms of default settings, presentation of output, and even calculations (see Desideratum E12 for an example). Different versions of a given software package may also differ in terms of these features. For this reason, researchers should provide the name and version of the package used.

## E7. Method of Extraction

As noted in the introduction to this chapter, our focus is on factor analysis rather than on component analysis. However, the two types of analyses are often confused; therefore, researchers and reviewers should ensure the analysis that is most appropriate for the goals of the study has been conducted. In general, component analysis is recommended for reducing a large number of variables to a smaller, more manageable number of components, whereas factor analysis is better suited for identifying the latent constructs underlying the variable correlations. Thus, although component analysis is often used for such purposes as scale development, if the researcher's intention is to interpret the components as latent dimensions or factors then factor analysis is the more appropriate analysis.

Extraction refers to the process by which the parameters of the factor solution, which include the factor pattern coefficients and, if appropriate, structure coefficients and factor intercorrelations, are estimated. A desirable solution is one that accounts for the correlations among the variables in the sense that these correlations can be accurately reproduced as functions of the estimated parameters. However, because there is an infinite number of combinations of pattern coefficients and factor correlations that will reproduce the variable correlations equally well, there is no one method of extraction that can be considered "best" in an absolute sense. Because of this, many methods of extraction have been proposed, each with a slightly different criterion regarding what is considered "best." Within the factor analysis model, principal axis (PA) and maximum likelihood (ML) methods are most commonly used. Other methods, such as generalized least squares, unweighted least squares, image analysis and alpha factoring, are also available, although not as widely used.

PA methods provide a least squares-type solution, attempting to minimize the residuals between the correlation matrix being analyzed and the matrix implied by the factor model (i.e., pattern coefficients and factor correlations). ML factor analysis explicitly takes into account the fact that a sample and not a population matrix is being analyzed, and seeks to obtain the solution that would best reproduce the population correlation values. ML factor analysis is thus an inferential method, and standard errors for model parameters as well as tests of the goodness of fit of the solution can be obtained. Although some researchers prefer ML for this reason, it should be noted that this method may not provide accurate estimates of pattern coefficients if the factors are weak and/or if the sample size is small (Briggs & MacCallum, 2003). In any case, however, the choice of estimation method should not be arbitrary or dictated by the defaults of the computer package used, but should be justified by the researcher.

## E8. Number of Factors

A major decision in EFA studies is that of determining the number of factors to retain. Methods of determining the number of factors can be classified as statistically based (Bartlett's test, Velicer's (1976) Minimum Average Partial [MAP] procedure, Parallel Analysis; Horn, 1965), mathematically based (the eigenvalue greater than one [K1] "rule"), or more heuristic (scree plot; retaining the number of factors that account for a pre-specified percentage of variance). In simulation studies, the Parallel Analysis and MAP procedures have been found to perform best, whereas K1 consistently performs worst. However, the simulations have been based on generated data in which the factors are known to be orthogonal. With correlated factors, it is likely that methods of determining the number of factors will be less accurate. In any case, decisions regarding the number of factors to retain should never be based on one criterion alone. Because EFA is an exploratory technique, it is expected that researchers will compare solutions that extract various plausible numbers of factors. These solutions should be obtained from several of the methods mentioned previously, but in making a final decision, more weight should be placed on solutions obtained from methods that are known to perform well (such as Parallel Analysis and MAP) relative to those known to provide biased estimates of the number of factors (such as K1). In addition, solutions on which the various methods converge are typically preferred to those for which only one method provides an optimum value.

In addition to these methods, researchers should use theory and/or previous research to inform decisions regarding the number of factors. For example, if previous research or theory suggests that there should be three factors, this solution should be obtained and compared to other possible solutions, even if it is not suggested by any of the methods used in determining the number of factors. Also, factors for which only one or two variables have strong pattern/structure coefficients should be carefully scrutinized, as such factors are relatively weak and are unlikely to replicate. In situations such as this, the researcher should obtain a solution with one fewer factor to determine whether the variables can be "forced" onto another factor. If this does not occur, it may be the case that the variables do represent a separate factor but that there are not sufficient variables to capture it. In any case, researchers should clearly state the criteria and logic used to determine the number of factors, and justify their choice of model.

### E9. Method of Rotation

Although there are many rotational methods, the primary distinction is that between those that produce orthogonal (uncorrelated) and oblique (correlated) factors. The choice between these should correspond to the researcher's theory regarding whether the dimensions of the construct being studied are correlated. In the absence of such theory, our view is that oblique rotations will generally result in more reasonable representations of the data, because the dimensions that underlie constructs in the social and behavioral sciences tend to be correlated. In addition, if an oblique rotation is applied to data in which the factors are not correlated, an orthogonal solution will result, so nothing is lost by such a specification. On the other hand, specifying an orthogonal solution when the structure is actually oblique will cause variables to load on more than one factor. This is because specification of an orthogonal solution in such a situation will force any cross-factor correlations to be manifested through the factor pattern/structure coefficients. However, keeping in mind that EFA is an exploratory procedure, it is acceptable to obtain both orthogonal and oblique rotations and compare the results. The solution that is more interpretable and theoretically justifiable can then be chosen. The important point is that the researcher must provide some justification for the rotational procedure chosen and, if choosing between different rotations, must explain the basis by which that decision was made.

Once the choice between an orthogonal and an oblique rotation has been made, the researcher has another choice among the various rotation methods in each category. In general, however, the latter choice is not as consequential as the former. Although certain rotation methods are more likely to obtain a general factor than others (e.g., quartimax), in our experience the different rotations within the orthogonal and oblique categories do not generally have a strong effect on the results. This will not always be the case, however, so researchers should provide the specific orthogonal (e.g., varimax) or oblique (e.g., oblimin, promax) rotation method used along with an explanation of why it was chosen.

### E10. Variable Elimination

It is not uncommon for researchers to eliminate variables from the model on the basis of low structure, pattern, or communality values or because the variable is strongly or equally influenced by multiple factors (i.e., high multidimensionality). While we are not against such practices in theory, we do feel that these decisions are often made in a somewhat cavalier manner. It must be kept in mind that the variables were presumably chosen for some reason and that eliminating some of them changes the definition of one's construct(s) to some extent. We therefore feel that researchers should carefully consider the decision to eliminate variables from the model with an eye toward the validity of the construct(s) being studied. Often, studies are conducted on samples that are not sufficient to support stable results. Because of this, the lack of saliency or problems with multidimensional variables may be the result of sampling error or a lack of stability. In such cases, if the variables have not been analyzed together previously our recommendation would be to retain any questionable variables in order to determine if their transgressions are repeated upon replication of the study.

If, after careful consideration, a researcher does choose to eliminate one or more variables from the analysis, the factor model must be re-analyzed with the remaining variables. This is because the elimination of even one variable can change the factor structure. Ideally, variables should be removed one at a time and analyses rerun after each removal. It may be the case that removal of one variable eliminates problems with others. Finally, the researcher should provide a summary of the decisions made regarding variable deletions and the justifications for these.

## E11. Parameter Values

After a factor model has been decided upon, the researcher should provide information on the model parameter estimates. There are potentially four sets of estimates authors should provide and discuss: communalities, structure coefficients, pattern coefficients, and factor correlations. As noted previously, communalities represent the proportion of a variable's total variance that is accounted for by the factor solution. Thus, low communality estimates can be used to identify variables that are not explained well by the factor solution. However, if communality values are moderate to high, the structure and pattern coefficients can be interpreted to more clearly understand the relationship between the factors and observed variables.

The values that represent the relationships between the factors and the observed variables are often called *loadings* in EFA. As mentioned previously, there are two different parameter estimates that are sometimes referred to as loadings: *structure coefficients* and *pattern coefficients*, and thus, the term is ambiguous. We recommend researchers use the terms structure and pattern coefficients and avoid the term loading when reporting EFA results. If an orthogonal rotation is used, or if there is only one factor, these two estimates are equivalent and represent the simple correlation between the factor and the variable (and range between $-1$ and $+1$). Researchers should interpret these coefficients as factor-variable relationships and may note that squaring these values represents the amount of variance in the variable that is explained by the factor. For oblique rotations, however, structure and pattern coefficients are not equivalent. Thus, with correlated factors both sets of coefficients should be presented and a clear distinction between them should be made: the pattern coefficient represents the unique relationship between a factor and variable, controlling for the other factors; the structure coefficient represents the simple, zero-order correlation between the factor and the variable. In other words, for obliquely rotated solutions, pattern coefficients are not correlation coefficients, but are analogous to standardized (beta) weights in multiple regression analyses (and can fall outside the range of $-1$ to $+1$). Structure coefficients are a function of the pattern coefficients and factor correlations and represent the total effect of the factor on the variable; that is, structure coefficients represent both the unique effect of a factor on a variable plus its effect via relationships with other factors. A factor may not have a strong unique effect on a variable (i.e., small pattern coefficient) but can still have a strong total effect on the variable (i.e., large structure coefficient) due to strong factor correlations. Thus, in order to accurately interpret the solution, the factor correlations should be reported and discussed for oblique solutions. If the factor correlations are very weak, the structure and pattern coefficients will be similar in magnitude; on the other hand, if the factor correlations are strong, the structure and pattern coefficients may be very different.

Although there are differing opinions as to which set of coefficients should be used to interpret the meaning of the factors, we agree with Gorsuch (1983) that "The basic matrix for interpreting the factors is the factor structure. By examining which variables correlate high with the factor and which correlate low, it is possible to draw some conclusions as to the nature of the factor" (p. 207). Other methodologists recommend that the focus should be placed on the pattern coefficients; these values tend to indicate a clearer structure, making it easier to find salient variables. However, one must realize that the pattern coefficients do not represent the complete relationship between the variable and the factor. Therefore, when interpreting the factor solution, we recommend attending to the structure coefficients first and then evaluating the pattern coefficients to understand the unique factor-variable relationships. Both sets of coefficients should be reported in a table to allow readers to best understand the factor- variable relationships.

### E12. Percentages of Variance

The amount of variance explained by the championed solution and by each factor should be reported. As noted previously (see Desideratum E8), the amount of variance explained is sometimes used to determine the number of factors to extract and rotate. In turn, researchers should and often do note the variance explained by the solution and by each factor *before rotation*. However, rotating the solution distributes the variance explained across the retained factors. Therefore, although the total percentage of variance explained by the solution before and after rotation remains the same, the amount of variance associated with each factor will be adjusted after rotation. These adjusted values are only calculated for orthogonal rotations and should be reported and discussed (e.g., do the retained factors explain nearly comparable amounts of variance or are certain factors associated with much more explained variance?). When using an oblique rotation the factors overlap, and each of the correlated factors is "credited" with any shared variance in the observed variables. Because of this, the total variance explained in a variable can appear to sum to over 1.00, therefore, the percentage of variance associated with each factor is not reported for oblique solutions. Instead, one can report the sum of squared structure coefficients associated with each factor after rotation.

Researchers and reviewers should realize that the two most commonly used statistical software programs (SPSS and SAS) compute the percentage of variance explained by each factor differently for EFA (thus, the importance of noting the software used; see Desideratum E6). Both compute the percentage/ proportion of variance explained when conducting an EFA, but the two packages use different denominators, resulting in values that can potentially be very different. SPSS calculates percentage of variance as the amount of variance explained out of the *total* variance, whereas SAS calculates percentage of variance explained as the amount of variance explained out of the total amount of *common* variance. Therefore, if one conducted an EFA in SPSS and SAS, the eigenvalues, communalities, and parameter estimates would be the same, but the percentage of explained variance would appear larger when computed by SAS (because the denominator would be smaller). It is critical that the amount of variance explained is interpreted with this in mind. In addition, if reviewers re-analyze the data to check the results, they may produce different values of explained variance than the author because of this software difference.

### E13. Interpretation of Factors

The factor solution should be interpreted using all of the relevant information. For an oblique solution, this includes the pattern coefficients, structure coefficients, and factor correlations, whereas for an orthogonal solution pattern coefficients provide sufficient information. In addition, knowledge of the variables being factored and the theory surrounding the construct should be incorporated in the interpretation process. Once factors are interpreted, the factor name is most often used to communicate the identity of the factor, rather than the observed variables themselves. Therefore, naming the factor is extremely important and the process of naming/interpreting the factor should be clearly communicated.

Researchers should note how they used the factor-variable relationships to interpret and name the factors. Interpreting the factor solution requires a determination of the value a coefficient must reach to be considered salient, or "high." Variables with coefficients that reach this value are used to name the factor. Given the difference in the interpretation of structure and pattern coefficients, one wouldn't expect the same value to be used for both coefficients. Common values for structure coefficients are .30 and .40. Although to some extent this choice is arbitrary, some thought should be given to choosing an appropriate value. Realize that, for uncorrelated factors, squaring the structure/pattern coefficient value represents the amount of variance explained in the variable by the factor. For correlated factors, the structure coefficients are affected by the factor correlations, and thus squaring these terms yields the amount of variance in the variable that the factor can explain uniquely and via relationships with other factors. Therefore, if approximately 10% is enough shared variance to deem a variable useful for factor interpretation, then a value of .30 or .40 could be used. One practice that should be guarded against is that in which the researcher chooses the cut-off value in a self-serving manner. For instance, we have seen published articles in which a value such as .42 was chosen. Although no reason for such a choice was provided, it appeared to be motivated by the fact that use of this value would allow the researchers to ignore variables having coefficients of .41 for more than one factor.

Interpretation and naming of the factor(s) is made easier if the solution exhibits simple structure. The key criteria for simple structure are that each variable has a large structure coefficient for one factor and small values for all other factors. This means that each factor has strong relationships with only some of the variables. Again, strong relationships indicate a high degree of overlap between the factor and the variables, which facilitates naming the factor. On the other hand, weak relationships between the factor and the variable or variables that have strong relationships with numerous factors make it difficult to determine what the factor represents.

It may be the case that simple structure is not achieved and this should be acknowledged and discussed. If the solution has many variables that are multidimensional (i.e., variables associated with multiple large structure coefficients that reflect the influence of multiple factors), this may be an indication of under-factoring. On the other hand, researchers should be cautious of solutions that extract too many factors in an effort to "eliminate" multidimensional variables. As noted previously, factors that are represented by only one or two variables may indicate over-factoring. Such possibilities underscore the importance of examining several different solutions (see Desideratum E8). When presenting the parameter estimates (see Desideratum E11), researchers should comment on the degree to which simple structure is achieved and note any variables that contribute to the deviation from simple structure. Authors should note that variables that have strong relationships with many factors are multidimensional and this finding should be discussed in the context of the theoretical conceptualization of the construct.

Finally, researchers should discuss how the factor structure (nature and number of factors) aligns with the current theoretical conceptualization of the construct. Again, it is important to note that obtained factors are completely driven by the variables that are factored (see Desideratum E3). Therefore, failure to include variables that cover the breadth of the construct will result in failure to represent the construct's "true" dimensionality. This must be addressed by authors (i.e., this is a potential problem or can be ruled out as a problem) when discussing "unexpected" or "interesting" findings, such as obtaining a simpler factor structure than expected. Finally, authors should note that a clear understanding of the factor necessitates replication and integration of the construct into its nomological net (see Desideratum E15) because, unfortunately, seemingly "interpretable" factors can emerge from random data.

### E14. Factor Score Determinacy

In many cases, researchers are interested in computing *factor scores* to use in other analyses. For example, a researcher may want to use the factor scores as variables in an analysis such as ANOVA or regression. Factor scores are weighted sums of the standardized variables. Factor scores can be either *exact* (also called *refined*) or *approximate* (also called *coarse*). The difference is simply that all of the variables are used to compute exact factor scores, whereas for approximate factor scores only the variables most associated with each factor are used. In the discussion that follows, we refer to exact factor scores. Several types of exact factor scores can be obtained; the differences among them have to do with differences in the weights used, which in turn result in factor scores with different properties. It should be noted that different procedures for obtaining *component* scores will produce the same results for component analysis. However, for common factor analysis the procedures will produce different *factor* scores. This is due to the fact that exact factor scores obtained from common factor analysis are indeterminate, meaning that there is an infinite number of ways in which factor scores could be obtained that would be consistent with a given pattern or structure matrix. The reason for this indeterminacy can be seen by considering the common factor model. This model posits that the observed variables $Z$ are functions of common factors as well as factors that are unique to each variable.

To take a simple example, it might be hypothesized that four standardized variables, $z_1$, $z_2$, $z_3$, and $z_4$ are functions of scores on two common factors, $f_1$ and $f_2$ and four unique factors, $u_1$, $u_2$, $u_3$, and $u_4$:

$$z_1 = w_{11}f_1 + w_{12}f_2 + x_1 u_1$$
$$z_2 = w_{21}f_1 + w_{22}f_2 + x_2 u_2$$
$$z_3 = w_{31}f_1 + w_{32}f_2 + x_3 u_3$$
$$z_4 = w_{41}f_1 + w_{42}f_2 + x_4 u_4$$

In these equations the $w$ and $x$ are weights that quantify the degree to which the variables are related to the common and unique factors. The problem with such a system of equations is that the number of common and unique factor scores to be estimated in $f$ and $u$ is greater than the number of equations. In our example there are four unique factor scores plus two common factor scores, but only four equations, a situation that is analogous to obtaining a solution for the equation $x + y = 10$. The problem is not that there are no values for $x$ and $y$ that will satisfy the equation, but that there is an infinite number of values that will do so. With regard to the factor scores, the problem is not that there is no set of factor scores that can be obtained from the variables scores; it is that there are many such sets of factor scores (see Grice, 2001, for a review of factor indeterminacy and various methods of computing factor scores). Note that a similar problem does not exist for exact scores in the component model, because there are no unique factor scores in this model.

The degree to which factor scores are indeterminate depends on the variable to factor ratio, level of communality of the variables, and the sample size, with increases in each of these leading to greater determinacy (Acito & Anderson, 1986; Gorsuch, 1983; Grice, 2001). Of these, the level of communality has been found to have the greatest effect, whereas the sample size has a relatively minor effect. We therefore recommend that measures of factor score indeterminacy be reported and discussed for situations in which factor scores are obtained from data characterized by low communalities and/or variable to factor ratios. Grice discusses several indices for evaluating the degree of indeterminacy, and provides SAS code to compute these. These indices include the multiple correlation between each factor and the variables (r) and the minimum possible correlation between two sets of factor scores computed in different ways ($2\rho^2 - 1$). The former index ranges from 0 to 1, with high values indicating greater degrees of determinacy. The second index ranges from $-1$ to $+1$ and represents the degree to which two sets of factor scores constructed in different ways will be correlated. Values at or below 0 for this index indicate that the two sets of scores are unrelated, or negatively correlated, thus higher values are desirable.

### E15. Factor Quality

Reliability estimates for the variables that represent each factor should be reported and interpreted. Although internal consistency of observed scores is often of interest, the type of reliability that is most relevant for a given scale will be dictated by its purpose. In many cases, researchers compute Cronbach's coefficient alpha for the complete set of variables even though a multidimensional solution emerged, which seems to defeat the purpose of obtaining multiple dimensions. In general, internal consistency estimates for multidimensional instruments should be obtained for the dimensional level at which the scale will be interpreted and used. If subscales representing separate dimensions of the construct are to be used independently, reliability coefficients should be calculated for these subscales. If, however, the total scale is conceptualized as a higher-order factor that incorporates all of the subscales, it may be useful to obtain reliability coefficients at both the total and subscale levels.

If reliability is low (less than .70), this should be discussed, rather than simply reported and ignored (see Lance, Butts, & Michels, 2006, for a discussion of acceptable levels of reliability). Although it is possible to obtain an interpretable factor structure along with low estimates of internal consistency, it is more likely that low reliability will result if the factor solution is unstable. For example, the variables being factored may have weak relationships, in which case the researcher should refer readers to the variable correlations and discuss this issue. Low internal consistency can also occur when there are few variables representing a factor. In such a case, the researcher should address the coverage of the construct's breadth. Whatever the cause, it is incumbent upon researchers to provide a rationale for the credibility of their interpretation when reliability of a factor is low.

In referring to factor quality, it should be noted that indexes such as Cronbach's alpha are, strictly speaking, only appropriate for composites obtained as simple sums of the variable scores in which each variable is weighted equally. Factor scores, on the other hand, are typically computed using weights based on the factor pattern or structure coefficients. For this reason, indexes of internal consistency more appropriate for factor models are often used in EFA and CFA (e.g., coefficient $H$; see Hancock & Mueller, 2001).

Whenever possible, external validity evidence should be gathered to support the proposed interpretation of the factors. This involves relating the construct under study to theoretically related variables and/or constructs. This can greatly facilitate the interpretation and naming of the factors because the meaningfulness of the factor is made apparent through its relationships with other variables (i.e., its nomological net). In fact, it is through this process that one really begins to understand what is represented by the factors (Benson, 1998). In addition, it can provide further evidence of the distinction or lack of distinction between factors (i.e., do the factors have differential predictive utility which would support their differentiation?). It is important that researchers note the construct validity evidence associated with the external variables; lack of such evidence severely limits the utility of these external variables in evaluating the factors under study. In sum, whenever possible reliability and validity evidence should be reported along with the factor structure results in order to fully interpret the findings.

## E16. Caveats and Limitations

Researchers should acknowledge that the factor structure championed in the study is only one possible representation of the relationships among the variables. Other models may represent the data just as well or better than the structure presented; therefore, language that suggests that this model represents "truth" or is "confirmed" should be avoided. Authors should also acknowledge the exploratory nature of the analysis. In EFA studies, it is typical to examine many solutions, and often variables are removed and data re-analyzed. This capitalizes on chance due to fitting the idiosyncrasies of the sample data. Therefore, authors should note that the results from the EFA represent the structure of the data for that particular sample and that there is a need for replication (cross-validation) in order to assess the stability of the factor structure across independent samples from the same population. Moreover, researchers should not imply that the structure championed will necessarily generalize to other populations. Further research is needed to support such generalizations.

## C1. Theory and/or Previous Research

As is the case in EFA, a solid understanding of the theory underlying the latent construct being modeled is essential in CFA studies, and the points made previously with regard to EFA are equally relevant for CFA (see Desideratum E1). In fact, because use of the CFA model requires the researcher to specify the model to be analyzed in more detail than is the case in EFA, knowledge of theory is even more important for these analyses.

Researchers should clearly specify the theoretical and/or empirical basis for the model, and to the extent possible should provide hypotheses about the expected sign (positive or negative) and magnitude of the coefficients to be estimated. Our preference is for researchers to present all models to be tested in the form of path diagrams (see Chapter 28, this volume, for more detail on path diagrams). Often, existing theory and research do not converge on a single plausible model. In such cases recommended practice is to specify alternative models corresponding to different theoretical perspectives (e.g., number of factors, factor relations) a priori, and to evaluate these against each other appropriately.

## C2. Confirmatory vs. Exploratory

The distinction between exploratory and confirmatory analyses was made previously in the context of EFA (see Desideratum E2), and will not be repeated here. Instead we use this desideratum to emphasize that the ability of CFA to evaluate and compare different a priori models developed on the basis of theory is the major strength of this method. Use of CFA should therefore be reserved for situations in which at least one theory-based model can be hypothesized. Although it is possible to use CFA in an exploratory manner, such usage often results in gross misfit of the model and it can be very difficult to "uncover" the structure that best represents the data through the use of CFA output. If theoretically derived a priori models cannot be specified, it will often be necessary to explore several different models in an attempt to determine which provides the best representation of the data; this is a task more suited to the use of EFA. As mentioned in previous desiderata, researchers should not conduct a CFA to "confirm" the EFA solution using the same sample; this practice results in capitalization on chance due to fitting the idiosyncrasies of the sample data. However, EFA is often used after CFA if the a priori specified model(s) results in extreme misfit. Using the same sample and moving to an exploratory approach (EFA) is completely justified and recommended; if the a priori specified model(s) do not fit, researchers can use the data to explore the structure that does underlie the observed variables via EFA. However, researchers should be clear that the analysis has changed from a confirmatory to an exploratory mode, documenting completely for the reader the different models and decision processes and including appropriate caveats.

## C3. Measured Variables

As noted for EFA in Desideratum E3, the observed variables under study must be clearly presented with respect to type and number; this speaks directly to the coverage of the breadth of the construct. Information pertaining to reliability and validity should be presented if available, and, whenever possible, examples of the variables under study should be presented. An additional consideration in CFA studies is that there must be a sufficient number of variables per factor to identify the model. In general, at least three variables per factor are required, although if there are two or more correlated factors, two variables per factor can be sufficient. However, it should be noted that these guidelines pertain only to model identification. More variables are typically required to encompass the scope of the constructs.

The relationship between the factors and the observed variables should be clearly presented when articulating the model(s) under study (see Desideratum C1). Up until now, we have only discussed models that specify factors as the causal agents of the observed variables. In such models the observed variables are hypothesized to be correlated with one another because they are a function of the same factor. Given this conceptualization of the factor-variable relationship, the factor is deemed *latent* and the direct paths flow from the latent variable to the observed variables. It is also possible to conceptualize a factor as being a function of the observed variables (e.g., overall stress is a function of work stress, spouse-related stress, and children-related stress). The direct paths flow from the observed variables to the *emergent* factor. A more detailed description of the emergent factor model and problems surrounding its estimation can be found in Chapter 28, this volume. When discussing the measured variables, researchers should clearly present how the variables are related to the factor (we recommend a figure) and explain why the particular model chosen (latent or emergent factor) is appropriate.

## C4. Sampling Method(s) and Sample Size(s)

Researchers should specify the type of sampling that was used to obtain the data. Most commonly used computer packages for structural equation modeling (SEM) analyses now allow the researcher to specify sampling weights for data obtained via stratified sampling methods, and allow for nested or hierarchical models that accommodate data obtained from clustered samples. Researchers using such sampling techniques should therefore employ the proper methodology.

With regard to sample size, the guidelines presented previously in the context of EFA also apply to CFA. However, because CFA is an inferential method, researchers should also consider issues of power and precision in addition to accuracy of parameter estimates when deciding on the necessary sample size. Power for CFA can be computed for both individual parameter estimates and for the model as a whole (see Hancock, 2006, and Chapter 28 of this volume, for more discussion of power in SEM). Finally, some estimation methods used with nonnormally distributed and/or categorized data (e.g., Asymptotically Distribution Free [ADF], Arbitrary Generalized Least Squares [AGLS], Weighted Least Squares [WLS]) require sample sizes that are much larger than those needed for normal theory-based methods.

## C5. Data Screening

Unlike EFA, CFA is an inferential methodology and commonly used estimation methods in CFA (e.g., maximum likelihood [ML], and generalized least squares [GLS]) are based on the assumption of multivariate normality. Violations of this assumption can result in underestimation of standard errors and inflation of chi-square values (fit indices based on chi-square will also be biased). Because of this, both univariate and multivariate normality should be assessed and reported. Univariate skewness and kurtosis values of less than |2.0| (some researchers suggest a more liberal standard of less than |7.0| for kurtosis) and values of Mardia's normalized multivariate kurtosis coefficient of less than 3.0 have been suggested as acceptable departures from normality. For levels of nonnormality outside these guidelines one of the estimation methods developed for nonnormally distributed data should be used (see Desideratum C6).

As with EFA, the scales of the observed variables can also affect results. Specifically, normal theory-based methods assume that variables are continuous in addition to being multivariately normally distributed. Variables with five or more scale points should not result in substantial bias, but researchers working with variables with fewer than five scale points should consider the use of estimation methods specifically designed for such data (see Desideratum C6).

Screening for univariate and multivariate outliers should be conducted prior to analysis. Univariate outliers can be identified as cases with large $z$-scores (e.g., +/–3 standard deviations from the mean). For multivariate outliers, Mahalanobis $D$ or $D^2$ can be used. Researchers should also determine the effect of outlying cases on the parameter estimates and fit indexes; with a large sample size these effects may be quite small. A common practice is to obtain estimates from data with and without the outliers; if these are similar there is arguably little reason to delete outlying cases.

Recent advances in missing data methodology have called into question the utility of more traditional missing data methods such as listwise and pairwise deletion. Currently, full information maximum likelihood (FIML) and methods based on Expectation Maximization (EM) algorithms are considered to be more acceptable methods for accommodating missing data, and most commercially available SEM software packages have incorporated at least one of these. In the FIML method, missing values are not imputed; instead parameter estimates are obtained from the information available from each case for the variables involved in the parameter being estimated. The cases contributing to the estimation vary across parameters because different cases may have missing values for different variables. EM methods, on the other hand, use a two-step process to impute missing values. In the first step, referred to as the Expectation or E step, regression-based methods are used to obtain the necessary information (variable sums of square and cross-products) to compute a complete covariance matrix. In the second, Maximation or M step, the information from the E step is used to compute a covariance matrix using Maximum Likelihood (ML) estimation. The new covariance matrix is then used in the next E step to obtain new estimates of the missing values, and the two-step process is repeated multiple times until some pre-set convergence criterion is met. The covariance matrix obtained at the last step can then be used as input into any SEM software package. Both methods assume that data are *missing at random* (MAR), meaning that missing values for each variable are independent of that variable. Researchers should identify the method used to handle missing data and address the assumptions underlying the method. In addition, the proportions of missing data across the variables in the study should be provided.

## C6. Estimation Method

Researchers should report the type of estimation that was implemented, along with a justification for use of the chosen method. Although ML estimation is the default in virtually every SEM computer package, this method assumes the data are continuous and multivariate normally distributed, as noted previously. Violations of either or both of these assumptions can result in underestimation of the standard errors and overestimation of the chi-square values. When data depart from normality, adjustments to standard procedures such as the Satorra-Bentler (SB) adjustment to the standard errors and chi-square values can be implemented. For data with fewer than five response categories, estimators such as the WLS, or mean and variance adjusted WLS (WLSMV) implemented in the Mplus program should be used. These procedures assume that there is a normally distributed continuous variable underlying the observed categories of the manifest variable. Based on this assumption, polychoric correlations and thresholds representing the estimated cut-points between the observed categories are computed from the raw data and used in subsequent parameter estimation.

## C7. Software

Commonly used computer packages for conducting CFA include AMOS, EQS, LISREL, Mplus, Mx, and SAS Proc CALIS. For basic CFA models, these packages should provide estimates that are essentially identical. However, the packages differ with regard to the estimators available, amount and type of output provided, and capabilities for advanced features such as incorporation of sampling weights, accommodation of nested data, and availability of modern missing data methodology. Researchers should therefore provide the name of the software package chosen. In addition, because software in this area is continually being upgraded, the specific version of the software package used should always be reported. If the LISREL package is used, researchers should also report the version of the pre-processor PRELIS (if used).

## C8. Estimation Problems

In some cases CFA estimation can fail. Non-positive definite matrices, convergence failures, and improper estimates are the most common reasons for such failures. Researchers should carefully examine their computer output for such problems, because they are not necessarily flagged by software packages. These problems are usually the result of model misspecification, collinearity, insufficient sample size, and/or a lack of identification, and it is incumbent upon the researcher to determine the cause of the problem and correct it. Failure to do so renders the parameter estimates and other statistical indexes suspect.

For CFA models, the most commonly encountered problem is the occurrence of negative error variances (Heywood cases). These can occur because of collinearity between variables, or somewhat paradoxically, because of a lack of correlation between variables intended to represent the same factor. Other improper values in CFA include factor correlations greater than 1.0. In addition, parameter values that are of the opposite sign from what was expected, or are much larger than expected (e.g., pattern coefficients of 20.0) are usually signs that something has gone wrong in the analysis. As noted previously, insufficient sample sizes, outliers in the data, underidentification, and model misspecification are all possible causes of both improper solutions and convergence failures.

If problems such as these arise during analyses, the researcher should briefly describe the problem and state what was done about it and why. In many cases, such difficulties in estimation are indicative of problems with model specification or identification. If this is the case, the model should be respecified. If other steps are taken to overcome estimation problems, these should also be clearly documented. Researchers sometimes ignore or gloss over such problems, so reviewers and editors should be mindful that these are potentially serious issues that require explanation and remediation.

## C9. Data-Model Fit

A thorough discussion of the plethora of fit indices that have been developed to measure the fit of CFA models is beyond the scope of this chapter. Instead, we provide here a general discussion of the assessment of data-model fit and focus on those indices that are recommended by methodologists in this area. CFA models are complex, and it is probably naïve to think that model fit can be properly assessed by a single index. Therefore, most methodologists agree that fit should be assessed through the use of several different criteria. Although the chi-square test of goodness of fit has been traditionally used to assess data-model fit, many methodologists feel that it is overly stringent because (1) the role of the null and alternate hypotheses is reversed in the logic underlying this test such that the desired outcome is a failure to reject the null hypothesis, (2) due to null and alternative hypothesis reversal and to the large sample sizes typically needed for CFA, the test is very powerful, and (3) the null hypothesis itself, that the model holds exactly in the population, is unrealistic. Despite these shortcomings, the chi-square test should be reported, along with its degrees of freedom and $p$-values; however, other fit indices should be reported as well. These are often categorized into three classes, which we present along with recommended exemplars and cut-off values in each class.

*Absolute* or *stand-alone indices* are measures of the discrepancy between the observed sample matrix and that implied by the CFA model being tested. One example of these is the chi-square value. Another is the standardized root mean square residual (SRMR), which is based on the average of the residuals between the observed and implied matrices. SRMR values of .08 or less are considered to be indicative of good fit.

*Parsimony-adjusted indices* also measure the discrepancy between the observed and implied matrices, but incorporate some type of penalty for model complexity. Because data-model fit improves as parameters are added to the model, these indices evaluate the improvement in fit resulting from the additional parameters relative to the number of parameters needed to obtain this improvement. One recommended index in this class is the root mean square error of approximation (RMSEA) and its associated confidence interval. This index (or its 90% confidence limits) should be .05 or below for a well-fitting model, or .08 or below for an "acceptable" model.

*Incremental fit indices* measure the fit of the model of interest relative to the fit of a null or baseline model. The latter model is typically one that posits no correlations among the variables. Recommended indices in this class include the comparative fit index (CFI) and the non-normed fit index (NNFI; also known as the Tucker-Lewis Index, or TLI). Values of both indices should be .95 or above.

Decisions regarding data-model fit should be based on an integration of all available information. Although evaluation of global fit indices is important, the matrix of standardized covariance residuals, which represents the discrepancy between corresponding elements of the observed and model-implied matrices, should be closely inspected to identify any local areas of misfit that were masked by the global fit indices. At a minimum, the range of these values should be reported when discussing fit. Large residuals should be taken as indications of areas in which the model does not account for the data; they should not be ignored but should be used to diagnose and better understand data-model misfit. Researchers should also examine parameter estimates to determine whether the signs and magnitudes are reasonable. In addition, convergence problems or excessively large standard errors can indicate model misspecification or collinearity problems, and should be considered indications of data-model misfit (see Desideratum C8).

Researchers should be aware that CFA models place very strict restrictions on the parameters. In general, CFA models represent a perfect simple structure in which variables represent one factor and have no direct relationship (i.e., zero pattern coefficients) with other factors. Such models are more representative of an ideal than of reality, and it is often the case that they do not fit well. In particular, CFA models with large numbers of variables often exhibit poor fit, because with more variables there are more idealized factor-variable relationships of zero that may not satisfy this standard. Another way of saying this is that the model becomes more falsifiable as the number of factor-variable relationships that are set to zero increases. We emphasize strongly here that we are not advocating relaxation of standards for fit of CFA models. Rather, we present these comments in the hope that they will motivate researchers to develop a better understanding of the structure of their data.

## C10. Model Comparisons

CFA provides an opportunity to examine the extent to which competing models explain the interrelationships among variables. In fact, CFA is most useful when a set of a priori alternative models are estimated and compared because the researcher is then able to make more informed decisions about the viability of a target model relative to competing models. This is because testing of alternative models provides support for a model not only through acceptable fit to the data, but also by the rejection of competing models. Authors should clearly discuss the utility of testing competing models and explain the manner in which these models will be compared.

Competing models may be nested or non-nested, and this influences the indices used to compare models. Models are nested if the simpler model can be derived from the more complex model but by fixing parameters. A chi-square difference ($\Delta\chi^2$) test can be used to compare nested models. If the $\Delta\chi^2$ is statistically significant, the model with additional parameters is inferred to be better than the simpler model. A common error in CFA is to treat models with different variables or different numbers of variables as nested models. However, nested models must have the same variables, as well as nested parameters. Non-nested models can also be compared, although not through the use of the chi-square difference tests. Instead, information criteria such as the Akaike Information Criterion (AIC) or its rescaled versions (ECVI, CAK, CAIC), which estimate how well the model would fit in future samples (cross-validate), are often used to compare non-nested models. Models with lower values of these indices are associated with better data-model fit and, therefore, are championed over models with higher values.

Competing models should only be compared to one another if they fit well in an absolute sense, both globally and locally (see Desideratum C9). If none of the competing models are viable representations of the data, comparing them can be misleading. For example, if authors make statements such as "Model A fits better than Model B," readers may then infer that Model A fits well in an absolute sense. Furthermore, we believe that if only one model fits the data, there is no need to compare this model with other models that do not represent the data well.

## C11. Post Hoc Model Modifications

There are two forms of post hoc model modification: the removal of nonsignificant paths from a well-fitting model ("trimming"), and the addition of paths to increase data-model fit for a poorly fitting model. The former involves simplifying the a priori specified model by removing paths that do not reach a particular level of statistical significance. We recommend against this practice for several reasons: (1) using the same sample to respecify and test a modified model capitalizes on sampling error and thus decreases the chance of obtaining replicable results; (2) the model no longer aligns with theory but instead is empirically based or data driven; and (3) respecified models are often presented as though they were a priori theoretically based models, thus misleading readers as to the initial models specified and tested. If post hoc model modification is undertaken authors should do the following at a minimum: (1) clearly present the results from the a priori model before any paths are removed (including fit indices, parameter estimates, and standard errors); (2) present the full set of results (fit indices, all parameter estimates, and standard errors) from the modified model, making a clear statement that fit index cutoffs and $p$-values associated with the parameter estimates do not apply to modified models estimated on the same data; (3) provide a thoughtful explanation for the lack of empirical support for that path (e.g., low variability associated with the variables due to the population under study; issues collecting the data that impacted its validity); (4) provide a clear statement regarding capitalization on chance and the possibility of lack of power (i.e., a path may not be significant because sample size was small but the same path could be found significant if a larger sample was used); and (5) make a call for replication given the exploratory nature of the model modification. Often researchers delete indicators that have non-significant or weak relationships with their intended factors, rather than simply deleting the path from the factor to the variable. The above recommendations apply to this practice as well. However, this practice is potentially more serious because it may impact the coverage of the breadth of the construct.

Most often, model modification involves the addition of paths to the model. A priori models often do not fit the data adequately. Researchers are then often tempted to add parameters based on *modification indices* (MIs); MIs provide estimates of the amount by which the chi-square value would decrease if the parameter were added. A particularly egregious practice is the addition of paths between measurement error variances. These parameters are often added to models in an attempt to improve fit. However, the need for such paths indicates that the associated variables have stronger correlations that can be accounted for by the factors. This may be due to method effects, similar wording, or other artifacts, or may indicate the need for more factors. Regardless of the source, the presence of unmodeled covariation is an indication that the hypothesized structure does not hold and should be interpreted as such. Unfortunately, many researchers lose sight of the purpose of CFA, which is to allow the testing of a priori models (see Desiderata C2 and C10). If a model does not fit the data, that information, along with a diagnosis of the source of the misfit, is useful and should inform the domain. On the other hand, thoughtlessly modifying a model post hoc in an attempt to make it fit the data is not the purpose of CFA and may simply lead to models that do not replicate due to fitting the idiosyncrasies of the sample data. Researchers and reviewers must keep in mind that the purpose of conducting a CFA study is to gain a better understanding of the underlying structure of the variables, not to force models to fit. The former is a useful scientific endeavor; the latter is not.

If a model does not fit the data, we recommend that this misfit be diagnosed using standardized residual values (see Desideratum C9) in conjunction with modification indices. Given the sample-specific nature of model misfit, we encourage replication studies to evaluate the stability of the misfit. If the same area of misfit is found upon replication, it should be taken seriously and possible theoretical explanations of the misfit should be presented. Given plausible and thoughtful reasons for the misfit, the model may be modified and treated as an a priori specified model in future studies.

Often researchers do not have multiple samples to evaluate the replicability of model misfit. In such cases researchers may choose to add parameters suggested by modification indices using the same data on which the model was originally estimated. Although we discourage such post hoc model modifications, if these are made researchers should do the following at a minimum: (1) report findings (fit indices, parameters estimates, and standard errors) from the a priori model and the modified model; (2) provide substantively meaningful justifications for the modifications; (3) clearly note that the modifications were done post hoc and explain the issues surrounding this practice (e.g., capitalization on chance; fit index cutoffs and $p$-values associated with parameter estimates do not apply); and (4) present the results as exploratory and make a call for replication of both the original model to assess the stability of model misfit and the newly proposed model. Until modified models are studied using independent samples, it is unknown if they will generalize and, in turn, if they are plausible. Thus, researchers should be very cautious when interpreting the results from modified models and avoid statements regarding the usefulness or plausibility of the model.

## C12. Parameter Estimates

If a model does not fit the data adequately, the parameter estimates may be biased and should not be reported. In such cases the focus should turn to diagnosing model misfit. When model-data fit has been deemed adequate, parameter estimates and their corresponding significance tests should be interpreted. At a minimum, the direct relationships between the factors and the observed variables (path coefficients) should be reported in standardized form. If an observed variable serves as an indicator to only one factor, the standardized coefficient can be squared to represent variance explained in the variable by the factor. Given that the variables were specifically chosen to indicate these factors, one would hope that the variance explained ($R^2$) would be high (at least .50). If these values are low, implications should be discussed (see Desideratum C13). The unstandardized coefficients, not standardized coefficients, are tested for significance. Therefore, the unstandardized coefficients should be presented with their corresponding standard errors. Reporting the unstandardized estimates and standard errors facilitates comparison of results across independent samples. There is no need to present the significance tests; readers can easily compute these values if needed. Instead, a statement regarding the statistical significance of the path coefficients can be made in the text or in the table note (e.g., "All unstandardized path coefficients were significant at $p < .05$"). However, it should be noted that significance of the path coefficients simply means that these are significantly different from zero, not that they are, in some sense, "good" indicators of the factors. Given that the variables were specifically chosen to represent the factors, statistical significance of the path coefficients would seem to be a minimum expectation. However, researchers often interpret significance as though it were evidence of the strength of the indicators. Lastly, if one has tested a multidimensional solution with factor covariances/correlations freely estimated, these values should be reported along with the corresponding significance tests. In general, parameter estimates can be reported in tables, on the path diagram, or a combination of the two. If several competing models fit the data well and are theoretically plausible, authors should present and interpret results from each model.

## C13. Factor Quality

As with EFA, the quality of the factor is assessed by the magnitude of the parameter estimates, reliability of scores, and available validity evidence. Given the confirmatory nature of the analyses, one expects the observed variables to relate strongly to the factor for which they serve as indicators; it is this assumption that led to the selection and use of the observed variables. Often authors conflate adequate data-model fit with strong relationships between the factors and corresponding observed variables. Weak factor-variable relationships can occur despite adequate model-data fit due to such things as low observed variable correlations. If the majority of the relationships between a factor and its indicators are weak ($R^2 < .5$), the author should acknowledge this and resist the temptation to label the factor as consistent with a priori expectations. Weak relationships between variables and the factors they are meant to measure indicate that the researcher's hypotheses about the variables are not supported. In addition, weak factor-variable relationships yield low internal consistency reliability, which will affect external validity (relationships with theoretically related constructs), making interpretation of the factor difficult.

If the factor-variable relationships are strong and reliability is adequate, external validity evidence should be gathered. The quality of the factor is ultimately dictated by how well observed relationships with other constructs align with theoretical expectations (e.g., through multitrait-multimethod analysis; see Chapter 22, this volume). If authors are unable to assess these relationships, the quality of the factors and what the factors represent remains in question; therefore, authors should refrain from making cavalier statements regarding the meaning of the factor and its utility until such relationships have been investigated.

## C14. Caveats and Limitations

Caveats and limitations for CFA are essentially the same as those for EFA (see Desideratum E16). In addition, as noted above, if post hoc model modifications are undertaken, the authors must clearly explain the effect of this practice on the interpretation of resulting model parameters. Furthermore, researchers may be tempted to use more far-reaching language related to the plausibility of the model when employing CFA compared to EFA. Finding adequate data-model fit does not imply that the model represents "truth" but instead that the model is one possible representation of the structure underlying the observed variables. Moreover, researchers should focus *equally* on the adequate fit of a model and the rejection of alternative models when interpreting results. It is the combination of rejecting competing models and failing to reject a model that provides the most useful insight into the dimensionality of the construct under study.

## Note

1  Although standard errors for factor analytic methods other than maximum likelihood have been derived, the computations are intensive and have not been programmed into most general-use statistical packages. However, the program CEFA (Browne, Cudeck, Tateneni, & Mels, 2008) does provide standard errors for coefficients.

# References

Acito, F., & Anderson, R. D. (1986). A simulation study of factor score indeterminacy. *Journal of Marketing Research, 23,* 111–118.

Arbuckle, J. L. (2007). Amos 5 [Computer software]. Chicago: Smallwaters.

Benson, J. (1998). Developing a strong program of construct validation: A test anxiety example. *Educational Measurement: Issues and Practice, 17,* 10–17, 22.

Bentler, P. M. (2006). EQS 6.1 for Windows [Computer software]. Encino, CA: Multivariate Software.

Bollen, K. A. (1989). *Structural equations with latent variables.* New York: Wiley.

Bollen, K. A. (2002). Latent variables in psychology and the social sciences. *Annual Review of Psychology, 53,* 605–634.

Bolt, D. M. (2005). Limited- and full-information estimation of item response theory models. In A. Maydeu-Olivares & J. J. McArdle (Eds.), *Contemporary psychometrics* (pp. 27–71). Mahwah, NJ: Erlbaum.

Briggs, N. E., & MacCallum, R. C. (2003). Recovery of weak common factors by maximum likelihood and ordinary least squares estimation. *Multivariate Behavioral Research, 38,* 25–56.

Brown, T. A. (2006). *Confirmatory factor analysis for applied research.* New York: Guilford Press.

Browne, M. W., Cudeck, R., Tateneni, K., & Mels, G. (2008). CEFA: Comprehensive Exploratory Factor Analysis, Version 3.00 [Computer software and manual]. Retrieved from http://faculty.psy.ohio-state.edu/browne/.

Byrne, B. M. (1998). *Structural equation modeling with LISREL, PRELIS and SIMPLIS: Basic concepts, applications and programming.* Mahwah, NJ: Erlbaum.

Byrne, B. M. (2001). *Structural equation modeling with AMOS: Basic concepts, applications, and programming.* Mahwah, NJ: Erlbaum. Byrne, B. M. (2006). *Structural equation modeling with EQS: Basic concepts, applications, and programming.* Mahwah, NJ: Erlbaum.

Comrey, A. L., & Lee, H. B. (1992). *A first course in factor analysis* (2nd ed.). Hillsdale, NJ: Erlbaum.

Finney, S. J., & DiStefano, C. (2006). Dealing with nonnormality and categorical data in structural equation modeling. In G. R. Hancock and R. O. Mueller (Eds.), *Structural equation modeling: A second course* (pp. 269–314). Greenwich, CT: Information Age Publishing.

Gorsuch, R. L. (1983). *Factor analysis* (2nd ed.). Hillsdale, NJ: Erlbaum.

Grice, J. W. (2001). Computing and evaluating factor scores. *Psychological Methods, 6,* 430–450.

Hancock, G. R. (2006). Power analysis in covariance structure modeling. In G. R. Hancock and R. Mueller (Eds.), *Structural equation modeling: A second course* (pp. 69–115). Greenwich, CT: Information Age Publishing.

Hancock, G. R., & Mueller, R. O. (2001). Rethinking construct reliability within latent variable systems. In R. Cudeck, S. Du Toit, & D. Sörbom (Eds.), *Structural equation modeling: Present and future* (pp. 195–216). Lincolnwood, IL: Scientific Software International.

Hogarty, K. Y., Hines, C. V., Kromrey, J. D., Ferron, J. M., & Mumford, K. R. (2005). The quality of factor solutions in exploratory factor analysis: The influence of sample size, communality, and overdetermination. *Educational and Psychological Measurement, 65,* 202–226.

Horn, J. L. (1965). A rationale and test for the number of factors in factor analysis. *Psychometrika, 30,* 179–185.

Jöreskog, K. G., & Moustaki, I. (2001). Factor analysis of ordinal variables: A comparison of three approaches. *Multivariate Behavioral Research, 36,* 347–387.

Jöreskog, K. G., & Sörbom, D. (2007). LISREL 8.80 [Computer software]. Lincolnwood, IL: Scientific Software International, Inc.

Kline, R. B. (2005). *Principles and practice of structural equation modeling* (2nd ed.). New York: Guilford Press.

Lance, C. E., Butts, M. M., & Michels, L. C. (2006). The sources of four commonly reported cutoff criteria. What did they really say? *Organizational Research Methods, 9,* 202–220.

Loehlin, J. C. (2004). *Latent variable models* (4th Ed.). Hillsdale, NJ: Lawrence Erlbaum.

MacCallum, R. C., Widaman, K. F., Preacher, K. J., & Hong, S. (2001). Sample size in factor analysis: The role of model error. *Multivariate Behavioral Research, 36,* 611–637.

MacCallum, R. C., Widaman, K. F., Zhang, S., & Hong, S. (1999). Sample size in factor analysis. *Psychological Methods, 4,* 84–99.

McDonald, R. P. (1985). *Factor analysis and related methods.* Hillsdale, NJ: Erlbaum.

McLeod, L. D., Swygert, K. A., & Thissen, D. (2001). Factor analysis for items scored in two categories. In D. Thissen & H. Wainer (Eds.), *Test scoring* (pp. 189–216). Mahwah, NJ: Erlbaum.

Moustaki, I. (2007). Factor analysis and latent structure of categorical and metric data. In R. Cudeck & R. C. MacCallum (Eds.), *Factor analysis at 100: Historical developments and future directions* (pp. 293–313). Mahwah, NJ: Erlbaum.

Muthén, L. K. and Muthén, B. O. (2007). *Mplus user's guide* (5th ed.). Los Angeles, CA: Muthén & Muthén.

Pett, M. A., Lackey, N. R., & Sullivan, J. J. (2003). *Making sense of factor analysis: The use of factor analysis for instrument development in health care research.* Thousand Oaks, CA: Sage.

Swygert, K. A., McLeod, L. D. & Thissen, D. (2001). Factor analysis for items or testlets scored in more than two categories. In D. Thissen & H. Wainer (Eds.), *Test scoring* (pp. 217–250). Mahwah, NJ: Erlbaum.

Velicer, W. F. (1976). Determining the number of components from the matrix of partial correlations. *Psychometrika, 41,* 321–327.

Velicer, W. F., & Fava, J. L. (1998). Effects of variable and subject sampling on factor pattern recovery. *Psychological Methods, 2,* 231–251.

Widaman, K. (2007). Common factors versus components: Principals and principles, errors and misconceptions. In R. Cudeck & R. MacCallum (Eds.), *Factor analysis at 100: Historical developments and future directions* (pp. 177–203). Mahwah, NJ: Erlbaum.

# 9
# Generalizability Theory

Amy Hendrickson and Ping Yin

Generalizability theory (*G theory*) is a powerful tool in educational measurement that can help researchers and educators conceptualize better and more efficient data collection efforts based on their needs and requirements. G theory is a statistical theory used to assess the consistency or dependability of scores over randomly parallel replications of a measurement procedure. In order to take full advantage of the potential of generalizability theory, one must constantly ask the question of what constitutes the measurement procedure. This is such an important question in any generalizability study that almost every piece of information in the analysis will depend on it.

G theory provides a conceptual framework and a set of statistical procedures and classical test theory and analysis of variance (ANOVA) can be viewed as its parents. In G theory, one identifies the various sources of error in observed scores and the relations among such sources and then estimates the error variance associated with each. The analysis of any generalizability study is important because there are various choices for procedures and methods for the analysis. The strengths of G theory are that the relative importance of multiple sources of error can be estimated separately in a single analysis for a given situation as well as be used to help the decision maker to design more efficient measurement procedures for the future. Valid generalization and interpretation of G theory results, however, depend on a carefully conceptualized design and well-executed analysis, and the authors should document and justify each of their decisions, as outlined in this chapter.

Software packages developed specifically for generalizability analyses are GENOVA (for balanced designs), urGENOVA (for unbalanced designs primarily), and mGENOVA (for multivariate designs). Other commercially available software packages such as SAS and SPSS may also be used for the analyses, but have limitations in their applicability.

For comprehensive descriptions of G theory we recommend texts by Brennan (2001), Cronbach, Gleser, Nanda, & Rajaratnam (1972), Shavelson and Webb (1991), and Traub (1994). Specific desiderata for applied studies that utilize G theory are presented in Table 9.1 and explained in detail subsequently.

## 1. Measurement of Reliability

Generalizability analyses are particularly appropriate when an investigator is concerned about reliability-related issues that involve generalizing over multiple tasks, raters, occasions, and so forth.

Table 9.1 Desiderata for Generalizability Theory

| *Desideratum* | *Manuscript Section(s)\** |
|---|---|
| 1. The need for measurement of reliability over replications of a measurement procedure is made clear. | I |
| 2. The measurement procedure is described in detail. | M |
| 3. The objects of measurement, universe(s) of admissible observations, universe(s) of generalization, and included facets are defined to aid in the understanding of the generalizability (G) and decision (D) studies. | M |
| 4. The design of each G and D study is described, including whether each design is univariate or multivariate and balanced or unbalanced, and including diagrams and/or tables to facilitate the understanding of each design. | M, R |
| 5. The name and version of the utilized software package is reported. | M, R |
| 6. Summary statistics (*n* per task, mean, SD) and variance components for all facets and their interactions are presented. | R |
| 7. Problems with estimation are reported and discussed. | R |
| 8. A table and discussion of G study variance components are included. | R |
| 9. A table and discussion of D study variance components are included. | R |
| 10. The D study error variances and generalizability and dependability coefficients are discussed and included in a table. | R, D |
| 11. The practical results are discussed, including (if applicable) a discussion of multivariate results and composite scores and including an evaluation of the current assessment design and of the ideal design for ensuring valid generalizations. | D |

\* *Note*: I = Introduction, M = Methods, R = Results, D = Discussion

Conceptually, it is useful to think of such analyses in terms of hypothetical replications of the measurement procedure. In the terminology of generalizability theory, replication is usually described in a *randomly parallel* fashion that involves different but similar instances of the measurement procedure. For example, a group of persons might be given a set of five items to complete. A randomly parallel replication of this measurement procedure would involve a different set of five items from the same item pool. The authors must discuss the need to measure variability (or consistency) over defined replications.

## 2. The Measurement Procedure

The authors must describe the measurement procedure, including any assessment instruments and the included tasks, as well as the administration and scoring processes. For example, for an analysis of rater and task variability over replications, the following questions should be answered: How many tasks were included? How many score points were used per task? What was the nature of the rubrics—was holistic or analytic scoring employed? How were tasks assigned to persons? How were tasks assigned to raters? How were persons assigned to raters? There should be enough detail included regarding the nature of the measurement procedure, including the assessment and administration, to support the indicated generalizability study (*G study*) design.

## 3. Universes, Objects of Measurement, and Facets

The *universe*(s) of admissible observations and universe(s) of generalization are part of the conceptual framework behind generalizability theory. The definition of these universes for a given study lends support to and aids in understanding the indicated generalizability (G) and decision (D) studies employed. As part of both types of universes, *facets*, or sets of similar conditions of measurement, must be identified and described. These facets often include raters, tasks, occasions, and so forth. The universe(s) of admissible observations defines the instances of the facets as well as the relations among the facets that are acceptable conditions of measurement. For example, all high school math teachers from a given state might constitute the universe of admissible raters, any word problem from a given pool of math items might constitute the universe of admissible tasks, and a pairing of any rater with any task might constitute an acceptable relationship between these facets (crossed, in this instance).

The *objects of measurement* represent the population for whom the admissible facets would be appropriate (often people or groups). Following the high school math test example above, all the students who might respond to the math items would be the objects of measurement.

While the universe(s) of admissible observations defines the facets and relationships as they exist in a given situation, the universe(s) of generalization defines the facets and relationships to which a decision-maker wants to generalize based on the results from the given situation. In this way, we can estimate the effect on score variability of modifying the numbers of and relations among the facets (e.g., raters, tasks, occasions) and thus improve future designs (for example, finding that we may use fewer raters or fewer items and still maintain acceptable levels of reliability or identifying ways to obtain more reliable results from the measurement procedure). Given a particular universe of generalization, a *universe score* is defined for each object of measurement in the population as their mean (or expected) score over all randomly parallel replications of the measurement procedure. Thus, the universe score is analogous to the true score in classical test theory. The purpose of a measurement is to accurately estimate this universe score based on a sample of observations.

The authors need to clearly identify and define these concepts as related to their study. Furthermore, specifics about the facets must be described, including whether they are fixed or random and crossed or nested. The identification of the facets and universe(s) of admissible observations and of generalization is crucial. They provide the framework for choosing G study and D study designs, and for interpreting generalizability results.

## 4. Study Designs

Once a population and universe(s) of admissible observations has been defined, the researcher collects and analyzes data to estimate the variances of the observed scores and of the facets in the universe(s) of admissible observations associated with the appropriate design. The design of this study, called a *generalizability study* (or *G study*), should match the facets and relationships included in the universe(s) of admissible observations. The variance of the facets in the universe(s) of admissible observations can be decomposed into several uncorrelated components based on the G study design. Each of the components is termed a *variance component*. The variance component estimates from the G study can then be used to estimate the variance components and reliability-like coefficients for one or more D studies. In any particular application of generalizability theory, there is usually only one universe of admissible observations and one G study (i.e., a univariate design). However, there are often multiple universes of generalization and multiple decision (D) studies that are of interest to an investigator. These multiple universes and D studies might differ in terms of sample sizes for facets, which facets are fixed and which are random, and/or the structure of the D study designs.

In cases where multiple universes of admissible observations, and thus multiple universe scores for each person, are of interest a multivariate generalizability study should be employed. In a multivariate G theory design, each universe score is associated with one level of a fixed facet and for each such level there is a random effects design. These various designs are *linked* through covariance components. An example of a multivariate design is the *table of specifications* model (Brennan, 2001; Yin, 2005). In this model, a test consists of items from several content areas: a different set of items is nested within each

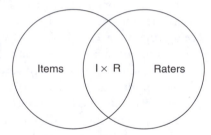

**Figure 9.1** Venn Diagram Depicting Items (I) Crossed with Raters (R).

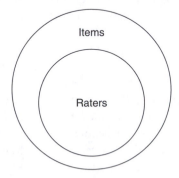

**Figure 9.2** Venn Diagram Depicting Raters (R) Nested within Items (I).

content area, and each content area is treated as a level of a *fixed* facet. The levels are linked because the same group of persons responded to all questions for each content area.

The authors need to clearly identify and describe the design of each G and D study, including whether each design is univariate or multivariate and balanced or unbalanced. The inclusion of Venn diagrams and/or tables that outline the various variance components help to facilitate the understanding of each design. For example, see Figures 9.1 and 9.2 for Venn diagrams that depict designs in which raters are either crossed or nested within items. The distinctions between the designs are easily discernible from the diagrams.

Furthermore, the authors need to be aware of and acknowledge that the power and flexibility of generalizability theory is associated with conceptual challenges in characterizing the designs, especially those with multivariate designs. Even though there are statistical methods associated with complex generalizability designs, the identification and conception of such designs is never clear-cut.

## 5. Software

The choice of computer software for estimating G theory results depends largely on the specific method desired for estimation. There are various methods for estimating variance components in a generalizability study. For example, one type of estimation method makes normality assumptions (e.g., maximum-likelihood and restricted maximum-likelihood procedures). Another type of procedure does not require such normality assumptions. For a detailed discussion of various estimation methods and their implementation in various computer software programs, see Brennan (2001, pp. 241–247).

Relatively few software packages exist for estimating G theory results. These are the suite of GENOVA programs, SAS, SPSS, and S-Plus. These packages tend to produce similar results for most basic designs, though they have various specializations and limitations.

Considerations to keep in mind when choosing and using software packages are: the estimation method employed, memory requirements, processing time, sample size, the design implemented (e.g., unbalanced vs. balanced) and the interrelatedness of these variables. For example, maximum-likelihood (ML) and restricted maximum-likelihood (REML) procedures necessarily make heavy demands on computer resources compared to the other procedures. When using the ML or REML procedures in combination with more than 10,000 observations, memory requirements and processing time are likely to be prohibitive. Furthermore, using SPSS for G theory analyses often results in an "insufficient memory" error message even with moderately large analyses. Finally, of these programs, only urGENOVA provides G study results for unbalanced designs.

The authors should identify the software program they used, the version of the software, the estimation method employed and any problems with estimation that they encountered.

### 6. Summary Statistics and Variance Components

Typically, various summary statistics are presented as part of the generalizability analysis related to the particular G study design. The most important G study statistics are the estimated variance components, because they can be used to design more efficient measurement procedures and to provide information in various decision studies (e.g., obtain D study variance components and subsequent D study statistics such as error variances and generalizability coefficients). The estimated variance components for all facets presented in the G study design are typically reported as part of the summary statistics.

### 7. Problems with Estimation

In theory, variance components cannot be negative. In practice, however, because sampling error is likely to be present, the estimates of variance components can sometimes be negative. In particular, when the number of facets in the design is large while the sample size for a particular facet is small, a possible consequence of sampling variability related to the facet with the small sample size could result in one or more negative estimated variance components.

One obvious solution to this problem is to increase the sample size for the facet with the small sample size. However, this might not always be feasible in practice, especially when time and resources are limited. Usually, negative estimates are simply set to zero. If an investigator wants to preclude the possibility of obtaining negative estimates, Bayesian estimation procedures can be used, but they are often challenging to implement. The negative estimate issue has theoretical implications because by definition, variance components cannot be negative. Also, because of the additive property of variance components, the negative variance component estimates for one facet may affect the variance components for other facets and some subsequent D study statistics if using certain estimation methods such as the *expected mean square* method (see Desideratum 8). However, the negative variance component estimates seldom make much difference from a practical point of view.

The authors need to be aware that such estimation problems do occur in generalizability analyses, especially when the sample size for a facet is small. The authors should also be aware that such estimation problems do not necessarily imply inappropriate study design or estimation procedure. The best solution for the estimation problem is to increase the sample size if possible. The authors should report any estimation problems encountered in initial runs and the strategies for solving these problems, including whether any and which negative estimates were set to zero.

### 8. G Study Variance Components

As discussed in Desideratum 4, the main purpose of a G study is to obtain variance component estimates associated with a universe(s) of admissible observations. Using the analysis of variance method, the variance of the observed scores over the population of objects of measurement and the facets in the universe(s) of admissible observations can be decomposed into several uncorrelated components, called variance components, based on the G study design.

Several procedures can be used to obtain the estimated variance components in a G study for balanced designs, and these procedures are discussed in detail in Brennan (2001). Only a brief summary is presented in this chapter. One method is called the *expected mean square* (EMS) procedure, in which a series of EMS equations are solved by replacing parameters with estimators (Cronbach et al., 1972). The most commonly used method is called the *ANOVA method*. It can be implemented using matrix procedures or via an algorithm described in Brennan (2001). The ANOVA method has two important characteristics: (1) it does not make any normality assumptions (which are often highly suspect in generalizability theory applications), and (2) the estimated variance components are unbiased.

For unbalanced designs, two general types of methods are available to obtain estimated variance components. One type of method (i.e., maximum likelihood) assumes normality and often involves computations with large matrices, which might not be appropriate when the normality assumption is not satisfied, or when the estimation process encounters problems related to the operation with large matrices. Another type of method does not make assumptions of normality and does not typically involve operations on large matrices. This method is called *Henderson's Method 1* or the *analogous-ANOVA method* by Brennan (2001). It is difficult to mount a compelling theoretical argument for one method over another when designs are unbalanced. From a practical perspective, however, Henderson's Method 1 is relatively simple and estimated variance components can be obtained quickly.

The authors should summarize the variance component results in a table, with each G study effect and estimated variance component clearly listed. For an example of such G study variance component tables, please see Table 9.3 in Brennan (2001). Information for the estimated G study variance components is typically combined with D study results (see Desideratum 9) and presented in one table.

## 9. D Study Variance Components

The specification of a universe of generalization is the most important aspect of a D study. It is very common for G and D studies to have the same structure. However, designs in G and D studies do not have to be the same. For example, different D study designs can be considered based on one G study for the purpose of designing more efficient measurement procedures and providing information for different decision studies. Therefore, D studies in generalizability analyses provide a unique and powerful tool in designing and improving measurement procedures.

The D study variance components are estimated for the universe of generalization. To obtain estimated D study variance component estimates, D study sample sizes need to be specified. Note that D study sample sizes do not need to be the same as the sample sizes for the G study. Also note that for the D study, the emphasis is on mean scores for facets considered in the design rather than individual conditions of facets.

The D study variance component for a facet can be interpreted as the variance of the distribution of mean scores for that facet, where each of these means is for the population of persons and the randomly parallel instances of the measurement procedure. Statistically, the variance of the mean score for an effect is simply the variance component for the individual effects divided by the sample size(s). Let $\alpha$ stand for an effect (e.g., the item effect); the D study variance component for the effect can be obtained using

$$\sigma^2(\bar{\alpha}) = \frac{\sigma^2(\alpha)}{n(\alpha)},$$ (1)

where $\sigma^2(\alpha)$ is the G study variance component for the effect, $\sigma^2(\bar{\alpha})$ is the D study variance component for the effect, and $n(\alpha)$ is the product of the D study sample sizes for all facets in the effect except for the object of measurement (e.g., persons). For instance, if the G study design involves persons, items, and the person × item interaction (or residual term), the D study variance component for the person × item interaction effect is simply the G study variance component for the person × item interaction effect divided by the D study sample size for items.

One of the advantages of generalizability theory is the flexibility in D studies. The authors should make an effort to specify the D study sample sizes and/or D study designs with likely desired scenarios (for example, using fewer raters while still achieving the same level of reliability) so that better and more efficient studies can be designed.

## 10. D Study Error Variances; Generalizability and Dependability Coefficients

Two types of error variances are typically reported in a D study, *relative* and *absolute*. These error variances are very useful for making norm-referenced and criterion-referenced interpretations of scores. Specifically, the absolute error variance is the error term involved in using an individual's observed mean score as an estimate of the individual's universe score, and for random models it is obtained by summing all variance components except the universe score variance in the D study. Absolute error is often associated with domain- or criterion-referenced interpretations of scores.

The relative error variance, on the other hand, is associated with using an individual's observed deviation score as an estimate of the individual's universe deviation score, which is the difference between a person's observed mean score and this person's universe or true score. For random models it is obtained by summing all variance components that include the universe score and at least one other facet in the D study. Relative error is often associated with norm-referenced interpretations of scores.

Based on the definitions of these two error variances, which indicate that at least the same variance components are summed in the calculation of absolute error variance, absolute and relative error variance is at least as large as relative error variance. Because the structure of the D study design often influences the magnitude of error variance, values of these absolute and relative error variances need to be considered or compared only in the context of the specific D study design.

D study coefficients are much like reliability coefficients in classical test theory, and have a range of zero to one. The two coefficients are *generalizability* and *dependability* coefficients. The generalizability coefficient is considered when making norm-referenced interpretations of scores:

$$E(\rho^2) = \frac{\sigma^2(p)}{\sigma^2(p) + \sigma^2(\delta)} \qquad (2)$$

where $E(\rho^2)$ is used to represent a generalizability coefficient that is considered a squared correlation between the universe and observed scores. In equation (2), $\sigma^2(\delta)$ is the relative error variance, and $\sigma^2(p)$ is the universe score variance. Similarly, the dependability coefficient is used when criterion-referenced interpretations of scores are to be made:

$$\Phi = \frac{\sigma^2(p)}{\sigma^2(p) + \sigma^2(\Delta)} \qquad (3)$$

where $\Phi$ is the dependability coefficient, and $\sigma^2(\Delta)$ is the absolute error variance.

The generalizability coefficient appears similar to the traditional reliability coefficients defined in classical test theory (e.g., test-rest reliability, Cronbach's alpha or internal consistency reliability coefficient). However, only in generalizability theory can a researcher specifically define the universe of generalization. It is also clear that multiple estimates of generalizability coefficients can be obtained based on including different facets, D study sample sizes, and even D study designs.

It can be noted from equations (2) and (3) that the only difference between the generalizability and dependability coefficients is in the error variances. If values for the relative and absolute error variances are the same, the two coefficients are the same as well. Typically, absolute error variance is larger than relative error variance because the former involves more variance components for the same design. Consequently, the dependability coefficient tends to be smaller than the generalizability coefficient.

When discussing different error variances and D study coefficients, the authors should be careful comparing the magnitude of the two error variances and the two D study coefficients. Depending on the G and D study designs, the two error variances and the two D study coefficients can be the same, similar, or different from each other. It is always advisable to discuss the differences instead of focusing simply on the values of these statistics.

## 11. Practical Results Are Discussed

The authors should pull together the results of their D study analyses and relate them back to the particular situation they are working with and the decisions that they set out to make. They should discuss the implications of the results for their measurement procedure and provide justification for using their current procedure or rationale for changes that they expect to make to the procedure.

## References

Brennan, R. L. (2001). *Generalizability theory*. New York: Springer-Verlag.

Cronbach, L. J., Gleser, G. C., Nanda, H., & Rajaratnam, N. (1972). *The dependability of behavioral measurements: Theory of generalizability for scores and profiles*. New York: Wiley.

GENOVA available on the web at: http://www.education.uiowa.edu/casma/computer_programs. htm#genova

mGENOVA available on the web at: http://www.education.uiowa.edu/casma/computer_programs. htm#genova

Shavelson, R. J., & Webb, N. M. (1991). *Generalizability theory: A primer*. Newbury Park, CA: Sage.

Traub, R. E. (1994). *Reliability for the social sciences: Theory and applications*. Newbury Park, CA: Sage.

urGENOVA available on the web at: http://www.education.uiowa.edu/casma/computer_programs. htm#genova

Yin, P. (2005). A multivariate generalizability analysis of the Multistate Bar Examination. *Educational and Psychological Measurement, 65*, 668–686.

# 10
# Hierarchical Linear Modeling

D. Betsy McCoach

In the social sciences, much of the data that we deal with are hierarchical in nature. Examples of naturally occurring hierarchies include students nested within schools, patients nested within hospitals, workers nested within companies, husbands and wives nested within couple dyads, and observations nested within people. Most traditional statistical analyses assume that observations are independent of each other. The assumption of independence means that subjects' responses are not correlated with each other. This assumption might be reasonable when data are randomly sampled from a large population. However, when people are clustered within naturally occurring organizational units (e.g., schools, classrooms, hospitals, companies), the responses of people from the same cluster are likely to exhibit some degree of relatedness with each other, given that they were sampled from the same organizational unit. Hierarchical linear modeling allows researchers to adjust for and model this non-independence.

With clustered data, traditional statistical analyses that assume independence will produce incorrect standard errors. In such a scenario, the estimates of the standard errors are smaller than they should be. Therefore, the Type I error rate is inflated for all inferential statistical tests that make the assumption of independence. In multilevel analyses, we explicitly estimate and model the degree of relatedness of observations within the same cluster, thereby correctly estimating the standard errors and eliminating the problem of inflated Type I error rates. These types of models are often referred to as *hierarchical linear models, multilevel models, mixed models,* or *random effects models.* These terms are generally used interchangeably, although there are slight differences in the meanings of the terms. For instance, hierarchical linear model is a more circumscribed term than the others, as it assumes a normally distributed response variable.

The advantages of hierarchical linear modeling, however, are not merely statistical in nature. Multilevel analyses allow us to exploit the information contained in cluster samples to explain both the between- and within-cluster variability of an outcome variable of interest. These models allow us to use predictors at both the individual (or lowest) level (level 1), and the organizational (or higher) level (level 2) to explain the variance in the dependent variable. We can also allow the relation between an independent variable and the dependent variable to randomly vary across clusters. If we find that the impact of the independent variable on the dependent variable varies across clusters, we can try to explain the variability in this relation using cluster-level variables. For example, we can allow the relation between students' SES and achievement to vary by school. If we find that the impact of SES does

Table 10.1 Desiderata for Hierarchical Linear Modeling

| Desideratum | Manuscript Section(s) |
|---|---|
| 1. Model theory and variables included in the model are consistent with the purposes of the study and the research questions or study hypotheses. | I |
| 2. The decision to include/exclude random effects should be justified theoretically. The number of random effects to include should be as realistic and yet as parsimonious as possible. If random effects are eliminated during the model-building process, this decision should be justified both empirically and theoretically. | M, R |
| 3. Statistical model is presented, preferably using equations. Otherwise, minimally, the statistical model is described in enough verbal detail to be replicable by other researchers, and for the reader to determine the fixed effects and the random effects at each level for each model. | M |
| 4. Sample size is specified at each level, and is sufficient for conducting the proposed analysis. Sampling strategy and mode(s) of data collection are identified and justified. If appropriate, weighting methods are described and justified. | M |
| 5. Measurement of the outcome/response variable is described and justified. Measurement of all explanatory variables is described and justified; evidence of reliability and validity is provided. | M |
| 6. Scaling and centering of predictor variables are described and justified. Coding of all categorical predictors is fully described. Special attention must be paid to the centering/coding of all lower-level independent variables and to the implications of these centering decisions for interpretation of the model results. | M |
| 7. Extent of missing data is clearly reported for all variables at all levels, and methods for accommodating missing data are described. Special attention is paid to the issue of missing data at higher levels of analysis, as higher level units with missing data are eliminated from the analysis by default. The final analytical sample is described. | M, R |
| 8. For longitudinal models, the shape of the growth trajectory is described, and the modeling of this trajectory is described and justified. | M, R |
| 9. The software or program used to run the models should be identified. Parameter estimation strategy (e.g., REML, ML) is identified and justified. | M, R |
| 10. Assumptions of the model are described and checked. This may include discussions of normality, outliers, multicollinearity, homogeneity or heterogeneity of variances, and residual diagnostics. | M, R |
| 11. The assumed error covariance structure should be described, and any plausible alternative error covariance structures should be described and tested. This is especially important for longitudinal models. | M, R |
| 12. Descriptive statistics for variables at each level of the analysis should be reported. These should include means, standard deviations, and correlations. | R |
| 13. The intraclass correlation coefficient for the unconditional model should be reported and interpreted. | R |
| 14. Generally, multilevel models are built sequentially, using a series of models: an unconditional model, a random coefficients model (containing lower-level predictors), and a full model (containing predictors at all level of the analysis). This series of models is described. | M, R |
| 15. The write-up includes a table that presents the results of the analysis. These results should include both fixed effect parameter estimates and variance components. | R |
| 16. Model fit issues are addressed. Deviance is reported for any estimated models. Additionally, other measures of model fit (e.g., AIC, BIC) are reported for all estimated models. Competing nested models are compared using the likelihood ratio/chi-square difference test. | R |

17. Some description/summary of the final model's predictive ability should be provided. This could include a proportion reduction in variance at each level/proportion of variance accounted for at each level.    R

18. Some measure of effect size or practical utility should be reported.    R, D

19. Language used in the presentation and discussion of results appropriately reflects the study design. Causal language is not used except when justified through study design.    R, D

*Note*: I = Introduction, M = Methods, R = Results, D = Discussion

vary by school, we can try to explain that variability using school-level predictors, such as type of school, school SES, or average per-pupil expenditures. If a level-2 variable, such as average per-pupil expenditure, moderates the relationship between a level-1 variable (SES) and the dependent variable (achievement), this is called a *cross-level interaction*. Thus, hierarchical linear modeling allows us to simultaneously model the impact of both individual (or lower-level) and institutional (or higher-level) variables on the dependent variable of interest, as well as to model the cross-level interactions between higher-level and lower-level variables on the outcome of interest.

Finally, growth curve and other longitudinal analyses can be reframed as hierarchical linear models, in which observations across time are nested within individuals. Using this framework, we can partition residual or error variances into those which are within-person (measurement error) and those that are between people. In such a scenario, between-person residual variance represents between-person variability in any randomly varying level-1 parameters of interest, such as the intercept (which is commonly centered to represent initial status in growth models) and the growth slope.

Contemporary expositions of hierarchical linear modeling include textbooks by Raudenbush and Bryk (2002), Hox (2002), and Snijders and Bosker (1999), and an edited volume by O'Connell and McCoach (2008). Table 10.1 presents specific desiderata for applied studies that utilize hierarchical linear modeling, and the remainder of this chapter is devoted to the explication of these desiderata.

## 1. Model/Theory Alignment

When using modeling techniques such as multilevel modeling, it is important that a coherent conceptual base informs and guides the statistical analyses. The model theory and the variables included in the model need to be consistent with the purposes of the study and the research questions or study hypotheses. Of course, any study should be guided by a coherent theory. However, given that excluding an important potential confounder creates the potential for bias in the estimates, the temptation with regression-type models is to try to add any variable that might be related to the outcome variable. While it is true that failing to include important potential confounders can create bias in the estimates of the effects of other variables, given the complexity of the error structure and the number of potential cross-level interactions, models that include large numbers of fixed and random effects can become unwieldy, difficult to interpret, and perhaps even impossible to estimate. Therefore, researchers should spend a great deal of time determining the variables for inclusion based on theory and relevant literature prior to undertaking the data analysis.

## 2. Random Effects

As in multiple regression (see Chapter 21, this volume), we estimate a within-cluster residual ($r$) that represents the deviation of a person's score from his or her predicted value. In a multilevel model, the intercept and the slopes for each of the level-1 variables can randomly vary across the level-2 units. In general, we allow the intercept to randomly vary across level-2 units. Therefore, we estimate a residual for each cluster ($u_0$). This is the deviation of a cluster's value from the overall intercept. It is this ability to partition variance into within-cluster variance and between-cluster variance that is the essence of the hierarchical linear model. For simplicity, imagine a model in which there are no predictors. Each person's score on the dependent variable is composed of three elements: the overall mean ($\gamma_{00}$), the deviation of the cluster mean from the overall mean ($u_{0j}$), and the deviation of the person's score from his/her cluster mean ($r_{ij}$). The $u_0$ term allows us to model the dependence of observations from the same cluster because $u_{0j}$ is the same for every student within school $j$ (Raudenbush & Bryk, 2002). The $u_0$ term is referred to as a *random effect* for the intercept because we assume that the value of $u_0$ randomly varies across the level-2 units (clusters) and that it has a mean of 0 and a variance of $\tau_{00}$.

Now, imagine a model in which there is one predictor at the lowest level. For this example, assume that we are predicting reading achievement ($Y_{ij}$) using SES. We continue to allow the intercept to randomly vary across schools. However, now we can allow the SES slope to randomly vary across schools as well by including $u_1$. By allowing the SES slope to randomly vary across schools, we are specifying a model in which the relation between SES and reading achievement is different for different schools. Therefore, in some schools, there could be no relation between students' SES and their reading achievement, whereas in other schools the relation between students' SES and their reading achievement could be quite strong. The set of equations for this model is

$$
\begin{aligned}
Y_{ij} &= \beta_{0j} + \beta_{1j}(SES)_{ij} + r_{ij} \\
\beta_{0j} &= \gamma_{00} + u_{0j} \\
\beta_{1j} &= \gamma_{10} + u_{1j}
\end{aligned}
\tag{1}
$$

Generally speaking, in multilevel modeling, fixed effects represent the average effect across an entire population and are expressed by the regression coefficient (Snijders, 2005). In contrast, a random effect varies randomly across the population of level-2 units and is estimated as a residual for each of level-2 units (Snijders, 2005). Generally, we are interested in estimating, modeling, and testing the variances (and covariances) for these random effects. The variances and covariances of the random effects are referred to as *variance components*. In the set of equations above (1), the $\gamma$ terms are the *fixed effects* and the $u$ terms are the *random effects*.

In a two-level model, the number of possible random effects is equal to the number of variables at level 1 plus 1 (the random effect for the intercept). Therefore, in a model that contains 10 different level-1 variables, there could be up to 11 random effects. The number of random effects included should be as realistic and yet as parsimonious as possible. At first glance, it might seem desirable to try to allow the slopes for all level-1 variables to vary randomly across the level-2 clusters and then to empirically eliminate any random effects that are not statistically significant. However, Raudenbush and Bryk (2002) cautioned against this practice: "If one overfits the model by specifying too many random level-1 coefficients, the variation is partitioned into many little pieces, none of which is of much significance" (p. 256). Instead, researchers should make the decision to include/exclude random effects based on theoretical grounds, rather than blindly allowing all level-1 slopes to randomly vary across level-2 clusters.

Even when using theory as a guide, sometimes analysts make changes to the random portion of the model during the model-building process. If random effects are eliminated during the model building process, this decision should be justified both empirically and theoretically. One common reason for eliminating random effects is that the level-2 variables in the model are able to explain the between-cluster variability in the slopes. For example, imagine a model in which prior vocabulary is a predictor of reading achievement and the prior vocabulary slope is allowed to randomly vary across schools. However, when the level-2 model includes the SES of the school and the proportion of English language learners as predictors of the vocabulary/reading achievement slope, the between-school variability in the slope is greatly reduced, and it is no longer statistically significant. In such a scenario, the variability in the impact of vocabulary knowledge on reading achievement is explained by school-level variables. Therefore, the slope of vocabulary on reading achievement is neither fixed nor randomly varying. Instead, it systematically varies as a function of the two level-2 variables.

## 3. Presentation of the Statistical Model

It is important for readers to be able to understand and potentially replicate the reported multilevel analyses. Therefore, the full hierarchical linear model must be specified clearly within the Methods section. There are many decisions that a researcher must make when building a multilevel model. Are the slopes of the level-1 coefficients allowed to randomly vary across the level-2 units? Which cross-level interactions between level-1 variables and level-2 variables are specified? Given the complexity of most multilevel models, the clearest and easiest way to communicate the exact specification of the model is by presenting the statistical model using equations. The equations for the multilevel model can be presented in one of two ways: using separate equations for the level-1 and level-2 variables or using a combined model.

To illustrate the multilevel and combined specifications, imagine a model in which the researcher wants to predict the reading achievement scores for students nested within schools. The level-1 independent variable is socio-economic status (SES), and the effect of SES is assumed to randomly vary across schools. The level-1 intercept is also allowed to randomly vary across schools. The level-2 independent variable is percentage of students receiving free lunch (FREELNCH), which serves as an indicator of School SES. The multilevel, multiple equation notation is:

$$
\begin{aligned}
Y_{ij} &= \beta_{0j} + \beta_{1j}(SES)_{ij} + r_{ij} \\
\beta_{0j} &= \gamma_{00} + \gamma_{01}(FREELNCH)_j + u_{0j}, \\
\beta_{1j} &= \gamma_{10} + \gamma_{11}(FREELNCH)_j + u_{1j}
\end{aligned}
\tag{2}
$$

The $\gamma_{00}$ term is the intercept of the school intercepts, indicating the predicted value of reading achievement when all other variables in the model are held constant at 0. The $\gamma_{01}$ term represents the unit change in the predicted value of the intercept per unit change in the free-lunch variable. The $\gamma_{10}$ term is the intercept of the SES slope, indicating the relationship between SES and achievement when FREELNCH = 0. Finally, $\gamma_{11}$ is the cross-level interaction between FREELNCH and SES, indicating the degree to which the percentage of free-lunch students in a school moderates the effect of SES on reading achievement. The $u_{0j}$ term indicates that the intercept ($\beta_{0j}$) is allowed to randomly vary across schools. The $u_{1j}$ term indicates that the slope of the SES variable ($\beta_{1j}$) is allowed to randomly vary across schools. The combined model is the same model and contains the same information as the multilevel, multiple equation notation. However, in the combined model, we substitute the expressions to the right of the equals sign for $\beta_0$ and $\beta_1$. Thus, the combined notation for the same model would be:

$$
\begin{aligned}
Y_{ij} &= \gamma_{00} + \gamma_{01}(FREELNCH) + u_{0j} + \gamma_{10}(SES)_{ij} + \gamma_{11}(FREELNCH)_j \\
&\quad (SES)_{ij} + u_{1j}(SES)_{ij} + r_{ij}.
\end{aligned}
\tag{3}
$$

Generally, these terms are regrouped so that the fixed effects are in the beginning of the equation and the random effects are at the end of the equation. So the standard combined form would be as follows:

$$
\begin{aligned}
Y_{ij} &= \gamma_{00} + \gamma_{01}(FREELNCH)_j + \gamma_{10}(SES)_{ij} + \gamma_{11}(FREELNCH)_j \\
&\quad (SES)_{ij} + u_{0j} + u_{1j}(SES)_{ij} + r_{ij}.
\end{aligned}
\tag{4}
$$

In reality, the model that is estimated is the combined model. Users of SAS and SPSS must specify the combined model, whereas users of the software package HLM (Raudenbush, Bryk, Cheong, Congdon, & du Toit, 2004) must use the multiple equation notation to estimate multilevel models. Thus, certain multilevel modelers prefer to use the combined notation while others prefer the multiple equation notation. Either convention is acceptable, as both sets of equations are equivalent and contain the same information.

There might be audiences who would be confused by the multilevel equations. In such a situation, a reasonable solution is to present the equations and then to explain them verbally within text. Occasionally, a researcher might present research findings to an audience who would be completely overwhelmed by the presentation of equations, and the editor may request that the equations be removed from the manuscript. In such a circumstance, the statistical model should be described in enough verbal detail to be replicable by other researchers based on the description. The reader should be able to determine the fixed effects and the random effects at each level and the cross-level interactions. Some models are so complex that describing them verbally might actually be more difficult than describing them using equations. Even so, the author should make every effort to present his or her multilevel models both verbally and through the use of combined or multilevel equations.

## 4. Sample Size Issues

Issues related to sample size are critically important in hierarchical linear models. Further, sample size issues are complicated by the multilevel nature of the data. In a sense, there are two sample sizes in a two-level model. Consider an organizational model in which people (level 1) are nested within organizations (level 2). The number of individuals represents the level-1 sample size, and the number of organizations represents the level-2 sample size. In a longitudinal model, on the other hand, observations across time are nested within people. Therefore, the number of observations across time (and people) is the level-1 sample size, and the number of people is the level-2 sample size. So, for example, in a longitudinal model where 100 people are each measured across four time points, the level-1 sample size would be 400 ($4 \times 100$), and the level-2 sample size would be 100.

Generally speaking, the overall sample size is less important than the number of level-2 units and average number of level-1 units within each of the level-2 units. Thus, it is important to report the sample size at each level. At a minimum, the researcher should report the number of level-1 units, the number of level-2 units, and the mean and standard deviation of the average cluster size. If there are a small number of clusters ($N < 50$), including a frequency table that shows the number of observations within each cluster can be useful (Ferron et al., 2008). In addition, the researcher should identify and justify the sampling strategy and the mode of data collection. When using sample weights, it is important to describe the method for weighting the data and justify the decision to use sample weights. For more information on the use of sampling weights see Stapleton and Thomas (2008).

There are two important considerations related to sample size. First, the number of units at each level of analysis must be large enough to estimate the multilevel model. Second, the analysis should be adequately powered to detect the effect of interest.

The average sample size at each level should be sufficient for conducting the proposed analysis. While small average numbers of level-1 units within level-2 clusters may limit the number of random effects that a researcher can estimate, the number of level-2 units is the more important sample size to consider when conducting or evaluating multilevel analyses. The sample size must be large enough to produce estimates, and these estimates must be reasonably free from bias. Maximum likelihood estimation is a large-sample technique that provides asymptotically unbiased estimates. However, in multilevel modeling, having a large overall sample size is not sufficient. The number of clusters (or the sample size at the highest level) must be large enough to support the estimation technique and to produce relatively unbiased estimates of the parameters and standard errors. What is the minimum number of clusters for a multilevel analysis? Certainly, it seems clear that multilevel analyses require a bare minimum of 10 clusters (Snijders & Bosker, 1999). However, such small sample sizes at level 2 might still produce biased estimates. The number of clusters impacts the estimates of the variance components and the standard errors, as well as the parameter estimates themselves.

Maas and Hox (2005) conducted a series of simulation studies to determine the smallest level-2 sample size that would produce unbiased parameter estimates and standard errors. With only 10 level-2 units, the regression parameters and the level-1 variance components exhibited little bias. However, the level-2 variance components were overestimated by approximately 25%, and the standard errors for all parameter estimates were underestimated. With at least 30 clusters, the parameter estimates for the regression slopes and both the level-1 and level-2 variance components tended to exhibit very little bias in samples. However, there were issues with the estimation of the standard errors. While the standard errors for the fixed effects and the level-1 variance components seemed to exhibit reasonable coverage with as few as 30 clusters, the standard errors for the level-2 variance components tended to be underestimated when there were fewer than 100 groups (Maas & Hox, 2005). This means that studies with small to moderate numbers of clusters might have a higher Type I error rate for the level-2 variance components, which could lead to concluding mistakenly that the between-group variance is more pronounced than it actually is. Therefore, while it is possible to produce unbiased estimates of the fixed effects with as few as 10 higher-level units, at least 30 clusters are required to produce unbiased estimates of the variance components and at least 100 clusters are necessary to have reasonable estimates of the standard errors of the level-2 variance components. In conclusion, while it may be possible to estimate a model with as few as 10 clusters, models with at least 30 clusters should provide reasonable estimates of variance components. Standard errors for higher-level variance components will be underestimated with small to moderate sized samples.

In addition, number of level-1 and level-2 units must be large enough to detect the effect of interest. In the simplest scenario, power in multilevel modeling is a function of the number of clusters, the number of units per cluster, the intraclass correlation coefficient (see Desideratum 13), and the effect size. Sample size at both levels increases power. However, increasing the number of clusters increases power more than increasing the number of units per cluster does. This effect is even more pronounced as the intraclass correlation increases. Several free software programs are available to conduct a priori power analyses for multilevel models. The Optimal Design software program and manual are freely available from http://sitemaker.umich.edu/group-based/optimal_design_ software.

## 5. Measurement Issues

As with any analysis, it is important to describe the scale of measurement of the outcome variable. Hierarchical linear models are appropriate for analyzing continuous, normally distributed outcome variables whereas hierarchical *generalized* linear models allow for the estimation of non-normal response variables (O'Connell, Goldstein, Rogers, & Peng, 2008; Raudenbush & Bryk, 2002).

In addition, the Methods section should include a description of the scale of measurement for all of the explanatory variables in the model. As with any statistical analysis, the researcher should provide evidence of reliability and validity of each of the variables in the model. Because hierarchical linear modeling is a regression-based technique, the assumptions of linear regression models continue to apply (see Chapter 21, this volume). One commonly overlooked assumption of linear regression is that the independent variables are measured reliably. When one or more predictor variables are measured with error, the results of the analysis might be misleading. The likelihood of Type II errors increases for the measures that exhibit low reliability, while the likelihood of Type I errors increases for the other variables in the model (e.g., Osborne & Waters, 2002). Therefore, it is especially important to provide evidence of reliability of scores for all of the continuous independent variables in the model.

## 6. Centering

In multilevel modeling, it is especially important to describe and justify the scaling and centering of all the predictor variables. Decisions about centering impact the interpretation of the parameter estimates. Centering decisions are especially important for the lower-level continuous independent variables because the choice of centering at the lower level(s) impacts the interpretation of both the lower- and higher-level parameter estimates. For organizational models, the two main centering techniques for lower-level independent variables are *grand mean centering* and *group mean centering*. In grand mean centering, the overall mean of the variable is subtracted from all scores. Therefore, the new score captures a person's standing relative to the full sample. In group mean centering, the cluster mean is subtracted from the score for each person in that cluster. As such, the transformed score captures a person's standing relative to his or her cluster. Whereas grand mean centering is a simple transformation of the raw score, group mean centering is not. There is some debate within the multilevel literature about whether grand mean centering or group mean centering is preferable from a statistical point of view. However, most experts in hierarchical linear modeling agree on three issues related to centering. First, the decision to use grand mean or group mean centering should be based on substantive reasons, not just statistical ones. For instance, if the primary research question involves understanding the impact of a level-2 variable on the dependent variable and the level-1 variables serve as control variables, grand mean centering may be the most appropriate choice. On the other hand, when level-1 variables are of primary research interest, group mean centering may be more appropriate. This is because group mean centering removes between cluster variation from the level-1 covariate and provides an estimate of the pooled within cluster variance (Enders & Tofighi, 2007). Second, it is important to explain the centering decision and to interpret the parameter estimates accordingly. Third, when using group mean centering, it is important to introduce an aggregate of the group mean centered variable (or a higher-level variable that measures the same construct) into the analysis. Without an aggregate or contextual variable at level 2, all of the information about the between-cluster variability is lost. See Enders and Tofighi (2007) for an excellent discussion of centering in organizational multilevel models.

In growth models, the time or age variable also needs to be centered so that the intercept represents an interpretable value. For linear growth models, the most common technique is to center time at initial status or age at the beginning of the study. When time is centered at initial status, then the intercept represents an individual's starting value. However, the time variable can be centered at any point in the data collection period. For certain research problems, analysts may prefer to center time at the final time point or at the middle of the data collection cycle. As Biesanz, Deeb-Sossa, Papadakis, Bollen, and Curran stated:

> The choice of where to place the origin of time has to be substantively driven. Because this choice determines that point in time at which individual differences will be examined for the lower order coefficients, the answer to which coding(s) of time to examine in detail lies with the researcher's specific substantive questions of interest. (2004, p. 37)

In addition to describing the centering and scaling of continuous variables, it is also important to describe the coding of all categorical predictors. Researchers should use the same conventions that they would use when conducting multiple regression to code the categorical variables in their models. Thus, researchers should use dummy coding, weighted or unweighted effects coding, or contrast coding for all categorical variables (see Cohen, Cohen, West, & Aiken, 2003, for an excellent discussion of coding for multiple regression analyses). The decision about the type of coding scheme is less important than the description of the coding scheme used and the correct interpretations of the parameter estimates that result from such a coding scheme. Finally, researchers need to create all same-level interactions among categorical and/or continuous variables in the same manner as they would if they were conducting a multiple regression analysis. Again, the interpretation of the interaction parameter estimates depends on the coding schemes used for the lower-order variables. See Aiken and West (1991) for an excellent discussion of creating and interpreting same-level interactions within a multiple regression framework.

## 7. Missing Data

The extent of missing data needs to be reported for all variables at all levels, and the author should describe the methods used to address the issue of missing data. Missing data is a problem for any analysis. However, in hierarchical linear modeling, missing data can be especially problematic at the higher levels of analysis. In most software programs, higher-level units with missing data on any of the covariates are eliminated from the analysis by default. For example, any school with missing data on any of the school-level covariates (e.g., percentage of free-lunch eligible students, average per pupil expenditures) is eliminated from the multilevel model. Thus, it is easy to see how even small amounts of missing data at the higher levels of analysis could drastically reduce the size of the sample as well as the generalizability of the results.

Several modern data techniques exist for dealing with the problem of missing data. These include *multiple imputation* (MI; Rubin, 1987, 1996), and *expectation maximization* (EM; Dempster, Laird, & Rubin, 1977; also see Enders, 2001), which are considered the best methods of dealing with missing data. Ideally, multiple imputation of either the dependent variable or lower-level covariates should take the clustered nature of the data into account (Black, 2008). Less desirable methods of dealing with missing data include listwise deletion and single-imputation techniques. Unfortunately, although both MI and EM are considered the most appropriate methods for handling missing data, these methods are not routinely employed in multilevel studies. When describing the sample, the author should explicitly describe the amount of missing data and justify his or her method of handling missing data.

## 8. Fitting Growth Trajectories

Fitting longitudinal growth models using hierarchical linear modeling techniques is becoming increasingly popular. In such a model, the person is at level 2, and observations across time are nested within person. The unconditional linear growth model is

$$Y_{ti} = \pi_{0i} + \pi_{1i} (TIME)_{ti} + e_{ti}$$
$$\pi_{0i} = \beta_{00} + r_{0i}$$
$$\pi_{1i} = \beta_{10} + r_{1i} \tag{5}$$

As the name implies, a linear growth model assumes a straight-line growth trajectory. However, many growth processes do not follow a linear trajectory. Assuming a linear growth trajectory is very limiting, and it may result in a serious misspecification of the model. Other shapes are accommodated easily using a variety of strategies. These include estimating piecewise models, polynomial models, or other non-linear models, as well as introducing time-varying covariates (see Singer & Willett, 2003, for an excellent discussion of these strategies). Therefore, the researcher should empirically examine the shape of the individual and average growth trajectories descriptively prior to fitting any statistical models. This information, in combination with the theory, can help guide decisions about the shape of the growth trajectory. When using multilevel modeling to fit longitudinal models, it is imperative that the researcher describe the shape of the growth trajectory, describe the level-1 model, and justify how the modeling procedure used at level-1 was able to capture the shapes of the growth trajectories for the sample.

## 9. Software and Parameter Estimation

The write-up should mention the program or software package used to conduct the analysis. Many computer programs have multilevel capabilities. Some of the most common programs used to conduct multilevel analyses include SAS, SPSS, HLM, MLWin, Stata, R/SPlus, and MPLUS. All of these programs handle straightforward two-level models with normal response variables with ease. Where the programs differ is in their ability to handle more complicated models such as cross-classified models, three-level models, multilevel meditational models, or models with non-normal outcome variables. For an overview and comparison of these different software programs, see Roberts and McLeod, 2008.

The two most common estimation techniques for hierarchical linear models with normal response variables are *maximum likelihood* (ML) and *restricted maximum likelihood* (REML). The two methods should produce similar results in terms of the fixed effects (regression parameters); however, they do produce different estimates of the variance components (Snijders & Bosker, 1999). In ML estimation the estimates of the variance and covariance components are conditional upon the point estimates of the fixed effects, whereas in REML they are not (Raudenbush & Bryk, 2002). Whereas REML estimates of variance-covariance components adjust for the uncertainty about the fixed effects, ML estimates do not. When estimating the variance components, REML takes "into account the loss of degrees of freedom resulting from the estimation of the regression parameters, whereas the ML method does not" (Snijders & Bosker, 1999, p. 56). Thus, the ML estimates are downwardly biased (Snijders & Bosker), especially when the number of level-2 units (clusters) is small. When the number of clusters is very large, REML and ML results should produce similar estimates of the variance components. However, when the number of level-2 units is relatively small, the ML estimates of the variance components ($\tau_{qq}$) are underestimated by a factor of $(J-F)/J$, where $J$ is the number of level-2 units and $F$ is the number of fixed effects (Raudenbush & Bryk, 2002). Thus, REML is the preferred estimation strategy for models with few level-2 units.

While REML may be preferable to ML for estimating the variance components, ML is often preferable to REML for testing model fit. The deviances of any two nested models that differ in terms of their fixed and/or random effects can be compared when using ML. However, REML only allows for comparison of nested models that differ in their random effects (Snijders & Bosker, 1999, p. 89). In addition, information criteria, such as the AIC and BIC, should be based on the ML estimates of the deviance (see Desideratum 16 for information about deviance and model fit.)

## 10. Assumptions and Residual Analyses

As with any statistical analysis, it is important to check the assumptions of the model and to describe any violations of the assumptions. Many regression diagnostics for single-level models are applicable within the multilevel framework as well. These may include discussions of normality, linearity, outliers, multicollinearity, homogeneity or heterogeneity of variances, and residual diagnostics. However, because the regression model is operating on multiple levels, tests of the assumptions become a bit more complex and time consuming. First, most residual analyses can and should be conducted at each level of the analysis. For example, in a two-level model where, say, students are nested within schools, it is possible to have an outlier at the student level. However, it is also possible to have an outlier at the school level. A school-level outlier would be a school "with implausible regression intercepts or slopes" (Raudenbush & Bryk, 2002, p. 252). Researchers should carefully check the assumptions of their models, and they should include a short description of the procedures that they used to check their assumptions. In addition, they should describe any violations of the assumptions and the procedures that they used to rectify those violations (e.g., Were any outliers deleted? Were any variables transformed?).

## 11. Error Covariance Structure

The researcher should briefly describe the assumed error covariance structure. Any plausible alternative error covariance structures should be described and tested. Generally, the assumed error covariance structure is quite reasonable for organizational models. The simplest error structure for a two-level model with a random intercept is depicted in equation (6). In this matrix, there are as many rows and columns as there are level-1 units. In this example, the first six level-1 units are shown. The first three level-1 units belong to cluster 1 and the second three units belong to cluster 2. The total residual variance for each person in the model is the sum of the within cluster residual ($\sigma^2$) and the between-cluster residual ($\tau_{00}$). The covariance between any two people who are members of the same cluster is accounted for by $\tau_{00}$, the between cluster residual. Finally, the residual covariance between 2 members of two different clusters is assumed to be 0.

$$\begin{bmatrix} \sigma^2 + \tau_{00} & \tau_{00} & \tau_{00} & 0 & 0 & 0 \\ \tau_{00} & \sigma^2 + \tau_{00} & \tau_{00} & 0 & 0 & 0 \\ \vdots & \tau_{00} & \tau_{00} & \sigma^2 + \tau_{00} & 0 & 0 & 0 & \cdots \\ 0 & 0 & 0 & \sigma^2 + \tau_{00} & \tau_{00} & \tau_{00} \\ 0 & 0 & 0 & \tau_{00} & \sigma^2 + \tau_{00} & \tau_{00} \\ 0 & 0 & 0 & \tau_{00} & \tau_{00} & \sigma^2 + \tau_{00} \end{bmatrix}. \quad (6)$$

Describing the error covariance structure is especially important for longitudinal models. Models that fail to adequately account for the covariances among repeated measurements may result in misleading inferences (Fitzmaurice, Laird, & Ware, 2004). On the other hand, when modeling these longitudinal covariances, the analyst's goal should be to "select the most parsimonious covariance structure that reasonably fits the data" (Wolfinger, 1996, p. 208).

The standard multilevel linear growth model imposes a very particular structure on the composite within-person/across time covariances (the composite of the covariances across waves). The structure is dependent on the number of random effects in the model. The maximum number of random effects that can be estimated in a repeated measures model is the number of waves of data minus 1. The standard multilevel linear growth model estimates a random effect for the intercept, a random effect for the linear growth slope, and a covariance between the intercept and the slope. Using the standard multilevel model, the model-implied variance-covariance matrix for a model with four waves of data is

$$\begin{bmatrix} \tau_{00} + \sigma^2 \\ \tau_{00} + \tau_{01} & \tau_{00} + 2\tau_{01} + \tau_{11} + \sigma^2 \\ \tau_{00} + 2\tau_{01} & \tau_{00} + 3\tau_{01} + 2\tau_{11} & \tau_{00} + 4\tau_{01} + 4\tau_{11} + \sigma^2 \\ \tau_{00} + 3\tau_{01} & \tau_{00} + 4\tau_{01} + 3\tau_{11} & \tau_{00} + 5\tau_{01} + 6\tau_{11} & \tau_{00} + 6\tau_{01} + 9\tau_{11} + \sigma^2 \end{bmatrix}. \quad (7)$$

Thus, all 10 unique elements of the variance covariance matrix for the four repeated measurements are estimated using four parameters: $\tau_{00}$, the between person variance in the intercept, $\tau_{11}$, the between person variance in the linear growth slope, $\tau_{01}$, the covariance between the slope and the intercept, and $\sigma^2$, the within person residual variance. Other options for estimating the covariance structure of the repeated measurements include fitting models with heterogeneous $\sigma^2$ across the time points, first order autoregressive models, first order moving average models, and unrestricted covariance matrices, to name a few. A complete treatment of this topic is beyond the scope of this chapter. However, researchers who are interested in learning more about covariance structures for repeated measures multilevel models should consult Hedeker and Gibbons (2006), Singer and Willett (2003), and Wolfinger (1996).

## 12. Descriptive Statistics

As in any research study, it is important to provide the reader with tables of descriptive statistics. Minimally, the author should provide a table of means and standard deviations for all of the continuous level-1 variables used in the analysis as well as a table of means and standard deviations for all of the continuous level-2 variables in the model. Dichotomous variables should be reported as proportions or percentages. In addition, the write-up should include a table of correlations corresponding to each level in the analysis. So, for a two-level model, one correlation matrix should detail the correlations among the level-1 variables, whereas another correlation table should provide the correlations among the level-2 variables.

## 13. Intraclass Correlation Coefficient

The *intraclass correlation coefficient* (ICC) is the proportion of variance that is between clusters, that is, the proportion of variance that can be explained by the clustering or grouping structure (Hox, 2002). Alternatively, one may interpret the ICC as the "expected correlation between any two randomly chosen units that are in the same group" (Hox, 2002, p. 15). The formula for the ICC is

$$\rho = \frac{\tau_{00}}{\tau_{00} + \sigma^2}. \tag{8}$$

where $\tau_{00}$ represents between cluster variance and $\sigma^2$ represents within cluster variance. The ICC is important to report because it indicates the degree of non-independence in the data. The higher the ICC, the more homogeneity there is within clusters (or the more heterogeneity there is between clusters). An ICC of 0 indicates independence of observations, and any ICC above 0 indicates some degree of dependence in the data. The smaller and more homogeneous the cluster is, the higher the expected ICC is. For example, in school effects research, ICCs typically range from .10 to .20. In dyadic research, on the other hand, ICCs above .50 are not uncommon.

The computation of the *design effect*, which indicates the degree to which the parameter estimates' standard errors are underestimated when assuming independence, utilizes the ICC ($\rho$) and the average number of units per cluster ($\bar{n}_j$):

$$\text{design effect} = \sqrt{1 + \rho\,(\bar{n}_j - 1)}. \tag{9}$$

Generally, design effects below 2.0 are considered fairly small. However, keep in mind that even with a design effect as low as 1.5, the standard errors in a model that assumes independence of observations are underestimated by a factor of 1.5. Therefore, the Type I error rate is already noticeably inflated, even with such a small design effect.

## 14. Model Building

Generally, multilevel models are built sequentially, using a series of models. First, researchers estimate an *unconditional* (or *null*) *model*, which contains no predictors. The purpose of this model is to obtain estimates of the level-1 and level-2 variance components for comparison to later, more parameterized models and to estimate the ICC. The second model estimated is a *random coefficients model*, which contains the level-1 predictors. Depending on the researcher's theoretical framework as well as the sample size at level-1, the slopes for some of the level-1 predictors may be estimated as randomly varying across level-2 units, or they can be estimated as fixed across all level-2 units. Raudenbush and Bryk (2002) cautioned against the "natural temptation," which "is to estimate a 'saturated' level-1 model … where all potential predictors are included with random slopes" (p. 256). Any level-1 slopes that do not have statistically significant variability across level-2 units should be fixed prior to conducting the full contextual analysis. The next model to be estimated is the *full contextual model*, which contains both level-1 and level-2 predictors. Level-2 predictors can be used to predict the intercept or the mean value of the dependent variable (when all of the level-1 variables are held constant at 0). Level-2 predictors also can help explain the variability of level-1 slopes across clusters. In such a scenario, the level-2 variable is used to predict the level-1 slopes, or the relationship of the level-1 predictor and the dependent variable across level-2 units. For example, imagine that SES is a level-1 predictor of math achievement. Sector, a level-2 variable that indicates whether a school is public or private, can be added as a predictor of the relation between SES and math achievement. This cross-level interaction indicates whether sector moderates how SES relates to math achievement. Finally, if any fixed or random effects are eliminated from the full model, a final contextual model should be estimated.

It is important to describe the process of building these sequential models. Analysts differ somewhat in their approaches to building multilevel models. Thus, authors must be sure to describe the model building process in enough detail that another analyst could replicate the entire model and decision sequence.

## 15. Tables

The Results section should include a table that presents the results of the analyses. If space allows, presenting the results of the series of models can be quite informative; however, minimally, the table should include the complete results from the final, full contextual model. These results should include the fixed effect parameter estimates, the random effect parameter estimates (the variances of the random effects), the standard errors for all parameter estimates, and tests of statistical significance for both the fixed and random effects. Also, the table may include covariances among the random effects.

## 16. Deviance and Model Fit

It is important to address model fit issues as part of the model building and testing process. The deviance compares the log-likelihood of the specified model to the log-likelihood of a saturated model that fits the sample data perfectly (Singer & Willett, 2003, p. 117). Specifically, deviance $= -2LL$, where LL is the log-likelihood of the current model minus the log-likelihood of the saturated model. Therefore, deviance is a measure of the badness of fit of a given model; it describes how much worse the specified model is than the best possible model. Deviance statistics cannot be interpreted directly since deviance is a function of sample size as well as the fit of the model.

When one model is a subset of the other, the two models are said to be "nested" (e.g., Kline, 1998). In nested models, "the more complex model includes all of the parameters of the simpler model plus one or more additional parameters" (Raudenbush, Bryk, Cheong, & Congdon, 2000, pp. 80–81). When two models are nested, their deviance can be compared directly using the chi-square difference test. The deviance of the simpler model $(D_1)$, which has $p_1$ degrees of freedom, minus the deviance of the more complex model $(D_2)$, which has $p_2$ degrees of freedom $(p_2 < p_1)$, provides the change in deviance $(\Delta D = D_1 - D_2)$. As the number of parameters in a model increases, the deviance value decreases. In sufficiently large samples, the difference between the deviances of two hierarchically nested models is distributed as an approximate chi-square distribution with degrees of freedom equal to the difference in the number of parameters being estimated between the two models (e.g., de Leeuw, 2004).

In evaluating model fit using the chi-square difference test, the more parsimonious model is preferred, as long as it does not result in statistically significantly worse fit. In other words, if the model with the larger number of parameters fails to reduce the deviance by a substantial amount, the more parsimonious model is retained. However, when the change in deviance $(\Delta D)$ exceeds the critical value of chi-square with $p_2 - p_1$ degrees of freedom, then the additional parameters have resulted in statistically significantly improved model fit. In this scenario, the more complex model (i.e., with $p_1$ degrees of freedom) is favored.

In ML, the number of reported parameters includes the fixed effects (the $\gamma$ terms) as well as the variance/covariance components. When using REML, the number of reported parameters includes only the variance and covariance components. To compare two nested models that differ in their fixed effects, it is necessary to use ML estimation, not REML estimation. REML only allows for comparison of models that differ in terms of their random effects but have the same fixed effects. Because most programs use REML as the default method of estimation, it is important to remember to select ML estimation to use the deviance estimates to compare two nested models with different fixed effects.

## 17. Predictive Ability of the Model

In single-level regression models, an important determinant of the utility of the model is the proportion of variance explained by the model, or $R^2$. Unfortunately, there is no exact multilevel analog to the proportion of variance explained. Variance components exist at each level of the multilevel model; therefore, variance can be accounted for at each level of the multilevel model. In addition, in random coefficients models, the relation between an independent variable at level 1 and the dependent variable can vary as a function of the level-2 unit or cluster. Consequently, there is no constant proportion of variance in the dependent variable that is explained by the independent variable. Instead, the variance in the dependent variable that is explained by the independent variable varies by cluster. Finally, because the variance components are estimated using ML estimation, the estimation of the variance can differ slightly from model to model. Therefore, it is impossible to compute an $R^2$ value for the entire model. However, both Raudenbush and Bryk (2002) and Snijders and Bosker (1999) have proposed multilevel analogs to $R^2$. In both cases, the authors provided two separate formulas: one to explain variance at level-1 and another to explain variance at level-2.

Perhaps the most common statistic used to estimate the variance explained is the *proportional reduction in variance* statistic (Raudenbush & Bryk, 2002). The proportional reduction in variance can be estimated for any variance component in the model. This statistic compares the variance in the more parameterized model to the variance in a simpler baseline model. To compute the proportional reduction in variance, subtract the remaining variance within the more parameterized model from the variance within a baseline model. Then divide this difference by the variance within the baseline model. That statistic is computed

$$\frac{\hat{\sigma}_b^2 - \hat{\sigma}_f^2}{\hat{\sigma}_b^2}. \tag{10}$$

where $\hat{\sigma}_b^2$ is the estimated level-1 variance for the baseline model and $\hat{\sigma}_f^2$ is the estimated level-1 variance for the fitted model (Raudenbush & Bryk, 2002). At level-2, population variance components estimates are represented by $\hat{\tau}_{qq}$ and are given for the intercepts ($\beta_{0j}$) and each slope estimate ($\beta_{1j}, \beta_{2j}, \ldots, \beta_{qj}$) that is allowed to randomly vary across clusters. The proportional reduction in the variance of a given slope, $\beta_{qj}$, is

$$\frac{\hat{\tau}_{qq_b} - \hat{\tau}_{qq_f}}{\hat{\tau}_{qq_b}} \tag{11}$$

where $\hat{\tau}_{qq_b}$ is the estimated variance of slope $q$ in the base model and $\hat{\tau}_{qq_f}$ is the estimated variance of slope $q$ in the fitted model.

It should be noted, however, that the proportion reduction in variance statistic does not behave like the familiar $R^2$. First, the proportional reduction in variance statistic proposed by Raudenbush and Bryk (2002) represents a comparison of one model to another model, and as such it cannot be interpreted as an explanation of the absolute amount of variance in the dependent variable. In addition, the proportion reduction in variance statistic can be negative. This actually happens with some regularity when comparing the level-2 intercept variance of a completely null model (a random effects ANOVA model which includes no predictors at level-1 or level-2) to the level-2 intercept variance of a model that includes a group mean centered predictor at level 1.

The second method of deriving a multilevel $R^2$ type statistic (Snijders & Bosker, 1994, 1999) produces measures of *proportional reduction in prediction error* for level 1 (the prediction of $Y_{ij}$) and level 2 (the prediction of $\bar{Y}_{.j}$). These statistics are only available for models that include random intercepts but not for random coefficients models, which include randomly varying slopes. Like the proportional reduction in variance static presented above, the proportional reduction in prediction error for level 1 (the prediction of $Y_{ij}$) compares the amount of residual variance in the more parameterized model to a simpler baseline model. However, this formula uses the total estimated variance, $\hat{\sigma}^2 + \hat{\tau}_{00}$, to compare the two models. The rationale is that $\hat{\sigma}^2 + \hat{\tau}_{00}$ provides a reasonable estimate of the total sample variance of the outcome variable $Y$ (Snijders & Bosker, 1994). Because $\hat{\sigma}^2 + \hat{\tau}_{00}$ is being used as a proxy for the total variance in the dependent variable, this formula is only appropriate for models without randomly varying slopes. Given a random intercepts only model, the prediction error for individual outcomes ($Y_{ij}$) is equal to the sum of the level-1 and level-2 variance components, $\hat{\sigma}^2 + \hat{\tau}_{00}$.

The proportional reduction of prediction error at level 1 compares the total residual variance of a fitted (or more parameterized) model, f, to that of a baseline (or less parameterized) model, b. The formula for $R_1^2$ is

$$R_1^2 = 1 - \frac{(\hat{\sigma}^2 + \hat{\tau}_{00})_f}{(\hat{\sigma}^2 + \hat{\tau}_{00})_b} \tag{12}$$

Where the fraction's numerator is the prediction error for the fitted model and the fraction's denominator is the prediction error for the baseline model.

With respect to level 2, Snijders and Bosker's (1999) explained proportion of variance at level 2 is the proportional reduction in the mean squared prediction error for the cluster mean "$\bar{Y}_{.j}$ for a randomly drawn level-two unit $j$" (p. 103). The prediction error for the group mean is

$$\frac{\hat{\sigma}^2}{n_j} + \hat{\tau}_{00}. \tag{13}$$

Thus, the level-2 proportional reduction in the prediction error, $R_2^2$, is

$$R_2^2 = 1 - \frac{\left(\dfrac{\hat{\sigma}^2}{n_j} + \hat{\tau}_{00}\right)_f}{\left(\dfrac{\hat{\sigma}^2}{n_j} + \hat{\tau}_{00}\right)_b}. \tag{14}$$

Where the fraction's numerator is the prediction error variance for the fitted model and the fraction's denominator is the prediction error variance for the baseline model. In this case, $\bar{n}_j$ is a representative value for average group size.

The various multilevel $R^2$-type statistics described above provide heuristics to compare models in terms of their ability to "explain variance." However, it is important to remember their shortcomings. First, these estimates do not provide unequivocal estimates of the variance explained by a model. Instead, they compare two models in terms of their ability to reduce some type of variance at one of the levels of the hierarchy. Second, when a model contains random slopes, $R^2$ does not have a unique definition (Hox, 2002; Kreft, deLeeuw, & Aiken, 1995). The relation between the level-1 predictor and the dependent variable varies across level-2 units, and the level-2 variance estimate is not constant in these models (Snijders & Bosker, 1999). Finally, these statistics can produce negative estimates, which effectively demonstrates that they are not actually proportions of variance explained. However, even given these shortcomings, multilevel $R^2$ analogs do help researchers to compare predictive ability of various multilevel models. Therefore, they should be reported within the Results section of a multilevel paper. When reporting their $R^2$ results, researchers should be sure to specify whether they used Raudenbush and Bryk's (2002) or Snijders and Bosker's (1999) method to compute these proportional reduction in variance estimates, and they also should clearly specify which model they used as the baseline model and which model they used as the fitted (or more parameterized model) for each of their computations.

## 18. Effect Size

As with any statistical analyses, it is important to report effect size measures for multilevel models. The $R^2$ analogs described above can help researchers and readers to determine the impact that a variable or a set of variables has on a model. In addition, researchers can compute Cohen's $d$-type effect sizes to describe the mean differences among groups (see Chapter 7, this volume). To calculate the equivalent of Cohen's $d$ for a group-randomized study (where the treatment variable occurs at level-2), use the following formula:

$$\delta = \frac{\hat{\gamma}_{01}}{\sqrt{\hat{\sigma}^2 + \hat{\tau}_{00}}} \tag{15}$$

(Spybrook, Raudenbush, Liu, Congdon, & Martinez 2006). Assuming the two groups have been coded as 0/1 or $-.5/+.5$, the numerator of the formula represents the difference between the treatment and control groups. The denominator utilizes the $\sigma^2$ and $\tau_{00}$ from the unconditional model. In the unconditional model, the total variance in the dependent variable is divided into two components: the between-cluster variance, $\tau_{00}$, and the within-cluster variance, $\gamma_{01}$.

To facilitate understanding among readers, researchers should consider including figures that illustrate cross-level interactions among variables. Just as plotting same level interactions facilitates an understanding of interaction effects (Aiken & West, 1991), similar visual graphics of interactions between two variables at different levels of the data hierarchy can effectively display cross-level moderation. In addition, researchers should include predicted values for prototypical participants. These predicted values also can help the reader to make sense of the magnitude of the effects that are being reported. Thus, they serve as a form of "unstandardized" effect size.

## 19. Causal Claims

Multilevel modeling solves certain statistical issues that arise from non-independent or clustered data, and it allows for more nuanced analyses of variables that occur at different levels of the hierarchy. However, any causal claims that can be made from a multilevel analysis are determined by the strength of the research design. As Kelloway (1995) stated, "No amount of sophisticated analyses can strengthen the inference obtainable from a weak design" (p. 216). It is common to refer to "effects" in multilevel modeling. In fact, the entire lexicon of the technique is replete with references to fixed effects, random effects, cross-level interaction effects, and so forth. However, none of these "effects" imply causation. When writing the Results and Discussion sections of a multilevel article, researchers should choose their language carefully so as not to imply causal claims that cannot be substantiated or defended given the design of the study.

## References

Aiken, L. S., & West, S. G. (1991). *Multiple regression: Testing and interpreting interactions*. Newbury Park, CA: Sage.

Biesanz, J. C., Deeb-Sossa, N., Papadakis, A. A., Bollen, K. A., & Curran, P. J. (2004). The role of coding time in estimating and interpreting growth curve models. *Psychological Methods, 9*, 30–52.

Black, A. C. (2008). *Maximum likelihood estimation and multiple imputation: A Monte Carlo comparison of modern missing data techniques for multilevel data*. (Unpublished doctoral dissertation). University of Connecticut. Storrs, CT.

Cohen, J., Cohen, P., West, S. G., & Aiken, L. S. (2003). *Applied multiple regression/correlation analysis for the behavioral sciences* (3rd ed.). Mahwah, NJ: Erlbaum.

de Leeuw, J. (2004). Multilevel analysis: Techniques and applications (Book review). *Journal of Educational Measurement, 41*, 73–77.

Dempster, A. P., Laird, N. M., & Rubin, D. B. (1977). Maximum likelihood from incomplete data via the EM algorithm. *Journal of the Royal Statistical Society, Series B, 39*, 1–38.

Enders, C. K. (2001). A primer on maximum likelihood algorithms available for use with missing data. *Structural Equation Modeling: A Multidisciplinary Journal, 8*, 128–141.

Enders, C. K., & Tofighi, D. (2007). Centering predictor variables in cross-sectional multilevel models: A new look at an old issue. *Psychological Methods, 12*, 121–138.

Ferron, J. M., Hogarty, K. Y., Dedrick, R. F., Hess, M. R., Niles, J. D., & Kromrey, J. D. (2008). Reporting results from multilevel analyses. In A. A. O'Connell & D. B. McCoach (Eds.) *Multilevel modeling of educational data* (pp. 391–426). Charlotte, NC: Information Age Publishing.

Fitzmaurice, G., Laird, N., & Ware, J. (2004). *Applied longitudinal analysis*. Hoboken, NJ: Wiley.

Hedeker, D., & Gibbons, R. D. (2006). *Longitudinal data analysis*. Hoboken, NJ: Wiley.

Hox, J. J. (2002). *Multilevel analysis: Techniques and applications*. Mahwah, NJ: Erlbaum.

Kelloway, E. K. (1995). Structural equation modeling in perspective. *Journal of Organizational Behavior, 16*, 215–224.

Kline, R. B. (1998). *Principles and practice of structural equation modeling*. New York: Guilford Press.

Kreft, I. G. G., de Leeuw, J., & Aiken, L. S. (1995). The effect of different forms of centering in hierarchical linear models. *Multivariate Behavioral Research, 30*, 1–21.

Maas, C. J. M., & Hox, J. J. (2005). Sufficient sample sizes for multilevel modeling. methodology. *European Journal of Research Methods for the Behavioral and Social Sciences, 1*, 86–92.

O'Connell, A. A., Goldstein, J., Rogers, H. J., & Peng, C. Y. J. (2008). Multilevel logistic models for dichotomous and ordinal data. In A. A. O'Connell & D. B. McCoach (Eds.) *Multilevel modeling of educational data* (pp. 199–244). Charlotte, NC: Information Age Publishing.

O'Connell, A. A., & McCoach, D. B. (2008). *Multilevel modeling of educational data*. Charlotte, NC: Information Age Publishing.

Osborne, J., & Waters, E. (2002). Four assumptions of multiple regression that researchers should always test. *Practical Assessment, Research & Evaluation, 8*(2). Retrieved May 15, 2008 from http://PAREonline.net/getvn.asp?v=8&n=2 .

Raudenbush, S., & Bryk, A. (2002). *Hierarchical linear models* (2nd ed.). Newbury Park, CA: Sage.

Raudenbush, S. W., Bryk, A. S., Cheong, Y. F., & Congdon, R. (2000). *HLM 5: Hierarchical linear and non-linear modeling.* Chicago, IL: Scientific Software International.

Raudenbush, S., Bryk, A., Cheong, Y. F., Congdon, R., & du Toit, M. (2004). *HLM6: Hierarchical linear and non-linear modeling.* Lincolnwood, IL: Scientific Software International.

Roberts, J. K., & McLeod, P. (2008). Software options for multilevel models. In A. A. O'Connell & D. B. McCoach (Eds.) *Multilevel modeling of educational data* (pp. 427–467). Charlotte, NC: Information Age Publishing.

Rubin, D. B. (1987). *Multiple imputation for nonresponse in surveys.* New York: Wiley & Sons.

Rubin, D. B. (1996). Multiple imputation after 18+ years. *Journal of the American Statistical Association, 91*, 473–489.

Snijders, T. A.B. (2005) Fixed and random effects. In B. S. Everitt and D. C. Howell (Eds.), *Encyclopedia of statistics in behavioral science.* Vol. 2, 664–665. Wiley.

Spybrook, J., Raudenbush, S. W., Liu, X., Congdon, R., & Martinez, A. (2006). *Optimal design for longitudinal and multilevel research. V1.77* [Computer software]. Retrieved May 20, 2008 from http://sitemaker.umich.edu/group-based.

Singer, J. D., & Willett, J. B. (2003). *Applied longitudinal data analysis: Modeling change and event occurrence.* New York: Oxford.

Snijders, T., & Bosker, R. (1994). Modeled variance in two-level models. *Sociological Methods & Research, 22*, 342–363.

Snijders, T., & Bosker, R. (1999). *Multilevel analysis.* Thousand Oaks, CA: Sage.

Stapleton, L. M., & Thomas, S. L. (2008). Sources and issues in the use of national datasets for pedagogy and research. In A. A. O'Connell & D. B. McCoach (Eds.), *Multilevel analysis of educational data* (pp. 11–57). Charlotte, NC: Information Age Publishing.

Wolfinger, R. D. (1996). Heterogeneous variance covariance structures for repeated measures. *Journal of Agricultural, Biological, and Environmental Statistics, 1*, 205–230.

# 11
# Interrater Reliability and Agreement

**William T. Hoyt**

When researchers make use of observer ratings (where *observer* may refer to acquaintances such as peers, family members, and teachers, or to trained observers not previously acquainted with the research participants), they provide evidence of dependability or *replicability* of ratings by reporting coefficients of reliability (for continuous scores) or agreement (for categorical ratings). Many methods have been recommended for quantifying dependability of ratings, and investigators (for whom this task is often only a peripheral issue) might not be aware of well-documented limitations of some of these approaches. Interrater reliability (for continuous rating scales) is best quantified as an *intraclass correlation coefficient* (ICC). Shrout and Fleiss (1979) provided a primer on the different types of ICCs and how to choose among them. For interrater agreement (for nominal scales) Cohen's (1960) *kappa coefficient* is recommended when there are exactly two raters, or Fleiss's (1971) extension for three or more raters. Tinsley and Weiss (1975) offered a helpful introduction to reliability and agreement, including critiques of inferior approaches to estimation. Hoyt and Melby (1999; see also Lakes & Hoyt, 2009) noted that multiple sources of error (e.g., instability of scores over time, internal inconsistency of rating scales, as well as rater variance) contribute to unreliability of ratings, and researchers may find it useful to report generalizability coefficients (Brennan, 2001; Shavelson & Webb, 1991) as a means of quantifying dependability with respect to multiple sources of error simultaneously. Schmidt and Hunter (1996; see also Schmidt, Le, & Ilies, 2003) offered a helpful discussion of the impact of measurement error on study findings, and the importance of reporting coefficients that reflect the relevant sources of error in scores. Hoyt (2000; see also Hoyt & Kerns, 1999) discussed issues for interpretation of findings in the presence of rater errors. Feldt and Brennan (1989) provided a technical treatment of the relation between reliability and generalizability coefficients.

I consider three possible contexts in which interrater reliability should be reported (Contexts A, B, C in Table 11.1). Investigators who are using established rating scales (Context A) will wish to report the dependability of scores for their own sample. Investigators who create a rating scale for a unique purpose (e.g., coding open-ended responses from participants; coding studies in a meta-analysis; Context B) will likewise wish to provide evidence that similar ratings would have been obtained using a different set of coders. Finally, investigators who are developing a new rating scale intended to be used in future substantive inquiries (Context C) should provide more detailed conceptual and psychometric information, to assist future users of this scale.

Table 11.1 Desiderata for Interrater Reliability and Agreement

| Desideratum | Context A | B | C | Manuscript Section(s)* |
|---|:---:|:---:|:---:|---|
| 1. Construct is clearly defined, with theory-based predictions related to establishment of reliability and validity of the measure. | | | • | I |
| 2. Justification is provided for use of ratings as a source of data, with reference to the nature of the construct and past research in the area. | • | • | • | I, M |
| 3. Procedures for generating items (or other rating instructions) are described. | | • | • | M |
| 4. Reports of interrater reliability in previous research include brief description of both targets and rater characteristics (including training procedures) in that study. | • | | | M |
| 5. Rater selection and training procedures for the current study are described. | • | • | • | M |
| 6. Procedures for computing coefficients of interrater reliability or agreement are clearly described. | • | • | • | M |
| 7. Reported interrater reliability (or generalizability) coefficients are congruent with the rating design (number of raters per target; raters crossed or not crossed with targets) used to produce the scores to be analyzed in Results section. | • | • | • | M, R |
| 8. Dependability of ratings is appropriately considered in interpretation of findings. | • | • | • | R, D |
| 9. Dependability of ratings based on current study leads to recommendations for appropriate use of scale in future research (e.g., rating design, rater training). | | | • | R, D |
| 10. Dependability of ratings is considered in discussing study limitations and suggestions for future research. Authors acknowledge that interrater errors are one of multiple sources of error that may affect ratings data. | • | • | • | D |

*Note*: Context A refers to investigations using established rating scales; Context B refers to "rough and ready" rating scales created uniquely for a particular study (e.g., coding systems for studies included in a meta-analysis); Context C refers to studies designed to establish reliability and validity for a new rating measure that will be used in future substantive research.

* I = Introduction, M = Methods, R = Results, and D = Discussion

## 1. Definition of Construct (Context C)

It is always desirable for investigators to provide clear definitions of latent constructs intended to be represented by measured variables, and theoretical linkages among these constructs that lead to the research hypotheses for the study. This consideration is particularly important in instrument development studies (Context C), because the nature of the construct determines the type of validity and reliability evidence that should be sought. For example, when a construct is conceptualized as a trait, it is expected to be relatively stable over time, so that high coefficients of stability will be one characteristic of a valid measure, whereas valid measures of psychological states can be expected to have more modest stability coefficients. Thus, a careful analysis of the nature of the latent construct provides a rationale for choosing among types of evidence that bear on the validity of the scale. In the language of generalizability theory (see Chapter 9, this volume), this analysis assists the investigator in identifying the *facets of measurement* (i.e., the types of error) relevant to assessment of dependability of measurement.

Of special concern for developers of rating scales are questions about the level of inference required to judge a target's standing on this construct, and the consistency with which this construct will be embodied in target behaviors over time and across situations. Constructs requiring low levels of inference (e.g., smiling; talking time in a conversation) can usually be reliably rated by a single rater, whereas higher level constructs (e.g., shyness; manipulativeness) will likely yield lower levels of consensus among raters. For such constructs, reliable ratings may be obtained by computing the mean of ratings provided by multiple raters, and an important question concerns the number of raters that will be necessary to provide scores that dependably reflect the construct.

Behavioral consistency varies from construct to construct, and may also vary from person to person for a given psychological construct. For example, consensus on target extraversion tends to be substantial even after brief acquaintance with the target person, and may be adequate even among raters who view the target person in different settings. So extraversion is a medium-inference construct (i.e., there are observable behaviors that many raters agree constitute cues about a person's level of extraversion), and people may be relatively consistent about the level of extraversion they display in different situations. Many psychological constructs, however, are likely to be context-sensitive, so that observation in multiple situations would be important for validity of scores. Relatedly, behavioral indicators of some constructs (e.g., cheating) have relatively low base rates, so that valid ratings may not be obtainable without protracted periods of observation.

## 2. Ratings as a Source of Data (Contexts A, B, and C)

For ratings to be a source of valid information on psychological states (or traits), these states (or traits) must be reflected in observable behaviors to which the raters will have access. Investigators should provide a theory-based rationale for the types of behavioral cues to which observers have access as a basis for judgments about the construct(s) of interest. These theoretical linkages between observable behaviors and psychological characteristics can assist readers to evaluate the face validity of the rating system. For low-inference rating scales, justification of ratings as a valid data source may include theoretical explanations for typical behavioral cues (e.g., talking time in group, for ratings of extraversion) to which raters will be referred in evaluating participants' status on the rating dimension of interest. For high-inference rating scales, researchers rely on raters' global judgments' of participants' status. A rationale for the use of high-inference rating scales might include theoretical considerations (e.g., an evolutionary argument that people are attuned to social cues signaling important interpersonal dimensions such as dominance and affiliation) and reference to past empirical findings that attest to the accuracy of these social perceptions.

## 3. Item Generation (Contexts B and C)

Investigators creating a new coding system should describe how the coding instructions were created, with attention to how these operationalize the behavior-construct linkages noted in Desideratum 2. When creating a rating scale for future substantive use (Context C), investigators might want to attend in more detail to content validity of the proposed scale, perhaps including an evaluation of proposed scale items for relevance and domain coverage by experts in the area, pilot testing, or similar procedures.

## 4. Previously Obtained Reliability or Agreement Estimates (Context A)

Users of existing rating scales (Context A) should report evidence of interrater reliability (for continuous scales) or agreement (for nominal scales) from past studies. (Evidence of the validity of scores on the rating scale should also be reported, as available.) Reports of dependability of ratings in past studies should be accompanied by a description of the populations under study, characteristics of the raters, and training of the raters. When multiple reliability or agreement estimates are available, it is desirable to select one from a study as similar as possible to the present rating context on these three dimensions.

One challenge that can arise regarding this desideratum is that previous users of the measure might have reported indicators of reliability or agreement that are not optimal estimators of rater consensus.

For example, past users of a nominal rating scale might have quantified rater agreement as percentage agreement (rather than reporting Cohen's kappa, which corrects for agreements expected due to chance). In such a case, investigators might wish to include a caveat about the limitations of available past data. To address comparability of current rater training procedures with those of the previous study, one could then report both percentage agreement and the kappa coefficient for the ratings made in the present study.

## 5. Rater Selection and Training (Contexts A, B, and C)

By describing salient characteristics of those selected as raters, and also procedures for training those raters, investigators provide important information for future potential users of the rating scale, as well as some indication of the generalizability of these ratings. Raters in some studies are selected because of their previous acquaintance with the target person (e.g., spouse ratings, parent ratings of children), whereas in other studies raters are initially unacquainted with targets. This feature has important implications for the rating design (discussed in Desideratum 6). Unacquainted observers might be selected for their special expertise with the construct being rated (e.g., experienced clinicians versus trained undergraduate research assistants as raters of participant diagnostic category). Description of training should include approximate hours of training and brief description of training procedures (e.g., rating of standardized stimuli, criterion for determining levels of accuracy or consensus sufficient for involvement with actual study data).

## 6. Estimating Dependability of Ratings (Contexts A, B, and C)

Investigators should select an index of agreement (e.g., kappa, reflecting the proportion of agreement among raters corrected for chance) or reliability (e.g., the intraclass correlation coefficient, or ICC, representing the proportion of consensual variance in scores) that is appropriate to the rating scale. Coefficients of agreement are appropriate for nominal rating scales but are generally less useful for ordinal or interval scales. Coefficients of reliability are appropriate for interval or ratio scales, and are also recommended for use with ordinal scales as long as there is no reason to believe that the scale grossly violates the assumption of equal intervals between scale points. Users of interval, ratio, and most ordinal scales also have a second set of options available for quantifying dependability of ratings by reporting one or more generalizability coefficients (see Desideratum 7).

Depending on the measure being evaluated, investigators will make some decisions about how to compute a coefficient of reliability or agreement. For example, there might be a choice about the level of reliability analysis (i.e., reliability of item scores, subscale scores, or full scale scores). For some measures, scores of a single rater are expected to demonstrate adequate reliability, and only a subset of targets will be coded by a second rater for purposes of demonstrating this reliability in sample data. The research report should indicate what choices the investigator has made about these and other issues that will affect the interpretation of the reliability or agreement coefficients. In general, it is essential to document the dependability of scores (or categorical ratings) that will actually be used in the analyses in the Results section, and computational procedures should be selected with this consideration in mind (see Desideratum 7).

When computing interrater reliability coefficients, a potentially confusing issue concerns whether the mean squares used to compute the ICC are derived from a one-way ANOVA (with targets as the lone factor) or a two-way ANOVA (including targets and raters as factors, as well as the target × rater interaction). As Shrout and Fleiss (1979) noted, the choice of models depends on the rating design. When raters are nested within targets, the one-way ANOVA should be used. For example, if each participant is rated on trustworthiness by three acquaintances, then each target has a unique set of raters (a nested rating design), and the ICC should be computed with mean squares derived from the one-way ANOVA.

In a crossed rating design (raters crossed with targets), one set of raters evaluates all targets in the data set, and a two-way ANOVA should be used to compute the ICC. For example, if a set of five trained observers rates videotapes of family interactions on dimensions of cooperation and hostility, then each target is evaluated by the same set of raters (a crossed rating design), and the ICC should be computed from mean squares derived from a two-way ANOVA. When raters are crossed with targets, but the one-way ANOVA is used to derive mean squares for the ICC, reliability is generally underestimated.

Some rating studies use a "mixed" rating design. For example, the investigator might have 10 trained raters available, but will randomly assign two of these to each target. This is not a nested design, because each rater judges multiple targets, but it is not crossed either, because different targets are judged by different sets of raters. If Shrout and Fleiss's (1979) formulas are used to determine inter-rater reliability in a mixed rating design, a one-way ANOVA should be used (similar to the nested design).

Finally, the interrater reliability coefficient estimates the proportion of observed score variance that is attributable to targets, rather than to biases of the rater(s) used in the study. It is not sufficient to report a different type of reliability coefficient (e.g., coefficient alpha, which assesses error variance attributable to differential interpretation of items) that tells nothing about rater-based errors. Further, to compute the ICC, it is necessary to have multiple targets in the reliability study (so that there will be target variance). Although novel procedures have occasionally been recommended for computing reliability or agreement of ratings for a single target person (e.g., Cicchetti, Showalter, & Rosenheck, 1997; James, Demaree, & Wolf, 1984), these coefficients are not comparable to those that examine agreement across multiple targets (see Schmidt & Hunter, 1989 for a detailed explanation). It is the latter type of coefficient that is preferred for almost all research applications in the social and behavioral sciences.

To summarize, there are many methods to compute coefficients reflecting interrater agreement and especially interrater reliability. For readers to be able to interpret reported coefficients, it is desirable that investigators be specific about how the data were derived (full sample or subsample, rating design, model for deriving mean squares and computing the coefficient). Although such a description conveys a lot of useful information, it can usually be relatively brief. An example is provided in the next section (Desideratum 7).

## 7. Dependability Coefficients and Rating Design (Contexts A, B, and C)

The material in this section applies to coefficients of reliability, and also to coefficients of generalizability. I discuss options for computing reliability coefficients first, then briefly present an argument for considering generalizability coefficients as useful global summaries of dependability of ratings. I then present some recommendations for computing and reporting generalizability coefficients for scores involving ratings. Further discussion of generalizability theory may be found in Chapter 9, this volume.

*Intraclass correlation coefficient (ICC).* Table 11.2 lists the six different ICCs discussed by Shrout and Fleiss (1979). The choice of the correct coefficient depends upon three questions. First, is the rating design nested (Case 1) or crossed (Cases 2 or 3)? Second, if the rating design is crossed, does rater variance count as error (Case 2) or not (Case 3)? Third, considering the actual scores to be analyzed (which will not necessarily be the same as those produced for the reliability study), will they be based on judgments of a single rater per target ($n'_r = 1$) or on $k$ raters per target ($n'_r = k$)?

The first question, which concerns the distinction between nested and crossed rating designs, is discussed in Desideratum 6. The second question (when raters are crossed with targets) asks whether

Table 11.2  Six Intraclass Correlation Coefficients and Comparable (One-facet) Generalizability Coefficients

| | $n'_r$ | ICC (Shrout & Fleiss, 1979) | Generalizability Coefficient |
|---|---|---|---|
| Case 1: raters nested within targets (mean squares derived from one-way ANOVA). Use also for mixed design. | Single rater per target $(n'_r = 1)$ | $ICC(1,1)$ $= \dfrac{BMS - WMS}{BMS + (k-1)WMS}$ | $\hat{E}\rho^2 = \dfrac{\hat{\sigma}_p^2}{\hat{\sigma}_p^2 + \hat{\sigma}_{r,pr,e}^2}$ $= \dfrac{MS_p - MS_{r,pr,e}}{MS_p + (n_r-1)MS_{r,pr,e}}$ |
| | Multiple raters per target $(n'_r = k)$ | $ICC(1,k)$ $= \dfrac{BMS - WMS}{BMS}$ | $\hat{E}\rho^2 = \dfrac{\hat{\sigma}_p^2}{\hat{\sigma}_p^2 + \hat{\sigma}_{r,pr,e}^2/n'_r}$ $= \dfrac{MS_p - MS_{r,pr,e}}{MS_p}$ |
| Case 2: raters ("random") crossed with targets (mean squares derived from two-way ANOVA) | Single rater per target $(n'_r = 1)$ | $ICC(2,1)$ $= \dfrac{BMS - EMS}{BMS + (k-1)EMS + k(JMS - EMS)/n}$ | $\hat{E}\rho_{abs}^2 = \dfrac{\hat{\sigma}_p^2}{\hat{\sigma}_p^2 + \hat{\sigma}_r^2 + \hat{\sigma}_{pr,e}^2}$ $= \dfrac{MS_p - MS_{pr,e}}{MS_p + (n_r-1)MS_{pr,e} + n_r(MS_r - MS_{pr,e})/n_p}$ |
| | Multiple raters per target $(n'_r = k)$ | $ICC(2,k)$ $= \dfrac{BMS - EMS}{BMS + (JMS - EMS)/n}$ | $\hat{E}\rho_{abs}^2 = \dfrac{\hat{\sigma}_p^2}{\hat{\sigma}_p^2 + \hat{\sigma}_r^2/n'_r + \hat{\sigma}_{pr,e}^2/n'_r}$ $= \dfrac{MS_p - MS_{pr,e}}{MS_p + (MS_r - MS_{pr,e})/n_p}$ |

| | | |
|---|---|---|
| Case 3: raters ("fixed") crossed with targets (mean squares derived from two-way ANOVA) | Single rater per target $(n'_r = 1)$ | $ICC(3, 1)$ $$\hat{E}\rho^2_{rel} = \frac{\hat{\sigma}^2_p}{\hat{\sigma}^2_p + \hat{\sigma}^2_{pr,e}}$$ $$= \frac{MS_p - MS_{pr,e}}{MS_p + (n_r - 1)MS_{pr,e}}$$ $$= \frac{BMS - EMS}{BMS + (k-1)EMS}$$ |
| | Multiple raters per target $(n'_r = k)$ | $ICC(3, k)$ $$\hat{E}\rho^2_{rel} = \frac{\hat{\sigma}^2_p}{\hat{\sigma}^2_p + \hat{\sigma}^2_{pr,e}/n'_r}$$ $$= \frac{MS_p - MS_{pr,e}}{MS_p}$$ $$= \frac{BMS - EMS}{BMS}$$ |

*Note:* For one-way ANOVA (Case 1), $BMS = MS_p$ = mean square between persons (targets); $WMS = MS_{r,pr,e}$ = mean square error (confounds variance attributable to raters, person × rater interaction, and error). For two-way ANOVA, $BMS = MS_p$ = mean square between persons (targets); $JMS = MS_r$ = mean square for raters (judges); $EMS = MS_{pr,e}$ = mean square error (confounds variance attributable to person × rater interaction and error). For the ICC, $k$ represents the number of raters per target used to estimate the ICC, whereas in G studies, this quantity is notated $n_r$ (so $n_r = k$ for all formulas); $n$ represents the number of targets used to estimate the ICC, whereas in G studies this quantity is notated $n_p$ (so $n = n_p$ for all formulas); $n'_r$ represents the number of raters whose scores are actually aggregated to determine participant scores, which may be different from $k$. Shrout & Fleiss (1979) provided formulas for computing the ICC for two possible values of $n'_r$: 1 and $k$. $\hat{E}\rho^2$ is a common notation for the generalizability coefficient, which represents the estimated expected value of the squared correlation between observed scores and universe (a.k.a. "true") scores. When raters are crossed with persons (targets), investigators need to decide whether rater variance ($\hat{\sigma}^2_r$) counts as error (Case 2) or not (Case 3). Cronbach et al. (1972) suggested that Case 2 coefficients ($\hat{E}\rho^2_{abs}$) apply when decisions will be based on absolute scores—e.g., if ratings determine scores on a qualifying examination on which examinees must exceed a predetermined cutoff score to pass. Case 3 coefficients ($\hat{E}\rho^2_{rel}$) apply when decisions will be based on relative scores (i.e., when the relative standing is important, but adding or subtracting a constant from all scores would not change the decision)—e.g., when scores will be correlated with scores on another variable to determine the degree of association between constructs.

rater variance should count as error. Shrout and Fleiss (1979) related this decision to the issue of whether raters are treated as fixed (Case 3; rater variance does not count as error) or random (Case 2; rater variance counts as error).

In generalizability theory, this decision is framed in terms of the type of decision (absolute or relative) that will be based on the scores. An *absolute* decision will be based on the actual score, such as when examinees must score above a predetermined cutoff in order to pass a qualifying examination. In this case (or in any other case in which obtained ratings will be compared to ratings derived from other sets of raters or to criterion scores), rater variance counts as error, and the ICC should be computed using Shrout and Fleiss's Case 2. A *relative* decision will be based on the participant's relative standing within the sample rather than the actual score. For example, when scores derived from the ratings will be correlated with scores on another variable, only the relative standing is relevant. (The correlation coefficient would not be altered if a constant value were added to all scores in the data set.) For relative decisions, rater variance does not count as error, and Shrout and Fleiss's Case 3 should be used to compute the ICC. I return to the question of whether to count rater variance as error below, as part of the discussion of generalizability approaches.

Finally, for constructs involving little or no inference, basing each target's score on the judgment provided by a single rater might yield scores that are adequately reliable. To estimate the reliability of these scores, it will of course be necessary to have at least a subset of targets evaluated by two or more raters. Nonetheless, if the scores to be analyzed in the study are based on only a single rater, then the relevant reliability coefficient is the estimate based on $n'_r = 1$: for example, ICC(3,1) in Table 11.2. For constructs that involve some inference on the part of raters, bias variance will be larger, and it is likely that scores based on a single rater will be relatively unreliable. To achieve adequate reliability, it will be necessary to aggregate scores across raters, using the mean or the sum of scores from multiple raters in the analysis. The reliability of this composite score will be greater than that of scores based on judgments of a single rater. Assuming that the interrater reliability analysis is conducted on data from all the raters whose data contribute to the composite scores (i.e., that $n'_r = k$), then Shrout and Fleiss's coefficient for multiple raters—e.g., ICC(3,2)—gives the correct reliability estimate.

When raters are crossed with targets, the RELIABILITY procedure in SPSS can be used to compute ICC(3,1) or ICC(3,$k$). The data file should be set up so that each row contains data from a single target, with ratings from the $k$ raters (which must be assigned different variable names) occupying $k$ columns. Using the pull-down menus, select Analysis → Scale → Reliability Analysis. The variable names corresponding to the ratings must be entered into the "Items" box, and "Intraclass correlation coefficient" must be checked as an option under "Statistics." For Case 1 and Case 2 coefficients, it is necessary to run the appropriate ANOVA model and compute the ICC from the mean squares provided in the output (see formulas in Table 11.2).

*Advantages of generalizability coefficients.* The reason for reporting an interrater reliability coefficient is to inform readers (and to remind the investigator) of the proportion of replicable variance in the scores derived from the ratings, which shows the likelihood that an independent set of raters, evaluating the same target behaviors, would arrive at a similar relative ranking of the targets (ICC Cases 1 and 3) or at similar absolute scores for all targets (ICC Case 2). The complement of the reliability coefficient $(1 - ICC)$ quantifies the proportion of measurement error, defined as rater error plus random error, in the scores used in the analysis.

In classical test theory, replicable variance was conceptualized as true score (valid) variance, so that the reliability coefficient estimated the proportion of score variance attributable to actual differences between targets on the construct of interest. Because at least some of the variance in observed scores is error variance, the correlation between these scores and other variables will be *attenuated* relative to (hypothetical) correlations between error-free measures of the same constructs (assuming independence of error terms for the measures of these constructs). Thus, the coefficient of reliability has important implications for interpretation of findings, in that it provides a means to estimate the magnitude of attenuation in effect size estimates (e.g., correlation coefficients, regression coefficients) derived from scores based on ratings. In many research contexts, it is useful to report effect sizes corrected for attenuation, to enhance comparability of these findings with effect sizes derived from other studies using different measurement methods.

By the 1940s, psychometricians were in widespread agreement that the idealized interpretation of reliability coefficients as estimates of the proportion of true score variance in scores represented a simplification, because of the difficulty of finding a replicated measure that is truly *parallel* (meaning that all covariance between replicated scores is attributable to true scores) in the classical sense. In practice, replicable variance in a reliability analysis usually includes variance attributable to one or more sources of error that were not varied in the reliability study.

To illustrate this principle for ratings measures, consider the case of a team of observers who rate videotaped family interactions, where the ratings will be used to derive a score on hostility for each family. The ICC for these hostility scores quantifies the extent to which raters who observe the same set of family behaviors agree in their judgments of hostility (i.e., the generalizability over raters of scores based on the same videotape), which provides an estimate of the effects of rater error on hostility scores. However, there are other sources of error that might contribute to the generalizability of ratings, depending on the use that will be made of these scores. For example, we could ask how scores derived from these videotaped interactions would compare with scores from other videotapes at a different time, or in a different setting (i.e., generalizability over occasions or settings). We might also be curious about errors attributable to the wording of items to which the raters responded (i.e., generalizability over items on the rating measure).

Because these other sources of error (occasions, settings, items) were held constant in our study of interrater reliability, any variance attributable to these sources is *replicable* variance in these studies. The ICC estimates the proportion of replicable variance in ratings, but this replicable variance is not all *true score* variance (i.e., valid variance between families in their levels of hostility). If the goal is to obtain a dependability coefficient that estimates the proportion of valid variance in ratings (and therefore can be used to estimate the degree of attenuation in effect sizes derived from these scores), then the investigator needs to think carefully about which sources of measurement error are important in this rating design. If the research hypothesis concerns the general level of hostility in family interactions, then it is important to know whether scores obtained from a given set of raters, using the specific items on the rating scale as applied to a videotaped family interaction on a single occasion, generalize to scores obtained from different raters, using different but still relevant items, applied to family interactions on different (but presumably temporally proximate) occasions.

Generalizability theory (GT; Cronbach, Gleser, Nanda, & Rajaratnam, 1972; see Chapter 9, this volume) is a set of techniques for simultaneously evaluating generalizability (dependability) of measurement across multiple sources of error, and yields generalizability (G) coefficients that are interpreted as ICCs representing the proportion of valid (i.e., generalizable) variance in ratings that would be expected in a number of possible future rating designs. Because multiple sources of error are likely to be a concern for users of observer ratings, GT may be preferable to the standard approaches for determining interrater reliability. G studies yield dependability coefficients that take multiple sources of error into account, and can provide important guidance for improving dependability in future research applications, as well as understanding the likely impact of measurement error in attenuating study findings.

*Computing and reporting G coefficients.* G studies are designed to estimate the proportion of variance attributable to differences among targets, and also variance attributable to other *facets* (sources) that contribute to variability in observed scores. The investigator determines which facets (usually sources of error variance) are of interest for a given measure. Typically, facets such as raters, items, and occasions all contribute to non-target (i.e., error) variance in scores. In general, interactions between these facets and the targets (e.g., target × rater interactions), along with random error, are the largest sources of error variance, although facet main effects may also be important for some rating designs.

Data for the G study often are collected using a crossed design. For example, each target could be judged by all raters, using the same set of items, on several occasions. The first step in data analysis is *variance partitioning*, in which observed score variance is partitioned into components attributable to targets, facet main effects, and all possible interactions. (The highest-level interaction component is always confounded with random error in a G study.) Providing a table of variance components is very helpful for readers, as it allows them to compute G coefficients relevant to a variety of different rating designs that could be used in future investigations. For example, a researcher might be interested in the consequences for score dependability of using a nested rating design, or of including more or fewer levels of some facet in a future study.

Once the variance components have been estimated, G coefficients can be computed for the scores that will be analyzed in the present study. Because many different G coefficients can be estimated from the same set of variance components (depending on how scores are computed), it is generally helpful to provide the formula used in computing this coefficient. In general, G coefficients are computed as a ratio of target variance [usually notated $\hat{\sigma}_p^2$ or var($p$)] to total variance. The total variance is computed as a sum of variance components that contribute to observed scores for that rating design, usually with multipliers to correct for aggregation (when scores are computed as means or sums across multiple levels of some or all facets of measurement). Which variance components appear in the denominator of the G coefficient, and which multipliers are used, will depend on how the scores are computed from the available ratings.

For many applications, the data from the study conducted to compute the G coefficient (the *G study*) are also the data that will be analyzed to test substantive hypotheses (the decision study, or *D study*). But sometimes investigators conduct a separate G study (perhaps on a subset of cases using a larger number of raters with a crossed design) to establish dependability for a different rating design (e.g., a smaller number of raters, and not crossed with targets) to be used in the D study. In such cases, the G coefficient for the G study rating design will be different (and higher) than that for the D study. Investigators should report the G coefficient for the rating design used in the D study (i.e., the estimate of dependability for the scores that will actually be analyzed), and should describe the computational procedures and justify them based on this design. For researchers who are creating rating measures to be used in future D studies (Context C in Table 11.1), GT is a particularly attractive approach, as they can use the variance components from the G study to forecast G coefficients for a variety of possible rating designs that might be used by future investigators. It can be helpful to tabulate the predicted coefficients for different designs (e.g., varying the number of raters and occasions) to illustrate how these choices on the part of future scale users are likely to impact dependability of scores.

*Notational issues: ICCs and G coefficients.* One challenge for researchers interested in using GT may be unfamiliarity with the standard notation developed by Cronbach et al. (1972) to elucidate this framework. A helpful reference point for gaining comfort with GT notation is provided in Table 11.2, where formulas derived from basic (one-facet) G study designs are compared to the equivalent formulas in the more familiar ANOVA notation employed by Shrout and Fleiss (1979). All of these coefficients are derived from one-facet G studies, because raters ($r$) are the only source of error that is varied in the G study. For example, Shrout and Fleiss's Case 1 corresponds to a G study where raters are nested within targets. In this design, var($r$), var($pr$), and var($e$) are confounded and cannot be decomposed into separate variance estimates. This is indicated by denoting the error component as var($r,pr,e$) or $\hat{\sigma}_{r,\,pr,\,e}^2$. The G coefficient for a single rater is computed as the ratio var($p$)/[var($p$) + var($r,pr,e$)], and the expression for this ratio in terms of the mean squares from the one-way ANOVA is identical to that provided by Shrout and Fleiss, except for notational differences.

When scores are computed as aggregates based on judgments of multiple raters, the contribution of rater variance to observed scores is reduced. This increases the relative contribution of target variance to observed score variance, which means an increase in dependability of measurement. This is represented in the G coefficient for multiple raters by a multiplier $(1/n'_r)$ for variance components involving raters, where $n'_r$ is the number of raters for each target in the D study. This procedure can be used to forecast the dependability of ratings for any value of $n'_r$. Table 11.2 uses the value of $n'_r = k$ (i.e., the number of raters in the D study will be the same as that for the G study), and shows how this ratio using mean squares is again identical to that provided by Shrout and Fleiss (1979), except for notational differences.

G coefficients for Shrout and Fleiss's (1979) Cases 2 and 3 are based on a crossed rating design. In this design, rater variance [var($r$), or $\hat{\sigma}_r^2$] can be estimated separate from the target × rater interaction [which is confounded with random error: var($pr,e$) or $\hat{\sigma}_{pr,e}^2$]. A GT perspective sheds some light on the choice between Cases 2 and 3, for studies with the crossed design. As Shrout and Fleiss pointed out, the choice between Cases 2 and 3 focuses on whether rater variance counts as error. The two main uses of Case 2 coefficients [which include var($r$) in the denominator of the G coefficient] are (a) when decisions for the study will be based on absolute scores (such as comparing observed ratings to a pre-established cutoff score) and (b) when raters are crossed with targets in the G study, but the investigator wishes to provide an estimated G coefficient for a future (D) study using a nested rating design. Thus, when investigators are mainly interested in providing an estimate of dependability for the current study (as in Contexts A and B from Table 11.1), it is usually appropriate to exclude var($r$) from the

Table 11.3  Recommendations for Reporting ICCs and G Coefficients

---

### ICCs

1. Note whether the rating design used to compute the ICC was nested (each target is judged by a different set of raters), crossed (all targets are judged by the same set of raters) or mixed.

2. If using Shrout and Fleiss's (1979) formulas (see Table 11.2), be sure to use the correct ANOVA model—one-way ANOVA for nested designs, two-way for crossed. SPSS RELIABILITY can also be used for crossed designs (Case 3).

3. If raters are crossed with targets, determine whether Case 2 (rater variance included in error term) or Case 3 (rater variance not included) is appropriate.

4. Report the ICC that reflects the level of aggregation for the scores you will be analyzing. For example, ICC(3,1) if scores are based on judgments of a single rater for each target, or ICC(3,$k$) if scores are composites based on $k$ raters.

### G Coefficients

1. Determine which facets of measurement contribute to error variance, based on the intended application of scores. For many rating scales, scores are sought that generalize over raters, items, and observation occasions.

2. In the G study, vary each of these facets (e.g., each target could be judged by multiple raters, using a set of multiple relevant items, on several occasions).

3. Tabulate the results of the variance partitioning, so that readers will have access to variance estimates for facet main effects and interactions.

4. Report a G coefficient appropriate for the rating design used in the D study (actual data to be analyzed); this may be a different design than that used in the G study. Specify which sources (facet main effects and interactions) contribute to the total variance estimate for this coefficient, and the level of aggregation (e.g., number of raters per target) for each facet.

5. If the rating measure is being created for use in future substantive (D) studies, consider tabulating estimated G coefficients for possible future rating designs (e.g., crossed versus nested, and different levels of aggregation for the facets), to inform future users about the likely effects of these choices on score dependability.

denominator of the G coefficient [i.e., to report ICC(3,$k$)], because var($r$) does not contribute to variability observed scores in a crossed rating design.

For investigators who are developing a new rating measure for use in future (D) studies (Case C in Table 11.1), it may be helpful to report both the Case 2 and Case 3 coefficients (either G coefficients or ICCs), so that future users of the scale will have an indication of what the consequences will be for dependability of measurement if a nested, rather than a crossed rating design is used. When raters are nested within targets, var($r$) contributes to error variance, which generally results in some decrease in dependability of measurement. The difference between the Case 2 and Case 3 coefficients shows the likely magnitude of this decrease.

*Summary.* Because there are many types of ICCs (see Table 11.2), investigators using this method for reporting interrater reliability need to be explicit about which coefficient they are reporting and why. Users of rating measures should be aware that other sources of variance (e.g., items, occasions) contribute to error of measurement in many applications. Thus, they might opt to report a G coefficient from a G study including several facets of measurement that may contribute to measurement error, to give a clearer indication of the impact of measurement error on the findings to be reported. For investigators developing new rating measures, this approach might be particularly attractive, because it allows for tabulation of predicted G coefficients under a number of different future measurement designs. Table 11.3 provides a summary of recommendations for studies reporting either ICCs or G coefficients as indices of interrater reliability.

## 8. Interpretation of Findings (Contexts A, B, and C)

Methodologists have offered rules of thumb for describing the degree of agreement or reliability, based on the magnitude of the obtained coefficients. Fleiss (1981, p. 218), following Landis and Koch (1977, p. 165) recommended that kappa coefficients in the ranges .75 and higher, .40 to .75, and below .40 be characterized as evidence of excellent, fair-to-good, and poor agreement, respectively. Typically, reliability coefficients of .80 and above have been considered to reflect good dependability of scores, with coefficients between .70 and .80 reflecting marginal dependability. Recall, however, that because multiple sources (e.g., raters, items, observation occasions) contribute to error in ratings, conventional reliability coefficients overestimate the dependability of ratings (see Desideratum 7). The rules of thumb just discussed can be useful in encouraging consistency in the labels applied to coefficients by different investigators, but ultimately the importance of these coefficients lies in what they reveal about the effects of error of measurement on the effect sizes obtained in the study.

The effects of error of measurement on study findings are complex. In general, however, when predictor and criterion variables are measured with less than perfect reliability, effect sizes involving these variables are attenuated (i.e., are smaller than they would have been under hypothetical error-free conditions). The smaller the dependability coefficient, the greater is the expected degree of effect size attenuation. Thus, interpretation of observed effect sizes should take account of the reliability of the measures that contributed to those effect sizes. This may be particularly important for comparison with effect sizes using different measurement procedures. For example, a psychotherapy process researcher may compare working alliance scores during the third session of brief psychotherapy with outcome measured at the end of treatment. If working alliance is measured via client reports, and outcome via judgments of trained interviewers, then error of measurement in each measure acts to attenuate the observed effect size. Suppose that this effect size is compared with that of an earlier study that used client report measures for both constructs and found to be smaller. It is important to ask whether this represents a substantive difference in findings, or whether it may be explained by the fact that measurement errors for the two constructs are uncorrelated in the present study, but correlated (i.e., method covariance between the two sets of client reports) in the earlier investigation. Thus, the present study may be viewed as stronger evidence for the hypothesis (despite its smaller effect size) because method covariance can be ruled out as a plausible alternative explanation for the observed correlation.

One option for quantifying the impact of measurement error is to publish effect sizes corrected for unreliability of measurement. As noted by Schmidt and Hunter (1996), it is important for investigators who use this correction to be sure that it is based on a dependability coefficient that quantifies the proportion of replicable variance taking all relevant sources of error into account (see Desiderata 7 and 10).

### 9. Recommendations for Future Users (Context C)

For investigators who are creating a new rating measure, it is important to provide guidance for future researchers about how to use the scale. Unlike self-report questionnaires, which are relatively standardized, rating measures offer multiple options for future users regarding the number of raters who will provide judgments of each target, the assignment of raters to targets (i.e., nested versus crossed design), and the procedures used to train raters. As noted above (see Desideratum 7), for continuous scales, a generalizability approach (reporting variance estimates and G coefficients for multiple possible rating designs) offers investigators maximum flexibility to provide guidance for future users.

### 10. Limitations and Suggestions (Contexts A, B, and C)

Finally, when reliability or agreement is weak or marginal, this fact should be noted as a limitation of the study in the Discussion section. Low dependability of measurement indicates that rather different scores would probably have been obtained under different rating conditions (and especially, in the case of interrater reliability or agreement, if a different set of raters had been used). As noted in Desideratum 8, the likely consequence of unreliability in a continuous predictor or criterion variable is attenuation in the obtained effect size, so it would be expected that a replication with more reliable measures would show a stronger association. However, the impact of measurement error in other (e.g., statistical control) variables in the equation is not straightforward to predict, so that weak measurement procedures generally reduce confidence in the obtained results.

For continuous measures, investigators who report interrrater reliability coefficients should acknowledge that these do not reflect the contribution of other sources of error that are relevant to evaluating dependability of measurement (see discussion of GT in Desideratum 7). Unless a generalizability analysis is undertaken, it is generally not possible to be very precise about the contribution of multiple sources of measurement error to distortion of study findings.

### References

Brennan, R. L. (2001). *Generalizability theory.* New York: Springer Verlag.

Cicchetti, D. V., Showalter, D., & Rosenheck, R. (1997). A new method for assessing interexaminer agreement when multiple ratings are made on a single subject: Applications to the assessment of neuropsychiatric symptomatology. *Psychiatry Research, 72,* 51–63.

Cohen, J. (1960). A coefficient of agreement for nominal scales. *Educational and Psychological Measurement, 20,* 37–46.

Cronbach, L. J., Gleser, G. C., Nanda, A. N., & Rajaratnam, N. (1972). *The dependability of behavioral measurements: Theory of generalizability for scores and profiles.* New York: Wiley.

Feldt, L. S., & Brennan, R. L. (1989). Reliability. In R. L. Linn (Ed.), *Educational Measurement* (3rd ed.) (pp. 105–146). New York: Macmillan.

Fleiss, J. L. (1971). Measuring nominal scale agreement among many raters. *Psychological Bulletin, 76,* 378–382.

Fleiss, J. L. (1981). *Statistical methods for rates and proportions* (2nd ed.). New York: Wiley.

Hoyt, W. T. (2000). Rater bias in psychological research: When is it a problem and what can we do about it? *Psychological Methods, 5,* 64–86.

Hoyt, W. T. & Kerns, M. D. (1999). Magnitude and moderators of bias in observer ratings: A meta-analysis. *Psychological Methods, 4,* 403–424.

Hoyt, W. T., & Melby, J. N. (1999). Dependability of measurement in counseling psychology: An introduction to generalizability theory. *The Counseling Psychologist, 27,* 325–352.

James, R. G., Demaree, R. G., & Wolf, G. (1984). Estimating within-group interrater reliability with and without response bias. *Journal of Applied Psychology, 69,* 85–98.

Lakes, K. D., & Hoyt, W. T. (2009). Applications of generalizability theory to clinical child psychology research. *Journal of Clinical Child and Adolescent Psychology, 38,* 144–165.

Landis, J. R., & Koch, G. G. (1977). The measurement of observer agreement for categorical data. *Biometrics, 33,* 159–174.

Schmidt, F. L., & Hunter, J. E. (1989). Interrater reliability coefficients cannot be computed when only a single stimulus is rated. *Journal of Applied Psychology, 74,* 368–370.

Schmidt, F. L., & Hunter, J. E. (1996). Measurement error in psychological research: Lessons from 26 research scenarios. *Psychological Methods, 1,* 199–223.

Schmidt, F. L., Le, H., & Ilies, R. (2003). Beyond alpha: An empirical examination of the effects of different sources of measurement error on reliability estimates for measures of individual difference constructs. *Psychological Methods, 8,* 206–224.

Shavelson, R. J., & Webb, N. M. (1991). *Generalizability theory: A primer.* Newbury Park, CA: Sage.

Shrout, P. E., & Fleiss, J. L. (1979). Intraclass correlations: Uses in assessing rater reliability. *Psychological Bulletin, 86,* 420–428.

Tinsley, H. E. A., & Weiss, D. J. (1975). Interrater reliability and agreement of subjective judgments. *Journal of Counseling Psychology, 22,* 358–376.

# 12
# Item Response Theory

R. J. De Ayala

Item response theory (IRT) is a psychometric modeling paradigm used for the measurement of psychological constructs that is typically based on continuous latent variables. In its simplest form an IRT model contains a single continuous latent variable that represents the construct of interest (e.g., mathematics proficiency, depression, social anxiety), which in turn is believed to determine persons' responses to a series of binary and/or polytomous questions. Both items and people are located on this continuous latent continuum. In general, a person's location on this latent continuum is estimated as a function of his or her responses to those questions and the items' location on the latent variable. From this basic model one can generalize to multiple latent construct variables and/or multiple item characterizations.

A typical IRT application produces an estimate of a person's location on the latent construct's continuum. However, IRT may also be used for other purposes, such as the design of an instrument with specific properties. For example, an instrument may be designed to provide highly accurate person proficiency estimates across a particular range of the latent continuum. IRT may also be used for creating an item bank for use with computerized adaptive testing or to facilitate creating alternate forms of an instrument. This latter use points toward another purpose of IRT, the equating of alternate forms of an instrument. In the desiderata and explications that follow, I distinguish between item-focused studies (e.g., item bank construction) and person-focused studies (e.g., diagnostic testing).

Although IRT typically requires larger sample sizes than those used for classical test theory (CTT) implementations, given satisfactory model-data fit IRT offers a number of advantages over the traditional CTT approach to measurement. For instance, the IRT person location estimate is not dependent on the specific instrument used for person measurement, the IRT item characterization(s) are independent of the sample of respondents, and the IRT model may be used to predict response behavior. Moreover, unlike CTT's global measure of observed score accuracy (i.e., the standard error of measurement), with IRT one knows the accuracy with which each person's location is estimated.

Because IRT models are nonlinear, obtaining estimates of the latent person and item parameters involves numerically intensive (and typically iterative) algorithms. Various software packages, such as BILOG-MG (Zimowski, Muraki, Mislevy, & Bock, 2003), MULTILOG (Thissen, Chen, & Bock, 2003), NOHARM (Fraser & McDonald, 2003), PARSCALE (Muraki & Bock, 2003), and WINSTEPS

Table 12.1 Desiderata for Item Response Theory

| Desideratum | Manuscript Section(s)* |
|---|---|
| 1. The construct of interest is defined. | I |
| 2. IRT is justified as the appropriate measurement approach (e.g., continuous latent variable vs. categorical latent variable). | I |
| 3. The specific model(s), with description and justification, are provided. | I, M |
| 4. The response data are fully described, including sampling, sample size(s), demographics, and testing environment (if appropriate). | M |
| 5. All instruments are fully described, including length, response format, and validity evidence (if appropriate). | M |
| 6. Software and estimation approach(es) are fully specified. | M |
| 7. Estimation problems are documented, as are details as to how they were addressed. | R, D |
| 8. A complete description is provided of how missing data were addressed. | R, D |
| *Item-focused studies* (linking, item bank construction, instrument construction) | |
| 9. Details regarding model fit analysis are provided, including those related to dimensionality, fit statistics, invariance, and model selection (if appropriate). | M, R, D |
| 10. Details regarding item fit analysis are provided, including those related to conditional independence, functional form, fit statistics, invariance, predicted vs. observed item response functions, and handling of misfitting items. | R, D |
| 11. Instrument calibration results are presented (item parameter estimates and/or summary statistics, total information function). | R, D |
| *Person-focused studies* (CAT, diagnosis, equating, vertical scaling) | |
| 12. Person fit analysis results are presented, including fit statistics and appropriateness measurement. | R, D |
| 13. Person location estimate results are described, including relevant standard errors. | R, D |
| 14. Methods of equating scores on different metrics are described in detail. | M, R |

* *Note*: I = Introduction, M = Methods, R = Results, D = Discussion

(Linacre, 2001) are available to provide parameter estimates. Technical treatments of IRT are provided by Baker and Kim (2004), Lord and Novick (1968), and van der Linden and Hambleton (1997). Readable introductions may be found in Hambleton, Swaminathan, and Rogers (1991), as well as De Ayala (2009). The desiderata for IRT applications are presented in Table 12.1 and are explicated in the remainder of this chapter.

## 1. Defining the Construct of Interest

In any application of IRT one needs to define the construct(s) of interest. Because IRT is a latent variable modeling approach the authors should make clear to the reader why they believe that one or more latent variables underlie the observed behavior. In some statistical contexts (e.g., exit polling) it is unnecessary to posit the existence of a latent variable; in other cases, however, convention or theory dictates that one or more latent variables are the most meaningful conceptualization of the research questions at hand. In these cases, the operational definition of the construct should be clearly specified. Stated in other words, the linkage between, for example, a latent variable and its observed manifestations needs to be explicated.

## 2. Appropriateness of IRT

Although a latent variable may be invoked to explain individuals' responses to binary and/or polytomous items, this latent variable may be conceptualized as categorical, continuous, or even some combination of the two. In the case of a categorical conceptualization of the latent variable, then the use of Latent Class Analysis (see Chapter 13, this volume) is the recommended psychometric technique. However, if the latent variable is conceptualized as continuous, then an IRT model is warranted. If the latent variable is believed to have categorical and continuous facets, then a mixture IRT model, such as the mixed Rasch model, may be used. Authors must make clear, by virtue of the hypothesized nature of the construct(s) of interest, that IRT is indeed the appropriate approach.

## 3. Specifying the Model(s)

After presenting the study's context, theory, purpose, and justification of why IRT is an appropriate technique, the researcher should present the IRT model(s) that will be used and a description of the model parameters. Because in some cases there may be multiple IRT models that may be applicable the researcher should justify the models selected and articulate the implication of the selection. To elucidate this statement I will first present a taxonomy of IRT models followed by a brief discussion of the implication(s) of selecting particular models over other models.

IRT models may be classified in multiple ways. One taxonomy uses the type of response to classify the models into two broad categories, those designed solely for dichotomous (binary) response data and those for polytomous response data (e.g., from Likert response scales or rater judgments). A second classification approach reflects differences in intent. Specifically, is the researcher's intent simply to model the data or to gauge if it is possible to construct an instrument in an attempt to measure the construct of interest? This latter purpose is associated with the Rasch family of IRT models, Guttman Scalogram, and Coombs Unfolding. In Table 12.2 these two classification approaches are used to cross-classify several commonly used IRT models. Specifically, the IRT models are classified into whether the researcher's intent is to describe or model the data as oppose to using the model to construct the instrument and whether the model is or is not restricted to dichotomous responses; note this is a non-exhaustive list of IRT models. This grid should not be interpreted as consisting of impermeable cells. For instance, all polytomous models simplify to one of the dichotomous models and all the models listed in the first row may be seen as special cases of the models in the second row. Nevertheless, the taxonomy is a useful organizational scheme and presents two questions that need to be answered in an IRT application: (1) is the researcher primarily concerned with modeling the data

Table 12.2  Example IRT Models

| Intent | Response Type | |
|---|---|---|
| | Dichotomous | Polytomous |
| Construct | Rasch/One-parameter model | Rating Scale model |
| | Linear Logistic Test Model (LLTM) | Partial Credit model |
| | Mixed Rasch model | |
| Describe Data | Two-parameter model | Generalized Rating Scale model |
| | Three-parameter model | Generalized Partial Credit model |
| | Multidimensional two-parameter model | Graded Response model |
| | | Nominal Response model |
| | Multidimensional three-parameter model | Multiple-choice model |

(i.e., the second row) or in using the model to determine whether or not it is possible to construct an instrument to measure the latent variable (i.e., the first row), and (2) what type or types of response data will the researcher be working with.

If the researcher adopts the perspective that the model determines whether it is possible to measure the construct (i.e., the first row), then there are a number of implications. Specifically, the models in the first row require that all items on an instrument to be approximately equally good at discriminating among respondents located at different points along the latent variable continuum. As a result, these models characterize an item only in terms of its location or, in terms of polytomous data, locations on the latent continuum. Therefore, when applying a Rasch model to empirical data one *might* have greater difficulty in obtaining model-data fit for some of the items because these items discriminate differently than other items and/or because of chance success on the item.

Another implication of using a model in the Rasch family involves parameter estimation. In general, with Rasch family models it is possible to obtain reasonably good item parameter estimates with smaller sample sizes than would be needed with non-Rasch models. Moreover, because there is less difficulty in estimating the item's location(s) than the other item parameter(s) (which are discussed below), convergence problems are less likely to occur with Rasch family models.

A third implication of using a Rasch family model is that results are comparatively easy to present and explain to a lay audience. For instance, two individuals with the same observed score (i.e., the sum of item responses on an instrument) will obtain the same location estimate on the latent variable. In contrast, with non-Rasch models these two individuals may receive different location estimates depending on their respective response patterns. Some individuals refer to the Rasch model as the one-parameter logistic (1PL) model, whereas others believe that because the Rasch model represents a philosophical approach to measurement not embodied in the 1PL model that the terms should not be interchanged.

The non-Rasch models in the second row may be viewed as focusing on describing the response data. To accomplish this objective these models contain one or more discrimination parameters and, in the case of the three-parameter models, an additional parameter representing the chance (so-called *guessing*) success on an item is called the pseudo-guessing or pseudo-chance parameter. These additional item parameters allow greater flexibility in modeling the data. One commonly seen multi-item parameter model is the three-parameter model (e.g., the three-parameter logistic [3PL] model). In the 3PL model each item is characterized by a discrimination parameter ($\alpha$), a location parameter ($\delta$), and a pseudo-chance parameter ($\chi$). If one constrains $\chi$ to be zero, then the 3PL model simplifies to another common model the two-parameter model (e.g., the two-parameter logistic [2PL] model). As can be seen from Table 12.2 these models may be extended to have multiple latent variables (e.g., the multidimensional two-parameter model).

In an application's write-up the researcher should always present the model and define its parameters rather than relying on the model's name because some models have multiple names. For example, some individuals refer to the *generalized partial credit* model as the *two-parameter partial credit* model. Moreover, presenting the model makes it clear whether a probit or a logit link function is being used. Although the difference in link functions does not affect model-data fit, which link function is used and whether any rescaling is done (e.g., the use of the $D$ scaling constant) affects the latent continuum's metric.

## 4. Describing the Response Data

In the Methods section the researcher should specify how the respondents were selected (i.e., random sampling, convenience sampling, matrix sampling, etc.) along with the sample's demographics and size. Regarding the latter, although there are a number of sample size guidelines (e.g., 100, 500, 1000, depending on the model, estimation method, and assumption tenability), these should not be interpreted as hard and fast rules. This is due, in part, to the number of the factors that need to be considered and are at times data specific; these include a researcher's model-data misfit tolerance (i.e., acceptable level of risk of failing to reject an inappropriate model), ancillary technique sample size requirements (e.g., principal component analysis), the amount of missing data, the model, the application context, the use of prior distributions for estimation as well as the estimation algorithm. An additional factor may be the number of items being calibrated. For instance, the joint maximum likelihood estimation procedure provides more accurate estimates when used with more than 25 items than with fewer items. As may be surmised, from this example some of these factors interact with one another. For instance, the greater the amount of missing data the larger the overall sample size needs to be in order to compensate for the missing data. In general, it behooves the researcher to provide a rationale for the sample size used.

Depending on the study's context the environment in which the instrument is administered should also be described. This includes any time constraints, administration medium (paper-and-pencil, computerized, etc.), and whether the administration was to an individual or a group.

## 5. Describing the Instrument(s)

Each instrument's purpose, the number of items, as well as any available validity information should be presented. Because IRT models do not require a particular item response format (e.g., multiple-choice, open-ended, ratings) it is customary to specify how the response data arose. For instance, if responses to open-ended questions are coded, a description of the rubric and an example of its application should be provided. In addition, if multiple raters are using the rubric to score the same item, then the researcher should provide a description of the rater training process, an assessment of inter-rater reliability, and how rating contradictions are resolved. If a Likert response scale is used, then the category labels should be specified along with whether items were reversed scored due to negative wording.

Although it is common to consider the dichotomous models as only applicable for proficiency assessment, the models are applicable for dichotomous data regardless of whether data represent correct/incorrect responses. For example, a response of 1 may be a correct response on an examination question, the successful completion of a task, a rater's judgment, or a response of TRUE on an attitude or personality item using a TRUE/FALSE response format. As such, the phrase "a response of 1" will be used herein rather than "correct response" to emphasize the generalizability of the models.

With some instruments multiple IRT models may be used to estimate the parameters (e.g., a dichotomous model for some items and the partial credit model for other items). In these cases the researcher should specify which items are calibrated by each model. Similarly, if an instrument contains subscales, then how the multiple subscales are treated should be presented. As an example, if a unidimensional model is used with an instrument containing multiple scales, then in general each subscale should be separately calibrated. However, if the subscales are highly interrelated, then it might not be appropriate to treat each subscale individually. The researcher should make clear the approach that was used and why.

## 6. Estimation Approach(es)

Several estimation algorithms are available to estimate item and person parameters. The researcher should specify the estimation approach for items and, if necessary, for respondents. Furthermore, the researcher needs to indicate whether the estimation used the program's default approach or whether the defaults were changed and, if so, what the changes were.

Some of the estimation approaches make assumptions of the data, whereas others do not make these assumptions but work best in certain situations. Moreover, because certain estimation programs allow the user to select from various estimation approaches one cannot simply rely on the estimation program's name to convey information about the estimation algorithm used. Therefore, both the program name and the estimation technique used need to be specified. For instance, three commonly used approaches are *joint maximum likelihood estimation* (JMLE), *conditional maximum likelihood estimation* (CMLE), and *marginal maximum likelihood estimation* (MMLE). These latter two approaches are available in the estimation program OPLM (Verhelst, Glas, & Verstralen, 1995). A fourth estimation strategy uses unweighted least squares to fit a polynomial that approximates a two-parameter normal ogive model (i.e., a two-parameter model using a probit link function). This strategy is typically used for estimating item parameters for a dichotomous multidimensional model, although it may also be used with dichotomous unidimensional models.

Although IRT itself does not make any distribution assumptions, MMLE makes an assumption about the respondent distribution. Typically, this assumption is that the respondents come from a normally distributed population. In contrast, JMLE does not make assumptions about the respondents' distribution. It should be noted, however, that JMLE has known inherent weaknesses that are exacerbated under certain conditions (e.g., short instruments and small samples) and, as a consequence, may encounter estimation difficulties with certain data sets and certain IRT models.

Some of the estimation strategies, including MMLE and unweighted least squares, only provide item parameter estimates. Therefore, if one needs estimates of the respondents' locations it would be necessary to perform a second step. Some of the commonly available person estimation approaches are *maximum likelihood estimation* (MLE) and Bayesian estimation such as *maximum a posteriori* (MAP) or *expected a posteriori* (EAP). The Bayesian approaches make a respondent distribution assumption and, as a result, the degree of regression of the person location estimate toward the mean will, in part, be dependent on the (prior) distribution's parameters. However, unlike using MLE for estimating an individual's location the Bayesian approaches will provide a person location estimate for each individual. Therefore, the estimated person locations will differ across the estimation procedures (although they will be highly linearly related with one another). Again, because some programs (e.g., BILOG-MG) provide multiple person estimation algorithms, simply specifying the program's name would be insufficient to inform the reader of which person estimation approach was used. As a result, both the estimation program and the estimation algorithm used for persons should be made clear in the Methods section.

The decision of which algorithm to use depends on a number of factors, including the model being used, the instrument length, available calibration software. For instance, CMLE is only an option with the Rasch family of models. With other models one could use JMLE, MMLE, or unweighted least squares. However, if one's instrument is 10 items long, then JMLE would not be the best approach to use because research has shown that instruments should be at least 25 items long. Moreover, when performing JMLE calibrations for the two- and three-parameter models there should be at least 1000 respondents to reduce bias in the parameter estimates. For the Rasch model family this bias can be ameliorated with certain JMLE programs. In general, most calibrations for non-Rasch family models are currently being performed using MMLE.

## 7. Addressing Estimation Problems

Estimation problems are more likely to occur with models that include a discrimination parameter and/or a pseudo-guessing parameter. Therefore, the following will apply primarily to non-Rasch models and the use of MMLE.

In some situations it is possible to experience difficulty in estimating an item's discrimination parameter ($\alpha$). For example, for some items the estimate of $\alpha$ may drift off to infinity; this is sometimes referred to as an example of a *Heywood case*. There are several approaches that one might use to aid in estimating an item's discrimination. One approach uses a prior distribution (e.g., a log-normal distribution) for the estimation of $\alpha$. Although, in general, the use of a prior distribution produces estimates that may be regressed toward the prior distribution's mean, the use of a prior distribution with discrimination parameter estimation has a less serious impact than in the case of person and item location parameters. Unless otherwise specified by the user, BILOG imposes a log-normal prior distribution when estimating $\alpha$ for the two- and three-parameter models. An alternative strategy is to impose an upper limit on the values that the estimated discrimination parameters may take on. This is the approach used in some JMLE programs.

Difficulty in estimating an item's lower asymptote or pseudo-guessing parameter ($\chi$) may occur because of (1) problems in estimating an item's other parameters, (2) because of the other parameters' estimates (e.g., a low estimated $\alpha$ parameter), (3) because the item is located at the lower end of the continuum (i.e., a very easy test item), and/or (4) because there are insufficient data at the lower end of the continuum with which to estimate an item's $\chi$. In this latter case, there may be several different combinations of an item's parameters that produce item response functions (IRFs) whose lower asymptotes are similar to one another even though the item parameter estimates may be vastly different from one another.

Different approaches for handling these estimation difficulties involve using a prior distribution or fixing $\chi$ to a specific value. In this latter case, this constant (common) value for $\chi$ may be set arbitrarily (e.g., $1/m - 0.05$, where $m$ is the number of item options), by averaging the non-problematic $\chi$ estimates, by averaging the $\chi$ estimates for items located at the lower end of the continuum (i.e., the easy items), or by fixing the lower asymptote to some nonzero value determined by inspecting the lower asymptote of empirical item response function. In addition, a "stability" criterion may be invoked to determine whether an item's $\chi$ parameter should be estimated at all (or assigned a constant value). That is, $\chi$ is estimated only when $\delta - 2/\alpha > -2.5$, where $\delta$ is the item location.

As mentioned above, we may also use a prior distribution for estimating the items' $\chi$ parameters. The use of a prior distribution for estimation of $\chi$ can lead to reasonable parameter estimates for the model. Moreover, the regression toward the mean phenomenon that typically occurs when using a prior distribution is not as problematic in estimating $\chi$ as it is when estimating person and item location parameters. The use of a prior distribution is recommended as the first strategy to facilitate the estimation of $\chi$. By default BILOG uses a beta distribution as a prior for estimating the IRF's lower asymptote, $\chi$, for the three-parameter model.

With polytomous data it is possible that one or more of an item's response categories may not be attractive and may never be chosen by respondents. These are sometimes referred to as *null categories*. In general, it is not possible to estimate the parameters for a category that does not have any observations. However, some software packages may provide "estimates" for the null category. If a null category occurs, then one should ignore the null category's parameter estimates and re-calibrate the item set specifying the appropriate number of categories actually observed for each item.

The preceding has been concerned with item parameter estimation problems. However, it is possible to experience problems when estimating person locations. If this occurs it is typically associated with using MLE. Although it is possible with MLE to obtain nonfinite person location estimates with poorly behaved likelihood functions, the most common problem to occur with binary response strings is an infinite person location estimate due to either responses of all 1s or responses of all 0s. There are several strategies that may be used to perform MLE with response strings consisting of all 1s (i.e., perfect response strings) or all 0s (i.e., zero response strings). The gist of these strategies is to modify the response strings to introduce some nonuniformity. One approach, the "half-item rule," assigns 0.5 to the item with the smallest location value for a uniformly 0 response vector and to the item with the largest location value for a uniformly 1 response vector. For example, assuming five items in increasing order of their locations, then a zero response string would become [0.5 0 0 0 0] and a perfect response string would be [1 1 1 1 0.5]. The strategy used for addressing perfect response or zero response strings should be specified in the write-up of the study. In contrast and as mentioned above, with EAP and MAP, person location estimates are available for all response vectors.

## 8. Missing Data

IRT is concerned with modeling *observed* responses. However, in working with empirical data one will, at times, encounter situations where some items do not have responses from all individuals in the calibration sample. Some of these missing data situations may be considered to be missing *by design* or *structurally missing*. For example, one may administer an instrument to one group of people and an alternate form of the instrument to another group. If these two forms have some items in common, then the calibration sample can consist of both groups. As a result, the data contain individuals who have not responded to all the items on both forms. In situations where the nonresponses are missing by design, these missing data may be ignored because of the IRT properties of person and item parameter invariance. However, when nonresponses are not structurally missing, then one needs to consider how to treat these nonresponses.

In general, missing data (e.g., omitted responses) may be classified in terms of the mechanism that generated the missing values: *missing completely at random* (MCAR), *missing at random* (MAR), and *nonignorable*. MCAR refers to data in which the missing values are statistically independent to the values that could have been observed as well as to other variables. In contrast, when data are MAR then the missing values are conditionally independent on one or more variable(s). Nonignorable missing values are data for which the probability of omission is related to what the response would be if the person had responded.

In the IRT context there are various reasons why an individual's response vector might not contain responses to each item. For instance and as mentioned above, in the missing by design case one has *not-presented* items. This missingness due to not-presented items arises in, for example, adaptive testing or the simultaneous calibration of multiple forms. These nonresponses represent conditions in which the missingness may be ignored for purposes of person location estimation. Therefore, the estimation is based only on the observed responses.

A second situation that will produce missing data occurs when an individual has insufficient time to answer the item(s). These *not-reached* items are (typically) identified as collectively occurring at the end of an instrument (this assumes the individual responds to the test items in a serial fashion) and represent *speededness*. Although IRT should be applied to unspeeded tests, if we knew which items the examinee did not have time to consider, then these not-reached items could be ignored for person location estimation because they contain no readily quantifiable information about the individual's proficiency (Lord, 1980). Therefore, when one has (some) missing data due to not-reached items, then the person's location can be estimated using only the observed responses. This should not be interpreted as indicating that IRT should be applied to speeded instruments, nor that the item parameter estimates for the not-reached item(s) are unaffected by being speeded. In fact, speeded situations may lead to violation of the unidimensionality assumption and biased item parameter estimates. For example, research has shown that the speeded items' $\alpha$ and $\delta$ parameters are overestimated and the $\chi$ parameters are underestimated. Because of the $\alpha$ overestimation the corresponding item information, and therefore the instrument's total information, are inflated. Identifying the speeded items as not-reached mitigates the bias in item parameter estimation.

The third situation that will produce missing data occurs when an examinee intentionally chooses not to respond to an item for which he or she does not have an answer. These *omitted responses* represent nonignorable missing data. Again, assuming that an individual responds in a serial fashion to an instrument, omitted responses may be distinguished from not-reached items by appearing throughout the response vector and not just at its end. Omitted responses may not be ignored because individuals could obtain, for example, as high a proficiency estimate as they wished by simply answering only those items they had confidence in correctly answering.

Research has shown that with dichotomous data omits should not be treated as responses of 0 nor should they be ignored. In these cases using a fractional value of 0.5 in place of omitted values leads to improved person location estimation compared to treating the omits as responses of 0 or using some other fractional value. By using this fractional value one is simply imputing a "response" for a binomial variable and thereby attempting to "smooth irregularities" in the likelihood function. Additionally, SAS's PROC MI or SPSS's MISSING VALUE ANALYSIS—EM may be used to impute values for the omitted responses and the resulting complete data calibrated.

The useful and accuracy of either of these approaches depends on the flexibility of the practioner's calibration software. That is, these approaches will yield decimal values and with some calibration software these values will need to be converted to integers. An alternative approach is to perform hot-decking. That is, for each case with missing data another case is found that is similar in characteristics to the case with the missing value(s), but has responses for the item(s) in question. The responses from this second case are substituted for those in the case with missing values. If one has multiple matching cases, then one selects the matching case at random. A third strategy that may be fruitful in some situations is to treat the omission as its own response category and apply a polytomous model such as the multiple choice or the nominal response models.

There are some issues in the treatment of omits that the practitioner should be aware of. For instance, in the context of proficiency assessment all the imputation procedures that produce "complete" data for analysis are, in effect, potentially giving credit (partial or otherwise) for an omitted response. With the substitution of a fractional value for the omitted response (e.g., 0.5) a second issue is the use of the same imputed value for all omits assumes that individuals located at different points on the latent variable continuum can all be treated the same. These issues are mentioned so that the practitioner understands the assumptions that are being made with some of the missing data approaches discussed above. However, they may or may not be of concern to an individual practitioner. Moreover, when IRT is used in personality testing or with attitude or interest inventories these may be nonissues. A third issue that should also be noted is that omits tend to be associated with personality characteristics and demographic variables as well as proficiency level. As such, in those situations where information on these variable(s) is available one might wish to use this information as covariate(s) in the imputation process; use of these covariate(s) might or might not have any meaningful impact on the person location estimates.

It is good practice when calibrating a data set to identify items without responses by some code. For instance, in the data file not-reached items may be identified by a code of, say, 9, not-presented items by a code of 8, omitted items by a code of 7. With certain calibration programs (e.g., BILOG 3, BILOG-MG, MULTILOG) any ASCII character may be used (e.g., the letters "R" for not-reached, "P" for not-presented, and "O" for omit). For BILOG omitted responses must be identified as such, whereas with other programs (e.g., MULTILOG) any response code encountered in the data file that was not identified as a valid response is considered to reflect an omitted item.

In practice, the user should report the strategy used to address nonignorable omitted responses as well as the rationale for why it was selected. As an aid in deciding on a strategy the practitioner may want to perform the estimation with and without the missing data strategy and compare the results to determine the impact of the strategy. Furthermore, this comparison (e.g., determining the impact of the strategy) may be facilitated by simulating data using the calibration model and the estimated item parameters

## 9. Model Fit

Ignoring the socio-political factors that in practice may determine which model is used, model selection would, in general, involve considering the latent structure (e.g., the number of latent variables, whether the latent variables are continuous and/or categorical), response data characteristics (e.g., dichotomous, polytomous), and the importance and necessity of modeling chance success as well as differential discrimination across items. Therefore, model selection would initially involve each of these factors as well as assessing the tenability of the corresponding assumptions (e.g., a model's dimensionality assumption). However, after selecting one or more models there remains the issue of model-data fit.

If the dimensionality analysis suggests that one or more models based on a single latent variable are appropriate, then one approach to assessing model-data fit is by using the likelihood ratio statistic, $G^2$. A statistically nonsignificant $G^2$ provides evidence supporting that the data are consistent with what a model would predict. In some cases the sample size and the typical instrument length will yield significant $G^2$ values. However, in these situations convention has been to still use $G^2$ to compare (nested) models of varying complexity. (Calibration programs typically produce the log likelihood value that would be used in calculating $G^2$.) For instance, because the one-parameter model is nested within the two-parameter model one may use the $G^2$ statistic to determine if the two-parameter model's varying discrimination parameter is necessary. If the $G^2$ is not statistically significant, then this would indicate that it is not necessary to provide each item with its own discrimination parameter and the simpler one-parameter model would be favored. Conversely, if $G^2$ is statistically significant, then the additional item parameter in the two-parameter model would be necessary. Similarly, the necessity of including a pseudo-guessing parameter for each item (i.e., the three-parameter model) could be assessed by comparing the model's fit with that of models without this item parameter (i.e., the one- or two-parameter models). The degrees of freedom for $G^2$ would be the difference in the number of item parameters between the two models.

With regard to polytomous response data, all the models listed in Table 12.2 (except for the graded response model) may be viewed as nested within the multiple-choice model. As such, the rating scale model is subsumed by the generalized rating scale, partial credit, and generalized partial credit models, the partial credit model is nested within the generalized partial credit model, the generalized partial credit model is nested within the nominal response model, and the nominal response model is nested within the multiple-choice model. Therefore, and as was the case with the dichotomous unidimensional models, the $G^2$ statistic can be used to compare various polytomous models; degrees of freedom associated with $G^2$ would be the difference in the number of item parameters between the compared models.

With multidimensional data the $G^2$ model comparison strategy is currently difficult to implement because the most popular software for parameter estimation does not produce a log likelihood value for the model solution. As a consequence, selecting between the multidimensional two- and three-parameter models would be based on a more heuristic approach. In general, instruments that contain items without a correct answer, such as, those found on a personality scale, should be adequately modeled using the multidimensional two-parameter model because there is no reason to believe that a respondent would guess at a response. However, for instruments that contain items on which chance success is a possibility, then one might use the *unidimensional* three-parameter model to get a sense of the degree to which chance success is evident by examining the $\chi$ parameter estimates. If this degree is very small, then using the multidimensional two-parameter model may be adequate for modeling the data; otherwise, a multidimensional three-parameter model would be called for.

## 10. Item Fit Analysis

Although one might have evidence of model-data fit this does not necessarily mean that one has evidence of fit for each item. Stated another way, if one has item-data fit for each item, then one will have model-data fit; however, the converse does not have to be true. Whether or not one or more misfitting items adversely affect the model-data fit analysis depends on the number of problem items and the instrument's length. For instance, one may have satisfactory overall model fit, but upon further investigation finds that a couple of items exhibit differential performance across the manifest groups of males and females. Therefore, after selecting a model and/or obtaining overall model-data fit one proceeds to perform item level analyses. In addition, in some cases, model-level *misfit* may be diagnosed by examining the item-level fit statistics (e.g., chi-square, INFIT, OUTFIT).

Item level fit analyses should involve both statistical indices and graphical analyses. The statistical analyses depend, in part, on the calibration program because different programs provide different statistics or none at all. The graphical analyses allow an examination of IRT's functional form assumption as well as certain model assumptions (e.g., constant discrimination). In the following I begin with some statistical indices and then proceed to graphical methods.

Most statistical indices are variants of a chi-square statistic. Programs such as WINSTEPS produce two fit statistics, INFIT and OUTFIT, for examining fit; these statistics may also be used for model-data fit analysis. Other programs such as BILOG-MG produce a chi-square fit statistic. I will first discuss INFIT and OUTFIT.

INFIT is a weighted fit statistic based on the squared standardized residual between what is observed and what would be expected on the basis of the model. These squared standardized residuals are information weighted and then summed across observations for the $j$th item; the weight is $p_j(1 - p_j)$ where $p_j$ is the probability of a response of 1 according to, for example, the Rasch model. OUTFIT is also based on the squared standardized residual between what is observed and what would be expected, but the squared standardized residual is not weighted when summed across observations. As such, OUTFIT is an unweighted standardized fit statistic. Both statistics are averaged to produce mean-square statistics. INFIT and OUTFIT each have a range from 0 to infinity with an expectation of 1; their distributions are positively skewed.

These two statistics differ in their sensitivity to where the discrepancy between what is observed and what is expected occurs. For instance, responses by persons located near an item's location estimate that are consistent with what would be expected according to the model produce INFIT values close to 1 (given the stochastic nature of the model). However, responses on items located near the item's location estimate that are inconsistent with what would be expected lead to large INFIT values. In short, INFIT is sensitive to unexpected responses near the item's location. In contrast, OUTFIT has a value close to its expected value of 1 when responses by persons located away from an item's location estimate are consistent with what is predicted by the model (again, given the stochastic nature of the model). However, unexpected responses by persons located away from an item's location estimate lead to OUTFIT values substantially greater than 1. Therefore, OUTFIT is sensitive to, say, a high ability person incorrectly responding to an easy item or a low ability person correctly responding to a hard item.

Although there are various interpretation guidelines, one guideline states that values from 0.5 to 1.5 are "acceptable" with values greater than 2 warranting closer inspection of the associated item. However, using a common cutoff value does not necessarily result in the fit statistic (either INFIT or OUTFIT) having correct Type I error rates. Some have suggested taking sample size ($N$) into account when interpreting INFIT and OUTFIT. Specifically, for INFIT and OUTFIT one would use $1 \pm 2/\sqrt{N}$ and $1 \pm 6/\sqrt{N}$ as cutoff values, respectively. Alternatively, given INFIT and OUTFIT's expectation and their range it is clear that there is an asymmetry in their scales. Therefore, INFIT and OUTFIT can be transformed to have a scale that is symmetric about 0.0. The result of this transformation is a standardized (0, 1) fit statistic, ZSTD. On this metric good fit is indicated by INFIT ZSTD and OUTFIT ZSTD values close to 0. Because the ZSTD values are approximate $t$ statistics, as sample size increases they approach $z$ statistics. As such, values of $\pm2$ are sometimes used for identifying items that warrant further inspection. See Linacre and Wright (2001) and Smith (2004) for more information on the INFIT and OUTFIT statistics and their transformations.

As mentioned above, BILOG-MG produces a chi-square fit statistic. This statistic is unreliable when calculated on instruments with less than 20 items (Zimowski et al., 2003). These chi-square statistics are based on combining individuals into, by default, 9 intervals on the basis of their Bayes location estimates; the user may change the number of intervals. The chi-square statistic tests the null hypothesis that an item's data are consistent with the model. In other words, item fit is associated with statistically nonsignificant chi-square values. As is generally true, failure to reject the null hypothesis does not imply that the model is correct, but only that there is insufficient evidence to reject the model.

In addition to using item fit statistics one should also compare the agreement between the empirical IRF with the model predicted IRF; for polytomous models one would compare the empirical and predicted option response functions (ORFs). This graphical examination should be viewed as a complementary approach to assessing item fit with fit statistics and not a replacement for fit statistics.

What defines agreement between the predicted and empirical IRFs is not absolute. One may use error bands to define reasonable agreement between the two IRFs. That is, if the empirical IRF falls within the error bands, then there is agreement between the empirical and the predicted IRFs (i.e., item-data fit). Because the width of these error bands is a function of the standard error one needs to decide on the number of standard error units that would indicate agreement. For example, the error bands might reflect two standard errors above and below the predicted IRF.

It should be noted that the agreement between the empirical and predicted IRFs is a matter of degree and only informs our judgment of fit. For instance, sometimes we find that the empirical IRF reflects an ogival pattern that shows close agreement with the predicted IRF for a substantial range of the continuum, but disagreement at, for example, the lower end of continuum (say below −2). Depending on the application this lack of fit at and below −2 may not be reason for concern. In short, different situations may be more amenable to or accepting of a certain degree of less than perfect fit. Another consideration is the number of intervals used in creating the empirical IRFs. For example, with a small number of intervals we might observe strong agreement between the empirical and the predicted IRFs, but with a larger number of intervals the degree of agreement is not as strong all other things being equal. Also, in making our judgment of fit we recognize that the choice of say, two, standard errors for defining the error band width is a reasonable, but arbitrary, decision. Again, all of this information is used to inform our judgment of fit along with the context (e.g., the number of items on the instrument, the number of items exhibiting "weak agreement," the number of respondents, the amount of missing data, the purpose of the application, and so forth).

So far we have been concerned with assessing item-level fit. However, we can also use item-level information to assess the conditional independence assumption that all IRT models make. The gist of this assumption is that responses to two items are statistically independent of one another after conditioning on the latent person variable(s). Although there are a number of different statistics that may be used for determining the tenability of this assumption, one simple approach is to use the $Q_3$ statistic. The $Q_3$ statistic is the correlation between the residuals for a pair of items. The residual for an item is the difference between the individuals' observed responses and their corresponding expected responses on the item. Therefore, after fitting the model the Pearson correlation coefficient is used to examine the linear relationship between pairs of residuals. For instance, with dichotomous data the observed response on the $j$th item $(x_{ij})$ is either a 1 or 0 and the expected response is the probability $(p_j)$ given by the IRT model. Symbolically, the residual for person $i$ for item $j$ is $d_{ij} = x_{ij} - p_j(\hat{\theta}_{ii})$ and for item $k$ it is $d_{ik} = x_{ik} - pk(\hat{\theta}_i)$; $\hat{\theta}$ is the person location estimate. The $Q_3$ statistic is the correlation between $d_{ij}$ and $d_{ik}$ across persons (i.e., $Q_{3jk} = r_{djdk}$).

If $|Q_3|$ equals 1.0, then the two items are perfectly dependent. A $Q_3$ of 0.0 is a necessary, but not a sufficient condition for independence because one may obtain a $Q_3 = 0$ because the items in an item pair are either independent of one another or because they exhibit a nonlinear relationship. Therefore, $Q_3$ is useful for identifying items that exhibit item *dependence*. Under conditional independence $Q_3$ should have an expected value of $-1/(L-1)$, where $L$ is the number of items on the instrument.

In some situations one may explain item dependence in terms of multidimensionality. That is, the dependency between two items is due to a common additional latent variable (e.g., test-wiseness in achievement testing). If two items are conditionally independent, then their interrelationship is completely explained by the latent structure of the model. However, it has been shown that if one applies a unidimensional model to bidimensional data, then items that are influenced by both latent variables will show a negative local dependence and items that are affected by only one of the two latent variables will show a positive local dependence. If only one of the latent variables is used, then the items that are influenced only by that underlying variable will show a slight negative local dependence. To obtain a large $Q_3$ value one needs to have similarity of items parameters for the items in question and the items need to share one or more unique dimensions. Therefore, similarity of parameters is a necessary, but not a sufficient condition, for obtaining a large $Q_3$ value. To summarize, if one determines that the various item pairs are exhibiting conditional independence, then one also has evidence support model-data fit.

Another facet of item-level fit analysis involves obtaining evidence of item parameter invariance. In the current context, invariance refers to one or more sets of item parameters that are interchangeable within a linear transformation. Although, theoretically, IRT item parameters are invariant, whether invariance is realized in practice is contingent on the degree of model-data fit. Therefore, obtaining evidence of invariance can be used as part of a model-data fit investigation. The quality of model-data fit may be assessed by randomly dividing the calibration sample into two subsamples. Each of these subsamples is separately calibrated and their item parameter estimates compared to determine their degree of linearity. This comparison can simply involve calculating the correlation coefficients between the subsamples' calibration results. One would have invariance evidence if the correlation coefficients are large (e.g., greater than 0.9). This approach would be applied across items for each of the items' parameters (i.e., a correlation coefficient for item discrimination across subsamples, another for item difficulty, etc.).

The correlation coefficient approach for invariance assessment has its disadvantages. One potential disadvantage is that the coefficient does not identify individual problem item(s). This issue may be addressed by obtaining the scatterplot for each coefficient. Another disadvantage is that it is not possible to simultaneously examine the interaction of the item's parameter estimates because the correlation is calculated for each item parameter. However, more sophisticated approaches for comparing an item's response functions based on each subsample's estimates allow one to simultaneously evaluate an item's multiple parameter estimates.

To simultaneously compare an item's multiple parameter estimates across subsamples one can use some of the differential item functioning (DIF) methods. DIF methods are used to detect items that are functioning differently across manifest groups of individuals. Typically, these manifest groups reflect majority/minority groups (e.g., males and females, African Americans and Caucasian, Hispanics and non Hispanics). However, these manifest groups can also be two randomly created subsamples. Some DIF approaches that could be used for this purpose are the likelihood ratio method of Thissen, Steinberg, and Wainer (1988), Lord's Chi-Square (Lord, 1980), and the Exact Signed Area and H Statistic methods (Raju, 1988, 1990).

## 11. Instrument Calibration

Which calibration results one presents and the completeness of the presentation depends on the study's purpose. In general, it is good practice to describe one's instrument statistically and/or graphically. At a minimum one should provide descriptive statistics for each type of item parameter. For example, one would provide the mean, range, the standard deviation, and so forth for the item parameter estimates. Moreover, one can graphically present the instrument's total (i.e., test) information function to show where on the continuum the instrument is expected to provide the most accurate person location estimates.

When the purpose of the study involves linking different metrics (see Desideratum 14) then the researcher should provide information about the metric transformation coefficients. Similarly, in those cases where person location estimates are transformed to the total score metric then the researcher should provide the total (i.e., test) characteristic function (TCF). In some cases, the item parameter estimates for all the items can be presented in a table either in the body of the paper or in an appendix; the corresponding standard errors should also be provided.

With the Rasch family of models it is possible to present an Item-Person Map (also known as a Variable Map). The Item-Person Map shows how the distributions of respondents and items relate to one another. By comparing the item locations to the person distribution one can obtain an idea of how well the items are functioning in measuring persons' latent trait(s). In short, this graph allows one to see not only how well the respondents' distribution matches the range of the instrument, but also provides an idea of how well the items are distributed across the continuum. Using this information one may anticipate where on the continuum one may experience greater difficulty in estimating person as well as item locations.

## 12. Person Fit Analysis

Analogous to item fit analysis, different calibration programs provide different person fit information. For instance, some of the calibration programs for the Rasch family of models produce the INFIT and OUTFIT statistics (Desideratum 10). In terms of person fit, these statistics are interpreted in a fashion analogous to their use with item fit. Specifically, responses on items located near the person's estimated location, $\hat{\theta}$, that are consistent with what would be expected produce INFIT values close to 1. However, responses on items located near the person's $\hat{\theta}$ that are inconsistent with what would be expected lead to large INFIT values. That is, INFIT is sensitive to unexpected responses near the person's $\hat{\theta}$. In contrast, OUTFIT has a value close to its expected value of 1 when responses on items located away from a person's $\hat{\theta}$ are consistent with what is predicted by the model. Conversely, OUTFIT values substantially greater than 1 arise because of unexpected responses on items located away from a person's $\hat{\theta}$.

Other fit statistics include, but are not limited to, the UB statistic (Smith, 1985) and Klauer and Rettig's (1990) chi-square statistic. The UB statistic may be standardized and a standard normal table can be used to provide screening values that would aid in identifying individuals that warrant further scrutiny. For example, a large UB statistic would indicate a person that is behaving inconsistent with the model (i.e., a misfitting person). The Klauer and Rettig chi-square statistic is asymptotically distributed as a chi-square. Therefore, the standard chi-square table of critical values would be used to identify a misfitting person. Klauer and Rettig have evaluated the significance of their statistic with a $\alpha$ level of 0.10. In general, if one or more persons are found to be misfitting, then at the very least they should be removed from the sample and the response data re-calibrated to determine their impact on the item parameter estimation. If it is determined that misfitting persons have minimal impact on the calibration results, then the researchers may choose to report the results that include the misfitting people.

A complementary graphical approach for person-fit assessment is the *person response function* (PRF). The PRF presents the relationship of the probability of a person's response pattern and the item locations. In addition to being used for identifying misfitting individuals, the PRF may be used to identify a particular item or set of items for which person-item fit is problematic as well as for providing diagnostic information, such as inattention, guessing, identifying copying, and so on (Trabin & Weiss, 1983). Typically, the PRF is assumed to be a nonincreasing function of the item locations. Departures from this monotonicity assumption are taken as indicators of person-model misfit for all or some subset of the instrument's items. To examine person fit, one compares a person's observed PRF (OPRF) with his or her expected PRF (EPRF). To obtain an individual's EPRF one uses his or her $\hat{\theta}$ and the item parameter estimates to calculate the person's probability of a response of 1. To obtain the individual's OPRF we first group the items in terms of the similarity of their locations and then determine the proportion of items in each group for which the individual has a response of 1. Large discrepancies between the individual's OPRF and EPRF reflect an individual who is behaving inconsistently with the model.

An alternative to the fit statistics' perspective (i.e., is the person behaving consistent with the model) is *appropriateness measurement*. In appropriateness measurement one asks, "What is the appropriateness of a person's estimated location as a measure of his or her true location ($\theta$)?" For example, assume that a person has a response pattern of missing easy items and correctly answering more difficult items. One might ask, "Did this pattern arise from the person correctly guessing on some difficult items and incorrectly responding to easier items or does this reflect a person that was able to copy the answers on some items?" Various statistically based indices have been developed to measure the degree to which an individual's response pattern is unusual or is inconsistent with the model used for characterizing his or her performance.

One appropriateness index, $l_z$, has been found to perform better than other measures. This index is based on the standardization of the person log likelihood function, and allows the comparison of individuals at different $\theta$ levels on the basis of their $l_z$ values. Although $l_z$ is purported to have a unit normal distribution, this has not been found to be the case for instruments of different lengths. As a result, it is not advisable to use the standard normal curve for hypothesis testing with $l_z$. Nevertheless, various guidelines exist for using $l_z$ for informed judgment. In general, a "good" $l_z$ is one around 0.0. An $l_z$ that is negative reflects a relatively unlikely response vector (i.e., inconsistent responses), whereas a positive value indicates a comparatively more likely response vector than would be expected on the basis of the model.

### 13. Person Location Estimate Results

In a classification or certification situation the decision about the individual is provided. For non-classification or non-certification contexts one typically presents the person location estimates on a transformed metric to eliminate negative person estimates and to make the scores more easily interpreted than they would be on the untransformed metric. This transformed metric may be a total score scale (e.g., a number correct scale) that ranges from 0 to the number of items on the instrument, or another metric that has been adopted (e.g., T-scores, College Board Score Scale, a proprietary scale). In other situations, particularly when not presenting the person estimate to the public, the person's location estimates may be left on the standard score-like metric that allows for both negative and positive location estimates. It is good practice to provide the estimate's standard error. How one transforms the $\theta$ standard metric to another metric is discussed next in Desideratum 14.

### 14. Metric Definition and/or Transformation

Because of the indeterminacy of the parameter estimate metric, programs typically use either person-centering or item-centering to identify the calibration model. The net effect of this approach is that the metric is defined relative to the sample used for the parameter estimation. Assuming acceptable model-data fit, then the administration of an instrument to two distinct samples will more than likely result in parameter estimates that are not identical because each sample defines its own metric. However, these two metrics are linearly related to one another (i.e., the metric is determined up to a linear transformation). As such, it is possible to transform one metric to another so that the interpretation of the estimates is freed from the particular sample used for estimation.

To linearly transform one metric to another involves using *metric transformation coefficients* (also known as *equating coefficients*). In general, the linear transformation from one metric to another for both person and item locations (or their estimates) is

$$\xi^* = \zeta(\xi) + \kappa, \tag{1}$$

where $\zeta$ and $\kappa$ are the unit and location coefficients, respectively. The $\xi$ term represents the parameter (or its estimate) on the untransformed or *initial metric* and $\xi^*$ represents the same parameter transformed to the *target metric*. The target metric (sometimes called the common metric) is the metric onto which all other metrics are transformed. For example, $\xi$ can represent the item location parameter, $\delta_j$ (or its estimate, $\hat{\delta}_j$), on the initial metric and $\xi^*$ is $\delta_j^*$ (or $\hat{\delta}_j^*$) on the target metric.

To transform the initial metric's item discrimination parameter, $\alpha_j$, to the target item discrimination parameter metric, $\alpha_j^*$, we use

$$\alpha_j^* = \frac{\alpha_j}{\zeta}. \tag{2}$$

(The discrimination parameters' estimates may be used in lieu of $\alpha$.) The IRFs' lower asymptote parameters (or their estimates) are on a common [0, 1] metric and do not need to be transformed.

In those cases where either a proprietary scale (e.g., the College Board scale) or a commonly defined scale (e.g., the T-score) is the target metric, then $\zeta$ and $\kappa$ are given by the target metric's definition. For instance, for the T-score scale, $\zeta = 10$ and $\kappa = 50$. However, when $\zeta$ and $\kappa$ are not given by the target metric then it is necessary to estimate $\zeta$ and $\kappa$. For example, this would be the case when we are linking two metrics to one another.

Multiple approaches for determining the values of $\zeta$ and $\kappa$ have been developed. One approach obtains the metric transformation coefficients by using the mean and standard deviations of the common items; this approach is sometimes called *linear equating*. Specifically, the coefficient $\zeta$ is obtained by taking the ratio of the target to initial metric standard deviations (*s*) of the locations:

$$\zeta = \frac{s_{\delta^*}}{s_{\delta}}, \tag{3}$$

where $s_{\delta^*}$ is the standard deviation of the item locations (or their estimates) on the target metric and $s_{\delta}$ is the standard deviation of the item locations (or their estimates) on the initial metric. Once $\zeta$ is determined the other coefficient, $\kappa$, is obtained by

$$\kappa = \overline{\delta}_j^* - \zeta\overline{\delta}_j. \tag{4}$$

where $\overline{\delta}_j^*$ is the mean of the item locations on the target metric and $\overline{\delta}_j$ is the mean of the item locations on the initial metric. Once the metric transformation coefficients are obtained, then the linking of the separate metrics is performed by applying equations (1) and (2) item by item to the item parameter estimates. To place the person location estimates onto the target metric we apply $\theta_i^* = \zeta(\theta_i) + \kappa$ to each individual's person location or its estimate.

In contrast to linear equating, a second approach, *total characteristic function equating*, uses all the item parameter estimates to determine the values of $\zeta$ and $\kappa$. The objective in this method (also known as *true score equating, test characteristic curve equating*) is to align as closely as possible the initial metric's total characteristic function with that of the target metric. The metric transformation coefficients are the values of $\zeta$ and $\kappa$ that satisfy this objective. This approach requires that the two metrics to be link have either all or a subset of items in common.

As mentioned above, another use of a metric transformation is to transform a metric to make it more meaningful or interpretable. One target metric that has intrinsic meaning for people is the total score metric. For instance, rather than informing a respondent that his or her $\hat{\theta}$ is 1.2, which may or may not have any inherent meaning to the respondent, we can transform the respondent's $\hat{\theta}$ to the more familiar total score metric. That is, a respondent with a $\hat{\theta}$ of 1.2 is told his or her score on the 20-item instrument is 15 (or 15/20 = 0.75). This transformation is performed through the total characteristic function. The gist of this approach is to sum, across an instrument's items, the response probabilities for a given person location estimate. In a proficiency assessment situation the total score metric indicates the expected number of correctly answered items. A variant of this approach divides this sum by the number of items on the instrument to obtain an expected proportion equivalence for $\theta$ (i.e., the proportion of responses of 1), which, in proficiency assessment, is the expected proportion of correct responses.

# References

Baker, F. B., & Kim, S.-H. (2004). *Item response theory: Parameter estimation techniques* (2nd ed.). New York: Marcel Dekker.

De Ayala, R. J. (2009). *The theory and practice of item response theory.* New York: Guilford Press.

Fraser, C., & McDonald, R. P. (2003). *NOHARM: A Windows program for fitting both unidimensional and multidimensional normal ogive models of latent trait theory* [Computer Program]. Niagara College, Welland, Ontario, Canada. Available at http://www.niagarac.on.ca/~cfraser/download/.

Hambleton, R. K., Swaminathan, H., & Rogers, H. J. (1991). *Fundamentals of item response theory.* Newbury Park, CA: Sage.

Klauer, K. C., & Rettig, K. (1990). An approximately standardized person test for assessing consistency with a latent trait model. *British Journal of Mathematical and Statistical Psychology, 43,* 193–206.

Linacre, J. M. (2001). *A user's guide to WINSTEPS/MINISTEPS.* Chicago, IL: Winsteps.com.

Linacre, J. M., & Wright, B. D. (2001). *A user's guide to BIGSTEPS.* Chicago, IL: MESA Press.

Lord, F. M. (1980). *Applications of item response theory to practical testing problems.* Hillsdale, NJ: Erlbaum.

Lord, F. M., & Novick, M. R. (1968). *Statistical theories of mental test scores.* Reading, MA: Addison-Wesley Publishing.

Muraki, E., & Bock, R. D. (2003). *PARSCALE* (Version 4.1) [Computer Program]. Mooresville, IN: Scientific Software, Inc.

Raju, N. S. (1988). The area between two item characteristic curves. *Psychometrika, 53,* 495–502.

Raju, N. S. (1990). Determining the significance of estimated signed and unsigned areas between two item response functions. *Applied Psychological Measurement, 14,* 197–207. (A correction may be found in *Applied Psychological Measurement, 15,* 352.)

Smith, R. M. (1985). A comparison of Rasch person analysis and robust estimators. *Educational and Psychological Measurement, 45,* 433–444.

Smith, R. M. (2004). Fit analysis in latent trait measurement models. In E. V. Smith & R. M. Smith (Eds.), *Introduction to Rasch measurement* (pp. 73–92). Maple Grove, MN: JAM Press.

Thissen, D. J., Chen, W.-H., & Bock, R. D. (2003). *MULTILOG* (Version 7.0) [Computer Program]. Mooresville, IN: Scientific Software, Inc.

Thissen, D. J., Steinberg, L., & Wainer, H. (1988). Use of item response theory in the study of group differences in trace lines. In H. Wainer & H. I. Braun (Eds.) *Test validity* (pp. 147–169). Hillsdale, NJ: Erlbaum.

Trabin, T. E., & Weiss, D. J. (1983). The person response curve: Fit of individuals to item response theory models. In D. J. Weiss (Ed.), *New horizons in testing: Latent trait test theory and computerized adaptive testing* (pp. 83–108). New York: Academic Press.

Van der Linden, W. J., & Hambleton, R. K. (Eds.). (1997). *Handbook of modern item response theory.* New York: Springer.

Verhelst, N. D., Glas, C. A. W., & Verstralen, H. H. F. M. (1995). *One-parameter logistic model* (OPLM). Arnhem, The Netherlands: CITO.

Zimowski, M., Muraki, E., Mislevy, R. J., & Bock, R. D. (2003). *BILOG-MG* (Version 3.0) [Computer Program]. Mooresville, IN: Scientific Software, Inc.

# 13
# Latent Class Analysis

**Karen M. Samuelsen and C. Mitchell Dayton**

Latent Class Analysis (LCA) is a statistical method for identifying unobserved groups based on categorical data. LCA is related to cluster analysis (see Chapter 4, this volume) in that both methods are concerned with the classification of cases (e.g., people or objects) into groups that are not known or specified *a priori*. In LCA, cases with similar response patterns on a series of manifest variables are classified into the same latent class although membership in latent classes is probabilistic rather than deterministic. In addition, LCA can be viewed as analogous to factor analysis (see Chapter 8, this volume), with the former examining categorical variables and the latter continuous ones; however, this comparison is less direct than in the case of cluster analysis. Although both LCA and factor analysis utilize manifest variables to gain insights into latent constructs, the focus of conventional factor analysis is on the structure of the variables as opposed to the structure of the cases. However, both LCA and factor analysis are based on the principle that observed variables are (conditionally) independent assuming knowledge of the latent structure. Finally, LCA is related to item response theory (see Chapter 12, this volume) and can be viewed as a generalization of discrete response models such as the Rasch model (Lindsay, Clogg, & Grego, 1991).

LCA has been applied in the health and medical fields for the identification of disease subtypes and for the assessment of diagnostic agreement (e.g., Albert, McShane, & Shih, 2001; Rindskopf & Rindskopf, 1986; Uebersax & Grove, 1990) as well as for studying dietary patterns (Patterson, Dayton, & Graubard, 2002). In addition to many other research areas, LCA has been applied to modeling jury verdicts (e.g., Gelfand & Solomon, 1974), market segmentation research (Wedel & Kamakura, 2000), and psychology (e.g., Kendler, Karkowski, & Walsh, 1998).

Overviews of LCA can be found in McCutcheon (1987), Dayton (1999), and Heinen (1996). Related and more complex models are discussed in Hagenaars (1993), Langeheine and Rost (1988), and Hagenaars and McCutcheon (2002). An excellent web-based resource is provided by Uebersax (2008). In addition, selected seminal papers on the subject include: Clogg and Goodman (1984), Dayton and Macready (1976), and Goodman (1974). For an extended bibliography, see the Uebersax website. Finally, as discussed in Desideratum 8, various software programs are available for estimation and analysis of latent class models.

Table 13.1 Desiderata for Latent Class Analysis

| Desideratum | Manuscript Section(s)* |
|---|---|
| 1. Substantive theories guiding the choice of models to be evaluated are synthesized. | I |
| 2. Theoretical connections among the manifest variables, covariates, and potential latent classes are explicated. | I |
| 3. The assumption of local independence is discussed and evidence is offered as to why the manifest variables would be independent of each other within latent classes. | I |
| 4. Manifest variables are defined and their appropriateness is justified. | M |
| 5. Categorical and continuous covariates used in the analysis are discussed and a rationale for their inclusion is provided. | M |
| 6. Sampling method(s) and sample size(s) are explicated and justified. The impact of sample size on cell frequencies and model fit statistics should also be discussed. | M |
| 7. The mathematical model(s) being considered are presented along with a substantive justification of the constraints (if any) placed on the allowable values of the parameters to be estimated. | M, R |
| 8. The name and version of the utilized software package are reported. The parameter estimation method is justified and its underlying assumptions are addressed. | M, R |
| 9. Problems with model convergence and model identification are reported and discussed. | R |
| 10. Recommended fit indices are presented and evaluated using literature-based criteria. | R |
| 11. For competing models comparisons are made using statistical tests and/or information criteria. | R |
| 12. Latent class proportions and conditional probabilities are reported. | R |
| 13. Evidence is provided that global versus local maxima have been reached. | R |
| 14. Boundary value parameter estimates are highlighted and implications are discussed. | R, D |
| 15. Meaningfulness of the latent class proportions is considered. | R, D |
| 16. Membership of the latent classes is discussed. | R, D |

\* *Note*: I = Introduction, M = Methods, R = Results, D = Discussion

## 1. Theories and Models

Like factor analysis and its generalized formulation, structural equation modeling, latent class analysis (LCA) uses manifest variables to provide information regarding hypothesized latent variables. In the case of LCA, one or more unobserved, or latent, categorical variables are assumed to be responsible for observed, or manifest, response patterns. LCA can be either exploratory or confirmatory. In the exploratory mode, LCA can be considered to be a special case of cluster analysis with categorical variables. In the confirmatory mode, LCA can be used to fit scaling models to categorical data and, in fact, to fit the equivalent of structural equation models (Hagenaars, 1993). In the absence of an explicit theory regarding the latent variables the researcher is still expected to discuss prior experiences that lead to the choice of latent class models to compare. When in a confirmatory mode, it is incumbent on the researcher to explain, in some depth, the choices made vis-à-vis the models chosen.

In either an exploratory or confirmatory mode, one issue that needs to be addressed is how the hypothesized latent classes represent qualitative differences among subjects. For example, it is possible that there are different latent classes of shoppers and that membership in those classes is predicated on the basis of motivational considerations behind those shoppers' purchases (e.g., fashion, price, and quality). What experience, literature, or theory supports the existence of these classes?

In the previous example, the latent classes simply represented different categories of shoppers. It is possible that in a different situation the latent classes could be ordered in some way. An example would be latent classes predicated on cigarette smoking behavior, where the latent classes are comprised of initiators, experimenters, regular users, and daily/dependent smokers (Flaherty, 2008). Once again, the question of what experience, literature, or theory substantiates that ordering must be addressed.

Finally, it is possible that one or more unscalable classes might be necessary to adequately model the data (Dayton & Macready, 1980; Goodman, 1975). For example, when young children are asked to select personality statements that best describe themselves, they may have neither a command of the vocabulary needed nor the cognitive understanding when questions are not concrete in nature (Meijer, Egberink, Emons, & Sijtsma, 2008). These children will tend to provide inconsistent response patterns because they resort to guessing or lack understanding. If one or more unscalable classes are included, the researcher must explain why certain respondents' responses could be inconsistent.

## 2. Linking Variables and Covariates to Potential Latent Classes

Many important questions in the social sciences involve comparisons among manifest groups. As an example, consider the data set concerning academic cheating reported by college students that were examined by Dayton and Scheers (1997). When considering academic cheating it may be interesting to compare groups such as: (1) males and females; (2) juniors and seniors; or (3) students in different academic programs. In the Introduction section of the manuscript, the researcher should discuss why the grouping or covariates included in the analysis should be linked to membership in the latent classes. What literature supports this linkage and if the literature is sufficiently detailed, what differences have been shown to exist between the manifest groups?

## 3. Local Independence

In LCA, it is assumed that: (1) the model correctly specifies the number of classes, (2) each respondent belongs to only one latent class, and (3) respondents within a class are homogeneous. Building on these, the fundamental concept of LCA is that of local (i.e., conditional) independence meaning that the observed manifest responses are independent given that latent class membership is known. Although, in theory, latent class models may include dependencies among residual terms (e.g., Hagenaars, 1988), this is quite uncommon in the research literature and it should only be done for substantive reasons, not to improve model fit (Flaherty, 2008). In either case, when local independence is assumed or when dependencies among residuals are allowed, the researcher should discuss these assumptions.

## 4. Manifest Variables

Relatively few assumptions underlie latent class analyses, particularly with regard to the manifest variables under consideration; however, these variables merit some discussion in the Methods section. The most obvious characteristic that should be highlighted is whether they are dichotomous, polytomous, or ordered polytomous in nature. If polytomous variables are employed there should be some justification for the number of scale points used. For example, if manifest variables were based on 5-point Likert scales, the researcher should verify and report that all scale points were chosen by respondents. If not, it might be appropriate to reduce the number of categories for some or all of the scales as well as to consider whether or not the variables are more appropriately modeled as dichotomous. Simple descriptive statistics can verify that scale points were used, but need not be shown in the document. If scale points are ordered (as for a Likert scale), then appropriate ordinal modeling methods should be employed (see Rost, 1988).

For confirmatory scaling models (and related models), permissible response patterns for a specific latent class model are also of interest. For example, the Lazarsfeld-Stouffer questionnaire data set (Lazarsfeld, 1950) was based on responses by non-commissioned officers to four dichotomous items regarding attitudes toward to the Army. These items were intended to express increasingly more favorable attitudes such that the permissible response vectors were {0000}, {1000}, {1100}, {1110}, and {1111}, where 0 is a negative response and 1 is a positive response. Thus, for example, the implication is that when a respondent agrees with the fourth item, one highly favorable toward the Army, the respondent also agrees with the three other items. Theoretically, all other response patterns are not permissible, however this does not mean that they will not occur since response errors and other factors may affect responses (additional discussion of these models is presented in Desideratum 7).

## 5. Covariates

Concomitant variables can be incorporated into LCA in two different manners: (1) as grouping (stratification) variables; or (2) as covariates that are modeled in a manner similar to logistic regression (see Dayton & Macready, 2002). In either case, some rationale should be provided for the choice of specific concomitant variables and their manner(s) of being included in the latent class model. When grouping variables are used, it is possible to explore homogeneous, partially homogeneous, and heterogeneous models. When fully heterogeneous models are not included, some rationale for this omission should be provided. When covariates are used, the form of the regression component of the model (e.g., logistic) should be described. In addition, the rationale for including, or not including, higher-order terms such quadratic functions of continuous covariates should be addressed. Finally, if grouping variables are used as covariates (e.g., by including indicator or dummy variables), the rationale for including, or not including, product terms for continuous and grouping covariates should be addressed. Whenever possible, some form of graphical display for the relation between the covariate(s) and latent class membership should be included.

## 6. Sampling Method(s) and Sample Size(s)

The inferences that can be drawn from LCA are limited by the sample available for analysis. Samples that are not representative of the population of interest to the researcher will severely limit the inferences that may be drawn from the analysis. Virtually all theoretical treatments of LCA have been based on the notions of simple random sampling, with, perhaps, the inclusion of manifest grouping (stratification) variables. In practice, however, data sets are often based on complex survey designs that involve clusters of respondents and disproportionate sampling. While it is relatively straightforward to incorporate sampling weights to make adjustments for disproportionate sampling, corrections for cluster sampling are more difficult. Failure to take clustering into account can result in severe underestimation of standard errors and can distort results from significance tests. Patterson, Dayton, and Graubard (2002) discussed these issues and provided some guidance for these situations. In any case, when complex sample designs are involved, these issues should be explicitly discussed in a manuscript.

Sample size is an important consideration in LCA, especially as it relates to observed cell frequencies. For the dichotomous case with $V$ manifest variables, the possible number of unique response vectors is $2^V$. While for four variables there are only 16 patterns, for 10 variables there would be 1024 possible response vectors. Thus, with larger numbers of variables, analyses based on frequencies of response vectors are not practical. In addition, goodness-of-fit tests are not applicable unless sample sizes are truly enormous given that frequency tables will be sparse (i.e., contain large numbers of 0 frequencies). In general, analyses with large numbers of manifest variables are possible using raw data (i.e., responses for individual cases rather than frequency data). Note that identification and convergence issues become more critical in these situations and need to be discussed in a manuscript.

## 7. Mathematical Model

In the notation formalized by Goodman (1974), manifest variables are denoted by capital letters ($A$, $B$, $C$, etc.) and, in the dichotomous case, the variables have levels $i = \{1,0\}$, $j = \{1,0\}$, $k = \{1,0\}$, and so forth, respectively. Latent variables are denoted $X$ and have levels $t = \{1, ..., T\}$. Therefore, for the $t^{th}$ latent class, the conditional probabilities are represented by $\pi_{it}^{\bar{A}X}$, $\pi_{jt}^{\bar{B}X}$, $\pi_{kt}^{\bar{C}X}$ etc. Assuming three manifest variables and a latent class $t$, the conditional probability of a response vector $\mathbf{y}$ for the $s^{th}$ case associated with the $t^{th}$ class can be written as:

$$P(\mathbf{y}_s \mid t) = \pi_{ijkt}^{\bar{A}\bar{B}\bar{C}X} = \pi_{it}^{\bar{A}X}\, \pi_{jt}^{\bar{B}X}\, \pi_{kt}^{\bar{C}X}. \tag{1}$$

Assuming local independence and latent class proportions for a total of $T$ latent classes, the probability of obtaining a response vector can be expressed as the weighted sum across classes:

$$P(\mathbf{y}_s) = \sum_{t=1}^{T} \pi_t^{X}\, \pi_{it}^{\bar{A}X}\, \pi_{jt}^{\bar{B}X}\, \pi_{kt}^{\bar{C}X}. \tag{2}$$

This simple model is generally presented in any article using LCA and provides a frame of reference from which to discuss any constraints placed on the model. These constraints depend upon the purpose of the analysis (exploratory or confirmatory), whether the models under consideration are restricted, and how response errors are modeled (e.g., Proctor, intrusion-omission), as described below.

Typically, exploratory LCA is conducted with no restrictions on the values of conditional probabilities or latent class proportions. In confirmatory analyses, however, the researcher must specify the hypotheses of interest and, typically, constraints are imposed that reflect these hypotheses. For example, the Proctor (1970) scaling model proposes permissible response vectors that represent "true types" in the population and incorporates equal rates of response error for all manifest variables. In a Proctor model with three dichotomous variables ($A$, $B$, $C$) and a linear (i.e., Guttman) scale, the permissible response vectors are {000}, {100}, {110}, and {111}, reflecting membership in the four latent classes of the latent variable $X$ (classes 1, 2, 3, and 4, respectively); the other possible manifest vectors, {001}, {010}, {011}, and {101}, reflect some sort of response error. Assuming equal rates of response error for the three variables, as in the Proctor model, the following constraints would be imposed:

$$\begin{aligned}
\pi_{11}^{\bar{A}X} &= \pi_{11}^{\bar{B}X} = \pi_{11}^{\bar{C}X} = \\
\pi_{12}^{\bar{B}X} &= \pi_{12}^{\bar{C}X} = \pi_{13}^{\bar{C}X} = \\
\pi_{02}^{\bar{A}X} &= \pi_{03}^{\bar{A}X} = \pi_{03}^{\bar{B}X} = \\
\pi_{04}^{\bar{A}X} &= \pi_{04}^{\bar{B}X} = \pi_{04}^{\bar{C}X} = \pi_{error}
\end{aligned} \tag{3}$$

The purpose of this brief explanation of the Proctor model is to show the necessity for discussing the general model in an article. In practice, more complex models may also be investigated that are generalizations of the Proctor model. For example, the intrusion-omission error model (Dayton & Macready, 1976) relaxes the assumption of equal error rates and allows for errors of intrusion (i.e., the occurrence of an observed response of 1 when a permissible vector calls for a 0 response) and errors of omission (i.e., the occurrence of an observed response of 0 when a permissible vector calls for a 1 response).

## 8. Software Package

There are many good software programs for LCA. These typically include Latent Gold (Vermunt & Magidson, 2005), LEM (a free program also by Vermunt, 1997), MPlus (Muthén & Muthén, 2006), PANMARK3 (van de Pol, Langeheine, & de Jong, 1996), and WINMIRA2001 (von Davier, 2001). SAS also has two procedures, PROC LCA and PROC LTA, to handle latent class analyses. Note that this is not meant to be an exhaustive list, but rather a sample of the widely used LCA programs. The name and version of the software package must be reported in any manuscript using LCA. This indirectly provides the reader with some information like whether standard errors are available for parameter estimates, multiple start values are automatically tested, and if local dependence can be handled. All other things being equal, we would recommend using programs that provide standard errors and multiple start values; however the choice of exactly which program to use depends upon many other factors as well (e.g., expense, type of model(s), user experience).

In addition, the parameter estimation method used within the chosen software program, along with its underlying assumptions, should be discussed. Parameters will generally be found using either maximum likelihood (ML) estimation or Bayesian methods such as those available in Latent Gold (Vermunt & Magdison, 2005). Markov Chain Monte Carlo (MCMC) methods can also be applied to latent class models; however, this methodology requires special software, original programming, and extensive knowledge of the techniques. Therefore, we would not recommend these methods for anyone but the most advanced users.

## 9. Model Convergence and Model Identification

Most LCA programs use ML estimation which is an iterative procedure that continues until the change in the log likelihood between successive iterations is less than some pre-set criterion. The value of that convergence criterion affects the likelihood that local, rather than global, maxima are identified, and thus this criterion should be reported (see Desideratum 13 for more on local and global maxima).

An identified model is one for which there is a single "best" solution. Non-identified models have multiple "best" solutions. For information on the reasons models may be non-identified, see Uebersax (2008). From a practical standpoint a researcher needs to know that the number of parameters to be estimated is limited by the number of unique response vectors minus one, which equals the degrees of freedom. Unfortunately, this is a necessary but not sufficient check for an identified model, because models may be non-identified due to the observed data and their match with a particular model. For this reason it is also common to report that the estimated variance-covariance matrix is of full rank. Note that this matrix contains estimated values for the sampling variances and covariances for the ML parameter estimates.

## 10. Fit Indices

The goodness of fit of a particular latent class model to the observed data can be assessed using some form of a chi-square test. Two familiar versions of this significance test are the Pearson statistic, $\chi^2$, and the likelihood ratio statistic, $G^2$. The former is based on the differences between observed and expected frequencies, while the latter is based on the logarithm of the ratio of the observed and expected frequencies. In theory, with dichotomous response variables for an unrestricted full-rank model, the degrees of freedom are

$$df = 2^V - m - 1, \tag{4}$$

where $V$ is the number of dichotomous manifest variables and $m$ is the number of independent parameters estimated in the model. This equation must be modified appropriately if the rank of the estimated covariance matrix is less than $m$. This occurs for models with restrictions on the conditional probabilities (e.g., scaling models), for models with conditional probabilities that approach 0 or 1, and for models that are not identified (see Dayton, 1999, for more information regarding these situations). Under most situations, $\chi^2$ and $G^2$ goodness-of-fit statistics yield similar numerical results and lead to the same conclusions. In general, both statistics are reported along with degrees of freedom. When some cell frequencies are small, both $\chi^2$ and $G^2$ tend to not follow the appropriate reference chi-square distribution and, thus, become inaccurate. Under this scenario the Read-Cressie statistic is recommended (see Dayton, 1999).

One problem with all goodness-of-fit chi-square tests is that, when fitting models based on large sample sizes, the null hypothesis of perfect fit tends to be rejected. This may happen even though the residuals (i.e., differences between observed and expected frequencies) indicate that fit seems satisfactory from a substantive point of view. To complement the limited amount of information provided by chi-square tests, other indices of model fit may be used. Two such indices are the index of dissimilarity, $I_D$ (Dayton, 1999), and $\pi^*$ (Rudas, Clogg, & Lindsay, 1994). For $I_D$, defined in terms of observed and expected frequencies, values that are less than 0.05 are generally considered satisfactory. For $\pi^*$, which is based on the number of cases that would need to be deleted to achieve perfect model fit, values must be interpreted with reference to the application although .10 is a commonly recommended value (Dayton, 1999). Typical LCA programs do not provide estimates for $\pi^*$; however, a spreadsheet procedure for estimating $\pi^*$ for two-class models with dichotomous response variables is described and illustrated in Dayton (1999). Given that programs do not provide $\pi^*$ we recommend that authors include $I_D$ with their results, supplementing with $\pi^*$ should they choose to.

An indication of how well the model fits the data can also be gained by examining the standard errors, and confidence intervals constructed from them, for the latent class proportions and conditional probabilities. Researchers should, when possible, utilize LCA computer programs that provide estimated standard errors for latent class proportions and conditional probabilities. The accuracy of these estimates has not been widely investigated, especially for small sample sizes; this is one reason for favoring large samples in LCA. If a program is used that does not provide standard errors, it is possible to use resampling methods, such as the jackknife or parametric bootstrap, to estimate standard errors from frequency data. This, however, requires original programming that can be carried out, albeit somewhat tediously, in a spreadsheet as illustrated in Dayton (1999).

## 11. Statistical Tests and Information Criteria

In addition to testing the absolute fit of a model, as discussed in Desideratum 10, it is often of interest to researchers to compare alternate models. For nested models the chi-square difference test can be used in certain circumstances. In those situations one computes the difference between the $G^2$ statistics for the two models and uses the difference between the degrees of freedom to evaluate whether the more complex model provides statistically significantly better fit. There are, however, technical issues surrounding the use of chi-square difference tests (see Dayton, 1999, for an overview), and for that reason it is often recommended that they are used in concert with measures of relative fit based on information criteria which do not require the models to be nested. Also, it should be noted that chi-square difference tests cannot be used to compare models based on differing numbers of latent classes. Although this comparison is almost always of interest in exploratory LCA, such tests are, in theory, invalid and in practice are known to yield misleading results.

The Akaike (1973, 1987) Information Criterion (AIC) includes terms for the log likelihood and the number of independent parameters to be estimated. This index, because it does not include a term for the sample size, lacks asymptotic consistency (Bozdogan, 1987). Schwarz (1978) included a penalty term that reflects sample size in his Bayesian Information Criterion (BIC) and Bozdogan (1987) proposed a Consistent Akaike Information Criterion (CAIC) that, in practice, performs virtually the same as BIC. For all of these information criteria, the decision procedure entails selecting the model with the lowest value of the criterion as the best fitting model. Often, the strategy is to decide on the preferred model based on multiple information criteria and to report those criteria for all models within a journal article. However, it should be noted that AIC tends to select more complex models (that is, models with more parameters) and BIC (or, CAIC) tends to select simpler models with the choice between them, to some extent, being a matter of philosophy with respect to the model-fitting enterprise (see Bozdogan, 1987, for more discussion).

## 12. Latent Class Proportions and Conditional Probabilities

When relatively few response patterns are possible, it is helpful to present those patterns, the conditional probabilities, and the latent class proportions in one table. The frequency of each response pattern and total sample size can also be reported. As shown in Table 13.2 (adapted from Dayton, 1999), arranging data in this manner provides the reader with a comprehensive summary of the analysis in question.

When the number of unique response patterns becomes large it is more convenient to show how the responses to individual items differ across the classes. In this example in Table 13.3 (adapted from Dayton, 1999), four dichotomous items ($A$, $B$, $C$, and $D$) were examined. The conditional probabilities for both the positive response (1) and negative response (0) are shown even though that information is redundant. In this case it is obvious that the first latent class (labeled LC 1) was much larger than the second (approximately 85% to 15%), that for LC 1 there are boundary value estimates (see Desideratum 14) for variable $B$, and that the conditional probabilities of a positive response differ substantially across classes for all of the items. The standard errors included in parentheses also indicate to the reader which parameters were estimated with greater precision. A similar table could be created for polytomous items with 0 through the number of response options shown for each item. This representation could also be used to show similarities and differences among manifest groups.

For exploratory analyses when models are estimated based on differing numbers of classes, a simple graphical display of the conditional probabilities can be included. A rationale for why these graphs are valuable can be seen in Figure 13.1 from a six-item dataset regarding attitudes toward abortion (Dayton, 2006). The first three items (labeled 1, 2 and 3 on the $x$-axis) deal with favoring abortions for reasons such as mother's health, rape or incest and birth defect whereas the last three items (labeled 4, 5 and 6) deal with favoring abortions for reasons such as being unmarried, being poor or having too many children already. The first class in the 4-class model includes individuals who are strongly pro-abortion without restriction. The individuals in the second and third classes put conditions on their

Table 13.2 Latent Class Structure for Two Hypothetical Items

| Response Pattern | Frequency | Conditional Probabilities | |
| --- | --- | --- | --- |
| | | Latent Class 1 | Latent Class 2 |
| {0 0} | 516 | 0.004 | 0.512 |
| {1 0} | 144 | 0.016 | 0.128 |
| {0 1} | 164 | 0.036 | 0.128 |
| {1 1} | 176 | 0.144 | 0.032 |
| Total | 1000 | 0.200 | 0.800 |

Table 13.3 Cheating Data for Males Only (Standard Errors in Parentheses)

|   |   | LC 1 | LC 2 |
|---|---|---|---|
|   |   | 0.8545 (0.0595) | 0.1455 (0.0595) |
| A | 0 | 0.9775 (0.0281) | 0.4299 (0.1702) |
| A | 1 | 0.0225 (0.0281) | 0.5701 (0.1702) |
| B | 0 | 1.0000 (0.0000) | 0.5486 (0.1943) |
| B | 1 | 0.0000 (0.0000) | 0.4514 (0.1943) |
| C | 0 | 0.9944 (0.0184) | 0.6817 (0.1341) |
| C | 1 | 0.0056 (0.0184) | 0.3183 (0.1341) |
| D | 0 | 0.8553 (0.0364) | 0.5457 (0.1425) |
| D | 1 | 0.1447 (0.0364) | 0.4543 (0.1425) |

support of abortion with relatively strong support for items 1–3 but weaker support for items 4–6. The fourth class includes individuals who express fairly negative opinions about abortion regardless of the situation. The 5-class model shows these same four classes plus another class that seems to include individuals who may be responding randomly to the items. Given that the responses do appear random, and that this class only represents 1% of the respondents, it would be appropriate to choose the 4-class model over the 5-class model even if the more complex showed better fit.

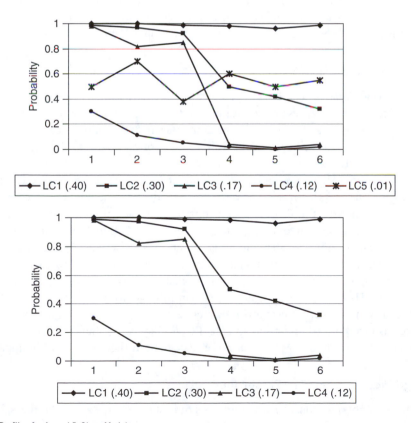

**Figure 13.1** Profiles for 4- and 5-Class Models.

## 13. Global vs. Local Maxima

Because of the mathematical form of the equations that must be solved to conduct maximum-likelihood estimation with latent class models, the ordering of latent classes is arbitrary. For example, for a two-class solution where the first latent class proportion is .68 there is an equivalent two-class solution where the first class proportion is .32 (and the item conditional probabilities are correspondingly switched). In the technical literature, this is known as "label switching." Although, in practice, this is merely a bookkeeping issue, it must be kept in mind when comparing solutions with varying number of latent classes or solutions with different sets of restrictions given that the classes may not directly correspond across different solutions.

Another issue is that local maxima might exist. Ideally, the algorithm being used to estimate model parameters should seek out the global maximum. However, existing algorithms using ML or Bayesian estimation cannot, in general, distinguish between the global maximum and locally optimal solutions. Thus, when reporting LCA solutions it is necessary to provide evidence that, in fact, a global maximum was reached. There are several types of evidence that are suitable. First, the LCA program should be executed several times with different sets of start values. If all of these runs result in the same solution, this is strong evidence that the global maximum has been reached. If those runs reach different maxima, it may be defensible to select an analysis with the largest log-likelihood but this requires explicit discussion as it is possible that the model is not identified. An article should also include the minimum convergence criteria used; stringent criteria (e.g., the change in the log-likelihood between iterations is $10^{-8}$ or less) will lend more credence that a global maximum has been reached when multiple LCA solutions identify that same maximum. Parameters whose estimates approach boundary values of 0 or 1 should be cause for suspicion. As discussed by Uebersax (2008), the greater the number of parameter estimates at the boundary values, the greater a researcher should be concerned that the solution is a local, rather than global, maximum. Desideratum 14 provides a more in-depth discussion of boundary values.

## 14. Boundary Value Parameter Estimates

When estimated conditional probabilities approach 0 or 1, computational issues arise with respect to estimated variances and covariances for the parameter estimates that render the associated confidence intervals and significance tests of questionable meaning (Garre & Vermunt, 2006). Therefore, in the Results section, one should be very specific regarding which conditional probabilities, if any, went to boundary values of 0 or 1.

Because estimates approaching boundary values might indicate that the data have been overfit, the researcher generally seeks to simplify or restrict the model in some way. One recommendation is to restrict these parameters to 0 or 1 (as appropriate), re-estimate the remaining parameters, and reduce the degrees of freedom for chi-square goodness-of-fit tests by 1 for each such restriction.

Certain LCA programs, notably Latent Gold (Vermunt & Magidson, 2005), incorporate Bayesian methods with Dirichlet prior probabilities for latent class parameters so that the boundary value issue is, for practical purposes, eliminated. Also, there is evidence that this assists in avoiding local maxima without distorting results (Uebersax, 2008). In general, Bayesian estimates will not be the same as ML estimates for identified models and, hence, justification for their use should be provided. For identified models with the weak priors as incorporated as defaults in Latent Gold, Bayesian parameter estimates tend to be "shrunken" relative to conventional ML estimates (i.e., tend to be further from the 0/1 boundaries).

## 15. Meaningfulness of Latent Class Proportions

Researchers need to not only consider which models fit best based on statistical indices, but on the sizes of the classes as well. This is especially true if one of the identified classes is extremely small (although the definition of "small" depends upon context). For example, with a sample size of 200, a latent class proportion of .05 represents only 10 cases, whereas with a sample size of 20,000 it represents 1000 cases. In essence, the question is whether or not a class is substantively meaningful or an artifact of the particular sample being analyzed. When a latent class proportion is extremely small, it is incumbent upon the researcher to offer a substantive rationale, based on past research or theory, for the inclusion of the class.

### 16. Latent Class Membership

A posteriori probabilities for each response vector and latent class can be computed and used to classify respondents into one of the latent classes. The specificity of the discussion regarding latent class membership will depend on the number of response vectors and how those correspond with the latent classes. If a few different response vectors were responsible for latent class membership it would be informative to state exactly what those vectors were. For example, with academic cheating data for four items, Dayton and Scheers (1997) found that all response vectors, except {0000}, {0010}, {0001}, and {0011}, resulted in classification in a latent class identified as representing persistent cheaters. It appears, then, that classification as a non-persistent-cheater hinged on responding "no" (i.e., 0) to the first two items. In situations with more items, on the other hand, if dozens of response vectors were associated with each class it would be of little value to enumerate those vectors. Instead, it would be more informative to discuss the items that seemed to be most predictive of class membership.

The interpretation of latent classes can be enhanced by post hoc analyses that involve concomitant variables. This may be viewed as a type of construct validation for the naming of latent classes. For example, the analysis of academic cheating data by Dayton and Scheers (1997) incorporated student grade point average as a covariate and resulted in an interpretation of the two classes as persistent cheaters and non-persistent-cheaters. Note that it is reasonable to expect that the probability of respondents being classified in these latent classes would be related to academic success in college.

### References

Akaike, H. (1973). Information theory and an extension of the maximum likelihood principle. In B. N. Petrov & F. Csake (Eds.), *Second International Symposium on Information Theory* (pp. 267–281). Budapest: Akademiai Kiado.

Akaike, H. (1987). Factor analysis and AIC. *Psychometrika, 52,* 317–332.

Albert, P. S., McShane, L. M., & Shih, J. H. (2001). Latent class modeling approaches for assessing diagnostic error without a gold standard: with applications to p53 immunohistochemical assays in bladder tumors. *Biometrics, 57,* 610–619.

Bozdogan, H. (1987). Model selection and Akaike's information criterion (AIC): The general theory and its analytical extensions. *Psychometrika, 52,* 345–370.

Clogg, C. C., & Goodman, L. A. (1984). Latent structure analysis of a set of multi-dimensional contingency tables. *Journal of the American Statistical Association, 79,* 762–771.

Dayton, C. M. (1999). *Latent class scaling analysis.* Thousand Oaks, CA: Sage.

Dayton, C. M. (2006). Latent structure of attitudes toward abortion. In S. S. Sawilowsky (Ed.), *Real data analysis* (pp. 293–298). Greenwich, CT: Information Age Publishing.

Dayton, C. M., & Macready, G. B. (1976). A probabilistic model for validation of behavioral hierarchies. *Psychometrika, 41,* 189–204.

Dayton, C. M., & Macready, G. B. (1980). A scaling model with response errors and intrinsically unscalable respondents. *Psychometrika, 45,* 343–356.

Dayton, C. M., & Macready, G. B. (2002). Use of categorical and continuous covariates in latent class analysis. In A. McCutcheon & J. Hagenaars (Eds.), *Advances in latent class modeling* (pp. 213–233). Cambridge, UK: Cambridge University Press.

Dayton, C. M., & Scheers, N. J. (1997). Latent class analysis of survey data dealing with academic dishonesty. In J. Rost & R. Langeheine (Eds.), *Applications of latent trait and latent class models in the social sciences* (pp. 172–180). Munich: Waxman Verlag.

Flaherty, B. P. (2008). Examining contingent discrete change over time with associative latent transition analysis. In G. R. Hancock & K. M. Samuelsen (Eds.), *Advances in latent variable mixture models* (pp. 299–316). Charlotte, NC: Information Age Publishing.

Garre, F. G., & Vermunt, J. K. (2006). Avoiding boundary estimates in latent class analysis by Bayesian posterior mode estimation. *Behaviormetrika, 33*, 43–59.

Gelfand, A. E., & Solomon, H. (1974). Modeling jury verdicts in the American legal system. *Journal of the American Statistical Association, 69*, 32–37.

Goodman, L. A. (1974). Exploratory latent structure analysis using both identifiable and unidentifiable mdoels. *Biometrika, 61*, 215–231.

Goodman, L. A. (1975). A new model for scaling response patterns: An application of the quasi-independence concept. *Journal of the American Statistical Association, 70*, 755–768.

Hagenaars, J. A. (1988). Latent structure models with direct effects between indicators: Local dependence models. *Sociological Methods & Research, 16*, 379–405.

Hagenaars, J. A. (1993). *Log-linear models with latent variables.* Thousand Oaks, CA: Sage.

Hagenaars, J. A., & McCutcheon, A. L. (Eds.) (2002). *Applied latent class analysis.* Cambridge, UK: Cambridge University Press.

Heinen, T. (1996). *Latent class and discrete trait models.* Thousand Oaks, CA: Sage.

Kendler, K. S., Karkowski, L. M., & Walsh, D. (1998). The structure of psychosis: Latent class analysis of probands from the Roscommon Family Study. *Archives of General Psychiatry, 55*, 492–499.

Langeheine, R., & Rost, J. (Eds.) (1988). *Latent trait and latent class models.* New York: Plenum Press.

Lazarsfeld, P. F. (1950). The logical and mathematical foundation of latent structure analysis. In S. A. Stouffer, L. Guttman, E. A. Suchman, P. R. Lazarsfeld, S. A. Star, & J. A Clausen (Eds.), *Measurement and prediction* (pp. 362–412). Princeton, NJ: Princeton University Press.

Lindsay, B., Clogg, C. C., & Grego, J. (1991). Semiparametric estimation in the Rasch model and related exponential response models, including a simple latent class model for item analysis. *Journal of the American Statistical Association, 86*, 96–107.

McCutcheon, A. C. (1987). *Latent class analysis.* Thousand Oaks, CA: Sage.

Meijer, R. R., Egberink, I. J. L., Emons, W. H. M., & Sijtsma, K. (2008). Detection and validation of unscalable item score patterns using item response theory: An illustration with Harter's self-perception profile for children. *Journal of Personality Assessment, 90*, 227–238.

Muthén, L. K., & Muthén, B. O. (2006). Mplus user's guide (4th ed.). Los Angeles: Muthén and Muthén.

Patterson, B., Dayton, C. M., & Graubard, B. (2002). Latent class analysis of complex survey data: application to dietary data. *Journal of the American Statistical Association, 97*, 721–729.

Proctor, C. H. (1970). A probabilistic formulation in statistical analysis of Gutttman scaling. *Psychometrika, 35*, 73–78.

Rindskopf, R., & Rindskopf, W. (1986). The value of latent class analysis in medical diagnosis. *Statistics in Medicine, 5*, 21–27.

Rost, J. (1988). Rating scale analysis with latent class models. *Psychometrika, 53*, 327–348.

Rudas, T., Clogg, C. C., & Lindsay, B. G. (1994). A new index of fit based on mixture methods for the analysis of contingency tables. *Journal of the Royal Statistical Society, Series B, 56*, 623–639.

Schwarz, G. (1978). Estimating the dimension of a model. *Annals of Statistics, 6*, 461–464.

Uebersax, J. (2008). A brief study of local maximum solutions in latent class analysis. Retrieved October 24, 2008, from http://ourworld.compuserve.com/homepages/jsuebersax/local.htm.

Uebersax, J. S., & Grove, W. M. (1990). Latent class analysis of diagnostic agreement. *Statistics in Medicine, 9*, 559–572.

van de Pol, F., Langeheine, R., & de Jong, W. (1996). *PANMARK 3. User's manual. Panel analysis using Markov chains—A latent class analysis program.* Voorburg, The Netherlands: Statistics Netherlands.

Vermunt, J. K. (1997). LEM: A general program for the analysis of categorical data. Department of Methodology and Statistics, Tilburg University

Vermunt, J. K., & Magidson, J. (2005). Technical guide for Latent GOLD 4.0: Basic and advanced. Belmont Massachusetts: Statistical Innovations Inc.

von Davier, M. (2001). WINMIRA user manual. Kiel, Germany: Institut für die Pädagogik der Naturwissenschaften.

Wedel, M., & Kamakura, W. A. (2000). *Market segmentation: Conceptual and methodological foundations* (2nd ed.). Dordrecht: Kluwer.

# 14
# Latent Growth Curve Models

Kristopher J. Preacher

Structural equation modeling (SEM) is one of the most flexible and commonly used tools in the statistical toolbox of the social scientist. *Latent growth curve modeling* (LGM), the subject of this chapter, is one application of SEM to the analysis of change. In LGM, repeated measures of a variable (hereafter, *Y*) are treated as indicators of latent variables, called *basis curves*, that represent aspects of change—typically intercept and linear slope factors. Values of the time metric (e.g., age, day, or wave of measurement) are built into the factor loading matrix to reflect the form of the hypothesized *trajectory*, or trend over time. There are many extensions of this idea, but these are the basic elements common to all applications of LGM. LGM contains elements of both variable-centered and person-centered approaches (Curran & Willoughby, 2003), in that a sample-level summary of change is provided, yet individual differences in initial status and change are also considered.

The advantages of LGM over rival techniques for modeling change are numerous. A primary advantage is that LGM affords the researcher the ability to model aspects of change as *random effects*; that is, the means, variances, and covariances of individual differences in intercepts and slopes can be estimated. Because LGM is a special case of SEM, all of the benefits of SEM apply to LGM as well. These include the ability to assess the fit of the model to data, the ability to assess change in latent variables, and the ability to examine the antecedents and sequelae of change. Missing data (assuming they are missing at random) pose no problem for LGM. Perhaps the greatest advantage of LGM is its flexibility. Cases need not be measured at the same occasions, or even at equally spaced intervals. Complex nonlinear trajectories can be modeled. LGM can be adapted in creative ways to address new problems.

There are currently four book references devoted exclusively or primarily to LGM (Bollen & Curran, 2006; Duncan, Duncan, & Strycker, 2006; Preacher, Wichman, MacCallum, & Briggs, 2008). In addition, there exist many accessible articles and chapters on the subject (e.g., Byrne & Crombie, 2003; Chan, 1998; Curran, 2000; Curran & Hussong, 2003; Hancock & Lawrence, 2006; Singer & Willett, 2003; Willett & Sayer, 1996). Any SEM software capable of accommodating mean structures and multiple groups (e.g., AMOS, EQS, LISREL, Mplus, Mx) may be used to specify these models.

Because LGM is a special application of SEM, the reader may notice some degree of overlap between the desiderata enumerated here and those described in Chapter 28 of this volume. Table 14.1 of the present chapter addresses some of these in the context of LGM, and includes *additional* desiderata specific to the case of LGM. To get the most out of this chapter, it should be read in conjunction with Chapter 28.

Table 14.1 Desiderata for Latent Growth Curve Models

| Desideratum | Manuscript Section(s)* |
|---|---|
| 1. Substantive theories motivating the model under scrutiny are described; a set of a priori specified competing models is generally preferred. | I |
| 2. The metric of time (or, more generally, the substrate of change) should be described. | I |
| 3. The functional form of the hypothesized trajectory of change is described. | I |
| 4. Path diagrams are presented to facilitate the understanding of the conceptual model of change and the specification of the statistical model. | I |
| 5. The scope of the study is outlined; if the study delineates a theory of change over time, enough time must be permitted to elapse for the phenomenon of interest to unfold. | I |
| 6. Repeatedly measured variables are defined and their appropriateness for inclusion in the study is justified. | M |
| 7. How any theoretically relevant control variables are integrated into the model is explained. | M |
| 8. The sampling method(s) and sample size(s) are explicated and justified. | M |
| 9. The treatment of missing data and outliers is addressed. | M, R |
| 10. The name and version of the utilized software package are reported; the parameter estimation method is justified and its underlying assumptions are addressed. | M, R |
| 11. Problems with model convergence, offending estimates, and/or model identification are reported and discussed. | R |
| 12. Summary statistics for measured variables are presented; if raw data were analyzed, information on how to gain access to data is provided. | R |
| 13. Recommended data-model fit indices from multiple classes are presented and evaluated using literature-based criteria. | R |
| 14. If incremental fit indices are used, an appropriate null model is specified and fit rather than relying on incorrect estimates provided by software. | R |
| 15. For competing models, comparisons are made using statistical tests (for nested models) or information criteria (for non-nested models). | R |
| 16. For any post hoc model respecification, theoretical and statistical justifications are provided and the model is fit to a new sample. | R |
| 17. Parameter estimates, together with information regarding their statistical significance, are provided. | R |
| 18. Appropriate language regarding model tenability and structural relations is used. | D |

*  *Note*: I = Introduction, M = Methods, R = Results, D = Discussion

## 1. Substantive Theories and Latent Growth Curve Models

LGM is intended as a way to test the a priori predictions of a theory of change against observed data. Therefore, it is critical that the researcher have a well-articulated theory of change before attempting to use LGM. The Introduction section of an article using LGM should build a case for testing specific hypotheses of change. Typically this involves stating a theoretical reason for specifying individual trajectories that are characterized by aspects of change (intercept, slope, and so on) that, in turn, are expected to vary across sampling units. The point of most applications of LGM is to obtain estimates of the means, variances, and covariances of these trajectories, and these parameters should have consequences for the theory under scrutiny. The lack of strong theoretical predictions can lead to the misuse of LGM to generate theory from data in an exploratory, inductive fashion. As with SEM in general, testing alternative models of change provides a more comprehensive survey of competing ideas—and is more scientifically sound—than testing a single model of change.

Given that the researcher has in mind a strong theory of change that makes LGM an appropriate analytic strategy, then attention must be given to the question "change in what?" Most applications of LGM involve modeling change in the same variable over time, but this is a questionable undertaking if the nature or meaning of the variable itself changes over time. Change in the fundamental meaning of the variable could be mistaken for change in mean level. For example, common age-appropriate aggressive behaviors in 6-year-old children may decrease in frequency over time not because levels of the aggression construct decline, but rather because other aggressive behaviors take their place. One way to address this problem is by invoking theory or past findings to support the stability in interpretation of the repeatedly measured variable. Another way is to replace directly observed repeated measures with repeated latent variables, each of which has multiple indicators at each measurement occasion. This approach permits tests of *longitudinal factorial invariance*, a way to assess stability in the meaning of a construct over time. Applications that use repeated latent variables should formally address longitudinal invariance.

## 2. The Metric of Time

The overwhelming majority of applications of LGM involve some metric of time as the substrate of change. Time can be measured in units ranging from milliseconds to decades. Quasi-time metrics such as wave of study, developmental stage, or school grade may be used. In fact, the data analyzed with LGM need not be longitudinal at all. For example, there is theoretically no hindrance to replacing the time metric with, for example, stimulus intensity, distance, or dosage, assuming these repeated measures are assessed or administered in a within-subject fashion and ordered in some logical way. In the case of stimulus intensity, the "origin" measure (typically time 0 in a longitudinal study) could represent the absence of a stimulus; it could represent the baseline dosage of a drug in a repeated-measures medical trial. For the remainder of this chapter I refer to the metric as "time," but this is not meant to exclude other substrates of change.

Two factors should be considered in any application of LGM: *origin* and *scale*. The origin of the time metric refers to the zero-point. The location of the zero-point (e.g., age 37, initial wave of measurement, or time of intervention) has implications for the interpretation of model parameters related to the intercept. For example, the first occasion of measurement is often chosen as the origin to permit interpretation of the intercept as "initial status," although other choices are feasible and more appropriate in different circumstances (e.g., "time of death" as the last measurement occasion). The scale of a metric refers to the unit of time (e.g., year, minutes since treatment, or developmental stage). Scale is important mainly for interpreting parameters related to the slope, because (linear) slopes are interpreted as the model-implied change in the outcome per unit increase in time. The choice of origin and scale either should be justified by the researcher or should be obvious from the context, and should not be chosen arbitrarily.

## 3. Functional Form

Most applications of LGM involve testing linear trends. That is, the researcher hypothesizes that scores on the repeated measures proceed upward or downward in a linear fashion. Individuals may vary around this mean linear trend if intercept and slope variances are also estimated, meaning that the basic growth curve model provides for variability in level and rate of change. In LGM, values representing the trend are entered into columns of a matrix of factor loadings ($\Lambda$) in the following way. The first column of $\Lambda$ always consists of a column of 1s to act as multipliers for the intercept factor. The remaining columns represent functions of the values of the time metric. For example, for a simple linear trend with four equally spaced repeated measurements, $\boldsymbol{\Lambda}$ could be represented in any of the following equivalent ways:

$$\boldsymbol{\Lambda}_A = \begin{bmatrix} 1 & 0 \\ 1 & 1 \\ 1 & 2 \\ 1 & 3 \end{bmatrix} \quad \boldsymbol{\Lambda}_B = \begin{bmatrix} 1 & -2 \\ 1 & 0 \\ 1 & 2 \\ 1 & 4 \end{bmatrix} \quad \boldsymbol{\Lambda}_C = \begin{bmatrix} 1 & -3 \\ 1 & -2 \\ 1 & -1 \\ 1 & 0 \end{bmatrix} \quad \boldsymbol{\Lambda}_D = \begin{bmatrix} 1 & 10 \\ 1 & 20 \\ 1 & 30 \\ 1 & 40 \end{bmatrix},$$

where the first column is always the constant 1 and the second column (containing actual values of the time metric) represents linear growth. The factor loadings in the $\boldsymbol{\Lambda}$ matrix often are represented in diagram form as path values on arrows connecting intercept and slope factors to the repeated measures of the outcome variable. In Figure 14.1, the loadings in $\boldsymbol{\Lambda}_A$, $\boldsymbol{\Lambda}_B$, $\boldsymbol{\Lambda}_C$, and $\boldsymbol{\Lambda}_D$ are depicted in simplified path diagram form (path diagrams are treated at greater length under Desideratum 4). Notice that the location of the zero-point, and thus the occasion at which the intercept is interpreted, changes from specification to specification, as does the metric of time. The choice of both the origin and scale of the time metric should be consistent with theory and with the research context. The origin should be chosen carefully to correspond with a theoretically important occasion—for example, the time of initial assessment, time of intervention, or time of death—so that time can be thought of as *time since* (or until) that event. In the first loading matrix ($\boldsymbol{\Lambda}_A$), the origin is placed at the first occasion of measurement. In the second ($\boldsymbol{\Lambda}_B$), the origin occurs at the second occasion of measurement, and so on. The scale is always chosen to correspond to a theoretically important metric. For example, change over time in $\boldsymbol{\Lambda}_B$ might be measured in two-month intervals, with an intervention imposed at the second occasion of measurement, whereas in $\boldsymbol{\Lambda}_D$ the unit of time is age in years, and the origin (i.e., birth) falls 10 years before the first assessment. It is usually not necessary to explicitly provide the $\boldsymbol{\Lambda}$ matrix, especially if an appropriately labeled path diagram is provided (see Desideratum 4), but it often can be helpful as an aid to understanding.

The basic linear LGM can be extended in numerous creative ways. For example, the first $\boldsymbol{\Lambda}$ matrix below ($\boldsymbol{\Lambda}_E$) specifies quadratic growth for four equally spaced repeated measurements—the first column provides multipliers for an intercept factor, the second for a linear factor, and the third for a quadratic factor. The $\boldsymbol{\Lambda}_F$ loading matrix provides for linear growth over four unequally spaced

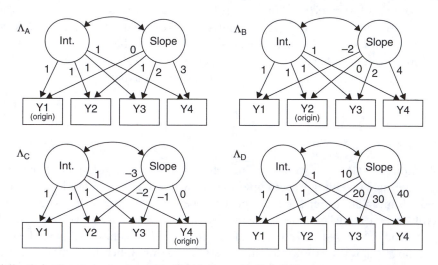

**Figure 14.1** How the Loadings in $\boldsymbol{\Lambda}_A$, $\boldsymbol{\Lambda}_B$, $\boldsymbol{\Lambda}_C$, and $\boldsymbol{\Lambda}_D$ Might Be Represented in Path Diagrams.

measurement occasions (the second occasion being 4 time units since the initial occasion, the third 5 units, and the fourth 7 units). The $\Lambda_G$ loading matrix represents an unspecified trajectory, in which the researcher has no specific trajectory in mind, but is willing to let the data determine the shape of change over time.

$$\Lambda_E = \begin{bmatrix} 1 & 0 & 0 \\ 1 & 1 & 1 \\ 1 & 2 & 4 \\ 1 & 3 & 9 \end{bmatrix} \quad \Lambda_F = \begin{bmatrix} 1 & 0 \\ 1 & 4 \\ 1 & 5 \\ 1 & 7 \end{bmatrix} \quad \Lambda_G = \begin{bmatrix} 1 & 0 \\ 1 & \lambda_{2,2} \\ 1 & \lambda_{3,2} \\ 1 & 1 \end{bmatrix}$$

There are two major points that should be addressed concerning functional form. The first is that neither the functional form of the hypothesized trajectory nor the measurement schedule need to conform to a rigid and limited set of options. Flexibility is a hallmark of LGM. The second point is that, regardless of what trend is hypothesized and fit, it must be explicitly justified on the basis of theory. It is rarely a good idea to use LGM in a theoretical vacuum, or to use it to approximate a trend of unknown shape for descriptive purposes. The option to approximate a functional form rather than test an a priori hypothesized one does exist, as the $\Lambda_G$ matrix above demonstrates, but this practice is exploratory, not confirmatory, so conclusions should be worded to reflect the partly atheoretical nature of such trends.

## 4. Path Diagrams

The use of *path diagrams* is explained in Chapter 28 of this volume. Everything said about path diagrams in Chapter 28 applies here as well, because LGM is a special case of SEM. Path diagrams are not

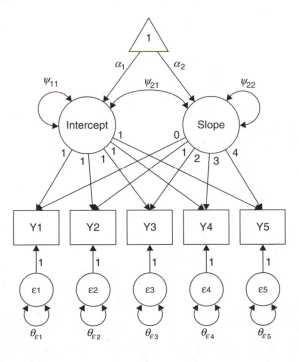

**Figure 14.2** Path Diagram of a Linear Latent Growth Curve Model with Random Intercepts, Random Slopes, and Unconstrained Residual Variances.

required in applications of LGM, but they almost always greatly facilitate interpretation, especially for readers unacquainted with the method.

Very spartan diagrams were used to illustrate loadings under Desideratum 3. An example of a full latent growth curve path diagram is given in Figure 14.2. As in other SEM path diagrams, circles represent latent variables, squares are measured variables (here, repeated measures), single-headed arrows are path coefficients (regression-type weights), and double-headed arrows are variances or covariances. Aspects of change (intercepts, slopes, and so on) are considered latent variables because they cannot be directly observed. They usually are permitted to vary across people and to covary with one another (e.g., initial status may covary with rate of change), so parameters representing those variances and covariances are often included in the diagram (therein labeled as $\psi$). Together, the Intercept and Slope factors comprise the *latent trajectory*. In addition to the information provided in Chapter 28 in this volume, there are several noteworthy features specific to diagrams used in LGM. The triangle represents a constant 1.0, and is otherwise treated as a variable. Therefore, the path coefficients labeled as $\alpha_1$ and $\alpha_2$ represent the means of the Intercept and Slope latent variables, respectively. Occasion-specific residual variances, labeled as $\theta_{\varepsilon(1-5)}$ in Figure 14.2, are included to represent the portion of the variance in the outcome not explained by the latent trajectory.

The other noteworthy feature of Figure 14.2 is the set of loadings connecting the latent trajectory factors with the outcomes. Unlike most applications of SEM or confirmatory factor analysis (see

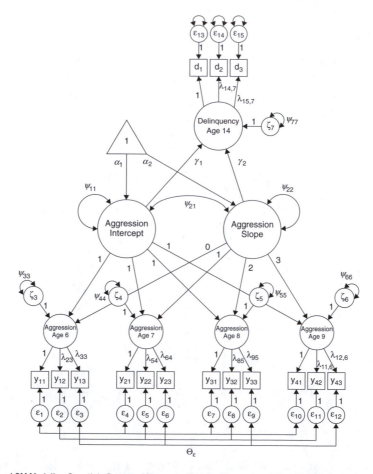

**Figure 14.3** A Linear LGM Modeling Growth in Repeated Measures That Are Themselves Latent Variables with Multiple Indicators.

Chapter 8), all of these factors are typically fixed to point values that reflect the hypothesized trajectory. In Figure 14.2, that trajectory is linear, but that can be changed by altering the elements of the $\Lambda$ matrix (see Desideratum 3). If any elements of $\Lambda$ are freely estimated, the researcher is approximating an unknown trend rather than testing a hypothesis about a specific trend.

Growth curve models do not have to conform to the linear model depicted in Figure 14.2. For example, more latent trajectory factors may be added (e.g., quadratic, cubic). The equality constraint on residual variances can be freed, within certain limits (see Desiderata 5 and 11). The observed repeated measures can be replaced with latent variables. Because LGM is a special application of SEM, Intercept and Slope factors may serve as independent or dependent variables in larger path diagrams. Multiple latent growth curves may be included in the same model. Figure 14.3 depicts an elaborate example of hypothesized linear growth in a latent aggression construct from ages 6 through 9, where the effects of Intercept and Slope on Delinquency at age 14 are of interest. Residuals of similar indicators (i.e., those of $y_{11}$, $y_{21}$, $y_{31}$, and $y_{41}$) are permitted to covary in this example.

## 5. Scope of the Study

*Scope* in this context refers to the number and range of repeated measurements. These models find their greatest use when there are only a few repeated measurements per case (but usually at least 4 or 5), and when the sample size is large (see Desideratum 8), although LGM is by no means limited to such situations. The three most important questions to consider about the scope of a study are:

1. *Were a sufficient number of occasions chosen to identify the model?* A model is *identified* if the type and number of constraints is sufficient to guarantee a unique solution for every model parameter. For time-balanced data (in which data are collected over a limited number of discrete occasions), there must be at least $k + 1$ repeated measures for the model to be identified, where $k$ is the number of basis curves (i.e., growth trajectory factors). This rule applies regardless of whether or not the residual variances are constrained to equality over time. For example, because a cubic trend requires four basis curves (intercept and linear, quadratic, and cubic slope components), there must be at least five repeated measures. If there are exactly five repeated measures and residual variances are constrained to equality, then $df = 5$. If residual variances are freely estimated, then $df = 1$. This rule is simply an elaboration of the assertion that two points define a line. At least two repeated measures are required in order for linear growth to be defined, but in order to test hypotheses about linearity, at least one more occasion is required. Identification rules become more complicated as the model departs from a basic growth curve. Fewer repeated measures may be required if some parameters are further constrained.

2. *Were a sufficient number of occasions chosen to adequately estimate the mean trend?* Even if the model is identified, $k + 1$ repeated measures are only barely enough to estimate parameters for a model with $k$ basis curves. More than the bare minimum are generally preferred. Three occasions are required to test a hypothesis of linearity, but six occasions provide a far superior test of linearity. A large number of repeated measures becomes particularly important as the complexity of the hypothesized trend increases. It is also important to concentrate measurements around the part of the trend likely to show the greatest changes. For example, if the hypothesized trend is a negative quadratic curve (in which levels of $Y$ are expect to start low, reach a maximum, and then decrease again), and the researcher has control over when measurements are scheduled, it is sensible to time the measurements such that they are more highly concentrated in the region of the maximum than in the extremes (although these are also necessary). Doing so generally reduces the likelihood of encountering estimation problems.

3. *Does the study span an interval sufficiently long to capture the process?* It is important to ensure that enough time elapses over the course of the study to adequately capture the process of interest. For

example, if growth in the height of elementary school children is measured on seven occasions spanning five weeks, clearly not enough time has elapsed to observe meaningful growth, and error of measurement likely will eclipse actual growth. If the trend under study is actually S-shaped (e.g., a learning curve) but data are collected during only a brief window of this process, the trend may appear to be linear. In such cases, a linear LGM might provide spuriously good fit, whereas a more appropriate nonlinear LGM may appear to overfit the data. By the same token, extending measurement beyond what is necessary to capture a trend can waste resources. If a simple linear trend is assessed across 18 occasions, the researcher may be better served to collect data from more cases at fewer occasions.

The key points regarding the scope of the study are that the number, spacing, and range of repeated measurements all need to be considered and justified on the basis of theory and minimum identification requirements. The number of repeated measurements should exceed the minimum, but should not be so numerous as to be wasteful. Measurements should be concentrated around regions of greatest expected change. Finally, the interval spanned by the first and last measurement must be sufficient to permit the process under study to unfold.

## 6. Repeatedly Measured Variables

As in any application of SEM, all measured variables in the model should be defined clearly, or else references should be provided. Some basic requirements of the repeatedly measured variables are that they be reliable and valid, and they should represent some attribute or characteristic that is able to change in *level* over time, but not *meaning*. The repeatedly measured $Y$ should not represent stable attributes for which there is no theory of systematic change. In addition, it is necessary to provide a theoretical rationale for expecting not only growth, but growth of a particular form in a particular direction. If intercept and slopes factors are expected to vary across individuals, this expectation should have its roots in theory. In short, it should be realistic and theoretically appropriate to expect change in the chosen $Y$ variable, and $Y$ should be demonstrably reliable and valid.

As in other applications of SEM, special estimation procedures are required for variables that are ordinal, binary, or censored. Application of standard LGM to discontinuous or nonnormally distributed variables violates key assumptions necessary for legitimate statistical inference. Special procedures must be invoked.

## 7. Control Variables

Often researchers may want to examine the effects of some variables on others after controlling for covariates. Chapter 28 in this volume outlines some rules that should be followed in including covariates in SEM, and these rules largely apply in LGM as well. In LGM, two kinds of covariates are distinguished based on their location in the model. *Time-varying covariates* (TVCs), as their name implies, are predictor variables modeled at the level of the repeated measurements. They may be included in the model either as distinct variables predicting $Y$ at each occasion of measurement, or as additional loading columns in the $\Lambda$ matrix in the same way that the variable *time* is included in the $\Lambda$ matrix. The first method is traditional practice in applications of LGM. The second is identical to how TVCs are included in hierarchical linear models (see Chapter 10, this volume). The LGM framework is flexible enough to permit either specification. Using the first approach, it is probably wise to permit TVCs to covary with other exogenous variables in the model (e.g., the latent trajectory factors).

*Time-invariant covariates* (TICs), on the other hand, are included at the subject level. TICs are exogenous predictor variables used to predict individual differences in aspects of change. For example, in Figure 14.3 we might introduce gender as a predictor of the Intercept and Slope factors. For both kinds of covariate (TVC or TIC), interest may be in controlling for the effects of covariates, that is, removing them from consideration so that effects of more substantive interest may interpreted more "purely." Or, interest may be in interpreting the effects of the covariates directly.

## 8. Sampling Method and Sample Size

The requirements regarding sampling method and sample size stated in Chapter 28 still hold when the model in question is a latent growth curve. If stratified or cluster sampling is used, it is crucial that this be considered in the modeling stage by employing sampling weights or multilevel SEM, or else the researcher should justify why these advanced procedures are not employed.

It is also essential that studies reporting latent growth curve analyses explicitly address the issue of sample size. As in non-LGM applications of SEM, there are several things to consider when choosing a sample size. First, $N$ needs to be large enough to support the estimation of potentially many free model parameters. Maximum likelihood (ML) is a large-sample technique, and alternative estimation algorithms may require sample sizes much larger than ML. Second, the sample must be large enough to achieve adequate power for rejecting poor models by some criterion of fit. The criterion most commonly used for this purpose is RMSEA. Applications of LGM tend to have high power, but this does not release the researcher from the obligation to demonstrate that power is adequate in a given application. Third, the sample must be large enough for parameters of interest to have small standard errors (and thus narrow confidence intervals and high power). This is a very important consideration in LGM, where interpretation of parameters is of central interest. Finally, longitudinal studies are typically characterized by missing data due to attrition, death, late entry into the study, and other causes. The total sample size must be large enough to accommodate the amount of missing data. Although missing data techniques such as imputation may be used to fill the gaps in a data set to permit easy analysis, they cannot create information lost to attrition.

Ideally, the researcher should not only meet the minimum sample size suggested by these considerations, but exceed it by a considerable margin. The sample size may be just large enough to exceed the minimum required to achieve adequate power for tests of individual parameters, yet still fall short of the $N$ required to yield usefully narrow confidence intervals. If the sample size is beyond the researcher's control, such as when samples of convenience or archived data are used for analysis, the researcher should still demonstrate that $N$ is large enough to support estimation of a growth curve model and valid interpretation and testing of parameters. The important point here is the sample size should be justified on reasoned grounds, not simply reported.

## 9. Missing Data and Outliers

Very few studies have complete data on all variables for all cases. In long-term longitudinal studies, where the same individuals are followed over time, there are particularly many reasons some observations may be missing for some cases, and missing data in turn may threaten the generalizability of results. These reasons may include attrition due to illness, incarceration, death, data management errors, late entry into the study, lack of subject motivation to follow through, and cancellation of research funding. Late entry and attrition are particularly dangerous in LGM studies, as data missing due to late entry or attrition are typically *not* missing at random. Longitudinal studies with nontrivial amounts of missing data at the beginning or end of the studied trajectory are subject to severe bias in intercept and slope mean and variance estimates.

Four broad strategies for dealing with missing data may be identified: *prevention, deletion, full-information*, and *imputation*. The best strategy to address missing data is to preemptively prevent data loss by design. Researchers should make every reasonable effort to minimize the proportion of missing data. Data deletion strategies (i.e., pairwise and listwise deletion) use only those cases with complete data for all or some of the variables, and are usually to be avoided. Full-information strategies involve estimating model parameters using all available information, even if some of that information comes from cases with incomplete data. Usable information can be gleaned even from cases with a single valid data point. Imputation strategies use information from existing data about the relationships among variables, and then fill in, or *impute*, reasonable values for the missing data. Complete-case methods are then applied to the imputed data set. Mean imputation, in which missing values of a variable are replaced with the sample mean, should be avoided because it often results in a distribution with unrealistically many cases at the mean value. If imputation is used, multiple imputation (in which several data sets are imputed and the results are averaged across imputations) usually gives the best results. There are other, less often used strategies but these account for the majority of them. Full-information and multiple imputation methods are those most often recommended by methodologists to deal with missing data. Pairwise and listwise deletion and mean imputation should not be used without extraordinarily compelling reasons.

In LGM, data may be missing by design. For example, in research that combines multiple overlapping cohorts, one cohort may be measured at occasions 1, 2, 3, and 4, while "missing" the fifth occasion of measurement, whereas another may be measured at occasions 2, 3, 4, and 5, while "missing" the first occasion of measurement. Combined, all five occasions are represented, but the first and fifth occasions are not as well represented as the other three. This kind of data collection strategy can save time, but it obliges the researcher to assume that it is legitimate to combine cohorts to form a single trajectory. This assumption can and should be tested rather than assumed. If multiple cohorts are included, data missing by design from one cohort should not be imputed because it can lead to the creation of extrapolated data that are unrealistically congruent with those from other cohorts.

In summary, some general guidelines for reviewing studies with missing data can be developed. The amount and kind of missing data should be explicitly addressed, as should the likely reasons missing data were missing, the steps taken to address missingness in the analysis, and the likely impact of missing data on statistical analyses, the study's conclusions, and generalizability. Full-information and multiple imputation strategies are usually the best choices for dealing with missing data. Regardless of the strategy chosen to address missing data, the researcher should justify that choice. In reporting missing data, it is helpful to report the percentage of missing data for each variable at each occasion. Reporting only the percentage of complete cases does not provide enough information.

## 10. Software and Estimation Method

For key software considerations the reader is referred to Chapter 28 in this volume. Not all SEM software packages can fit growth curve models. Because the key parameters in LGM include the means of latent variables, the software must be capable of modeling means. Currently, the major SEM software packages capable of employing LGM are AMOS, EQS, LISREL, Mplus, Mx, and OpenMx. Because SEM software is regularly updated and improved, it is important to list the name and version of the software used to fit models and obtain parameter estimates.

## 11. Problems with Convergence, Estimates, and Identification

All of the advice in the corresponding section of Chapter 28 applies to LGM. Errors of identification, convergence, and estimation occur routinely in specifying and fitting models. Accurate documentation of these problems, and the steps taken to remedy them, is essential in reporting results.

It takes a fair amount of skill to properly specify a standard structural equation model, but LGM often requires even greater facility with software and knowledge of the mathematics behind the model. It is easy to make mistakes in specifying a model, and often these mistakes go unnoticed because the software is incapable of distinguishing sensible models from nonsensical ones, or because software will cheerfully provide reasonable-looking results despite serious problems. Here three examples of problems that sometimes occur in the application of LGM, and which may go unnoticed by inexperienced researchers, are described.

First, if a model is underidentified, some SEM software applications (e.g., LISREL and Mplus) may automatically and unobtrusively add constraints to some parameters to render the model identified. Occasionally the researcher is unaware that this automatic identification occurs, and parameters that should not be interpreted are interpreted nevertheless.

Second, negative or boundary values of variance parameters may be reported. Latent growth curve models are fairly robust to estimation problems, but if the model is severely misspecified some very strange things may happen. For example, residual variances may sometimes be estimated as negative (or, depending on the software, constrained to a boundary value of zero). This kind of result is a serious and common estimation error known as a *Heywood case*, and is usually indicative of serious model misspecification or, sometimes, a sample that is too small. If a residual variance of zero is reported, it is likely a Heywood case rather than a true zero. This kind of error may go unnoticed and unreported because the parameters of central interest in LGM are those related to the intercept and slope factors, not typically the residuals. Thus, if residual variances are not reported, they should be.

Third, covariances among aspects of change (intercept, slopes) may correspond to correlations that lie outside of the logical bounds of $-1.0$ and $+1.0$. This problem may not be immediately recognizable if only variances and covariances are reported. For example, both covariance matrices below, labeled as $\mathbf{\Psi}_A$ and $\mathbf{\Psi}_B$, may appear to be legitimate covariance matrices at first glance, but only $\mathbf{\Psi}_A$ is acceptable or "proper." The covariance in $\mathbf{\Psi}_A$ equates to a reasonable correlation of $.28/(.19^{1/2} \times .64^{1/2}) = .80$, but the covariance in $\mathbf{\Psi}_B$ equates to an impossible correlation of $.38/(.19^{1/2} \times .64^{1/2}) = 1.09$.

$$\mathbf{\Psi}_A = \begin{bmatrix} .19 & .28 \\ .28 & .64 \end{bmatrix} \qquad \mathbf{\Psi}_B = \begin{bmatrix} .19 & .38 \\ .38 & .64 \end{bmatrix}.$$

If an improper matrix is reported, this indicates that an undetected estimation error has probably occurred. Such errors sometimes can be addressed by correcting coding errors, removing outliers from the data, or providing better starting values for the estimation procedure. Sometimes the problem cannot be solved, which usually indicates that specified growth model is not appropriate for the data.

## 12. Data Display and Accessibility

Fitting latent growth curve models requires access either to raw data or to a covariance matrix and mean vector. In line with recommendations endorsed in Chapter 28 of this volume, the data should be reported or made available to permit other researchers to verify or reexamine reported results. Whenever possible—given the constraints of the study, journal space, and proprietary issues—summary information in the form of a covariance or correlation matrix, mean vector, and standard deviations for complete data typically are sufficient to permit reanalysis. If some data are missing or if journal space does not permit reporting summary data, authors should provide instructions informing readers how they may obtain the data.

## 13. Data-Model Fit

A primary advantage of SEM is that it permits the assessment of fit between the model and data. The advice offered in Chapter 28 of this volume is reiterated here. The $\chi^2$ statistic by itself has only limited usefulness as a fit index because of its sensitivity to sample size and trivial departures from perfect fit. It is usually wise to report multiple (at least three) fit indices drawing from the three broad classes (*absolute indices, parsimonious indices*, and *incremental indices*). SRMR, RMSEA, and TLI (NNFI) are perhaps the best representatives of these classes, but individual researchers may choose not to limit themselves to these, or to include all of them. If RMSEA is reported, its associated 90% confidence interval should also be reported and interpreted. Models may also be compared on the basis of relative fit.

## 14. Appropriate Null Model

Incremental fit indices like those discussed in Chapter 28 (NFI, NNFI/TLI, and CFI) constitute an important class of fit indices. They express the fit of a substantive model as falling somewhere between the fit of a highly restrictive "null" model and that of a saturated (perfectly fitting) model. The appropriate null model must satisfy two requirements: (a) it must be nested within the most restrictive substantive model to be tested and (b) it must constrain all covariances among observed variables to zero (Widaman & Thompson, 2003). The null model typically used in SEM is one in which only the means and variances of the observed variables are estimated, constraining the covariances to zero. In ordinary applications of SEM, this null model is perfectly appropriate. LGM, however, requires a different null model. Specifically, if the model to be tested is any growth curve model in which the residual variances are all freely estimated, the appropriate null model is an intercept-only model in which the only free parameters are the intercept mean and the free residual variances—a model with $p + 1$ free parameters that implicitly constrains the variables' means to equality but permits them to have different variances, and constrains their covariances to zero. On the other hand, if the model to be tested is any growth curve model in which the residual variances are constrained to equality, then the appropriate null model is an intercept-only model in which the only free parameters are the intercept mean and the constrained-equal residual variance (only two parameters). Thus, the incremental fit indices reported by default by most SEM programs are incorrect for latent growth curve models. They must be computed by hand by explicitly fitting the appropriate null model, obtaining the resulting $\chi^2$ statistic, and manually computing the desired incremental fit index.

## 15. Model Comparisons

Investigating alternative models of change is a sound and recommended approach to theory evaluation. Compared to the evaluation of individual models in isolation, the comparison of rival models of change has a better chance of eliminating some theories from consideration. For recommendations regarding model comparison, the reader is referred to Chapter 28, Desideratum 14. Model comparison proceeds exactly the same way in LGM as in the more general SEM. Examples of the kinds of models one might wish to compare in LGM could include linear vs. quadratic models, or models in which residual variances are constrained to equality vs. freely estimated.

## 16. Model Respecification

Latent growth curve models are notoriously poor-fitting by traditional criteria. This poor fit arises not because LGM is unrealistically restrictive, but rather because most other applications of SEM have relatively many free parameters and show unrealistically *good* fit. The trajectories specified in LGM are highly constrained and are not likely to arise by chance in nature. Like any model, growth curve models are merely approximations to reality, and cannot be expected to fit perfectly. However, because tradition and publication pressures have made good fit a necessary component of publishing applications of SEM, the researcher may be tempted to counter instances of poor fit by freeing parameters identified by modification indices (see Chapter 28, Desideratum 15) and fitting the modified model to the same data, resulting in better fit. This temptation should be resisted. A good rule of thumb is that a model may be modified to any degree on the basis of modification indices, but (a) the modifications must be theoretically appropriate and (b) the modified model should be fit to new data to avoid the possibility of capitalizing on chance characteristics of the sample.

Using modification indices is especially discouraged in LGM because relaxing constraints on the model can severely compromise the interpretation of the model as a specific trajectory. For example, if the model in Figure 14.2 were fit to data and the software reports a large modification index for the loading connecting *Y4* to the Slope factor, freeing the loading may improve fit, but the resulting model can no longer be interpreted as a linear growth curve. Furthermore, the model no longer represents a test of the original, theoretically prescribed hypothesis. Even when not fit to new data, modified models nevertheless have use as descriptive tools. The main point of this desideratum is that modified models may be useful, but tests of such models should not be treated as confirmatory or as strict tests of hypotheses about growth over time.

## 17. Parameter Estimates and Significance

The end product of model-fitting is a collection of parameter estimates and fit indices. Assuming that model fit is reasonable and that no convergence, estimation, or identification problems persist, it is important to report the magnitude and significance of all model parameters—not only those that are of central interest, but *all* of them. In typical applications of LGM this number is not large. In the basic linear LGM with homoscedastic residuals, for example, there are only six parameters to report: the mean intercept and slope, intercept and slope variances, their covariances, and a common residual variance. More complicated models result in more parameter estimates to report.

A simple way to report parameter estimates is to place the point estimates in the appropriate locations in a path diagram (see Desideratum 4) along with some indication of significance or precision, such as confidence intervals, standard errors, $p$-values, or a system of asterisks indicating levels of significance. The method by which significance is determined is an important and often overlooked consideration. For some parameters—path coefficients and latent variable means—the traditional $z$-tests (dividing the point estimate by its standard error) usually are appropriate. However, for variance and covariance parameters, $z$-tests are not appropriate because such tests require the assumption that the tested parameter is normally distributed over repeated sampling, an untenable assumption for variances. A better test for variances and covariances is the *likelihood ratio test* (see Chapter 28, Desideratum 14), in which the fit of a model that fixes the parameter to zero is compared to the fit of a model in which the parameter is freely estimated. Depending on the parameter's role in the model, this test may not always be possible or appropriate. Alternatives are to obtain bootstrap confidence intervals for parameter estimates (an option available in several SEM software applications) or likelihood-based confidence intervals (available in Mx).

## 18. Interpretive Language

Several important points from Chapter 28 regarding interpretive language are reiterated here. First, latent growth curve models, and the theories of change they represent, are never literally "true," nor can they ever be confirmed empirically. Such statements are extremely misleading and should be discouraged at every opportunity. Models represent formal hypotheses, and hypotheses may fail to be rejected for many reasons, among them low statistical power. In fact, any structural equation model (LGM included) can be made to fit perfectly by freeing a sufficient number of parameters, but perfect fit does not imply a confirmed model. The outcome variable $Y$ is not observed as a continuous function of time, so there is no foolproof way to confirm that $Y$ values corresponding to unobserved values of time would similarly conform to the trend followed by the observed values, even if perfect fit is observed in the sample. Similarly, growth curve models with zero or negative degrees of freedom would fit *any* data equally well (i.e., perfectly), regardless of what process generated the data. In neither case can perfect fit be taken as support for the researcher's theory of growth. Even identified models with good fit can be described only as tenable in light of the data.

Second, causal language should be used sparingly if at all. A theory's predictions may be causal in nature, but a model's results can never completely support a conclusion that a process is causal, regardless of how well-designed the study may be. Alternative explanations can always be devised. However, confidence that a process is causal may be strengthened by experimental manipulation (e.g., treatment vs. control), temporal precedence (causes must always precede effects in time), and strong theory that prohibits or limits plausible alternative explanations (e.g., kindergarten may "cause" growth in verbal knowledge, but never the reverse). In applications of LGM, the data are naturally longitudinal, but this does not grant a license to use causal language with impunity. At most, findings may be supportive of a causal process, but can never definitively demonstrate causality.

## References

Bollen, K. A., & Curran, P. J. (2006). *Latent curve models: A structural equation perspective.* Hoboken, NJ: John Wiley & Sons, Inc.

Byrne, B. M., & Crombie, G. (2003). Modeling and testing change: An introduction to the latent growth curve model. *Understanding Statistics, 2,* 177–203.

Chan, D. (1998). The conceptualization and analysis of change over time: An integrative approach incorporating longitudinal mean and covariance structures analysis (LMACS) and multiple indicator latent growth modeling (MLGM). *Organizational Research Methods, 1,* 421–483.

Curran, P. J. (2000). A latent curve framework for the study of developmental trajectories in adolescent substance use. In J. S. Rose, L. Chassin, C. C. Presson, & S. J. Sherman (Eds.), *Multivariate applications in substance use research.* Mahwah, NJ: Erlbaum.

Curran, P. J., & Hussong, A. M. (2003). The use of latent trajectory models in psychopathology research. *Journal of Abnormal Psychology, 112,* 526–544.

Curran, P. J., & Willoughby, M. J. (2003). Reconciling theoretical and statistical models of developmental processes. *Development and Psychopathology, 15,* 581–612.

Duncan, T. E., Duncan, S. C., & Strycker, L. A. (2006). *An introduction to latent variable growth curve modeling: Concepts, issues, and applications* (2nd ed.). Mahwah, NJ: Erlbaum.

Hancock, G. R., & Lawrence, F. R. (2006). Using latent growth models to evaluate longitudinal change. In G. R. Hancock & R. O. Mueller (Eds.), *Structural equation modeling: A second course* (pp. 171–196). Greenwich, CT: Information Age Publishing, Inc.

Preacher, K. J., Wichman, A. L., MacCallum, R. C., & Briggs, N. E. (2008). *Latent growth curve modeling.* Thousand Oaks, CA: Sage.

Singer, J. D., & Willett, J. B. (2003). *Applied longitudinal data analysis: Modeling change and event occurrence.* New York: Oxford University Press.

Widaman, K. F., & Thompson, J. S. (2003). On specifying the null model for incremental fit indices in structural equation modeling. *Psychological Methods, 8,* 16–37.

Willett, J. B., & Sayer, A. G. (1996). Cross-domain analyses of change over time: Combining growth modeling and covariance structure analysis. In G. A. Marcoulides & R. E. Schumacker (Eds.), *Advanced structural equation modeling: Issues and techniques* (pp. 125–157). Mahwah, NJ: Erlbaum.

# 15
# Latent Transition Analysis

### David Rindskopf

Social science theories refer to characteristics (and associated measured variables) that are either continuous (or nearly so) or categorical. For categorical characteristics, we may then distinguish among theories that refer to measurement at a single time point (static) or measurements at two or more time points (dynamic). For dynamic theories, interest centers around the initial distribution of people across categories, and how people transition from a category at each time point to a category (either the same or different) at another time point. The measurements at different time points might be of the same characteristic or of different characteristics (e.g., how does personality type measured at an early age relate to whether a person ends up in a white-collar or blue-collar job later). To complicate the situation further, observed measurements may be thought to be imperfect indicators of an underlying (latent) characteristic. We may now define *latent transition analysis* (LTA) as a statistical model in which (i) latent categorical constructs are defined at two or more time points, (ii) parameters are included that assess initial status and transition probabilities from time $i$ to time $i + 1$ (for most models; for others, we predict further into the future), and (iii) observed variables are imperfect indicators of the hypothesized latent variables. As a simple example, we might theorize that at each age, children either can or cannot conserve volume (in accordance with Piaget's theory). We could give a three-item test to each of a large number of children at ages 3, 4, 5, 6, and 7, and then see whether the data are consistent with the theory, and if so, assess the rate at which children transition from being nonconservers to conservers at each age. The distinguishing characteristics of this study are that (i) the model allows children to be in one of two true states at each time, (ii) the observed test items are categorical (scored right/wrong), and (iii) each child is measured several times (here, over a four-year period).

The roots of latent transition analysis are in (i) latent class analysis (see Chapter 13, this volume), conceptually originated by Lazarsfeld (1950a, 1950b, 1959) and systematically developed by Goodman (1974a, 1974b) and Haberman (1974, 1977), and (ii) panel analysis, developed originally by Lazarsfeld and expanded on by Wiggins (1973). The ideas of latent transition analysis then grew in several independent but closely related strands, sometimes under different names. General references include Collins and Flaherty (2002), Collins and Wugalter (1992), and Lanza, Flaherty, and Collins (2003). More advanced works include Böckenholt (2005), Humphreys and Janson (2000), Langeheine and van de Pol (1994), Molenberghs and Verbeke (2005), Mooijaart (1998), van de Pol, and Langeheine (1990), and Vermunt, Langeheine, and Böckenholt (1999). Example applications include: Chung, Park, and Lanza (2005), who studied substance use among females as they went

Table 15.1 Desiderata for Latent Transition Analysis

| Desideratum | Manuscript Section(s)* |
|---|---|
| 1. A theoretical structure is presented that hypothesizes | I |
|    (i) discrete latent variables occurring at more than one time/age; | |
|    (ii) conditional/predictive relations of earlier to later times/ages; | |
|    (iii) discrete observed measures of latent variables. | |
| 2. Explicit consideration is given to plausible alternative structures. | I |
| 3. Diagrams of models, if useful for communication of structures, or comparisons among them, are included. | I |
| 4. Equations representing the model(s) are included. | I |
| 5. Parameter identification (unique estimation) is proved or demonstrated. | I |
| 6. A rationale is provided for any restrictions (e.g., equality constraints) used to make model(s) identified. | I |
| 7. Software used to estimate parameters is described. | M |
| 8. Fit statistics used to evaluate model(s) are described. | M |
| 9. A tabular presentation is included of model fit statistics for each model tested, and (where feasible) cell frequencies are provided for possible reanalysis. | R |
| 10. Tabular or graphical presentation of parameter estimates and standard errors (where appropriate) are provided. | R |
| 11. Models retained (as plausible) and rejected (as implausible) are discussed, and a rationale based on fit statistics is provided. | D |
| 12. Any implausible or unusual results in any models not rejected on statistical basis (fit statistics) are discussed. | D |
| 13. Parameter estimates for each retained model are discussed. | D |
| 14. Implications of each retained model are discussed, and a comparison among them is provided. | D |

* *Note*: I = Introduction, M = Methods, R = Results, D = Discussion

through puberty (ages 12–15); Graham, Collins, Wugalter, Chung, and Hansen (1991), who studied alcohol and tobacco use among adolescents and the effects of a substance abuse prevention program on that use; and Reboussin, Reboussin, Liang, and Anthony (1998), who studied health risk behavior (in this case, carrying weapons) of schoolchildren over a five-year period.

Computer programs available for estimation of latent transition models include WinLTA (Collins, Lanza, Schafer, & Flaherty, 2002), LEM (Vermunt, 1997), and Panmark (Van der Pol, Langeheine, & De Jong, 1991), all of which are either free or inexpensive. Commercial packages such as Mplus (Muthén & Muthén, 2007) and LatentGOLD (Vermunt & Magidson, 2000, 2003, 2005a, 2005b) and SAS PROC LTA and LCA (Lanza & Collins, 2008; Lanza, Collins, Lemmon, & Schafer, 2007; Lemmon, Bray, & Chung, 2007) are also available.

Table 15.1 lists desirable characteristics of studies that use latent transition analysis. These are discussed in more detail in the corresponding sections of this chapter.

## 1. Theoretical Structure

The theoretical structure delineation begins with a consideration of the possible discrete states in which a person might be at each time of measurement. For a single construct measured repeatedly, these states (and the nature of the latent variable) will be the same at each time point. For example, developmental psychologists might want to know whether a child has reached the stage where s/he can conserve volume, and might measure each child at three ages. In other cases, the constructs might be different at different ages: A reading theory might indicate that a child who develops certain word decoding skills by the time of entry into kindergarten will be more likely to be able to read aloud fluently by the end of second grade.

Part of the theory might concern possible versus impossible transitions from one age to another. Without illness or injury, one might expect that a child who can conserve volume at one age will not lose that skill at a later age, so the transition from a more advanced to a less advanced stage would be constrained to have probability zero.

It is the authors' responsibility to make sure that all assumptions are explicitly discussed. These include the number of latent classes, which observed variables measure which latent variables (if there is more than one latent variable), and any restrictions on parameter estimates.

## 2. Plausible Alternative Structures

As reasonable as our favorite theory might seem, one must always entertain the possibility that it is wrong in some minor (or even major) aspect. Therefore, alternative plausible theories should be explicitly considered. The ideas of "multiple working hypotheses" and "strong inference" should be in every empirical researcher's repertoire; see Chamberlain (1890/1965) and Platt (1964) for a detailed explanation and rationale.

Some alternative structures will also be in the form of latent transition models, and should be tested as described here. Others might have a different structure; most commonly this will be due to some variables (either latent or observed) being continuous or count instead of discrete with a few categories. The researcher should therefore explicitly discuss all plausible alternative structures for the data.

## 3. Model Diagrams

Diagrams of models can often indicate the main qualitative features in a more easily comprehensible fashion than can equations, although equations will contain the full quantitative specification of the model. Diagrams are also useful when comparing features of two or more models. For latent transition analysis there are no standard methods of representing models, but two methods are generally used. One such method uses the conventions of path analysis models; in this type of diagram there is no explicit differentiation between latent and observed variables. Curved lines with arrowheads at both ends are used to represent unexplained relations among variables; straight lines with an arrowhead on one end are used to represent hypothesized causal inferences.

The other common method is similar to structural equation model diagrams (see Chapter 28, this volume); in these diagrams, circles are used to represent latent variables, and rectangles are used to represent observed variables. The other aspects of path diagrams (curved and straight lines) are retained in these diagrams. When feasible, a diagram should be presented for each distinct type of model tested.

## 4. Equations Representing the Model(s)

The simplest latent transition models can generally be represented notationally by three categories of parameters. First, there are the (unconditional) probabilities of being in each category of the latent measure at the first time point. For example, at the first time of measurement, children might be conservers or nonconservers of volume; the probability of being in each category must be estimated. Next are the transition probabilities, which are the probabilities of being in particular states at each future time point given the state at the previous time points. Quite often each state is hypothesized to depend only on the immediately preceding state, which results in a *latent Markov model*. Finally, parameters are needed to account for the probability of being in each category of each observed variable as a function of a person's state in the latent variable underlying the observed variable. Sometimes these latter probabilities are considered to characterize the measurement properties of the observed variables, much as factor loadings indicate how well continuous observed variables measure factors (see Chapter 8, this volume).

In some cases, LTA models are written in terms of equations related to logistic regression or loglinear modeling (see Chapters 17 and 18, this volume). These models are then translated back after estimation into probabilities. Although a model written in the logistic form is perfectly proper and valid, it is often not easily understood even by specialists, let alone researchers in subject matter areas who are not methodologists. For this reason, it is desirable to translate results of applications of LTA into probabilities (either in text, tables, or figures) when feasible.

The basic LTA model can be extended in various ways. The most frequent such extension is to multiple populations. For example, one might theorize that males and females either have different distributions among latent statuses at the first time of measurement, or that they have different measurement properties or transition probabilities. More than one such characteristic may be included in a model (e.g., gender, race, and SES, along with possible interactions), and should be when there are strong theoretical grounds for doing so.

## 5. Identifiability of Model Parameters

Some statistical techniques never (or rarely) encounter problems in estimating parameters. For example, multiple regression weights are always estimable if there are enough data, and if predictors are not overly (multi)collinear. Models with latent variables, however, are not the same; sometimes parameters are not estimable no matter how many subjects one has in a sample. Such parameters are not *identified*, to use technical terminology. A nonstatistical example might give the general idea: Consider the equation $x + y = 10$. One cannot solve uniquely for $x$ or $y$; there are too many unknowns and not enough (independent) equations.

The rules for determining whether each parameter in a model is identified, which would result in the whole model being identified, are not simple to apply. Luckily, there are numerical techniques that in most cases will discover for any particular data set whether each parameter is estimable. Also, many special cases have already been explored, and if one stays within these cases one always knows that (in principle) the model is identified.

Some models that are not identified can be made so by imposing restrictions on parameters. For example, if the same latent variable is measured at several times by the same observed variable, it is sometimes reasonable to presume that the measurement properties of the observed variable do not change over time. In this case, one would impose restrictions of equality on the conditional probabilities of responses at each time. By imposing restrictions, one is estimating fewer parameters, and in many cases this will be enough to make identified an otherwise unidentified model.

Another common type of restriction occurs when there are several times of measurement, and the conditional probabilities of changing from time $i$ to time $i + 1$ are restricted to be the same for all transition times. This type of restriction is less often theoretically justifiable than measurement restrictions, although if one can argue that a process has reached a steady state then there is more hope for it to be true.

Because most researchers will not be able to prove algebraically that a model is identified, they will have to rely on software to establish empirical identification. They should examine the standard error of each parameter; if any is suspiciously large (under most circumstances, bigger than 3 for the log-linear version of a model), they should suspect problems. If this occurs, and if any observed frequency is zero, one can add .5 to all cells and retest for identifiability. If all standard errors now look reasonable, the model can be considered identified.

## 6. Restrictions/Constraints Used for Identification

Just because one can impose restrictions to make a model identified does not mean that it is right to do so. In the case discussed in Desideratum 5, one might not be justified in assuming that an observed measure is equally good at every age. For example, at lower ages there might be more chance of misunderstandings by the child, which results in greater likelihood of errors. Such possibilities should always be considered, rather than just accepting model restrictions merely because they achieve identification. At the same time, if such restrictions are plausible (or of theoretical interest) they should be tested. Presuming both restricted and unrestricted versions of the model are over-identified, one can compare the fit of the models with and without restrictions of probabilities across groups or times of measurement.

In other cases, restrictions might be more plausible. For example, in comparing males and females, it will frequently be the case that conditional response probabilities to items will be equal for both groups, but unconditional probabilities of class membership or for transition probabilities may differ for the two groups. One can frequently test this assumption by relaxing restrictions one item at a time, searching for what is called *differential item functioning* (DIF) in the psychometrics literature. In the end, it is incumbent upon the authors to justify whatever restrictions or constraints were used to achieve identification of the model(s) being investigated.

## 7. Software

Although each computer program should provide the same estimates, not all programs will use the same terminology and structure. Authors of research manuscripts should describe the program and its use in a manner that aids those not familiar with it to fully understand what the program calculates. The major differences among programs are in the notation they use. Not only will most substantive researchers be unfamiliar with the concepts of latent transition analysis (which should be explained in the Introduction and Methods sections), but even more so they will be unfamiliar with notation used by different researchers who have developed these models. Therefore one should thoroughly explain the notation that was chosen. As mentioned in Desideratum 4, additional differences might occur in how the model is parameterized (in terms of probabilities or in terms of logistic or loglinear models).

## 8. Fit Statistics

Fit statistics are used for two main purposes. The first is to test whether a model is in reasonable agreement with the observed data patterns. This is the typical application, which is an extension of the usual chi-square tests of independence in two-way tables with which most researchers are familiar. The problem with such an extension is that if there is a large number of observed variables, the number of people in some (or possibly most) cells of the cross-tabulation will be so small that the usual test statistics will not follow a chi-square distribution. Some programs contain procedures for constructing bootstrap tests that circumvent this problem.

A second use of fit statistics is to compare the fit of different models. When one model is a special case of another (obtained by imposing restrictions on the more general model), then the likelihood-ratio fit statistics may be subtracted, and referred to a chi-square distribution with degrees of freedom equal to the number of restrictions made. This test has a chi-square distribution even though neither of its constituent fit statistics does. Although it would appear that models varying only in the number of latent classes would be nested (e.g., the two-class model is a special case of the three-class model), the comparison of such models is not straightforward, as the difference in likelihood-ratio statistics is not a chi-square distribution. Research on the comparison of such models is discussed in Nylund, Asparouhov, and Muthén (2007).

When models are not nested, one may compare any of a number of fit statistics that are adjusted for model complexity. These include the AIC (Akaike Information Criterion), BIC (Bayesian or Schwarz Information Criterion), and other modifications of these. For these statistics, like the chi-square statistics, smaller values indicate better fit. But unlike chi-square statistics, which always decrease as more parameters are added, the AIC and BIC can increase for more complex models due to the penalty that is added. For these statistics, the lowest value indicates the best fit.

Often the AIC will favor more complex models than the BIC, and one should inspect not only the overall fit statistic (to see if the BIC is penalizing complexity too much, or the AIC too little), but also the parameter estimates. Sometimes it will be obvious that a model may seem to fit well by one of these criteria, but not be easily interpretable (due to odd parameter estimates), and therefore can be rejected.

One must also take into account that sample size can influence some fit statistics; all unadjusted goodness-of-fit statistics (Pearson, Likelihood-Ratio, and Cressie-Read) will become large when sample sizes are large, even when a model is incorrect in only a minor way. Like all statistical tests, increasing sample size will result in high power to detect small differences from a null hypothesis (in this case, the hypothesis that a model fits the data in the population). The penalty for a BIC is also a function of sample size, and therefore heavily penalizes the addition of parameters (removal of degrees of freedom); this is why it tends to favor simpler models. As much as we would prefer to have absolute standards for model selection, it remains an art: One must consider the sample size, the fit statistics, the adjusted fit statistics, and the parameter estimates for plausible models. With small sample sizes, more than one model may remain plausible, and all such plausible models should be discussed.

## 9. Tables of Model Fit Statistics

When several models are tested, a table is the most useful way to display the fit statistics for each model. Degrees of freedom for each model should always be included, and if the sample size is large enough (and the observed table has few enough cells) one should include a $p$-value for each model. Similar rules apply here as for the usual cross-tabulation; if all expected cell frequencies are greater than five, there is no problem. If most are bigger than one or two, then the chi-square distribution should apply reasonably well. Some programs will print out various forms of the statistic (Pearson, Likelihood Ratio, and Cressie-Read); if all are reasonably similar, then most likely the fit statistic and associated $p$-value can be taken seriously.

## 10. Retained and Rejected Models

Referring to the table of fit statistics, the author should discuss why some models are retained as plausible, and other models are rejected as implausible. Here, overall fit statistics and comparisons of these statistics are used; later, more refined (though idiosyncratic) decisions can be made on the basis of specific aspects of a model (see Desideratum12).

When it is reasonable to do so, one might use a combination of judgments about the fit of each model in isolation and the comparison of fits of sets of models. Sometimes only comparisons are reasonable due to small expected frequencies. As in logistic regression, with large cross-tabulated tables many small cell frequencies occur, in which case the usual fit statistics do not have a chi-square distribution even when the model is correct. Comparisons of nested models are still valid under these conditions. Using only $p$-values from fit statistics (without comparisons) is never wise; one may find that Model A (just barely) fits according to some criterion, and Model B (just barely) does not, but there is no way to decide statistically that Model A fits better than Model B without a direct comparison. One can make qualitative comparisons of AIC and BIC statistics, even for nonnested models, but there is no statistical test comparing the AIC or BIC values for different models.

## 11. Tabular/Graphical Presentation

Typically only a small number of models fit the data (sometimes we are lucky just to get one!); in this case either tabular or graphical display of the results will be reasonable. Standard errors are usually produced, along with $z$-statistics (parameter estimates divided by standard errors), by most software. But one must be cautious with these: If standard errors are large, and the parameter estimates are also large, never take the ratio seriously because it will be wrong. Unlike linear models (including factor analysis and structural equation models), nonlinear models such as logistic regression (of which latent transition models can be considered a special case) can have parameters that go to boundaries. If the parameters are probabilities, zero and one are the boundaries, and the parameter estimates do not have a normal distribution (so standard errors do not tell us anything useful); if the parameters are on the logistic scale, probabilities of zero and one (or other, more complicated, effects) can make some parameters go to infinity, and again standard errors are not useful.

## 12. Implausible or Unusual Model Results

If there is reason to believe that an observed measure is not perfect (and in most cases this is true), then one should suspect any results that show no error of measurement, even if the overall model fit is excellent. Similarly, if one has good reason to believe that no one should go from state A at time $i$ to state B at time $i + 1$, but a large proportion are estimated to do so (e.g., children who can conserve volume at time 1 but cannot at time 2), then one should suspect the model's truth. Another tipoff is if two constructs (e.g., measures at two adjacent times) should be strongly related, and they are only weakly related. In the end, there are only a few good general rules here. Knowing which results are plausible and which are not is mostly a matter of knowing (or thinking you know) the subject matter and the model characteristics very well.

## 13. Parameter Estimates

Even sophisticated quantitative models are often best summarized in simple qualitative terms. For example,

> The observed measures are quite accurate reflections of the latent variables, as indicated by conditional probabilities greater than .8 for each item that a person in a mastery latent class will answer that item correctly, and that a person in a nonmastery latent class will answer the item incorrectly.

One need not discuss each individual parameter (which, in any case, would not be a summary). Instead make broad general statements and list exceptions to these generalities.

In most models, two general aspects are important. First, how well do the observed measures tap the latent classes? Second, what are the relations in the latent structure? This would include initial probabilities of being in each class, as well as transitional probabilities from each time point to the next. In models that have multiple groups, a comparison of the parameter estimates across groups should be made for those parameters that are not constrained equal across groups. These estimates should make sense in terms of the original theory from which the model was derived.

## 14. Implications and Comparisons of Retained Models

If there is more than one model that is consistent with the data, authors should discuss the ways in which their implications are similar, and the ways in which they are different. In some cases, it will be apparent what data would be necessary to distinguish the plausible models (refer again to Platt, 1964); if so, next steps for research that would help resolve remaining issues should be suggested. In some cases, larger sample sizes may be necessary to distinguish the fit of alternative models. In other cases, perhaps more times or different spacings of measurement points may be needed to elicit different predictions of competing models. Another possibility is that more measurements are needed at some (or all) time points.

## References

Böckenholt, U. (2005). A latent Markov model for the analysis of longitudinal data collected in continuous time: States, durations, and transitions. *Psychological Methods, 10*, 65–83.

Chamberlain, T. C. (1890). The method of multiple working hypotheses. *Science, 15*, 92–96. Reprinted as: Chamberlain, T. C. (1965). The method of multiple working hypotheses. *Science, 148*, 754–759.

Chung, H., Park, Y., & Lanza, S. T. (2005). Latent transition analysis with covariates: Pubertal timing and substance use behaviors in adolescent females. *Statistics in Medicine, 24*, 2895–2910.

Collins, L. M., & Flaherty, B. P. (2002). Latent class models for longitudinal data. In A. L. McCutcheon & J. A. Hagenaars (Eds.) *Applied latent class analysis* (pp. 287–303). Cambridge: Cambridge University Press

Collins, L. M., Lanza, S. T., Schafer, J. L., & Flaherty, B. P. (2002). *WinLTA User's Guide Version 3.0*. State College, PA: The Pennsylvania State University, The Methodology Center.

Collins, L. M., & Wugalter, S. E. (1992). Latent class models for stage-sequential dynamic latent variables. *Multivariate Behavioral Research, 27*, 131–157.

Goodman, L. A. (1974a). Exploratory latent structure analysis using both identifiable and unidentifiable models. *Biometrika, 61*, 215–231.

Goodman, L. A. (1974b). The analysis of systems of qualitative variables when some of the variables are unobservable: A modified latent structure approach. *American Journal of Sociology, 79*, 1179–1259.

Graham, J. W., Collins, L. M., Wugalter, S. E., Chung, N. K., & Hansen, W. B. (1991). Modeling transitions in latent stage-sequential processes: A substance use prevention example. *Journal of Consulting and Clinical Psychology, 59*, 48–57.

Haberman, S. J. (1974). Log-linear models for frequency tables derived by indirect observation: Maximum likelihood equations. *Annals of Statistics, 2*, 911–924.

Haberman, S. J. (1977). Product models for frequency tables involving indirect observation. *Annals of Statistics, 5*, 1124–1147.

Humphreys, K., & Janson, H. (2000). Latent transition analysis with covariates, nonresponse, summary statistics and diagnostics. *Multivariate Behavioral Research, 35*, 89–118.

Langeheine, R., & van de Pol, F. (1994). Discrete-time mixed Markov latent class models. In A. Dale & R. B. Davies (Eds.), *Analyzing social and political change: A casebook of methods* (pp. 171–197). London: Sage.

Lanza, S. T., & Collins, L. M. (2008). A new SAS procedure for latent transition analysis: Transitions in dating and sexual risk behavior. *Developmental Psychology, 44*, 446–456.

Lanza, S. T., Collins, L. M., Lemmon, D., & Schafer, J. L. (2007). PROC LCA: A SAS procedure for latent class analysis. *Structural Equation Modeling: A Multidisciplinary Journal, 14*, 671–694.

Lanza, S., Flaherty, B. P., & Collins, L. M. (2003). Latent class and latent transition analysis. In J. A. Schinka & W. F. Velicer (Eds.), *Handbook of psychology: Vol. 2. Research methods in psychology* (pp. 663–685). Hoboken, NJ: Wiley.

Lazarsfeld, P. F. (1950a). The logical and mathematical foundations of latent structure analysis. In S. A. Stouffer, L. Guttman, E. Suchman, P. F. Lazarsfeld, & J. Clausen (Eds). *Measurement and prediction; Volume IV of The American soldier: Studies in social psychology in World War II* (pp. 362–412). Princeton, NJ: Princeton University Press.

Lazarsfeld, P. F. (1950b). The interpretation and computation of some latent structures. In S.A. Stouffer, L. Guttman, E. Suchman, P.F. Lazarsfeld, & J. Clausen (Eds). *Measurement and prediction; Volume IV of The American soldier: Studies in social psychology in World War II* (pp. 413–472). Princeton, NJ: Princeton University Press.

Lazarsfeld, P.F. (1959). Latent structure analysis. In S. Koch (Ed.), *Psychology: A study of science, Vol. 3* (pp. 476–543). New York: McGraw Hill.

Lemmon, D. R., Bray, B. C., & Chung, H. (2007). *PROC LTA: Latent transition analysis for the SAS System*. Paper presented at the 15th Annual Meeting of the Society for Prevention Research, Washington, DC.

Molenberghs, G., & Verbeke, G. (2005). *Models for discrete longitudinal data*. New York: Springer-Verlag.

Mooijaart, A. (1998). Log-linear and Markov modeling of categorical longitudinal data. In C. C. J. H. Bijleveld & T. van der Kamp (Eds). *Longitudinal data analysis: Designs, models, and methods* (pp. 318–370). Newbury Park, CA: Sage.

Muthén, L. K., & Muthén, B. O. (2007). *Mplus user's guide* (5th ed.). Los Angeles, CA: Muthén & Muthén.

Nylund, K. L, Asparouhov, T., & Muthén, B. O. (2007). Deciding on the number of classes in latent class analysis and growth mixture modeling: A Monte Carlo simulation study. *Structural Equation Modeling: A Multidisciplinary Journal, 14,* 535–569.

Platt, J. R. (1964). Strong inference. *Science,* 146, 347–353.

Reboussin, B. A., Reboussin, D. M., Liang, K. Y., & Anthony, J. C. (1998). Latent transition modeling of progression of health-risk behavior. *Multivariate Behavioral Research, 33,* 457–478.

van de Pol, F., & Langeheine, R. (1990). Mixed Markov latent class models. In C. C. Clogg (Ed.), *Sociological Methodology* (pp. 213–247). Oxford: Blackwell.

van de Pol, F., Langeheine, R., & De Jong, W. (1991). *Panmark user manual; panel analysis using Markov chains.* Version 2.2. Voorburg, CBS, Department of Statistical Methods.

Vermunt, J. K. (1997). *LEM 1.0: A general program for the analysis of categorical data.* Tilburg: Tilburg University. Available at http://www.uvt.nl/faculteiten/fsw/organisatie/departementen/mto/ software2.html

Vermunt, J. K., Langeheine, R., & Böckenholt, U. (1999). Latent Markov models with time-constant and time varying-covariates. *Journal of Educational and Behavioral Statistics, 24,* 178–205.

Vermunt, J. K., & Magidson, J. (2000). *Latent GOLD user's manual.* Boston, MA: Statistical Innovations.

Vermunt, J. K., & Magidson, J. (2003). *Addendum to Latent GOLD user's guide: Upgrade for Version 3.0.* Boston, MA: Statistical Innovations.

Vermunt, J. K., & Magidson, J. (2005a). *Technical guide for latent GOLD 4.0: Basic and advanced.* Belmont, MA: Statistical Innovations.

Vermunt, J. K., & Magidson, J. (2005b). *Latent GOLD 4.0 user's guide.* Belmont, MA: Statistical Innovations.

Wiggins, L. M. (1973). *Panel analysis.* Amsterdam: Elsevier.

# 16
# Latent Variable Mixture Models

**Gitta Lubke**

Latent variable mixture models serve to investigate heterogeneous populations consisting of two or more clusters of subjects. For instance, a population may consist of subjects who master a study topic and are therefore prepared to take an exam, and subjects who are not well-prepared. Similarly, in a longitudinal setting, there may be groups of subjects who differ with respect to their developmental trajectories. The observed data from a potentially heterogeneous population are modeled using a *mixture distribution* rather than a single distribution. A mixture distribution is a weighted sum of component distributions, and each of the component distributions is usually assumed to correspond to a cluster of subjects. Model parameters can be specific for each component distribution, and can therefore be used to model differences between the clusters.

Due to the complexity of latent variable mixture models, researchers need to make a large number of decisions during the specification of a model. Decisions include those regarding distributional assumptions, class-specific vs. class-invariant parameters, and/or estimating variance parameters within class rather than constraining variances to zero. These choices should ideally be theory driven, but in practice they are often related to practical considerations such as model convergence. In order to permit an evaluation of the quality of a given mixture analysis it is essential that a paper covering a mixture analysis includes a description of the decisions as well as the arguments that support the choices. Such a description places considerable emphasis on the methodological details of a given analysis; however, it is a mandatory aspect of using complex statistical methods for the analysis of empirical data.

Literature describing basic and more complex latent variable mixture models include Arminger, Stein, and Wittenberg (1999), Bartholomew and Knott (1999), Dolan and van der Maas (1998), Heinen (1996), Jedidi, Jagpal, and DeSarbo (1997), McCutcheon (1987), Muthén and Shedden (1999), Vermunt and Magidson (2003), and Yung (1997). The current chapter focuses on general guidelines for conducting and evaluating latent variable mixture modeling analyses. Details related to analyses using specific types of mixture models can be obtained by consulting the corresponding literature.

It should be noted that mixture modeling is a rapidly evolving area, and that numerous aspects of model estimation and model performance are still under investigation. The table of desiderata is based on the current state of knowledge, and summarizes the different issues that reviewers and researchers should consider in a substantive study using mixture models. Details are provided in the subsequent explications.

Table 16.1 Desiderata for Latent Variable Mixture Models

| Desideratum | Manuscript Section(s)* |
|---|---|
| 1. A conceptual description of the models is given and their theoretical underpinnings are explained. | I |
| 2. Information regarding the exploratory and/or the more confirmatory parts of the used models is provided as are the general assumptions. | I |
| 3. The description of observed measures emphasizes item (or subscale) properties. | M |
| 4. Arguments for a stepwise approach are given when fitting more complex mixture models. | M |
| 5. A detailed description and justification of within-class measurement models is provided, especially with respect to class-specific and class-invariant parameters. | M |
| 6. Model equations are provided (and at least briefly explained) and path diagrams may be included to illustrate within class models. | M |
| 7. Explanation is provided as to how and why covariates and class-predicted outcomes (if any) are integrated into the fitted models as proposed. | M |
| 8. Sufficient sets of random starting values are provided to afford the replication of the likelihood of the accepted solution with different sets. | M |
| 9. A priori expectations of the power to distinguish between alternative models are described taking into account expected class proportions, class separation, sample size, and the difference between models in the number of estimated parameters. | M |
| 10. Model fit assessment includes tests to decide on the number of classes, information criteria (ICs) and a discussion of relevant parameter estimates. | M/R |
| 11. Assumptions regarding missing data should be clearly stated. | M/R |
| 12. For competing models, results are summarized in tables showing likelihood values, number of estimated parameters, ICs, relevant parameter estimates, and their standard errors. | R |
| 13. If applicable, a justification of additional models is provided. | R |
| 14. If possible, an attempt to validate the class structure using external criteria is provided to strengthen confidence in the results. | R |
| 15. Post hoc class comparisons are prone to assignment errors, and should be interpreted with great caution. | R |
| 16. Input files and information how to access the data (if possible) are provided. | R |
| 17. The interpretation of results should acknowledge the fact that the number of classes may not reflect the number of distinct groups in the population. | D |
| 18. A section detailing the limitations includes a discussion of specific alternative explanations of the results and the potential of lack of power to distinguish between models. | D |

* *Note*: I = Introduction, M = Methods, R = Results, D = Discussion

## 1. Conceptual Description of Models

Following the usual outline of the theoretical context of a study, the Introduction of a paper using a latent variable mixture model (LVMM) should provide a clear description how the particular research questions translate into the set mixture models that the researcher proposes to fit to the data. LVMMs permit the specification of a wide variety of submodels that can be related to quite different conceptual interpretations. For example, traditional latent class models (see Chapter 13, this volume) are based on the assumption that the latent classes represent typologies and that members of a given class only vary with respect to random error. Models that specify continuous latent factors within classes permit structural variation, which in turn can be used (for instance) to model severity differences of a disorder within a class, or differences in initial status in a growth model. Due to the conceptual differences of different LVMMs, it is helpful if the Introduction of a paper using LVMM provides clear arguments supporting the choices regarding the type of mixture models used in a particular analysis.

## 2. Exploratory and Confirmatory Aspects of Model(s)

Mixture models require specification of the number of latent classes as well as specification of the within-class model structures. Mixture analyses of empirical data commonly compare models with an increasing number of classes, and are therefore exploratory regarding the cluster structure of the sample. In addition, the within-class structure (or aspects thereof) may be unknown a priori. For instance, from an exploratory perspective, in a growth mixture analysis it might be part of an investigation to evaluate the necessity of including quadratic effects to model the shape of the growth trajectories, and/or to assess measurement invariance over time or across classes. In more confirmatory applications, questionnaires with a known factor structure are used, and choices regarding the measurement part of the mixture model are more theory driven. The Introduction of a manuscript describing a mixture analysis should indicate which aspects of the model have a more exploratory or a more confirmatory character, and which aspects of the model are based on assumptions or on prior research.

## 3. Description of Measures and Measurement Properties of Items

The basic part of any mixture model used in behavioral research is the measurement of the behavior of interest. The observed measures are usually indicators of traits and/or typologies that are not directly observable. As in any other latent variable analysis, distributional assumptions regarding the observed measures should be clearly described, and ordered categorical data should be handled appropriately. Apart from the distributional assumptions, it is important to realize that the psychometric properties of the scales have a direct impact on the results of a mixture analysis. The basic requirements regarding the measurement model for latent variables in mixture models are similar to those in other latent variable models. In order to adequately cover the construct of interest, factors should have more than the minimum number of indicators necessary to identify the model, and it is advantageous to use highly reliable items. In the mixture context, it is important to realize that item means or thresholds are directly related to class detection. This is illustrated in Figure 16.1.

Figure 16.1a shows a mixture distribution of a latent factor or trait with two components. The dotted curves correspond to a set of items that measure the trait, and each of the individual curves depicts the probability of answering that item correctly. The set of items in panel A covers the whole range of

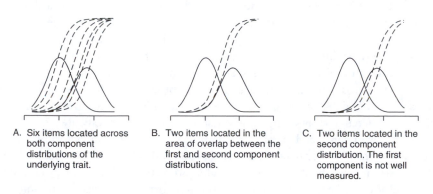

A. Six items located across both component distributions of the underlying trait.

B. Two items located in the area of overlap between the first and second component distributions.

C. Two items located in the second component distribution. The first component is not well measured.

**Figure 16.1** Probability of Answering Items Correctly Conditional on a Trait or Factor Following a Mixture Distribution.

the underlying trait, that is, the items are located such that there is variability in item scores *within* each class. Figure 16.1b shows two items in the overlap between the two component distributions. These items may be especially important for the estimation of class proportions. Figure 16.1c shows items that are located in the second class. This situation might be representative of a scale, questionnaire, or behavior checklist that is designed to distinguish between more or less affected individuals (e.g., checklists used to identify individuals with psychiatric disorders). The depicted items primarily discriminate between the affected individuals. If such a scale is used in a sample drawn from the general population, then unaffected individuals will score zero on these items, and estimation of variance parameters in the unaffected class is going to be problematic. Conversely, if item locations cover mainly the part of the distribution of the unaffected majority, then the detection of differences in the affected class may be problematic.

In sum, when describing the measures used in a particular analysis it is important to mention for which type of population the scales were originally designed, and whether it can be expected that the items actually measure the full range of the trait in the population that is investigated.

## 4. Stepwise Model Building

Estimating complex models can be problematic for at least two reasons. First, as the complexity increases, there is usually an accompanying increase in the number of potential misspecifications. Misspecifications can lead to convergence problems and/or to biased results. Second, fitting complex models is computationally intensive, and unusable results are especially unpleasant after a time-consuming estimation attempt.

It is therefore preferable to start a mixture analysis with simpler models that are easier to estimate but that might nonetheless reveal parts of the structure that can be incorporated in a final set of more complex models. For instance, second order growth mixture models (i.e., multivariate growth mixture models) have the advantage that measurement invariance can be tested across time. However, fitting an unconstrained second order growth mixture model as a first step to obtain a baseline model might be impractical. A model where class loading, factor variances and covariances, and residual variances are specified to be class-specific may not converge or may result in unacceptable parameter estimates. In such a case it is useful to evaluate the measurement model at each time point separately before combining the time points. The separate results from the different time points should give an idea whether the measurement model is appropriate, and also which parameters may be constrained to be equal across time points when combining time points in a more complex model.

A similar stepwise approach can be followed when fitting factor mixture models (FMMs). In case questionnaire data are analyzed that serve to indicate multiple factors, it might be useful to carry out an initial exploratory factor analysis (EFA; see Chapter 8, this volume). EFA is based on assuming a single homogeneous population, which is incorrect if the population is in fact clustered. It is therefore important to keep in mind that the EFA results are based on the *unweighted pooled* covariance matrix, that is, covariances between items may in part be due to mean differences between classes and not adequately reflect a factor structure within classes. However, an EFA can still provide some indication of the item properties. Items with very low reliabilities can be detected and removed from the main analysis. In addition, if the EFA shows that the items have a simple structure, one might consider fitting models to the different factors separately as a first step in order to investigate the cluster structure without having to invest considerable computation time trying to fit multifactor models. Especially with ordered categorical observed data, fitting multifactor models is computationally intensive.

A manuscript describing a mixture analysis should outline the analysis strategy and the steps the lead to the specification of the set of models used in the main part of the analysis. Providing the results of initial analyses will often enhance the confidence in the final results.

## 5. Measurement Models, and Class-Specific vs. Class-Invariant Parameters

One aspect of model building that deserves special consideration is the choice of class-specific and class-invariant parameters. In the absence of prior knowledge, it would be desirable to fit models with the majority of parameters specified to be class-specific, especially given that incorrectly imposing equality constraints across classes may results in accepting models with too many classes. However, there is a bit of a Catch-22. It might be necessary to impose constraints in addition to those that are mandatory to identify the model simply because there are practical limitations to estimating class-specific parameters. Models without any equality constraints across classes on loadings, factor (co)variances, residual variances or thresholds might not converge, or result in unacceptable parameter estimates.

The choice of the additional constraints may be theory driven, or based on the results of the stepwise model building described above. It is important to realize that most constraints correspond to some level of measurement invariance, and have direct conceptual implications. Consequently, it is essential to clarify the decisions made in any given analysis.

As an example, consider the following set of models fitted to observed variables that are multivariate normally distributed within classes.[1] Although it might be subject to discussion whether to start a series of models with the most lenient or the most restrictive model, in the context of mixture models the practical limitations mentioned above often do not leave much choice but to start with constrained models. An example of a highly constrained model is the latent profile class model,[2] which has one of the most simple within-class structures. The within-class model specification reflects the assumption that observed continuous variables are uncorrelated within classes (e.g., local independence). This means that non-zero covariation of the observed variables in the total sample (i.e., in the unweighted pooled covariance matrix) is entirely due to mean differences between the classes. On a conceptual level, this is equivalent to assuming that there is a latent variable underlying the observed variables, but that subjects *within* a class do not differ (have zero variance) with respect to the latent variable. The model parameters of the latent class model are the residual variances, the observed variable means within each class, and the class proportions. Although this model is usually unproblematic and very fast to estimate, it is a highly constrained model, and might be a too crude approximation in case subjects within a cluster actually vary with respect to the latent variable.

A smoother approximation of the true data generating process in the population can be obtained by allowing non-zero factor variances within classes. One can choose to specify a full factor model within classes, in which case one has to decide whether factor variances, loadings, residual variances, and the means of observed variables are class-specific or class-invariant. Instead of class-specific means of the observed variables, one can also chose to estimate factor mean differences between classes. Following the strategy of fitting increasingly lenient models, one can first fit latent class models, and then proceed to fit models that have class invariant observed variable means, loadings, and residual variances. Such a model accounts for non-zero factor variance within a class, but it also assumes that observed variables are measurement invariant across classes.

To test whether measurement invariance is tenable, one can proceed to free the observed variable means, loadings, and residual variances in a stepwise fashion. Note that it is useful to include models with fewer classes in the comparison given that more lenient models usually require fewer classes. Note also that it is informative to check the change in the parameter estimates across models. The process of successively freeing parameters to be class-specific can stop whenever it can be shown that class-invariance is tenable. In case the process is limited by convergence problems or unacceptable parameter estimates, this should be reported to permit the reader to put the results into an appropriate context. For instance, if fitting a model with class-specific residual variances in addition to class-specific loadings does not converge, then it is apparently not possible to test invariance of residual variances. Results would have to be interpreted taking into account that residual variances were assumed to be invariant. Given the direct impact on the conceptual interpretation, choices regarding class-specific and class-invariant parameters have to be described in detail.

## 6. Model Equations and Path Diagrams

Although the narrative of a mixture paper should provide a clear description of the models that are fitted to the data, due to the complexity of mixture models it is helpful to include the precise model equations. One should pay particular attention to the subscripts of the parameters given that subscripts provide the necessary information regarding class specificity or class invariance of the parameters. The parts of the paper describing the different models that are fitted to the data can then be linked to the model equations. The combination of equations and narrative can provide an unequivocal explanation of the analysis that is carried out. In addition to the equations, researchers may choose to include path diagrams depicting the within-class measurement models. Path diagrams provide a quick visual reference and can enhance the readability of the manuscript. As with the equations, path diagrams should clarify which parameters are class-specific and which are class-invariant.

## 7. Covariates and Class Predicted Outcomes

In general, the inclusion of covariates with substantive effects is helpful because it increases the separation between classes. As a thought experiment, suppose we have two classes and a binary covariate that perfectly predicts class membership. All subjects scoring, say, zero on the covariate will have zero probability of belonging to one of the classes and a probability of 1 to belong to the other class. This means that the classes are completely separated on the covariate. As a result, the multivariate distance between the latent classes computed for the observed variables and the covariate jointly will be larger than the distance computed for the observed variables only. Increased separation between classes is directly related to improved class assignment and increased power to distinguish between alternative models.

A second advantage of integrating covariates is that the model is embedded in a larger conceptual context. Latent classes can be characterized in reference to covariate effects, which may enhance confidence in the results. The same argument holds for integrating distal outcomes that are predicted by class membership and/or within-class factors.

The selection of covariates should be theory driven. Covariates can be specified to have effects on class membership, and/or on factors and observed variables within classes. It is important to realize that omission of direct covariate effects on within-class factors or variables can lead to biased results, which is very similar to the omission of direct effects in measured or latent variable path analysis. Class proportions, as well as even the direction of covariate effects, can be incorrect if important direct effects are omitted. It is therefore necessary to provide arguments supporting the choice of covariate effect incorporated in the set of mixture models. In case of lack of prior knowledge, direct effects should be tested. When reporting the results, it should be mentioned whether covariate effects or effects on distal outcomes are based on a priori expectations or whether integration of the effects is exploratory.

## 8. Random Starts

It is well-known that the likelihood surface of latent variable mixture models has numerous local maxima. As a result, it is necessary to start the estimation using different sets of random starting values. The number of starting values that is necessary to obtain stable results can depend heavily on the psychometric properties of the items (e.g., reliability, variance, location with respect to the trait), the complexity of the fitted model, and the degree of misspecification. It is useful to start with a lower number of sets of starting values, and check the likelihood and the stability of parameter estimates when increasing the number of sets. Replication of the likelihood using different sets of starting values enhances confidence in the solution; however, replication of the likelihood value is neither a sufficient nor a necessary requirement to ensure that a global rather than a local maximum has been found.

Some software packages first compute a limited number of iterations for a given number of sets of starting values, subsequently select a user-specified number of solutions with the best-ranked likelihoods, and then iterate those until a convergence criterion is met. Software packages differ in how starting values can be manipulated by the user, hence it is necessary to clearly mention the software package and version number in addition to how starting values were used.

## 9. Power to Discriminate Between Alternative Models, and Sample Size

Not all mixture analyses are exploratory and consist of comparisons of models with alternative within-class structures. Most analyses, however, compare models with an increasing number of classes. Model comparisons are usually based on information criteria such as the BIC that penalize for the number of parameters in the fitted model. Model choice therefore depends, in part, on the number of additional parameters that are estimated when fitting a model with an additional class. This is especially important when models are compared that use ordered categorical data. For instance, cases where it is unrealistic to assume that the item thresholds are class-invariant, models with class-specific thresholds are compared with an increasing number of classes. If items have a 5-point Likert response format, for example, then adding a class can imply a substantive increase in the number of parameters, and can have the consequence that a model with fewer classes is selected (Lubke & Neale, 2008). On the other hand, a researcher might want to compare a model with measurement invariance constraints to more lenient models with class-specific measurement parameters. In this case the measurement-invariant model with $k$ classes might actually have fewer parameters than the non-invariant model with $k-1$ classes (see the empirical example in Lubke & Neale, 2008), making the model with more classes more likely to be selected.

In addition to the number of parameters, sample size is clearly an important factor in class detection and discrimination between alternative models. Unfortunately, it is difficult to provide rules of thumb given that sample size requirements depend on class separation, model complexity, response format, and, as illustrated in Figure 16.1, on item properties. Depending on these factors, analyses for very simple latent class models may be carried out probably with as few as 30 subjects, whereas other analyses require thousands of subjects.

Bootstrapping provides an indication of the expected power to discriminate between alternative models. A large number of data sets may be generated under a given model, say Model A. Next, Model A and an alternative Model B, for instance a model with one additional class, are fitted to the data sets. Model comparisons are carried out for each pair of Model A and Model B fitted to an individual data set, and then the relative proportions may be determined in which Model A and Model B are preferred. Bootstrap options are integrated in some software packages for mixture analyses.

An important caveat of bootstrapping methods is that results are based on the fact that at least one of the fitted models (i.e., the data generating model) is a true model. In practice, the set of fitted models is unlikely to contain the true model. The true data generating process in the population underlying most human behavior is obviously very complex, and the data generation for bootstrapping is a crude simplification. Researchers should be aware that model comparisons between two models that are fitted to real data where both models contain various degrees of misspecifications do not necessarily have the same properties as bootstrap comparisons.

## 10. Model Fit Assessment

Model fit should be assessed in the context of the power to detect potentially small classes and to discriminate between alternative within-class models. Model comparisons can be based on information criteria and other indices, such as the bootstrapped likelihood ratio test statistic. In exploratory settings models with more or less constrained within-class structures should be compared. A difficulty in choosing a "best-fitting model" is related to the fact that that less constrained models (i.e., model with a large number of class-specific parameters) may be incorrectly rejected when compared more constrained models due to lack of power (Lubke & Neale, 2008). Given the fact that issues of power, class detection, and detection of class-specific parameters are highly dependent on the particular data and models under consideration, it may be desirable to refrain from narrowing the choice to a single best-fitting model, and rather present a small set of models that may be equally adequate to describe the structure of the data at hand.

## 11. Missing Data

Literature covering the effects of data not missing (completely) at random in mixture models is sparse. Missingness in data from heterogeneous population may be class-specific, for instance, drop-out in longitudinal studies may be higher for high-risk trajectories. Ongoing research aims at providing more insight in bias of class proportions and within-class parameters in growth mixture models in case missingness depends on class membership and/or growth factors (Yang, Lu, & Lubke, in preparation). While clear results regarding bias of parameter estimates for non-random missingness are not available, assumptions regarding missingness should be clearly stated.

## 12. Presentation of Results

The standard errors of the parameter estimates provide an indication of the stability of a given fitted model. This is especially true for the estimates of factor and residual variances within classes. Parameter estimates should be reported together with their standard errors.

Results of model comparisons should be summarized in tables. Preferably, one table provides the information criteria together with the number of estimated parameters and the log-likelihood values. In addition, a table showing class proportions and parameter estimates of interest is usually a useful reference.

## 13. Justification of Additional Models

Based on the results of a priori planned models, an analysis is often extended and additional models are fitted to the data. As in any other statistical analysis these additional models should be presented as post hoc exploratory analyses. Additional models can be a very useful source of information, for instance, planned analyses can be extended with additional covariates, or the necessity may arise to fit models with within factor structures that differ from the planned models. However, it is helpful to provide a clear justification why these models are included in the analysis.

## 14. Validation

As mentioned in Desideratum 7, effects of class predicting covariates and distal outcomes provide a conceptual context for a given mixture model, and can be useful to validate the class structure. When designing a study, one might therefore consider collecting data on potentially interesting covariates or outcome variables, and generating hypotheses about how latent classes are related to these supporting variables. Validation in a different sample drawn from the same population is of course desirable. Unless the original sample size is extraordinarily large, splitting the sample for validation purposes is discouraged given that sample size plays a crucial role in class detection and discrimination among models with alternative within-class structures. The sensitivity of mixture models to sampling fluctuation has not been thoroughly investigated, and generalizations of a given cluster structure may be highly sample-specific. Unless a validation is carried out, interpretation of results has to acknowledge this limitation.

## 15. Post Hoc Class Comparisons

Published studies sometimes report post hoc tests using posterior probabilities. Subjects are assigned to their most likely class based on the highest posterior class probability, thus transforming the latent classes into groups, and subsequent group comparisons are carried out with respect to variables that were not included when fitting the mixture model. This type of post hoc class comparisons can be problematic due to the error rates in assigning subjects to classes. This is especially true in the case of unbalanced class size. As illustrated in Figure 16.2, small classes have a high prior probability of incorrect assignment even if classes are relatively well separated. The left panel shows the distribution of a factor in two classes of equal size with a Mahalanobis distance of 1.5. The expected probability of incorrect assignment is symmetric. However, if one class contains only 25% of the subjects, as shown in the right panel, then the a priori probability of incorrect assignment in the small class is substantially higher, namely .5.

In addition to expectations related to class size, it should also be noted that posterior class probabilities are computed using parameter *estimates*, and therefore contain the accumulated uncertainty from those estimates. Studies have shown that assignment error rates can be considerable (Tueller & Lubke, in press), and studies examining the effect of assignment error on post hoc class comparisons are underway. Given the current uncertainty concerning the validity of post hoc testing, results should be interpreted with caution.

## 16. Input Files and Data

It is usually helpful to provide annotated input files of at least one of the fitted models in an appendix. If possible, one should provide (a link to) the empirical data so other researchers can repeat the analyses or fit alternative models. In the event that the empirical data may not be made public, a (link to the) full set of parameter estimates should be provided so that other researchers can simulate data.

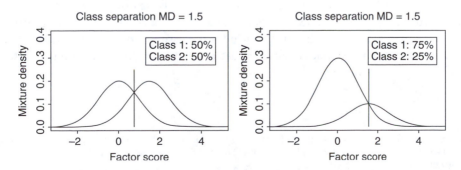

**Figure 16.2** The Effect of Class Size on Expected Incorrect Class Assignment.

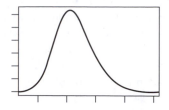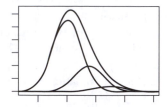

**Figure 16.3** Approximation of a Skewed Distribution Using a Mixture of Three Normal Component Distributions with Equal Variance.

## 17. Interpretation of Results

The interpretation of any given mixture analysis should be placed into the context of the psychometric properties of the items or scales, sample size, class separation, and response format of the items, since all of these factors influence the power to detect classes, to detect class-specific parameters, and to obtain stable results.

Mixture models are often used to detect qualitatively different clusters of subjects. However, it is important to note that the latent classes correspond to the components of a mixture distribution, and that the component distributions are used to approximate the joint distribution of the observed variables. Figure 16.3 illustrates the approximation of a distribution of a single continuous variable that is slightly skewed. The three normal mixture components in the right-hand panel have equal variance. It is easy to visualize that two components with unequal variance might provide a similarly good approximation.

Translating the idea to multivariate observed variables, suppose the observed distribution is multivariately skewed. Depending on the degree of multivariate skewness, a mixture of two or more multivariate normal component distributions can approximate the skewed distribution better than a single multivariate normal distribution. In other words, mixture components correspond to areas under the joint distribution of observed variables that contain subjects with similar response patterns. An interpretation along this conceptualization of latent classes is desirable.

## 18. Limitations

A number of journals have the standard of requesting a paragraph detailing the limitations of the current study. Such a section should be a staple especially in the context of latent variable mixture models due to the number of assumptions and potentially subjective decisions involved in the model building and model fitting process. It is useful to discuss the reasonableness of meeting model assumptions, the tenability of imposed constraints, and the potential detrimental effects of these decisions. The section on the limitations can also include a discussion of potential alternative interpretations of the current results (e.g., lack of power to detect small classes, overextraction of classes due to skewness).

### Notes

1 Measurement invariance for categorical data has been discussed in the multi-group context by, for instance, Muthén and Asparouhov, 2004, and Millsap and Tein, 2004. Some aspects that are specific to the mixture settings are described in Lubke and Neale (2008).
2 Latent profile models are latent class models for continuous outcome variables.

# References

Arminger, G., Stein, P., & Wittenberg, J. (1999). Mixtures of conditional mean and covariance structure models. *Psychometrika*, *64*, 475–494.

Bartholomew, D. J., & Knott, M. (1999). *Latent variables models and factor analysis* (2nd ed.). London: Arnold.

Dolan, C. V., & van der Maas, H. L. J. (1998). Fitting multivariate normal finite mixtures subject to structural equation modeling. *Psychometrika*, *63*, 227–253.

Heinen, T. (1996). *Latent class and discrete latent trait models: Similarities and differences*. Thousand Oaks, CA: Sage.

Jedidi, K., Jagpal, H. S., & DeSarbo, W. S. (1997). Finite mixture structural equation models for response based segmentation and unobserved heterogeneity. *Marketing Science*, *16*, 39–59.

Lubke, G. H., & Neale, M. C. (2008). Distinguishing between latent classes and continuous factors with categorical outcomes: Class invariance of parameters of factor mixture models. *Multivariate Behavioral Research*, *43*, 592–620.

McCutcheon, A. L. (1987). *Latent class analysis*. Thousand Oaks, CA: Sage.

Millsap, R. E., & Tein, J. Y. (2004). Model specification and identification in multiple-group factor analysis of ordered-categorical measures. *Multivariate Behavioral Research*, *39*, 479–515.

Muthén, B. O., & Asparouhov, T. (2002). Latent variable analysis with categorical outcomes: Multiple-group and growth modeling in Mplus [Mplus Webnote #4]. Los Angeles, CA: http://www.statmodel.com.

Muthén, B. O., & Shedden, K. (1999). Finite mixture modeling with mixture outcomes using the EM algorithm. *Biometrics*, *55*, 463–469.

Tueller, S., & Lubke, G. H. (in press). Evaluation of structural equation mixture models: Parameter estimates and correct class assignment. *Structural Equation Modeling: A Multidisciplinary Journal*.

Vermunt, J. K., & Magidson, J. (2003). Latent class models for classification. *Computational Statistics & Data Analysis*, *41*, 531–537.

Yung, Y. F. (1997). Finite mixtures in confirmatory factor analysis models. *Psychometrika*, *62*, 297–330.

# 17
## Logistic Regression

Ann A. O'Connell and K. Rivet Amico

Logistic regression (LR) is a type of regression analysis in the family of models more broadly known as *generalized linear models* (GLMs). LR provides a versatile and flexible modeling strategy for the analysis of binary data in the form of dichotomous outcomes, typically designated as either $Y = 1$ for success or $Y = 0$ for failure. Binary data appropriate for LR may also be grouped to represent the proportion of successes across multiple trials. The LR model is used to predict the probability of success, also known as the *response probability*, conditional on one or more predictors. Letting $x_i$ represent the collection of predictors for the $i^{th}$ person in the sample, we can write this response probability $P(Y = 1 \mid x_i)$ as $\pi(x_i)$. LR uses a logit link function to transform these conditional response probabilities into the natural log of their odds, called *logits*, where the odds are a quotient comparing the probability of success to the probability of failure. Thus,

$$\text{logit}(\pi(x_i)) = \ln\left(\frac{\pi(x_i)}{1-\pi(x_i)}\right).$$

Logits are useful in regression modeling because they form a continuous measure that spans the real line, unlike probability which is bounded between 0 and 1, or the odds, which have a lower bound of zero. The logits serve as the outcome being modeled in logistic regression, and a model's estimated logits can be easily back-transformed into estimated probabilities. Like standard linear regression models for continuous outcomes, LR models use single or multiple predictors that may be categorical or continuous, allow for polynomial terms or interactions between predictors, permit user-driven entry decisions or iterative methods (e.g., forward or stepwise), and provide model fit diagnostics and residual analyses.

Examples of LR are readily found in nearly all research areas in the social sciences, as variables that are appropriately understood or defined as dichotomies exist in virtually any substantive area of study—dropping out of, or persisting in, school; presence or absence of a chronic condition or risk behavior; attendance; pregnancy; incarceration; recidivism; completion of a program; and intervention success versus failure. LR modeling also can readily be extended to address other kinds of discrete or polytomous responses such as those obtained from ordinal (Clogg & Shihadeh, 1994; O'Connell, 2006) or multinomial (Long, 1997) variables, or discrete time-to-event data (Snyder & O'Connell, 2008).

The book by Hosmer and Lemeshow (2000) is widely regarded as the classic text on LR. We also recommend the texts by Agresti (2002, 2007), Allison (1999), Collett (2003), and McCullagh and Nelder

Table 17.1 Desiderata for Logistic Regression

| Desideratum | Manuscript Section(s)* |
|---|---|
| 1. Measurement of the outcome/response variable as dichotomous, multinomial, or ordinal is described and justified. | I |
| 2. Literature supporting design of the study as well as the proposed model of the outcome/response is reviewed and summarized; model theory is consistent with the purposes of the study and the research questions or study hypotheses. | I |
| 3. Core elements of generalized linear model are presented: sampling model (distribution of interest); link function (transformation linking predicted values to observed values); structural model (expression of transformed outcome as a function of a set of predictor variables). | M |
| 4. Choice of link function is described and justified: logit, probit, complementary log-log, multinomial, cumulative logit, etc. | M |
| 5. Sample size is identified and justified; sampling strategy and mode(s) of data collection are fully described. | M |
| 6. Software package is identified; response to be modeled is clarified; weighting methods, if necessary, are described and justified. | M |
| 7. Predictor variables are identified, and their measurement is described and justified; evidence for reliability and validity is provided. Selection process explained/justified. | M |
| 8. Coding of all categorical predictors is fully described. | M |
| 9. Parameter estimation strategy is identified. | M, R |
| 10. Extent of missing data is clearly reported for all variables, and methods for accommodating missing data are described. | M, R |
| 11. Choice for standardization of predictors (none, partial, full) is explicated and justified; limitations to any standardization strategy must be noted. | M, R |
| 12. Overdispersion is investigated and adjusted for as necessary. | M, R |
| 13. Preliminary model assessment strategies are provided, including investigation of multicollinearity, linearity in the logit, and existence of outliers; in situations where separation or complete separation occurs, offending variables are identified and corrective measures undertaken. | R |
| 14. Results of residual diagnostics are reported; impact of outliers and extreme or unusual observations are clarified. | R |
| 15. Choice for hypothesis testing for variable effects is justified (Likelihood Ratio Tests, Wald's $\chi^2$); if appropriate for the study design, the justification for trimming of variables is provided and the resulting competing models are statistically compared. | R |
| 16. Multiple summary statistics of model fit are presented. These include results of the Hosmer-Lemeshow test; Model Deviance; Chi-square difference tests for Goodness of Fit and comparisons of competing models; and pseudo $R^2$ values. | R, D |
| 17. Categorical assessment of model fit is provided, via classification tables and associated statistics. Percent of correct predictions cannot be the only criteria for classification accuracy; stronger supporting statistics include, for example, $\tau_p$ or $\lambda_p$. | R, D |
| 18. Final model presented is credible, addresses the research questions/hypotheses, and is supported through literature and theory. Causal language is not used except when justified through study design. | M, R, D |

* *Note*: I = Introduction, M = Methods, R = Results, D = Discussion

(1989). Three relatively brief but excellent descriptions of applied LR can be found in Harlow (2005), Tabachnick and Fidell (2007), and Wright (1995). More advanced treatment on the GLM can be found in Dobson (2002) and McCulloch and Searle (2001). In each section below we cover specific desiderata for applied studies utilizing traditional logistic regression models for binary phenomena, but we note in appropriate sections where adjustments can be made to accommodate extensions of the model.

## 1. Nature of the Response Variable

The response variable in LR refers to the outcome dichotomy of interest. Discussions regarding the response variable in LR must occur early in the manuscript and address whether the variable's dichotomy is a natural binary phenomenon (e.g., voting in the next election or not) or the result of an artificial dichotomization of an underlying scale or measure (e.g., categorizing children as overweight or not based on a cutoff BMI percentile rank, using median splits, or basing cutoff decisions on a certain number of standard deviations above or below the mean). Creating artificial dichotomies is an approach that is poorly regarded in most areas of inquiry. In fact, a recent examination of this practice upholds what historical reviews have long argued: forced dichotomization yields a crude categorization of a potentially useful continuous measure, subsequent loss of information, and a flawed understanding of the phenomena being studied (MacCallum, Zhang, Preacher, & Rucker, 2002). In terms of the impact of this process on logistic regression models, Taylor, West, and Aiken (2006) documented the loss of statistical power that occurs under different categorizations of a continuous response variable, and found that the loss was actually greatest when the outcome was dichotomized. Thus, from a substantive as well as a statistical perspective, the response variable for logistic regression should be based on a true underlying binary phenomenon. The application of LR to coarsely derived dichotomies needs to be accompanied by a carefully articulated justification and understanding of the limitations and consequences of this practice.

In addition to a qualitative description of the response variable, the Introduction section of manuscripts based on results of LR should contain a discussion of the prevalence or expected prevalence of the target response within the population of interest. This base-rate information is critical for informed understanding of the complex interplay between sample size, the number and measurement of covariates, the expected relations between covariates and response probability, and the use of the model for explanation or classification (or both). While the sample-based distribution of the dichotomy and the independent variables should be provided in the Methods and Results sections (see Desiderata 5, 7, and 13, below), the introduction of research using LR must address the anticipated distribution of events based on an articulated theory or previous research in the content area. As the targeted response probability becomes more extreme (closer to 0 or 1), there is an increased likelihood of the model being affected by numerical problems in the data including complete or quasi-complete separation or the presence of zero-cells (see Desideratum 13). Both of these numerical problems tend to be readily identifiable by extremely large standard errors for variable effects. Thus, because the conceptual and operational definition of the dichotomous outcome and its anticipated and real distributions given a set of hypothesized predictors have critical implications in research design and the statistical and substantive interpretations of results, these aspects of LR modeling should be clearly addressed beginning in the Introduction section of the manuscript.

## 2. Support for Model Theory

Models fit through LR should be guided by the same philosophical framework as other regression or prediction models. Thus, the capacity to inform the field and the credibility of the model depends on the theoretical underpinning of the model in relation to the investigated phenomena. A thorough introductory discussion of the LR model will clearly define the theoretical basis of the operationalization of the variables under study and the manner in which the constellation of variables modeled coexist. Moreover, because LR models, like standard regression models, evaluate the effect of predictors relative to other variables specifically included in the model, careful consideration of the literature and theory of the phenomenon in question must also address inclusion criteria for variables in the LR model and why some may be excluded.

Interestingly, LR deviates from multiple regression in a critical way regarding decisions on inclusion or exclusion of variables in the final regression model. In most purely exploratory prediction models, an initial analysis might include all available predictors, targeting for elimination from the model any variables that are not statistically significant, generally at an alpha level of .05 or, more conservatively, at .01. Use of this strategy in LR may lead to erroneous elimination of covariates that confound the relationship between a critical risk factor and the outcome. Consequently, elimination of model variables on the basis of their level of significance can lead to flawed interpretations of variable effects, particularly if a confounder is inappropriately removed (Tabachnick & Fidell, 2007).

Hosmer and Lemeshow (2000) detailed a careful model-building approach that begins with identification of a collection of theoretically or scientifically relevant variables, and then uses results from univariate analyses between each predictor and the outcome as an initial variable inclusion strategy. They recommended using a liberal level of significance in both univariate and a series of multivariable models to protect against removal or non-identification of confounders. Overall, their argument was for the researcher to use his or her best discretion as to the development of the model and the inclusion of all scientifically relevant explanatory variables. Clearly, these decisions can be made only with strong theoretical support drawn from existing literature or through adequately conceptualized studies framed around a body of previous research. The support developed in the Introduction of a study should make specific reference to the variables anticipated to be relevant in understanding the dichotomous outcome and in answering the research questions or study hypotheses. Later decisions to modify the resulting LR model should be articulated and reasonably justified.

## 3. Elements of the Generalized Linear Model

Generalized linear models have been used to represent the behavior of a wide variety of discrete outcomes in practice, and their theoretical connection with standard linear regression models simplifies their application. This simplicity is evident by considering the general structure of the GLM (McCullagh & Nelder, 1989; Nelder & Wedderburn, 1972), formally identified through three specific features: (1) a random component describing the anticipated distribution of the response variable and based on the exponential family (e.g., Normal, Binomial, Poisson); (2) a linear component describing how a transformation of the expected value of the response variable can be written as a linear predictor based on a given collection of covariates or explanatory variables; and (3) a link function specifying the connection between the original and the transformed responses.

In LR, the binomial distribution is typically used to describe the behavior of binary data or proportions and thus forms the random component. The distribution of a binary random response variable, $Y_i$, in logistic regression is generally expressed as $Y_i \sim B(1, \pi_i)$, where B indicates the binomial distribution, 1 indicates the number of trials (which equals 1 because each individual forms his or her own trial), and $\pi_i$ represents the probability of a successful outcome on the $i^{th}$ case or trial. The mean and variance of this Bernoulli (single-trial binomial) random variable are given by $\pi_i$ and $\pi_i^*(1 - \pi_i)$, respectively. Thus, the probability of success is heteroscedastic across cases—for each observation, the variance is different and depends on the expected value.

The linear component of the LR model describes how a transformation of the expected values can be written as a linear function of a set of predictors (covariates):

$$\eta_i = \beta_0 + \beta_1 X_{i1} + \beta_2 X_{i2} + \ldots \beta_p X_{ip}. \tag{1}$$

The corresponding link function, $\eta_i$, describes the connection between the observed responses and the transformed responses. If we were to base this connection on the identity link by letting $\eta_i = \pi(\underline{x}_i)$, where $\pi(\underline{x}_i)$ represents the success probability for the $i^{th}$ case given the set of $\underline{x}$ covariates, the model's predictions could potentially fall well outside the 0,1 range for probability. Further, the errors for binary data are non-normal and heteroscedastic, invalidating this linear probability model on several grounds. Thus, a suitable transformation of the expected values is required in order to construct the linear component of the model and honor the bounded nature of the binary responses. Several transformations are possible, but the logit link is often selected due to the simple and straightforward interpretation of model results in terms of odds and odds ratios (McCullagh & Nelder, 1989).

The logit link function is the natural log of the odds of success, or $\mathrm{logit}(\pi(\underline{x}_i))$, where the odds of success is a quotient comparing the probability of success to the probability of failure. Thus, the linear component of the logistic regression model can be written as:

$$\eta_i = \mathrm{logit}(\pi(\underline{x}_i)) = \ln \frac{\pi(\underline{x}_i)}{1 - \pi(\underline{x}_i)} = \beta_0 + \beta_1 X_{i1} + \ldots \beta_p X_{ip}. \tag{2}$$

The logit maps the expected values onto the real line from $-\infty$ to $+\infty$. Within the logistic regression model, each slope (holding all other effects constant) represents the change in the logit that is expected to occur given a one-unit change in the predictor. Exponentiating a regression coefficient (i.e., $e^\beta$) yields an odds ratio (OR) that describes the association between that variable and the outcome in terms of odds. An OR of 1.0 implies that the predictor has no associative effect on the odds of positive response. Small values of the OR ($< 1.0$) indicate that the odds of success tend to decrease as the predictor increases by one unit; larger values of the OR ($>1.0$) indicate that the odds of success tend to increase as the predictor increases by one unit. Odds ratios are routinely reported by statistical packages, and represent a measure of association between each predictor and the binary outcome; they are non-negative and range from 0 to $\infty$. The ORs can be interpreted directly, or they can be used to calculate a percentage change in the odds given a one-unit increase in a predictor based on the following formula: $100\% * (OR - 1)$.

In LR, the linear model describes how the log of the odds of success varies by the set of predictors. For predictions based on the logit link, the antilog (or inverse) will provide a prediction for the odds conditioned on that set of predictors:

$$\exp(\mathrm{logit}(\pi(\underline{x}_i)) = \ln \frac{\pi(\underline{x}_i)}{1 - \pi(\underline{x}_i)} = \exp(\beta_0 + \beta_1 X_{i1} + \ldots \beta_p X_{ip}). \tag{3}$$

Using this expression to solve for probability shows how the probability of success tends to vary as an inverse logistic function of the collection of covariates (McCullagh & Nelder, 1989):

$$\pi(\underline{x}_i) = \frac{\exp(\beta_0 + \beta_1 X_{i1} + \ldots \beta_p X_{ip})}{1 + \exp(\beta_0 + \beta_1 X_{i1} + \ldots \beta_p X_{ip})} = \frac{1}{1 + \exp(-(\beta_0 + \beta_1 X_{i1} + \ldots \beta_p X_{ip}))}. \tag{4}$$

Although LR modeling has many nuances in addition to the primary features of GLMs, these three aspects are the basic building blocks for understanding the LR model. Appropriate use and presentation of LR modeling strategies need not make extensive detailed description of the theory involved in

GLM, but reference to the key features (random component, linear component, and link function) should be present in a manuscript's Methods section.

## 4. Declaration of Link Function

The logit link function for dichotomous data yields what is commonly called the *binary logistic regression model*. Given the simple interpretation of this model in terms of odds and probability, it is often a researcher's preferred choice. Another frequently used model for binary data is the *probit regression model*. Probit regression uses a transformation of the probability of success based on the normal cumulative density function rather than log-odds; the inverse of the probit link is the cumulative standard normal distribution. Parameter estimates and standard errors will be different between probit and logit models given that they are based on different transformations of the data, but their overall interpretations of substantive effects will tend to be similar. Given their similarity, choice between logit or probit is often based on a researcher's familiarity with one of the two approaches. Fox (2008) described several advantages of the logit model over the probit model, emphasizing the simplicity of the logit model. However, a pragmatic investigator may want to compare and contrast goodness-of-fit statistics across competing options. A third link option for binary data is the complementary log-log (*clog-log*) transformation of the success probabilities. The inverse of the clog-log function is the extreme value distribution. Unlike the logit or probit transformation, the clog-log link is asymmetrical which becomes advantageous when analyzing discrete time-to-event data. Such data can be considered discrete and dichotomous when for each case in the sample it is known whether an event occurred or did not occur within a specific interval. Analyses for multinomial or ordinal data utilize these as well as other link options, such as cumulative logits or adjacent category logits (Long, 1997; O'Connell, 2006). Regardless of the choice for link function, it is incumbent upon the researcher to clearly identify the decision process used in making their choice, as well as consider how alternative approaches might impact on the substantive interpretation of the data.

## 5. Sample Details

Parameter estimation for GLMs and thus LR are based on maximum likelihood estimation, a large-sample methodology. Consequently, larger sample sizes are typically required for logistic regression than might be expected based on a standard linear regression. In addition, there are several interrelated factors that can impact the sample size necessary for reliable estimation of model coefficients or detection of an experimental effect. These include the base rate or response probability within the population of interest (rareness of the event), the difference in sample size between the two response categories (success versus failure), the number of observations per covariate pattern (sparseness of the data), the type of covariates included in the model (continuous versus categorical), and the expected number of events per covariate. Freeware or commercial sample size and power analysis packages are available that can estimate a desired sample size for given effects in LR. However, it is the rare case in which the desired statistical model can be adequately represented by the assumptions imposed by power software. Hosmer and Lemeshow (2000) reported that the type and number of covariates covered in power programs for LR are often limited, and the situation has not changed very much in recent years. As a first step in determining sample size, however, researchers could use power software or hand- calculations based on straightforward assumptions regarding factors related to sample size, and adjust the results as necessary to mimic the actual design.

As an event becomes increasingly unusual or rare within a given population, larger samples must be taken to ensure an adequate capture of sufficient cases to submit to a logistic regression analysis. Hosmer and Lemeshow (2000) and Allison (1999) described how sampling on the dependent variable in LR is protected against the selection bias inherent in standard linear regression models when sampling disproportionately on the outcome. Oversampling can help ensure sufficient numbers of events without biasing the odds ratios, although the intercept needs to be adjusted for the sampling fraction within both outcome groups. For analyses of extremely rare events (say, that occur with probability in the range of .05 or smaller), King and Zeng (2001a, 2001b) described a related oversampling selection adjustment that enhances the data collection process as well as improves understanding of uncommon occurrences.

When classification is the focus of the logistic regression procedure, additional attention must be paid to the relative sizes of the two outcome groups, given that LR tends to place participants into the larger group (Finch & Schneider, 2006). As a result, the misclassification rate for smaller groups can become extremely large when the two groups are very unequal in size.

Data sparseness is a problem that frequently plagues LR. The reliability of estimation weakens when sample sizes are small or when there are few observations per similar covariate pattern in the model. Sparse data patterns are particularly likely to occur when the LR model includes continuous covariates, given that there is expected to be a different covariate pattern for each individual in the sample. Evidence of estimation problems due to sparseness of data from small samples or in samples with continuous covariates are reflected in very high standard errors.

Some researchers have recommended minimum sample sizes of at least 50 observations per predictor for logistic regression (Aldrich & Nelson, 1984). Hosmer and Lemeshow (2000) discussed an extension of the minimum sample size criteria suggested by Peduzzi, Concato, Kemper, Holford, and Feinstein (1996) and recommended that for multivariable logistic models the sample size of the smallest response group be at least as large at $10(p + 1)$, where $p$ is the number of predictors in the model. Thus, the sample size of the smaller response group, and not the overall sample size, should be the main consideration in issues pertaining to power, including the number of appropriate parameters for a given model in a given sample or, alternatively, determining sufficient sample sizes to approach a priori models of a given complexity.

For readers to fully understand the quality of the research study, the Methods section of a manuscript also must provide information on the mode of data collection (mail survey, face-to-face interview, observation) and the sampling strategy utilized (simple random, cluster, etc.). In summary, manuscript authors should always provide sufficient methodological detail to demonstrate the utility of their desired and obtained sample sizes, carefully describing the impact of sample construction on generalizability of results and the impact of sample size on the capacity to detect statistically significant effects.

## 6. Software and Weighting

Under the assumption that data are drawn using a simple random sample, software packages are relatively comparable in their logistic regression procedures. However, analysis of complex survey data involving stratification, multi-stage, or cluster sampling often requires software capable of a design-based rather than a model-based approach. A design-based approach incorporates sample design features in the form of weights for each observation's primary sampling unit (PSU), strata, and cluster, into the analysis and adjusts the standard errors of variable effects accordingly. Sampling weights provide protection against potential bias inherent in the unweighted estimator and allow researchers to make valid inferences from the sample to the population the sample was designed to represent. Sudaan, Stata, AM, and SAS (Proc SURVEY LOGISTIC) can conduct logistic regression analyses based on complex survey data. There are some differences across these packages in estimation or statistical testing procedures and in how missing data are accommodated; thus, the software and version used should be reported by manuscript authors.

Software packages also may differ in how they internally refer to the dichotomous responses (e.g., 1 versus 2, or 1 versus 0), or in the designation of either outcome as the success response. Given that the odds for the complement of an event is simply the inverse of the odds for that event, the regression weights would retain the same magnitude but the signs would be reversed if the outcome codes were reversed (e.g., for modeling failure rather than modeling success). Thus, the choice has little impact on how the model is interpreted, although the researcher should clearly indicate which outcome is being modeled through the LR.

Finally, LR models based on data from case-control studies, in which the sampling process depends on the response variable, can proceed as if the data were collected through a simple random sample. Case-control studies are generally undertaken to ensure sufficient numbers of a rare event; consequently, event cases are oversampled at a rate much higher than that of controls. While the intercept is affected by sampling on the response variable, the remaining regression coefficients in the logistic model are not. Breslow's (1996) review of case-control studies remains one of the critically important ones in this field and should be read and cited by researchers employing this design. Extensions to the simple case-control design, including complex surveys involving stratification or one-to-one matching, induce several estimation and inferential complexities (Scott & Wild, 2003). Manuscript authors should justify and explain their sampling methods and adjustments in sufficient detail to allow critical examination of the validity of published results.

## 7. Identification and Measurement of Predictors

As with any type of statistical modeling, careful examination of each predictor or covariate is essential. Identification of potential predictor variables should be predicated primarily on theoretical grounds and supported through the literature review (see Desideratum 2). Procedures for investigating bivariate relationships between variables as well as between each potential predictor and the dependent variable should also be described. Decisions to consider interaction terms (products of predictor variables) should be justified based on prior knowledge of relationships among variables within the domain being studied. All relevant study variables including background and demographic variables should be clearly described; a summary of these variables for the study sample can be succinctly presented in a table or other summary form within the Methods section of the manuscript.

In LR models, predictors can be continuous, dichotomous, or categorical. The description of classification and coding systems used for dichotomous or categorical predictors must be complete (see Desideratum 8). The measurement of continuous predictors should be clearly articulated in the Methods section, including previous support for validity of the measurement process and careful evaluation and description of the predictor within the sample data in terms of distribution and reliability. As with most modeling strategies, unreliability in measures of predictor variables introduces error into the model and such measurement errors can attenuate estimated LR slope coefficients (Stefanski, 2000).

In research studies where a large number of variables may be considered for their potential as predictors, the procedures for determining variable selection for inclusion or exclusion during the modeling process need to be described and justified in sufficient detail for other researchers to be fully informed regarding validity of the intended approach and the resulting regression model. Variables may be included a priori based on defined or anticipated relationships with the outcome, or they may be selected through computer-driven approaches (stepwise, backwards elimination, forward selection, etc.). Reliance on computer-driven approaches, however, does not guarantee a credible final model. In all selection procedures, care must be taken so that inclusion or exclusion decisions are not made solely on the basis of a bivariate relationship between each potential predictor and the dependent variable, since assessment of individual contribution of a given predictor within a multivariable regression model is established relative to all other predictors in the model. Including irrelevant predictors that share association with other predictors can inflate the standard errors associated with these regression coefficients, and excluding relevant predictors can lead to cases where "third variable" effects are inappropriately attributed to the predictor set included. Proposed methodology for building the LR model must include consideration of complex variable relationships, the theoretical justification for each variable's expected contribution as a potential predictor within a multivariable model, and evidence of strong psychometric properties supporting the measurement of each variable.

## 8. Coding of Categorical Predictors

Within the Methods section, authors should include a clear description regarding their treatment of all categorical variables. For example, a description of a categorical variable for "education level" of a study participant must identify the specific education levels assigned to each category. In addition, the coding scheme corresponding to each categorical predictor when included in the regression model must be explicit.

Coding schemes are used within all regression-based analyses to completely identify group membership on a given categorical predictor. Depending on the software package used for analysis, there are quite a few options available for the coding of categorical predictors in LR, mirroring similar strategies for inclusion of categorical variables in standard linear regression. Some of the major statistical software packages have default coding schemes along with choices for alternative systems built into their logistic regression procedures that will automatically recode categories of a predictor when prompted by subcommand requests. Different coding schemes can easily be selected and applied to categorical data based on the needs of a particular analysis; alternatively, researchers can choose to do their own coding of categorical data within their data file based on a system that represents critical questions of interest regarding differences between categories. Most often, researcher-constructed schemes involve the creation of a series of dummy (or indicator) codes, effect codes, or a series of orthogonal contrast codes. As with all coding decisions it is incumbent upon the researcher to ensure that the final result for representation of category differences clearly represents the intended questions of interest.

## 9. Estimation of Parameters

The parameters of interest in most LR models are the regression weights and their associated significant departure from zero. The method used most often for estimation of model parameters in LR is maximum likelihood (ML). The ML estimates are the values for the parameters that maximize the likelihood function and thus provide the largest probability of producing the observed data. Mathematically, it is more convenient to maximize the log-likelihood (LL) rather than the likelihood itself; multiplying the LL by ($-2$) creates a quantity called the *deviance* that can be used for hypothesis-testing purposes to compare competing models (Hosmer & Lemeshow, 2000). The larger the deviance, the less well the fitted model reproduces the actual data; thus, smaller deviances are preferred. Statistical tests for model comparison are discussed in Desiderata 15 and 16.

ML estimation is based on large-sample theory, and estimates will typically be biased in small samples. Non-convergence of the iterative ML process generally occurs in cases where separation or complete separation is encountered, when zero-cells are present in the data, or when the data are sparse (see Desideratum 13). These problems become more likely when sample sizes are small. Another cause for non-convergence involves overfitting by including more parameters in the model than the data can adequately support. Alternatives to ML include exact methods, which can be applied in small samples or when non-convergence prevents parameter estimation. Hosmer and Lemeshow (2000) described two additional alternatives: iterative weighted least squares, and the use of discriminant analysis (see Chapter 6, this volume) which was relied on in early work on logistic regression. However, ML is the method predominantly used today.

In terms of manuscript preparation, the Methods and Results sections should identify the estimation strategy used for determination of parameter estimates. If alternatives to ML estimation are used, the selected procedure should be explained and justified. It may be helpful, but not necessary, for researchers to identify the numerical strategy used during the ML estimation process. Commercial statistics packages differ in the numerical procedures used to iterate and converge on the ML estimates. For example, SAS uses a Fisher Scoring algorithm (equivalent to iteratively reweighted least squares), while SPSS uses the Newton-Raphson technique. However, for logit models on binary data, these procedures provide equivalent parameter estimates.

## 10. Missing Data

Logistic regression is susceptible to the same potential biasing effects of missing data on model covariates as are other statistical models. The default method used in statistical packages for cases with missing data is listwise deletion, affecting the sample size as well as the sample's validity. For the most part, generalizability of statistical results hinges on the match between the obtained sample and the population from which it was drawn. Even when samples are drawn at random, missing data interferes with this match—particularly if the missingness follows a non-random pattern. Several excellent sources offer detailed information on analysis options in the presence of missing data; a historical review of missing data procedures and a comprehensive evaluation of approaches can be found in Schafer and Graham (2002).

Missing data on an entire case—often referred to as *unit non-response*— is often addressed at the analysis stage through the application of sample-weights for non-response or through post-stratification on known population characteristics such as age or gender distributions. Neither method, however, guarantees that the effects of non-response have been eliminated (Korn & Graubard, 1999). Thus, efforts to eliminate unit non-response should begin at the sample design and data collection stage, rather than at the analysis stage. The strategies taken to limit unit non-response (e.g., repeat callbacks, or the use of proxy-based information) should be clearly described in the Methods section, and if the research calls for sample weighting or post-stratification approaches to compensate for unit non-response, this process also must be articulated in detail within the Methods section. During the presentation of results, the degree of unit non-response (e.g., percentage refused, percentage unavailable) must be provided, and results of any analytic procedures (weighting schemes, post-stratification methods) or design-based approaches (percentage completed after callbacks, percentage completed by proxy) to adjust for unit non-response must be described with sufficient information to allow readers to evaluate the research findings in light of the approaches taken and to anticipate and understand the limitations involved in these strategies.

A second kind of non-response occurs when individual cases are missing responses on one or more items. Item non-response could occur for predictors or for the dependent variable of interest. Schafer and Graham (2002) pointed out that the procedures for dealing with missing dependent variables do not differ significantly from procedures for missing predictors; however, missing dependent variables in a study utilizing LR could seriously affect the patterns observed in the data and consequently our understanding of the phenomenon under study—particularly if the missingness is related to that dependent variable. For dependent variables that might be considered rare or hidden within a population (under-age drinking and driving; thoughts of suicide), care must be taken to prevent missing occurrences as much as possible. Disproportionate sampling (Desideratum 5) may be utilized here to assist in replacing cases with missing outcomes, as well as in attempts to model why certain persons in the sample offer differential response (present versus missing) on the outcome of interest.

There are many imputation methods available for estimating a replacement value or substitute for a missing item, some more reasonable than others. For example, mean-substitution is generally considered one of the weakest forms of imputation and is not recommended (nor is retention and analysis of cases containing only complete information on all variables). The text by Little and Rubin (1987) is likely the definitive resource on the analysis of incomplete data, but the treatment of missing data remains a rich and evolving research area. The current "state of the art" for working with studies involving missing data, according to Schafer and Graham (2002), assumes the data are missing at random (or completely at random) and include maximum likelihood procedures in missing data problems such as the EM algorithm (Dempster, Laird, & Rubin, 1977; Little & Rubin, 1987) or multiple imputation methods. In the case that items are *not* missing at random, sensitivity analyses can be used to gauge the bias that might result under certain prescribed and relevant situations.

Strategies for protecting against or limiting the effects of unit or item non-response need to be clearly described and detailed in the Methods and Results sections of a research manuscript. Sufficient information is required so that readers can evaluate the results and potential biases in model estimates, and anticipate any limitations to the research findings based on the approaches taken by the researcher.

## 11. Standardized Regression Coefficients

In LR, the raw (or unstandardized) regression coefficient, $b_j$, for a predictor variable, $X_j$, can be interpreted as the expected change in the log odds of success given a one-unit change in $X_j$, controlling for other variables in the model. Similar to linear regression, comparing the absolute size of the raw regression weights across multiple predictors and using the size of these raw weights as a marker for relative influence is not a reasonable practice, particularly when the predictor variables are measured in different scales or metrics. Mirroring concerns regarding the use of standardized regression coefficients in ordinary least squares (Bring, 1994), options for creating and interpreting standardized regression coefficients in LR continues to be a topic of interest among researchers. Specifically, coefficients in LR can be either partially or fully standardized, or be derived based on information theory (Menard, 2002, 2004a, 2004b). Insufficient knowledge or awareness about the estimates produced by different standardization processes can lead to their misuse and inappropriate substantive conclusions. Statistical packages vary in terms of options for standardization methods. For example, SAS offers a partially standardized regression coefficient, while SPSS no longer reports a standardized result. In any case, if a researcher decides to standardize regression coefficients, sufficient justification for the selected approach as well as a summary of the advantages and disadvantages of that approach for identification of relative importance of a predictor must be adequately presented in the Methods and Results.

An alternative use of the unstandardized coefficients in LR is to interpret them in terms of their odds ratios. Menard (2002) recommended basing substantive results for categorical variables or variables with definitive units of measure (length, cost, counts, etc.) on unstandardized regression coefficients or their corresponding odds ratios, and reserving the use and interpretation of fully standardized LR coefficients for other kinds of variables, such as those measured on a scale like self-concept or attitudes. Likewise, Hosmer and Lemeshow (2000) emphasized interpretation of variable effects in terms of clinical and theoretical importance, rather than in terms of relative contribution to a prediction model. Overall, in situations where researchers report and interpret standardized coefficients, principled reasons supporting the selected standardization method must be described in the Methods and used to clarify interpretation of results.

## 12. Overdispersion

The LR model for binary response data models the dispersion of the dichotomous outcome, $Y$, as a binomial random variable, or more simply, as following a Bernoulli distribution, given that the outcome of success or failure is observed on only a single trial for each participant in the study. The general form of the binomial variance is $\sigma^2 = m(\pi)(1 - \pi)$, where $\pi$ is the probability of success, and $m$ refers to the number of trials. If the response $Y$ is observed on a single trial, $m = 1$ and the mean, $\pi$, alone determines the variance. Overdispersion can occur in situations where $m > 1$; these are models for which the response probability for the $i^{th}$ unit is determined by summing the number of successes observed over $m$ trials. When the linear logistic regression model is applied to such proportion or percentage data, the data often exhibit more variability than would be expected based on the binomial variance. This overdispersion is sometimes referred to as *extra-binomial variation*, and its presence suggests heterogeneity in the data that is not accounted for by the model.

Overdispersion is quite common in practice, and can also occur when there is unaccounted for clustering or correlation within the data. Other causes include the presence of extreme observations or outliers, or when the underlying probability of success across each of the $m$ trials is not constant. Unfortunately, contributions to overdispersion cannot easily be distinguished (Collett, 2003). The impact of overdispersion is revealed in standard errors for LR coefficients that are smaller than they should be, leading to increased Type I errors for tests of variable effects and subsequent flawed understanding of the relationships being examined.

Dobson (2002) reported that overdispersion may be present if the deviance from a fitted model exceeds its degrees of freedom (generally determined as $J - (p + 1)$, where $J$ is the number of replicated covariate patterns and $p$ is the number of predictors in the model); however, this appraisal assumes that the model is correctly specified and that suspected outliers or other issues have been appropriately addressed. McCullagh and Nelder (1989) described how the covariance matrix for the regression coefficients can be rescaled based on the ratio of the deviance to degrees of freedom. The scaling factor or dispersion parameter, $\alpha$, reflects the degree to which the model variance should be inflated relative to the binomial variance: $\text{Var}(Y) = \alpha[m(\pi)(1 - \pi)]$. This same strategy can be used for overdispersed multinomial or ordinal models as well.

Tests for the scaling factor are available through statistical software packages, and models can be adjusted by incorporating $\alpha$ as a weighting factor on the variance estimates to compensate for the degrading effects of overdispersion. Alternatively, models that assume a specific form for the overdispersion may also be applied, such as the beta-binomial model for the analysis of proportions (Collett, 2003; McCullagh & Nelder, 1989). However, McCullagh and Nelder argued against the selection of a statistical model based on a specific assumed form for observed overdispersion on the grounds that mathematical convenience should not take precedence over the scientific plausibility of a chosen model. In support of this argument, they reported that models including an adjustment based on the dispersion factor generally perform better than beta-binomial models.

Overdispersion must be investigated for any logistic regression models of proportions or percentages (i.e., when $m > 1$). However, all manuscripts utilizing LR should include at least a brief discussion within the Methods section describing how overdispersion is to be investigated and identifying potential factors that might contribute to potential overdisperion (e.g., sampling design, omitted variables). If scaling is required, the Results section of the manuscript should report the size of the estimated scaling factor, and results should be clearly interpreted in light of any adjustments for overdispersion.

## 13. Preliminary Model Assessment

Preliminary assessments should focus on investigating linearity in the logit for continuous model predictors, the extent of multicollinearity, the presence of outliers or influential observations (see

Desideratum 15), and on locating zero-cells as well as evidence for separation or complete separation. The goal of the preliminary assessments is to develop support for the specification and validity of the model, and these issues should be investigated prior to determination of model fit and interpretation of effects for the final model. A summary of the results of these preliminary assessment procedures as well as decisions for dealing with any problems that were identified should be presented in the Results section of the manuscript.

Linearity in the logit refers to the relation between continuous predictors and the log-odds of the dependent variable. Departure from linearity affects statistical power and tends to underestimate the relationship between a predictor and the outcome. Linearity in the logit can be assessed graphically, or through orthogonal polynomials for categorical predictors. For continuous predictors the Box-Tidwell test might be used or orthogonal polynomials might be applied to a categorization of that predictor, in addition to graphical assessment (Hosmer & Lemeshow, 2000; Menard, 2002). If non-linearity in the logit is evident, adjustments include combining categories, or categorizing continuous variables so that separate parameter estimates might be obtained for different levels of the predictor.

Multicollinearity among predictors tends to inflate the standard errors for the estimated regression coefficients, which in turn affects the validity of statistical tests of these estimates. Multicollinearity also can lead to increased size of the regression coefficients. Multicollinearity is only a property of the predictors; thus, investigations for the presence of multicollinearity proceed as they would with standard linear regression. There is currently no single best practice in terms of dealing with multicollinearity in LR, other than its detection followed by review and potential elimination of redundant variables within a multivariable model. Thus, a correlation matrix for the predictors should be included in the results section so that the degree of collinearity can begin to be assessed. A brief discussion of statistical tolerance for the predictors in a multivariable regression model should form part of the Results section, as well. Menard (2002) suggested that as tolerance drops towards .10 (or less), the multicollinearity may be unreasonably large.

The problem of zero-cells occurs when all responses for a particular category of a nominal predictor are exactly the same; this invariance means that one of the two response categories does not occur in the sample data. Depending on the pattern of responses (either all success, or all failure), the estimated logit would be infinitely large or infinitely small, and the standard errors for the estimated logits will be extremely large as well. Options for dealing with zero-cells include collapsing categories of the predictor, eliminating the zero-cell category completely, weighting the data and assigning a particularly small weight to the zero-cell category, or rescaling the nominal variable in some fashion to represent an ordinal variable which can then be included in the model as if it were continuous (with one degree of freedom). These strategies may improve the numerical processes of the model, but will also affect the substantive interpretation of the predictors. An alternative approach uses exact logistic regression methods to determine the parameter estimates based on a conditional likelihood function rather than the standard approach utilizing the ML function (Collett, 2003).

*Separation* (also referred to as *quasi-separation*) and *complete-separation* are conditions that refer to near perfect or perfect predictions, respectively, of the response variable. As with zero-cells and multicollinearity, standard errors and the coefficients themselves will be extremely large and tend towards infinity. Separation and complete separation tend to occur for smaller sample sizes, and particularly when the number of participants experiencing the event of interest is small (Hosmer & Lemeshow, 2000). The risk of separation increases for increasing numbers of covariates and whenever the number of covariates becomes close to the sample size. Separation is not likely with continuous predictors, but complete separation can occur with any type of data (So, 1995). Although perfect prediction may seem like a great result for any model, separation hinders understanding of the effects of variables within a model since researchers may wish to determine the effect of a particularly strong variable and under conditions of separation the ML estimate for that regression coefficient no longer exists (the parameter estimate will be infinite). Adjustments for separation or complete separation mirror some of the options for zero-cells such as collapsing or eliminating categories; more advanced strategies are reviewed by Heinze and Schemper (2002), including exact logistic regression. However, they caution that there may be situations in which the requirements of the exact approach may not work well.

Overall, multicollinearity, zero-cells, and separation or complete separation manifest as the same problems within a logistic regression model: inflated standard errors and, often, inflated regression coefficients. Thus, the ability to distinguish among these issues and adequately adjust the analysis for their problematic effects requires generous preliminary assessments of the data. In addition, non-linearity in the logit affects the parameter estimates and thus the interpretation of effects. Thus, it is essential for statistical validity of the final model that any offending variables or combinations of variables be identified, and that all corrective measures are made explicit in the Results section of the manuscript.

## 14. Residual Diagnostics

Any regression model should be evaluated not only in terms of overall model fit (Desiderata 16 and 17), but also in regard to the distribution of fit (and ill-fit) of the model's predicted values to the values actually observed. These analyses are referred to as residual diagnostics in that they focus on aspects of that which remains "unpredicted" or in error after accounting for the model's predictions. In terms of methodology for residual diagnostics in LR, Pregibon (1981) remains the most influential reference in this field. Additional practical strategies for residual diagnostic evaluation in LR can be found in Hosmer and Lemeshow (2000) and Collett (2003).

Plots of residual statistics and corresponding visual assessment of extreme cases is the evaluation approach that is strongly relied upon during LR, due to the fact that the distributional properties of many of the LR residual statistics are largely unknown (Hosmer & Lemeshow, 2000). Thus, casewise or index-plots of residual statistics can often present a persuasive visual assessment of cases that are poorly fit by a model or that have undue influence in the model's parameter estimates.

In LR, residuals can be defined in terms of a single case or groups of cases that share the same covariate pattern. Software documentation should be reviewed to clarify the method used to calculate case residuals. In addition, standardized or Studentized residuals from a LR may not appropriately follow a normal or approximately normal distribution, and researchers should be cautious about interpreting the size of these residual statistics in the same way as one might interpret them from an OLS regression. Similarly, influence statistics such as leverage values or Cook's Distance should be evaluated in terms of relative size given values for other cases in the sample. This process of comparing residual statistics across cases can be used to locate extreme or unusual cases based on change statistics, such as the change in deviance or in the regression coefficients when a case (or cases with the same covariate pattern) is removed.

Due to the intensity of residual analyses, specific results or graphs are often not included in a manuscript but the findings from these analyses are discussed more generally. Of import when presenting the results of an LR analysis is the specific mention of having attended to these types of diagnostics and whether or not such evaluations led to manipulation of the data set. Decisions to delete cases on the basis of large residuals or unusual influence on the fitted model should be well-justified. Deletion of extreme cases will necessarily improve the model fit for the sample but unjustified or unexplained deletion on the basis of an extreme residual is unwarranted and tantamount to "stacking the deck" in favor of the given model's fit to that sample's data. Justification of deletion of extreme cases can include suspected error in data entry or in measurement completion, or conceptual impossibility or improbability of an observed value, but should never rest solely on the size of the residual or influence statistic. A detailed explanation in the Results section should always accompany any manipulation to the raw data. At minimum, results should make specific reference to having completed residual diagnostics, their general results, and any subsequent steps taken.

## 15. Variable Effects and Hypothesis Testing

For $j = 1$ to $p$ independent variables, the regression weights in the LR model represent the change in the logit for each one-unit increase in $X_j$, controlling or adjusting for the effects of the other independent variables in the model. The regression weights can be exponentiated to yield the odds ratio for each variable ($OR = \exp(\beta_j)$). Strong associations between independent variables and the outcome typically are represented by ORs further from 1.0, in either direction (see Desideratum 3). Statistical significance of an OR typically is assessed by testing if the regression coefficient, $\beta_j$, is statistically different from zero through one of three tests: Wald, score, or likelihood ratio.

In the Wald test, the parameter estimate for the effect of each independent variable in a logistic model is divided by its respective standard error, and the results are squared to represent a value from the chi-square distribution with one degree of freedom under the null hypothesis of no effect. Most major statistical packages report Wald chi-square statistics for each variable in the fitted model. However, Wald statistics can be problematic in small samples, samples with sparse cells, or samples with many covariate data patterns including samples with continuous independent variables. In these situations, statistical tests based on the likelihood function are preferred. The score test for the contribution of an independent variable in the model relies on derivatives of the likelihood function but is not directly available in many statistical packages. However, SPSS does use a score test in stepwise procedures to determine when variables enter or exit a developing model. Finally, the likelihood ratio test is generally regarded as the most reliable test for the contribution of an independent variable to a model, and is based on the difference in deviances between a model which contains that variable and a model that does not.

The general form for the likelihood ratio test is derived through comparisons of deviances of nested models (see Desideratum 16). The difference in deviances between nested models that differ only in the addition of a single variable approximates a chi-square distribution with one degree of freedom. While the likelihood ratio test is arguably the strongest test of a predictor's statistical contribution to a model, it may be time consuming to fit the appropriate one- variable-added models for each variable in a multivariable LR. The potential intensity of this process is the primary reason why results of the Wald test are often reported and interpreted in manuscripts and reports. However, for studies that do contain continuous predictors or that are based on small samples, results of the Wald tests should be supplemented by the likelihood ratio tests and appropriately included in the Results section of the manuscript.

Given the ready availability of Wald's test statistics, they are sometimes relied on for decisions on trimming non-statistically significant variables from a LR model. Support for these decisions rests strongly on the adequacy of the data structure as well as on the assumption of appropriate model specification, including the absence or incorporation of salient interactions. That is, all appropriate variables and interactions prior to trimming—from a substantive or theoretical perspective—have been considered. As in the development of all statistical models, variables scientifically critical to the research topic, including relevant interactions, should always be included in the final model, regardless of statistical significance. In situations where the data structure is questionable, decisions on trimming should be supplemented with the likelihood ratio test. Once a final model is decided on, the trimmed model can be compared to the final model using the general likelihood ratio test to assess model fit, described below.

## 16. Model Fit and Measures of Association

As described in Desideratum 9, the deviance for a fitted model is defined as $D = -2LL$ and can be thought of as representing "poorness" of fit. A perfect fitted model would have a likelihood of 1 and thus a deviance of 0; values of the deviance further from 0 represent worse fit. With grouped data (when specific covariate patterns can be grouped together into $J$ finite sets by their frequency of occurrence within the data), the deviance $D$ can be compared to a chi-square test statistic with $J - (p + 1)$ degrees of freedom, where $p + 1$ refers to the number of parameters in the model including the constant. With grouped data or a relatively small number of replicated covariate patterns, the deviance provides a test of goodness of fit. A good fitting model closely reproduces the observed data. In general, if $D$ exceeds the critical chi-square statistic, this suggests that the model does not provide a reasonable representation of the data.

Unfortunately, when the number of unique covariate patterns in the data becomes close to the sample size, which typically occurs when continuous covariates are present, $D$ cannot be assumed to follow a chi-square distribution, not even approximately. However, the difference in deviances between two nested models forms a general likelihood ratio statistic which will follow a chi-square distribution, and model comparisons can be approached from that perspective. Thus, $G = D_{reduced} - D_{full}$ is a quantity that represents improvement in fit, where the reduced model contains a subset of variables included in the full model. The degrees of freedom for this general likelihood ratio test is the difference in the number of estimated parameters between the two models. When the reduced model is the null or constant only model, this test is an omnibus test of the fitted model containing $p$ predictors. However, comparing nested models only provides information on whether the model with more parameters yields a statistically significant improvement (reduction) in the deviance relative to the reduced model. Statistical significance could occur even if a better model might still be found. Thus, additional options for describing model fit should be used to supplement the general likelihood ratio test.

The Hosmer-Lemeshow (H-L) test can be used to approximate a goodness-of-fit test when sparse cells are present in the data (which will nearly always occur when continuous variables are included in the model). The H-L test is based on formation of several groups referred to as "deciles of risk" that represent ordinal groupings of the estimated probabilities from the model. For most samples, ten groups are formed, but there may be fewer groups depending on the similarity of estimated probabilities across different covariate patterns. The frequencies of cases within the deciles are compared to expected frequencies using a Pearson Chi-square statistic with degrees of freedom equal to the number of groups (deciles) minus two. If the model fits well, there will be agreement between the observed and expected frequencies, and the null hypothesis of a good fit between the fitted values of the model and the actual data is retained.

There have been some concerns voiced in the literature regarding the power of the H-L test, but Hosmer and Lemeshow (2000) pointed out themselves that decisions on the adequacy of a model should be supported through a combination of criteria rather than on just a single statistical test. In addition to the likelihood ratio test and the H-L test, strategies for considering the quality of a model also include measures of association similar to $R^2$ values, and categorical fit measures representing predictive efficiency (Desideratum 17).

There are several logistic regression analogs to the familiar model $R^2$ from ordinary least squares regression that may be useful for informing about strength of association between the collection of independent variables and the outcome. However, there is some disagreement among researchers regarding which of these pseudo $R^2$ measures is best. The likelihood ratio $R^2$ ($R_L^2$; also called McFadden's $R^2$) seems to provide the most intuitive measure of improvement in fit and is determined by computing the proportion reduction in deviance obtained from the fitted model ($D_m$) relative to the null (or empty) model $(D_0)$ : $R_L^2 = 1 - (D_m/D_0)$. This statistic is not routinely reported by commercial statistics packages but is easily computed from available output which will contain deviances for both the null and the full model. Two other measures of association that also are based on model likelihoods and commonly reported in statistical output are the generalized $R^2$, also called the Cox and Snell $R^2$, and the Nagelkerke $R^2$, which in SAS is labeled Max-rescaled $R^2$. Menard's (2000) article comparing six different coefficients of determination for logistic regression models is an excellent resource on these pseudo $R^2$ statistics. As there is not agreement among researchers favoring one statistic over another, results of several of these measures of association should be reported and interpreted in the results section of manuscripts.

Finally, information criteria such as Akaike's Information Criterion (AIC) or Schwarz's Bayesian Information Criterion (BIC) provide model fit information through different adjustments to the $-2LL$ of a fitted model, based on sample size and the number of predictors. As with the deviance, lower values are more acceptable. These statistics are particularly useful when comparing non-nested models for a set of data. As noted above, however, good model fit in logistic regression is best assessed through a collection of evidence rather than relying on a single criterion.

## 17. Classification

In addition to measures of association and statistical tests for model fit, quality of a model can also be gauged through classification accuracy. Classification is based on the probabilities estimated from the model, such that if the estimated probability of response for a particular case is greater than .50 (for example), the case is assigned to the "success" outcome; otherwise, it is assigned to the "failure" outcome. Most software packages allow users to choose different cutpoint options for the classification probabilities; two-way tables showing correspondence between actual and predicted outcomes based on various cutpoints can offer additional information about quality of the model. If classification is the goal of the analysis, justification of these decisions must be clearly identified in the Results section. Often, this justification is empirical and based on the relative severity of Type I (classifying a non-event individual as having the event) versus Type II (classifying an event case as a non-event) error. The impact of cutpoint decisions on evaluation of the overall model should be included as part of the Discussion.

Accuracy of classification depends on the match between the classification frequency and the observed frequencies. Note that there are many situations where model fit is considered good but classification may be poor, particularly if there is a low observed percentage for one of the outcome variable groups. Thus, appropriate language should be adopted in presentation and discussion of results. In cases where classification is not reported or is poor, language suggestive of a model's ability to identify or predict group membership should be avoided in preference for language that describes model fit to the sample data.

Percentage correct is, by default, reported most often in most major statistical packages, yet is the least effective way in which to describe accuracy of classification because it does not accommodate either base rate or chance classification. Two excellent resources for information on alternative indexes of predictive efficiency are Menard (2000, 2002) and Long (1997). In particular, Menard reviewed and compared an extensive collection of classification measures. Among these are $\tau$ for prediction tables, $\tau_p$; the *adjusted count $R^2$* or $R^2_{adjCount}$ (also called $\lambda_p$ due to its similarity to the Goodman-Kruskal $\lambda$ as applied to prediction tables); and, for selection models, $\phi_p$. All three of these measures can be interpreted as proportional reduction in error statistics and can be tested for statistical significance. Given a researcher's extensive choice for predictive efficiency measures, it is important that the selected measure or measures be explicitly defined and that the choice matches the nature of the model. For classification models, Menard (2002) suggested that $\tau_p$ is the most appropriate measure, and $\phi_p$ is recommended for selection models. Unfortunately, these statistics are not directly programmed into commercial LR statistics packages and must be calculated from the classification table provided by the software. Whatever choice is made for determination of classification accuracy, sufficient detail and data must be presented in the results section to ensure reasonable and fair interpretation during discussion of the overall effectiveness of the model.

## 18. Credibility of the Model

The strength and utility of a statistical model rests on more than just global assessments of model fit. To effectively contribute to research and practice, model development must be guided by relevant research questions matched to rigorous methods that are appropriately designed to address those questions. The credibility of all statistical models and the research contributing to their development is, in large part, based on this connection whether for attempting to gather support for a hypothesized causal link or for identifying and describing patterns of associations between variables and an outcome. Thus, reviewers of LR studies should be cautious of methods or results that do not sufficiently address the proposed research questions or hypotheses, or that ignore or brush aside critical issues and factors affecting a LR analysis such as those discussed here. Long (1997) argued that measures of fit provide "only partial information that must be assessed within the context of the theory motivating the analysis, past research, and the estimated parameters of the model being considered" (p. 102). Accordingly, it is the researchers' responsibility to situate their work within a well-defined body of literature or preliminary research, to accurately describe their methodology and the measures/variables used, to responsibly justify their statistical decisions when addressing research questions or hypotheses, and to honestly interpret the results and resulting implications in support or refinement of relevant theory. This includes identifying and discussing impacts of limitations to the design, sample, and statistical methodology. Finally, as stated several times throughout this chapter, causal language should only be used when the research design supports such claims.

## References

Agresti, A. (2002). *Categorical data analysis* (2nd ed.). New York: John Wiley & Sons.
Agresti, A. (2007). *An introduction to categorical data analysis* (2nd ed.). New York: John Wiley & Sons.
Aldrich, J. H., & Nelson, F. D. (1984). *Linear probability, logit, and probit models.* Newbury Park, CA: Sage.
Allison, P. D. (1999). *Logistic regression using the SAS system: Theory and application.* Cary, NC: SAS Institute.
Breslow, N. E. (1996). Statistics in epidemiology: The case-control study. *Journal of the American Statistical Association, 91,* 14–28.
Bring, J. (1994). How to standardize regression coefficients. *The American Statistician, 48,* 209–213.
Clogg, C. C., and Shihadeh, E. S. (1994). *Statistical models for ordinal variables.* Thousand Oaks, CA: Sage.
Collett, D. (2003). *Modelling binary data* (2nd ed.). Boca Raton, FL: Chapman & Hall/CRC.
Dempster, A. P., Laird, N. M., & Rubin, D. B. (1977). Maximum likelihood from incomplete data via the EM algorithm (with discussion). *Journal of the Royal Statistical Society. Series B (Methodological), 39),* 1–38.
Dobson, A. J. (2002). *An introduction to generalized linear models* (2nd ed.). Boca Raton, FL: CRC Press.
Finch, W. H., & Schneider, M. K. (2006). Misclassification rates for four methods of group classification: Impact of predictor distribution, covariance inequality, effect size, sample size, and group size ratio. *Educational and Psychological Measurement, 66,* 240–257.
Fox, J. (2008). *Applied regression analysis and generalized linear models.* Thousand Oaks, CA: Sage.
Harlow, L. L. (2005). *The essence of multivariate thinking: Basic themes and methods.* Mahwah, NJ: Erlbaum.
Heinze, G., & Schemper, M. (2002). A solution to the problem of separation in logistic regression. *Statistics in Medicine, 22,* 2409–2419.
Hosmer, D. W. & Lemeshow, S. (2000). *Applied logistic regression* (2nd ed.). New York: John Wiley & Sons.
King, G., & Zeng, L. (2001a). Explaining rare events in international relations. *International Organization, 55,* 693–715.
King, G., & Zeng, L. (2001b). Logistic regression in rare events data. *Political Analysis, 9,* 137–163.
Korn, E. L., & Graubard, B. I. (1999). *Analysis of health surveys.* New York: Wiley.
Little, R. J. A., & Rubin, D. B. (1987). *Statistical analysis with missing data.* New York: Wiley.
Long, J. S. (1997). *Regression models for categorical and limited dependent variables.* Thousand Oaks, CA: Sage.
MacCallum, R. C., Zhang, S., Preacher, K. J., & Rucker, D. D. (2002). On the practice of dichotomization of quantitative variables. *Psychological Methods, 7,* 19–40.

McCullagh, P., & Nelder, J. A. (1989). *Generalized linear models.* Monographs on statistics and applied probability, *37.* Boca Raton, FL: Chapman & Hall/CRC.

McCulloch, C. E., & Searle, S. R. (2001). *Generalized, linear, and mixed models.* New York: John Wiley & Sons, Inc.

Menard, S. (2000). Coefficients of determination for multiple logistic regression analysis. *The American Statistician, 54,* 17–24.

Menard, S. (2002). *Applied logistic regression analysis* (2nd ed.). Thousand Oaks, CA: Sage.

Menard, S. (2004a). Six approaches to the calculating standardized logistic regression coefficients. *The American Statistician, 58,* 218–223.

Menard, S. (2004b). Correction. *The American Statistician, 58,* 364.

Nelder, J. A., & Wedderburn, R. W. M. (1972). Generalized linear models. *Journal of the Royal Statistical Society. Series A (Statistics in society), 135,* 370–384.

O'Connell, A. A. (2006). *Logistic regression models for ordinal response variables.* Thousand Oaks, CA: Sage.

Peduzzi, P., Concato, J., Kemper, E., Holford, T. R., & Feinstein, A. R. (1996). A simulation study of the number of events per variable in logistic regression analysis. *Journal of Clinical Epidemiology, 49,* 1373–1379.

Pregibon, D. (1981). Logistic regression diagnostics. *Annals of Statistics, 9,* 705–724.

Schafer, J. L., & Graham, J. W. (2002). Missing data: Our view of the state of the art. *Psychological Methods, 7,* 147–177.

Scott, A., & Wild, C. (2003). Fitting logistic regression models in case-control studies with complex sampling. In R. L. Chambers and C. J. Skinner (Eds.), *Analysis of survey data* (pp. 109–121). New York: Wiley.

Snyder, L. & O'Connell, A. A. (2008). Event history analysis for communications research. In M. D. Slater, A. F. Hayes, & L. B. Snyder (Eds.), *Sage sourcebook of advanced data analysis methods for communication research* (pp. 125–158). Thousand Oaks, CA: Sage.

So, Y. (1995, April). A tutorial on logistic regression. *SAS Conference Proceedings: SAS Users Group International 20.* Retrieved on February 13, 2009: http://support.sas.com/rnd/app/papers/ logistic.pdf.

Stefanski, L. A. (2000). Measurement error models. *Journal of the American Statistical Association, 95,* 1353–1358.

Tabachnick, B. G. and Fidell, L. S. (2007). *Using multivariate statistics* (4th ed.). Boston: Allyn and Bacon.

Taylor, A. B., West, S. G., & Aiken, L. S. (2006). Loss of power in logistic, ordinal logistic, and probit regression when an outcome variable is coarsely categorized. *Educational and Psychological Measurement, 66,* 228–239.

Wright, R. E. (1995). Logistic regression. In L. G. Grimm & P. R. Yarnold (Eds.), *Reading and understanding multivariate statistics* (pp. 217–244). Washington, DC: American Psychological Association.

# 18
## Log-Linear Analysis

Ronald C. Serlin and Michael A. Seaman

Log-linear analysis is a technique for both exploratory and confirmatory analysis of variable relations when all the variables of interest are classification (i.e., categorical) variables. Researchers who are familiar with contingency-table analysis will recognize log-linear analysis as a tool for assessing higher-order relations among two or more classification variables that cannot be achieved with traditional two-variable chi-square analysis. The term *log-linear* refers to the use of the logarithms of frequencies to create linear models that parallel the linear modeling in analysis of variance (see Chapter 1, this volume). In analysis of variance the response variable, measured on a quantitative scale, is viewed as a function of effects of explanatory classification variables. For log-linear analysis, the response is actually a log of expected frequencies associated with a specified model. Rather than effects, the explanatory variables are the classification variables of interest, so that the test of the model is one of structural relations, rather than effect. A special case of log-linear analysis, logistic regression (see Chapter 17, this volume), has a more direct parallel to analysis of variance, in that there are established response and explanatory variables, as well as a linear function of effects to explain the response. Log-linear analysis involves testing of hypotheses about the interaction among variables of interest. Results provide an understanding of relations and partial relations among these variables. Computations for log-linear analysis can be performed on most major statistical software packages, such as S-Plus, SAS, and SPSS, though the level of user input required varies among the packages. Default settings in software subroutines often do not address hypotheses of interest, so that software use might require a relatively high level of user sophistication. We recommend texts by Agresti (1990), Christensen (1997), and Fienberg (1980). Specific desiderata for applied studies that include log-linear analysis are presented in Table 18.1 and explained in the following sections.

## 1. Identification of Categorical Factors

The literature review should identify factors in previous studies that are clearly suited for classification of the units of analysis. Often the researcher has a choice of measurement scale, and arbitrary classification of intervals on a quantitative scale can result in categorical analysis that is less powerful for identifying relations than if the original scale was maintained. When the classification is not natural and obvious, the researcher should justify the use of a categorical variable. Similarly, log-linear analysis does not take advantage of the full information when the classification can be ordered, such as with

Table 18.1 Desiderata for Log-Linear Analysis

| Desideratum | Manuscript Section(s)* |
|---|---|
| 1. A literature review and statement of purpose refer to factors that are best represented as categorical variables. | I |
| 2. Categorical variables and the levels of these variables are clearly defined. | I, M |
| 3. Hypotheses are proposed regarding relations (or lack of relations) among the categorical variables. | I, M |
| 4. The methods of sampling and data collection are described. | M |
| 5. Log-linear models are developed to correspond to hypotheses of interest. | M |
| 6. Tables are constructed to display frequencies and proportions of cross-classified variables. | R |
| 7. Expected values, odds, and odds ratios are computed. | R |
| 8. Hypotheses are tested by comparing model fit statistics. | R |
| 9. Partial tests of association are conducted to avoid Simpson's paradox. | R |
| 10. The name and version of the utilized software package is reported, along with subroutine choices and justification. | R |
| 11. Planned and post hoc contrasts are tested using log-odds ratios, where relevant. | R |
| 12. Methods for dealing with cells with sparse or missing data are explicated. | R |
| 13. Effect size measures are reported to assess substantive importance of effects. | R, D |
| 14. Model performance is discussed in the context of theoretical understanding of the factors of interest. | D |
| 15. Results of hypothesis tests are discussed, including unexpected outcomes. | D |

\* *Note:* I = Introduction, M = Methods, R = Results, D = Discussion

Likert scales, but is best suited for unordered categorical data. For log-linear analysis to be appropriate, all variables of interest should be unordered classification variables.

To examine relations among the identified factors there must be cross-classification. That is, the researcher should focus on multiple factors that can be identified with a single sample or one or more factors identified for multiple samples (see Desideratum 4). The utility of log-linear analysis is most obvious when there are more than two of these factors. Prior to the development of log-linear methods, the common technique used for studying relations among two categorical variables (contingency table analysis with chi-square statistics) was often used for larger numbers of variables by studying multiple bivariate relations. Studies referenced in the literature review might have used this technique. The statement of purpose in the manuscript should identify a research question about the relations among two or more of the identified factors. The question(s) can be about full or partial associations (see Desiderata 8 and 9). Exploratory studies will be conducted for the purpose of identifying the structural nature of the relations, but confirmatory studies should clearly identify the nature of the relations.

## 2. Defining Categorical Variables

Once the factors of interest have been specified, the specific operational variables must be identified. These are unordered categorical variables (see Desideratum 1), so that the operational definitions consist of the classes or categories that will be identified for each unit in the sample. It is important that every unit fits into one and only one category for a single variable. That is, the categories are mutually exclusive and exhaustive. All units must also be amenable to cross-classification on the other factors to accommodate a complete study of the relations and partial relations. If this is not possible, full study is confined to those units that can be completely cross-classified, and other units can contribute to the study of a more restricted set of questions. If there are too many categories relative to the number of units in the sample, this can result in sparsely populated cells in the cross-classification (see Desiderata 6 and 12). The number of units classified in each category or cell (i.e., cross-classification) can be increased by combining categories, though this will decrease the information that can be deduced from the analysis. For example, collapsing low-frequency categories together into an "other" or "miscellaneous" category will increase the number of units within the category, but some of the information available with the original data is hidden. As with other types of analysis, larger numbers of units at each level of the variable will lead to a more powerful study of those levels, so the researcher must strike a balance between the specificity of the categories and the power available for studying these categories.

## 3. Proposing Hypotheses Regarding Relations

The manuscript should include hypotheses about the relations among the variables of interest. These hypotheses are of several types, some of which are usually trivial: (1) a hypothesis of equal (or unequal) proportions across all cross-classifications of the factors, (2) a hypothesis of equal (or unequal) proportions across all categories within a single factor (which can be repeated for some or all of the variables), (3) a hypothesis about independence (or association) among two of the factors, (4) a hypothesis about higher-order associations (i.e., more than two factors), and (5) a hypothesis about partial associations (i.e., an association among two or more factors within some or all levels of a separate factor or factors; see Desideratum 9). The most common of these is the hypothesis of association between two factors. Traditional analysis involved chi-square tests of two-factor data, but even when multiple two-factor relations are of primary interest, log-linear analysis provides error control that is typically ignored when multiple two-factor tests are conducted. In exploratory analysis, the hypotheses might not be explicit, but rather the focus is on ascertaining whether relations exist among the factors. For this reason, some common software packages have default settings to conduct all tests or conduct stepwise testing based on hypotheses with statistically significant results. Stepwise approaches are data driven and thus unreliable and sample-size dependent. If these methods are accepted, cross-validation and replication should be encouraged, if not required. Explicit hypotheses are preferred so that model testing can focus on the questions of interest (see Desideratum 5). Hypotheses are frequently stated in terms of research hypotheses, rather than statistical hypotheses.

## 4. Describing Sampling and Data Collection

There are three common sampling models that result in data collection best suited for log-linear modeling: (1) multinomial sampling, (2) product-multinomial sampling, and (3) Poisson sampling. Multinomial sampling is the process of selecting a sample of individuals from a single population and then cross-classifying every member of the sample on two or more factors. Product-multinomial sampling is the process of selecting a fixed number of units from each of two or more populations that differ in terms of classification on a single factor. Each unit within each of the multiple samples is then classified on the basis of one or more additional factors. The result of data tabulation from these two sampling methods looks identical: cross-tabulated frequency data for multiple factors, with each unit represented in one and only one cross-classification. The Poisson sampling model, though less common in social research, also results in data that can be analyzed using log-linear analysis. In this model, the possible cross-classifications, but not the number of observations, are known in advance. Observations during a fixed time period yield cell frequencies. These three sampling models do not form an exhaustive list, but they do encompass most of sampling types that lead to cross-tabulated data in published research.

A serendipitous property of these sampling models is that all three types asymptotically lead to the same log-linear analysis (Fienberg, 1980). More advanced so-called "exact methods," which derive confidence sets or critical values from permutations of observations, differ for each sampling method, but these methods are rarely found in the applied literature and will not be discussed here. Asymptotic methods are reasonable when the standard rule of thumb is observed that no more than 20% of the cross-classified cells have expected values less than 5 (see Desideratum 12 for more precise sample size rules).

Researchers often do not discuss the sampling method, presumably because the common sampling models asymptotically lead to the same statistical results. This is problematic. It is important to explicate the method, because the interpretation of the findings is directly linked to the sampling method. For example, in one model (product-multinomial sampling) the findings relate to the homogeneity (or lack thereof) of multiple populations on one or more factors. By contrast, when the sampling is multinomial the findings refer to the relationship of factors for a single population.

An advantage of these types of data collection is that there is often little or no ambiguity regarding the classification of the unit on the factors of interest. Thus, validity of the "measurement" is not an issue. If classification is ambiguous, construct validity is an issue that must be addressed. The researcher must clearly define classifications that are not widely known and accepted. If units are classified by judges, interrater reliability must be established (see Chapter 11, this volume). If classifications are created using intervals on a scale or otherwise ordered categories, the researcher must defend the choice of unordered classification, because typically this choice results in less specific results and/or a loss of power.

## 5. Log-Linear Models and Hypotheses of Interest

As with all research, the proposed hypotheses' truth or falseness is assessed by examining how well the collected data accord with them. The relationship (or lack of relationship) among measures of theoretical constructs that is specified in a hypothesis posits that a nonzero (or zero) value, respectively, of a log-odds or log-odds ratio exists in the population sampled. Analogous to analysis of variance, log-linear analysis is used to evaluate whether the data seem unlikely to have been sampled from a population in which the null hypothesized relations exist; if the data seem unlikely, given the hypotheses, then the null hypotheses are declared false. The hypothetical relations among the variables that are examined in the study must be clearly delineated in the manuscript.

The determination of the likelihood of the data, given the relevant hypotheses, is achieved through the use of a linear model. As in analysis of variance, the model includes an intercept (or grand mean) and main effect and interaction terms that are consistent with the specified hypotheses. For instance, in analysis of variance, one can test whether a population's mean on a continuous dependent variable is equal to a theoretically specified value. In log-linear analysis, because the variables are categorical, the analogous test would examine whether the log-odds, the logarithm of the ratio of the probabilities of the population falling into one category or another, is equal to a theoretically specified value.

In both instances, the data are speculated to arise as if generated at random according to an underlying linear model that includes terms whose magnitudes are theoretically specified by the hypotheses of interest. This model must be clearly delineated in the research report. In the example of analysis of variance, a null hypothesis regarding a population mean is specified as

$$H_0: \mu_t = \mu_0,$$

where $\mu_t$ denotes the true expected value of the population from which the sample was drawn, and $\mu_0$ denotes the theoretically specified expected value of the population. Equivalently, in log-linear analysis the null hypothesis concerns log-odds, the ratio of probabilities of members of the population falling into one or another category of the dependent variable, leading to the null hypothesis

$$H_0: \ln O_{t i_1 i_2} = \ln O_{0 i_1 i_2},$$

where $\ln O_{t i_1 i_2}$ denotes the ratio of the true probability of members of the population falling into category $i_1$ or $i_2$, and $\ln O_{0 i_1 i_2}$ denotes the theoretically specified ratio of the probability of members of the population falling into category $i_1$ or $i_2$.

With a specified sample size, $N$, the null hypothesis can equivalently be written in terms of *expected frequencies*, $F_p$, where $F_i = N p_i$, where $p_i$ denotes the probability of a member of the population falling into category $i$. In particular, log-linear analysis deals with models that specify logarithms of expected frequencies in terms of values of parameters that are determined by the hypothesized relations.

Most generally, if the hypotheses specify relations among variables $i, j, k, \ldots$, then the log-linear model would be written

$$\ln p_{ijk\ldots} = \mu + \alpha_i + \alpha_j + \alpha_k + \ldots + \gamma_{ij} + \gamma_{ik} + \gamma_{jk} + \ldots + \gamma_{ijk} + \ldots$$

where in direct analogy with analysis of variance, $\mu$ denotes a grand mean, $\alpha$ denotes a main effect, and $\gamma$ denotes an interaction. As main effects, the $\alpha$ terms represent differences in logarithms of probabilities, equivalent to the logarithms of ratios of probabilities, or log-odds. As interactions, the $\gamma$ terms represent differences of differences of logarithms of probabilities, equivalent to the logarithms of ratios of odds, or log-odds ratios. The hypothetical relations among the variables must be clearly specified in terms of parameters in the log-linear model, whereby those log-odds and log-odds ratios that are nonzero according to the hypothesized relations must correspondingly have nonzero effects in the model, and those log-odds and log-odds ratios that are zero according to the hypothesized relations must have the corresponding effects set to zero, thereby not having those terms included in the model.

## 6. Cross-Classification Tables

A *cross-classification table*, also known as a *crosstabs* or *contingency table*, is a display of the frequencies or proportions for all of the possible cross-classifications. The number of cross-classifications is

| Factor C | Factor A, Category 1 | | Factor A, Category 2 | |
|---|---|---|---|---|
| | Factor B, Category 1 | Factor B, Category 2 | Factor B, Category 1 | Factor B, Category |
| Category 1 | $f_{111}$ | $f_{121}$ | $f_{211}$ | $f_{221}$ |
| Category 2 | $f_{112}$ | $f_{122}$ | $f_{212}$ | $f_{222}$ |
| Category 3 | $f_{113}$ | $f_{123}$ | $f_{213}$ | $f_{223}$ |
| Category 4 | $f_{114}$ | $f_{124}$ | $f_{214}$ | $f_{224}$ |

**Figure 18.1** A $2 \times 2 \times 4$ Cross-Classification Table.

simply the product of the numbers of categories identified for each factor. Figure 18.1 contains a sample table for three factors, with two categories for two of the factors and four categories for the third factor.

It is more useful, alternatively or additionally, to display proportions or percentages, where the proportion is defined as $f / N_h$. Here $f$ is the frequency for a category or a cross-classification cell and $N_h$ is the appropriate sample size for the hypothesis of interest. It is unfortunately common for researchers to ignore the hypothesis (and therefore the research question) of interest when creating a cross-classification table. Proportions should be clearly defined and match interest. For example, using the total sample size for $N_h$ is misleading when the question is one of homogeneity of multiple populations on a response variable. In this case, $N_h$ should be the individual sample sizes for the samples drawn from each of the populations. Researchers should be clear about defining the proportions in the table, rather than requiring the reader to infer this by looking for the set of proportions that sum to one. The researcher should also make use of the descriptive information provided by the cross-classification table prior to focusing on the model-testing results. A well-constructed cross-classification table clearly leads to preliminary results, such as homogeneity (or heterogeneity) of samples, independence (or association) of factors, and partial independence (or association) of factors.

## 7. Expected Values, Odds, and Odds Ratios

Tests of log-linear models are based on the values one might expect to appear in cross-classification tables if the model is correct. Various terms of the model suggest contributions to the expectation of frequencies within categories or cross-classifications of categories. Essentially, a log-linear model should be retained or rejected depending on how close a match exists between the expectations provided by the model and the actual frequencies observed in the data. The calculation of an expectation is based on the marginal frequencies in the data and the hypothesis of interest, as reflected by terms in the model. For example, the common hypothesis of independence of two factors suggests that the expectations for the cross-classification of these factors can assume factor independence. Thus, $E(f_{ij}) = f\, p_{i\cdot}\, p_{\cdot j}$, where $E(f_{ij})$ is the expectation of the frequency for the $i$th category of Factor A and the $j$th category of Factor B given the total sample size ($f$) and the proportion of the units that are in category $i$ of Factor A ($p_{i\cdot}$) and category $j$ of Factor B ($p_{\cdot j}$). Similar calculations are available for both simpler and more complex hypotheses, though the complexity of the calculation for higher-order interactions and multiple interactions might require iterative processes.

Interactions should be descriptively examined using odds and odds ratios. An estimate of the odds of appearing in one category is given by $p_1/p_0$, where $p_1$ is the proportion of units in the category and $p_0$ is the proportion of units not in the category. With count data, interactions among factors are examined with odds ratios using contingent odds. That is, the odds of a unit appearing in a category for one factor might be contingent on the category the unit is in for a second factor. Ratios of these odds for different categories of one of the factors provide an estimate of the strength of association of the two factors.

Figure 18.2 illustrates a simple two-factor design with two categories in each factor. The estimate of odds of being in Category 1 of Factor B for units in Category 1 of Factor A is $f_{11}/f_{21}$. Similarly, the estimate of odds of being in Category 1 of Factor B for units in Category 2 of Factor A is $f_{12}/f_{22}$. The odds ratio is then $(f_{11}/f_{21})/(f_{12}/f_{22})$. If the odds of being in Category 1 of Factor B are not contingent on the categories of Factor A, then the odds will be the same and the ratio will be 1. Departures from 1 indicate a relation between Factors A and B, and the distance from 1 indicates the strength of this relation. With more factors and categories, contrasts of odds ratios should be used to study specific interactions related to the research questions.

| | | Factor A | | |
|---|---|---|---|---|
| | | Category 1 | Category 2 | |
| Factor B | Category 1 | $f_{11}$ | $f_{12}$ | $f_{1.}$ |
| | Category 2 | $f_{21}$ | $f_{22}$ | $f_{2.}$ |
| | | $f_{.1}$ | $f_{.2}$ | $f_{..}$ |

**Figure 18.2** Frequencies in a Two-Factor Study.

## 8. Hypothesis Tests and Model Comparison

In order to test hypotheses of interest, test statistics are computed to assess how well the observed frequencies, $f_{ijkl...}$, match the expected frequencies, $F_{ijkl...}$. The two statistics most commonly used for this purpose are Pearson's chi-square statistic,

$$\chi^2 = \sum_{i,\,j,\,k,\,...} \frac{(f_{ijk...} - F_{ijk...})^2}{F_{ijk...}},$$

and the likelihood ratio statistic,

$$G^2 = 2 \sum_{i,\,j,\,k,...} f_{ijk...} \ln\left(\frac{f_{ijk...}}{F_{ijk...}}\right),$$

both of which approximately follow (with large samples) a chi-square distribution. Research has shown that in most small-sample conditions, the Pearson statistic controls the Type I error rate better than the likelihood statistic. Nevertheless, when testing the effects of specific hypothesized relations, as reflected by tests of model parameters, the additive property of the likelihood statistic suggests that tests based on $G^2$ should be conducted and values of $G^2$ and associated $p$-values should be reported, as is typically done in log-linear analysis.

Both $\chi^2$ and $G^2$ are known as *goodness of fit* statistics, in that they quantify how well a model seems to fit the data, but in log-linear analysis they might better be referred to as "badness-of-fit" statistics, in that they get larger (and $p$-values smaller) as the fit between model and data becomes worse. This is especially salient in log-linear analysis, because there are many potential models (e.g., with 3 variables there are 19 models, not counting the model containing only the grand mean), and many of these models yield a statistically significantly bad fit between observations and expectations. If one focused solely on the fit between models and data, one would somehow have to determine that one significantly bad model is better than another significantly bad model. Furthermore, one would be searching for a model whose $G^2$ indicates a lack of significantly bad fit, which would lead a researcher to attempt to draw a conclusion on the basis of the lack of significance of a null hypothesis test, well known to be logically invalid.

Finally, it is only by focusing on the comparison in pairwise fashion of a much smaller set of models that one is able to draw conclusions about the original hypotheses of interest. For example, consider an investigation that involves only two variables, Factors A and B, and assume that one's hypotheses of interest involve both of the main effects and the interaction. Then there are five models possible:[1]

$$\ln F_{ij} = \mu$$
$$\ln F_{ij} = \mu + \alpha_A$$
$$\ln F_{ij} = \mu + \alpha_B$$
$$\ln F_{ij} = \mu + \alpha_A + \alpha_B$$
$$\ln F_{ij} = \mu + \alpha_A + \alpha_B + \gamma_{AB}$$

If the fit of the second model is statistically significantly better than the first, then one would conclude that the main effect of Factor A is nonzero in the population, because the fit of a model in which this term is nonzero has fit the data better than one in which the term is zero. It is here that the additive property of the $G^2$ statistic becomes essential, because the difference in the two models' $G^2$ statistics is itself distributed as chi-square. By comparing the reduction in $G^2$ to a critical value obtained from the chi-square distribution, one can test the null hypothesis that the A main effect is zero, that is, $G_A^2 = G_\mu^2 - G_{\mu+\alpha_A}^2$, where $G_A^2$ is a sample statistic testing the null hypothesis $H_0$: $\alpha_A = 0$, and $G_\mu^2$ and $G_{\mu+\alpha_A}^2$ are likelihood ratio statistics assessing the goodness of fit of log-linear models including only the grand mean, $\mu$, or including both the grand mean and the main effect of Factor A, $\alpha_A$, respectively.

One can test the main effect of B and the interaction of Factors A and B in a similar fashion. The last of these models, which includes all of the parameters, is known as the *saturated* model. In a saturated model the expected frequencies are equal to the observed frequencies, and so the $G^2$ statistic is equal to zero. The test of the highest-order interaction is identical to the test of the "goodness" (badness) of fit of the model in which all lower-order terms are included. The $G^2$ statistics used for testing the hypotheses of interests and their associated *p*-values should be reported in the manuscript.

## 9. Marginal and Partial Tests and Simpson's Paradox

Referring again to the two-factor example and the five associated models of interest (see Desideratum 8), notice that the main effect of Factor A can be tested by comparing the fit of the second and first models, as described, but also that the main effect of Factor A could be tested by comparing the goodness of fit of the fourth and third models, the former containing the main effect of A and the latter not. In the case of testing main effects, both approaches yield identical results, regardless of how many factors are included in the design. The same cannot be said when testing interactions in designs involving three or more factors. In such cases, one must distinguish between two approaches.

For one of these approaches, a *marginal* test of the interaction under examination, the researcher compares the $G^2$ statistic for a model in which only the interaction in question and the requisite lower-order terms are included to the $G^2$ statistic for a model that includes the same lower-order terms but not the interaction under examination. This test is conducted as if all factors not included in the interaction do not exist in the design, and the test is performed as if the data table had been collapsed across all other factors. For example, consider testing the hypothesis that the AB interaction is zero in a design that includes three factors, A, B, and C. The marginal test statistic would be calculated as

$$G_{AB}^2 = G_{\mu+\alpha_A+\alpha_B}^2 - G_{\mu+\alpha_A+\alpha_B+\gamma_{AB}}^2,$$

comparing the goodness of fit of models that do not include Factor C.

The other approach, a *partial* test of the interaction, includes all of the factors in the design. The partial test is conducted by comparing the $G^2$ of the model in which all interactions of the same order as

|  |  | $B_1$ | $B_2$ |
|---|---|---|---|
| $C_1$ | $A_1$ | 30 | 180 |
|  | $A_2$ | 70 | 220 |

|  |  | $B_1$ | $B_2$ |
|---|---|---|---|
| $C_2$ | $A_1$ | 350 | 90 |
|  | $A_2$ | 150 | 10 |

**Figure 18.3** Example Illustrating Simpson's Paradox.

the one of interest are included, to the $G^2$ of the model in which all interactions of the same order as the one of interest, but not the interaction of interest itself, are included, or

$$G^2_{AB.C} = G^2_{\mu + \alpha_A + \alpha_B + \alpha_C + \gamma_{AC} + \gamma_{BC}} - G^2_{\mu + \alpha_A + \alpha_B + \alpha_C + \gamma_{AB} + \gamma_{AC} + \gamma_{BC}},$$

where $G^2_{ABC}$ represents the likelihood ratio test of the AB interaction from which Factor C has been partialled, $\gamma_{AB.C}$.

Partial tests should always be conducted and reported, because marginal tests may fall prey to what is known as Simpson's paradox, resulting in erroneous conclusions due to the confounding influence of other variables that the marginal tests exclude from the analysis. As an example, consider again the design that includes Factors A, B, and C, and assume that each factor has two levels and that the data are as shown in Figure 18.3. In level $C_1$, the odds of falling in level $A_1$ versus level $A_2$ equal 0.4286 for level $B_1$ and 0.8182 for level $B_2$, and similarly in level $C_2$, the corresponding odds equal 2.3333 for level $B_1$ and 9 for level $B_2$; that is, in both levels of C, the odds are greater in level $B_2$ than in level $B_1$, and they are statistically significantly so. Yet, if one collapsed the table across levels of C, one would find that the corresponding odds in level $B_1$, 1.7273, are greater than the odds in level $B_2$, 1.1739, and statistically significantly so! Thus, one would draw a conclusion from the collapsed, marginal table that is opposite those found in the separate levels of variable C; this is an instance of Simpson's paradox.

The explanation for such an occurrence is that variable C is associated with both variables A and B, and this so-called third-variable influence can result in a spurious result if it is not accounted for. When the odds are compared in the separate levels of variable C, this variable's influence is controlled, or *partialled*, and the relationship between A and B thus cannot be due to variations in C. Because it is desirable that our conclusions be unconfounded, the tests reported in log-linear analysis should examine partialled effects.

## 10. Computer Programs

Researchers should specify which statistical computer program was used to perform the analyses and which default and custom options were selected, in order to indicate how estimates and test statistics were obtained. Most statistical packages perform log-linear analysis, with the notable exception of Minitab, and some programs offer several routines that do so. For instance, the SPSS routines GEN-LOG, HILOGLINEAR, and LOGLINEAR can all be used in this endeavor (although LOGLINEAR can only be accessed through a syntax window), and in SAS both PROC CATMOD and PROC GENMOD can yield log-linear analyses.

If a researcher is interested in testing partialled effects, then the procedure to use in SPSS is HILOGLINEAR, selecting "Model Selection" from the Analyze > Loglinear tabs and choosing the "Association Table" option with a default saturated model. SPSS adds a constant, denoted delta and equal to 0.5, to all cells when performing this analysis, and the researcher is encouraged to set delta equal to a very small number, say 0.00001. These options should be delineated in one's manuscript. To test partialled effects in SAS, PROC GENMOD should be used. A saturated model should be specified with a Poisson distribution, a LOG link, and a TYPE3 analysis should be requested. Again, these specifications should be indicated in the manuscript.

## 11. Planned and Post Hoc Contrasts

In analysis of variance, a statistically significant $F$ statistic whose numerator degrees of freedom exceed unity leads one to infer only that the complex null hypothesis under examination is false—that the means, for example, differ but not in precisely what way. Similarly, a statistically significant $G^2$ test of a main effect or interaction with degrees of freedom of two or more does not indicate the manner in which odds ratios differ from unity or from one another. In both kinds of analyses, planned or post hoc comparisons are examined to provide more detailed information regarding the model parameters and, consequently, about the research questions that the analysis is intended to address.

Analogous to analysis of variance interaction contrasts, written in terms of differences of mean differences, contrasts in log-linear analysis constructed to examine interactions among design factors are written in terms of differences in logarithms of population proportions and are tested in terms of corresponding sample statistics. As an example, say a design contains two factors, Factor A having two levels and Factor B having three levels. Then the complex null hypothesis would be written as $H_0$: $\gamma_{AB}$ = 0. This hypothesis is equivalent to $H_0$: $\gamma_{(1,2)(1,2)} = \gamma_{(1,2)(1,3)} = \gamma_{(1,2)(2,3)} = 0$, where the first pair of subscript values indicates that both levels of Factor A are involved in the interaction contrast, and the second pair of values indicate that particular levels of Factor B are involved. For example, we would write the first of these interactions as

$$\gamma_{(1,2)(1,2)} = (\ln p_{A_1B_1} - p_{A_1B_2}) - (\ln p_{A_2B_1} - \ln p_{A_2B_2}).$$

It has been shown (Gart & Zweifel, 1967) that this interaction is best estimated in terms of sample cell frequencies as

$$\hat{\gamma}_{(1,2)(1,2)} = \ln(f_{A_1B_1} + .5) - \ln(f_{A_1B_2} + .5) - [\ln(f_{A_2B_1} + .5) - \ln(f_{A_2B_2} + .5)],$$

and that the variance of this contrast is best estimated as

$$\sigma^2_{\hat{\gamma}_{(1,2)(1,2)}} = \frac{1}{f_{A_1B_1} + .5} + \frac{1}{f_{A_1B_2} + .5} + \frac{1}{f_{A_2B_1} + .5} + \frac{1}{f_{A_2B_2} + .5}.$$

According to Goodman (1964), in its most general form the omnibus null hypothesis is equivalent to writing that all linear combinations of tetrad interaction contrasts are equal to zero.

Most commonly, interaction contrasts are most interpretable when they involve a fourfold table. Consider again the data presented in Figure 18.3. The contrast that would be used to test the partialled AB interaction would be written as

$$\hat{\gamma}_{AB.C} = \frac{1}{2} [\hat{\gamma}_{(1,2)(1,2)(1)} + \hat{\gamma}_{(1,2)(1,2)(2)}]$$

$$= \frac{1}{2} [\ln(30.5) - \ln(180.5) - \ln(70.5) + \ln(220.5) + \ln(350.5) - \ln(90.5)$$
$$- \ln(150.5) + \ln(10.5)]$$

$$= \frac{1}{2} [(-0.6377) + (-1.3086)]$$

$$= -0.9731$$

The variance of this contrast would be calculated as

$$\sigma^2_{\hat{\gamma}_{AB.C}} = \frac{1}{4} [\sigma^2_{\hat{\gamma}_{(1,2)(1,2)(1)}} + \sigma^2_{\hat{\gamma}_{(1,2)(1,2)(2)}}] = 0.0432,$$

and the test of this interaction, which is asymptotically normally distributed, would equal

$$z_{\hat{\gamma}_{AB.C}} = \frac{-0.9731}{\sqrt{0.0432}} = -4.6816.$$

Standard multiple comparison procedures can be applied to tests of log-linear contrasts. Goodman (1964) showed that Scheffé's method is applicable, with the associated contrast tests having appreciably lower power than tests of the same contrasts using other multiple contrast procedures such as the planned contrast methods due to Dunn, Holm, or Shaffer. Dunn's procedure provides less power than the Holm sequentially rejective method, which in turn has lower power than the improved sequentially rejective method due to Shaffer. Except in those cases for which all but one of the factors possess two levels, the Shaffer method is much more difficult to apply than the Holm procedure, and so in general one would do best to rely on the Holm method to control Type I error rate. Regardless of choice of multiple comparison procedure, interpretable log-odds ratios should be tested and reported.

## 12. Sparseness

Traditionally, and correctly, concern has been expressed about the adequacy of the chi-square distribution to approximate tail probabilities of statistics when sample sizes are not large. Over the last 70 years, a number of suggestions have been offered regarding the minimum expected cell frequencies required for the approximation to be reasonably good (with suggested minima ranging between 1 and 20). More recently, however, researchers have found that both the magnitude of the expected frequencies and the ratio of total sample size to the number of cells involved in the interaction under study seem to relate to the adequacy of the approximation to the distribution of either the Pearson chi-square or the likelihood ratio statistics in sparse tables (i.e., those in which there are a number of cells with small frequencies). For instance, Agresti and Yang (1987) suggested that the chi-square approximation to the distribution of the likelihood ratio statistic performs poorly when testing the fit of a log-linear model when tables have cells with small expected frequencies, and that the chi-square approximation to the distribution of the Pearson statistic is adequate when $N \geq 10\sqrt{K}$, where $N$ is the total sample size and $K$ is the number of cells in the table. When testing hypotheses regarding partialled effects via subtraction of likelihood ratio statistics, however, these authors found that the chi-square distribution worked well in an $r \times c \times k$ table in approximating the distribution of the likelihood statistic for partialled tests when $N > 5[\max(rc, rk, ck)]$. On the other hand, the chi-square approximation to the distribution of the difference in Pearson fit statistics was inadequate throughout. The results also showed that adding a constant to all cell frequencies, typically 0.5, made both the Pearson and likelihood ratio statistics overly conservative for larger tables in testing both the fit of the log-linear model and the statistical significance of partialled effects (recall, however, that adding 0.5 to cell frequencies in the estimation of log-odds ratios and their variances yields unbiased estimators). Authors should comment on the likely adequacy of the chi-square approximation when testing the fit of the log-linear model, and they should note whether a constant has been added to cell frequencies in calculating the test statistics.

## 13. Power Analysis and Measures of Effect Size

In addition to reporting the results of statistical tests of parameters of interest, an assessment of the substantive importance of the design factors should be provided. There are a number of such measures available, but it might be most useful to provide measures that are directly analogous to the more familiar measures of effect size used to describe results of an analysis of variance or a multiple regression analysis. When assessing the importance of an effect in analysis of variance, the measure summarizing the size of mean differences among two or more groups is $\eta^2$, the ratio of the between-group and total variances. When assessing the strength of the relation between two variables, one often uses Pearson's correlation coefficient, and the natural extension to the relation between a set of predictors and a dependent variable yields the squared multiple correlation coefficient, $R^2$. These measures have analogs in log-linear analysis. Indeed, the measures in log-linear analysis have the same use in calculating prospective power of a test as do their counterparts in analysis of variance, so that the results of the power calculation can be reported in the manuscript.

The underlying conceptualization of the measures of effect size can be presented in terms of the Pearson chi-square statistic and the measure of the effect size for a contingency table, Cramer's $V$, whereby $\chi^2 = N V^2$. The measure analogous to $\eta^2$ can be defined similarly in terms of the likelihood ratio statistic, namely, $G^2 = N \lfloor 2 \Sigma \hat{p}_{1k} (\ln(\hat{p}_{1k} / \hat{p}_{0k})) \rfloor = N\hat{\eta}^2$, where $\hat{p}_{1k}$ is the observed value of a cell proportion and $\hat{p}_{0k}$ is the value of the cell proportion specified under the truth of the null hypothesis. Clearly, the easiest way to calculate this effect size measure is to divide the $G^2$ test of a factor by $N$.

In similar fashion, a measure analogous to the Pearson contingency coefficient can be obtained from the same relationship for a four-fold table. More particularly, the specific relationship between a one-degree of freedom $\chi^2$ variate and the square of a standard normal variate is $\chi_1^2 = z^2 = N\phi^2$, where $\phi$ is the Pearson contingency coefficient, and $z = \hat{\gamma} / \sigma_{\hat{\gamma}}$. From this, an analog to $\phi$ is given by

$$\phi_\gamma = \frac{\ln[(p_{11}p_{22}) / (p_{21}p_{12})]}{\sqrt{\dfrac{1}{p_{11}} + \dfrac{1}{p_{21}} + \dfrac{1}{p_{12}} + \dfrac{1}{p_{22}}}}$$

These measures of effect size relate in a natural way to power calculations, allowing a researcher to calculate the sample size required to detect an effect of a particular magnitude with sufficient power. In the case of the test of a main effect or interaction via $G^2$, one would need to specify the true probabilities in each cell of the design, $p_{1k}$, and the probabilities expected in each cell under the null hypothesis, $p_{0k}$, resulting in a noncentrality parameter $\lambda$ where

$$\lambda = N \left\lfloor 2 \sum p_{1k} \ln(p_{1k} / p_{0k}) \right\rfloor.$$

Similarly, when performing a power analysis for a test of a log-odds ratio in a fourfold table, one would calculate $\lambda$ as

These parameters would be used to perform the power analysis using a computer program such as G*power or NCSSCALC.

## 14. Model Performance Linked to Theory

The factors of interest are originally conceived as theoretical constructs that represent broad ideas and definitions about what underlies observable outcomes. These constructs are not themselves observable, but are inventions that define the researcher's understanding. The choice of specific variables moves the study from theory to practice and the definitions of mutually exclusive and exhaustive categories for each of the variables gives full operational form to the study. Log-linear models posit relations among the variables. When models do not fit well with the observations it is because terms are absent from the model that would have changed the expectations for cell frequencies. The relative importance of missing terms can be examined by determining how the addition of a term changes the fit of a model. These salient terms indicate operational variable relations that translate into relations among theoretical constructs. Researchers should describe both model performance and what this suggests about the factors of interest. This link between the outcomes of the study and theoretical understanding can be strengthened through reference to similar outcomes in other studies, but is restricted in a single study by both sampling and ecological considerations, primarily because of the inherent specificity that must be associated with the operational form of a single study.

## 15. Discussion of Hypotheses, Outcomes, and Expectations

The discussion about study findings will differ for exploratory and confirmatory studies. In exploratory studies, most or all models will be tested and the emphasis will be on terms that most contribute to fit of the model. A hypothesis of model fit will be rejected when these terms are absent from the model, thus suggesting a specific interaction among some or all of the variables. The exploratory nature of the study compels the researcher to discuss both statistically significant and non-significant contributions to the model. Statistical significance is a function of sample size and the reliability of the categorizations, so it is important for the researcher to also consider effect size (i.e., strength of associations) when discussing the results (see Desideratum 13). Statistically non-significant terms in model fit do not confirm the lack of relations, nor do statistically significant terms establish importance without the assessment of the size of relations. In confirmatory studies the focus is on specific models that highlight relations of interest. The researcher should test those models that highlight relations of interest and should discuss test outcomes regardless of whether results are anticipated or anomalous. In the case of anomalies, the researcher should provide further discussion and suggestions for future studies that would address the theoretical problems posed by the findings. For all relations of a higher order than two factors, the researcher should be careful to distinguish among non- contingent and contingent relations (see Desideratum 9).

## Note

1  It might seem that there are other models possible, such as $\ln F_{ij} = \mu + \alpha_A + \gamma_{AB}$, but log-linear models are *hierarchical* in nature, so that an interaction term such as $\gamma_{AB}$ cannot enter a model unless all lower level terms, here both main effects, are also entered, because an interaction cannot be defined without reference to the associated main effects.

## References

Agresti, A. (1990). *Categorical data analysis.* New York: Wiley.

Agresti, A., & Yang, M.-C. (1987). An empirical investigation of some effects of sparseness in contingency tables. *Computational Statistics & Data Analysis, 5,* 9–21.

Christensen, R. (1997). *Log-linear models and logistic regression* (2nd ed.). New York: Springer.

Dunn, O. J. (1961). Multiple comparisons among means. *Journal of the American Statistical Association, 56,* 52–64.

Fienberg, S. E. (1980). *The analysis of cross-classified categorical data* (2nd ed.). Cambridge, MA: MIT Press.

Gart, J. J., & Zweifel, J. R. (1967). On the bias of various estimators of the logit and its variance with applications to quantal bioassay. *Biometrika, 54,* 181–187.

Goodman, L. (1964). Simultaneous confidence limits for cross-product ratios in contingency tables. *Journal of the Royal Statistical Society, Series B, 31,* 486–498.

Holm, S. (1979). A simple sequentially rejective multiple test procedure. *Scandinavian Journal of Statistics, 6,* 65–70.

Shaffer, J. P. (1986). Modified sequentially rejective multiple test procedures. *Journal of the American Statistical Association, 81,* 826–831.

# 19
# Meta-Analysis

S. Natasha Beretvas

Meta-analysis entails a set of analytical techniques designed to synthesize findings from studies investigating similar research questions. While meta-analysis includes narrative integration of results, the current chapter will focus only on quantitative meta-analysis. Meta-analysis permits summary of studies' results and is designed for scenarios in which the primary studies' raw data are not available. The meta-analytic process involves summarizing the results of each study using an effect size (ES), calculating an overall average across studies of the resulting ESs, and exploring study- and sample-related sources of possible heterogeneity in the ESs. The overall average ES provides a single best estimate of the overall effect of interest to the meta-analyst. Meta-analysis can be used to explore possible differences in ESs as a function of study or sample characteristics. In the seminal article in which the term *meta-analysis* was coined, Smith and Glass used meta-analysis to summarize results from studies that had assessed the effectiveness of psychotherapy (1977). Thus, treatment effectiveness results provided the first type of ES to be synthesized using meta-analysis. Since the 1970s, the field of meta-analysis has grown to include methods for conducting the synthesis of other types of ESs including correlations, transformations of odds-ratios, validity coefficients, reliability coefficients, and so forth.

Many textbooks provide detailed descriptions of the meta-analytic process. Texts by Lipsey and Wilson (2001) and Rosenthal (1991) provide excellent introductions to meta-analysis. Hunter and Schmidt's (1990) textbook provides the seminal resource for meta-analysts interested in correcting ESs for artifacts (see Desideratum 11). Books by Cooper and Hedges (1994) and Hedges and Olkin (1985) are recommended for readers with more technical expertise. Meta-analysts interested in a text devoted to description of ways to assess and correct for publication bias should refer to Rothstein, Sutton, and Borenstein (2005). Desiderata for studies that involve use of meta-analysis are contained in Table 19.1 and thereafter they are discussed in further detail.

## 1. Theoretical Framework and Narrative Synthesis

As with any manuscript, a summary of past research must justify the selection of the study's research question. Similarly, a meta-analysis must be prefaced by a narrative synthesis summarizing results found in previous studies that are to be integrated in the meta-analysis. The narrative synthesis must clarify the specific research question associated with the effect size (ES) that is being synthesized. The narrative synthesis summarizes in words what previous research has found in terms of the patterns of

Table 19.1  Desiderata for Meta-Analysis

| Desideratum | Manuscript Section(s)* |
|---|---|
| 1. A theoretical framework is provided that supports the investigation of the effect size (ES) of interest and includes a narrative synthesis of previous findings. | I |
| 2. Type of ES of interest in the study is specifically detailed (e.g., correlation, standardized mean difference). | I, M |
| 3. Databases searched and keywords used to find relevant studies are listed, as well as criteria for deciding whether to include a study in the meta-analysis. | M |
| 4. Formulae used to calculate ESs are provided or referenced, and any transformations used (e.g., to normalize or stabilize ES sampling distributions) are made explicit. | M |
| 5. The coding that is used to categorize study and sample descriptors is provided. | M |
| 6. Estimates are provided that describe the inter-rater reliability of the information coded in each study. | M, R |
| 7. If study quality is assessed, a description is provided detailing how it is assessed and how study quality is incorporated into the meta-analysis. | M |
| 8. For weighted analyses, the type of weights used is provided. | M |
| 9. Methods used to handle within-study ES dependence (e.g., multiple ESs per study) are described. | M |
| 10. Methods used to access, assess, and handle missing data are detailed. | M |
| 11. If relevant, the method used to correct for artifacts is described. | M, R |
| 12. Homogeneity of ESs is assessed. | M, R |
| 13. Statistics describing the resulting meta-analytic dataset that was gathered and including pooled estimates of the effect size of interest are provided along with associated standard errors (and/or confidence intervals). | R |
| 14. Inferential statistics describing the relation between the study and sample descriptors and the effect size are presented. | R |
| 15. Interpretation is offered describing the practical significance of the ES magnitude and direction and the relation between moderators and the ES. | D |

* *Note*: I = Introduction, M = Method, R = Results, D = Discussion

results relevant to the ES of interest. While many studies might investigate the same basic research question, the studies can be distinguished by various sample and study composition descriptors. Examples of descriptors include demographic variables such as gender, ethnicity and age, and characteristics of the study's design such as the type and duration of an intervention, the outcome measure used, the research context, and the experimental design. The review of previous literature should clarify and identify the importance and relevance of these descriptors to the ES of interest. This then lays the groundwork for investigation of relations between these descriptors (termed *moderators*) and the ES in the ensuing meta-analysis.

## 2.  Effect Size

The fundamental unit of any meta-analysis is the effect size (ES). An ES provides a parsimonious descriptor containing information about the direction and magnitude of the results of a study. The most commonly used meta-analytic ESs include the standardized mean difference, the correlation (representing the relation between two variables), and the odds ratio. A meta-analytic ES describes the relation between a pair of variables. Operationalization of each of the two variables should be clarified and justified. For example, if student achievement is one of the pair of variables of interest in a meta-analysis then the sorts of test scores that qualify as student achievement should be clarified. Description of the research question of interest in the meta-analysis should clarify the ES being investigated both in terms of the statistical type as well as the operationalization of each of the relevant two variables.

### 3. Study Inclusion Criteria

The Introduction (see Desiderata 1 and 2) should have clarified the components necessary for deciding to include a study's results in the meta-analysis. A section in the Methods section of a meta-analysis must detail how the relevant studies and results were found. The databases (e.g., PsycInfo, ERIC) that were searched, the types of studies (e.g., peer-reviewed publications, dissertations, conference presentations) and the keywords used must be identified. Any additional means used for finding relevant studies that were not initially identified should also be described (e.g., using the References section in studies that had been identified in the database search). In addition to emphasizing the acceptable operationalizations of the constructs relevant to the ES, the population of interest should be described. For example, a researcher might solely be interested in an ES for adults and thus data would be excluded from any study that had investigated the relevant variables for adolescent respondents.

Meta-analysts must also decide on the types of study designs that qualify for inclusion. Some meta-analysts include only results from studies employing purely experimental designs while others also include quasi-experimental studies' results. Some studies necessitate the use of single subject designs for which there is still controversy in terms of how to meta-analytically synthesize the results. If a more general inclusion strategy is used, then meta-analysts should code the relevant design features and summarize descriptively or inferentially the potential differences in resulting ESs (see Desiderata 13 and 14).

### 4. Calculation of Effect Sizes

The (statistical) type of ES being synthesized should have been clarified in the Introduction (see Desideratum 2). Results reported in the primary studies being synthesized are not all in the same format. For example, a meta-analyst might be interested in synthesizing a treatment's effectiveness using a summary of standardized mean differences across studies. Some studies might provide the treatment and control groups' means and standard deviations for the relevant outcome. Other studies might instead provide the results of an independent samples $t$-test comparing the treatment and control groups on the outcome score. Results in both formats can be converted into a standardized mean difference ES metric. Authors should clarify any conversion formulas they use to convert studies' results into a common ES metric.

In addition, some estimators of the most commonly used ES (the standardized mean difference) have been found to be biased. There are a number of ways that this ES is calculated (including, most commonly, Cohen's $d$, Glass's $\Delta$, and Hedges' $g$). The meta-analyst must clarify and justify which estimate of the standardized mean difference is being used.

Sampling distributions of most of the typical untransformed ESs (e.g., standardized mean difference, correlation, odds ratio) have been found to be non-normal. One of the purposes of quantitative meta-analysis is to use statistical tests of the ES and its relation with sample and study descriptors. Thus, it is important to use the transformations that normalize (and stabilize the variances) of the sampling distributions of these ESs. Meta-analysts should detail the formulas that are used to transform the resulting ESs estimates for ensuing statistical analyses.

## 5. Coding of Study and Sample Descriptors

A host of variables typically distinguish the studies and samples being synthesized and might be related to the resulting ESs. Sources of the possible heterogeneity in ESs across studies can and should be explored using these variables. When gathering primary study data to be used for calculating the ESs, meta-analysts should also gather information associated with the samples in each study. Sample size is an essential variable that must be coded as it provides information about the precision of each study's ES and can be used as a weight in resulting ES analyses (see Desideratum 8). Demographic information (such as age, gender, and ethnicity composition of the sample) can also be coded and used in the meta-analysis. Characteristics of each of the two variables whose relation is being synthesized should also be coded and captured. For example, in a study summarizing a family-based treatment's effectiveness in reducing internalizing disorders, the meta-analyst might have multiple constructs such as depression and anxiety that qualify as internalizing disorders. Each type of outcome could be coded to explore possible differences in the treatment's effectiveness for the more specific kinds of internalizing disorders. This can lead to multiple ES estimates being gathered per study and thus some dependence that must be handled (see Desideratum 9). There might also be characteristics of the implementation of the treatment that distinguish the primary studies and define the resulting ESs. In the current internalizing disorders example, interventions might be designed to involve both parents and children or they might be designed only for parents. Thus, categories distinguishing interventions could also be coded and collected. In addition, and specifically for intervention effectiveness meta-analyses, some studies might report results for more than one intervention. As with a study reporting multiple outcomes, the dependence resulting from multiple ESs per study needs to be appropriately handled (see Desideratum 9).

Facets of a study's design can also be gathered and included in the meta-analysis (as described in Desideratum 14). As mentioned in Desideratum 3, a study's design should be coded as it can later be used to explore potential differences in ESs resulting from differing experimental designs. Some meta-analysts code "study quality" and evaluate its relation to the ES values. Some meta-analysts correct their ESs for artifacts. They correct their ESs to match what the ESs would be for a perfect study that used an infinitely large sample with access to perfectly reliable and valid test scores. If interested in correcting for artifacts, the meta-analyst would gather relevant information including, for example, the reliability of scores on the measures of interest (see Desideratum 11). Additional selection of study and sample descriptors should be founded in the meta-analysts' research questions in terms of what they hypothesize might explain variability in ESs.

Values for some of the descriptors might differ for samples within a study. Group sample size in a meta-analysis of a treatment's effectiveness (i.e., using the standardized mean difference ES to summarize the difference in means between a treatment and control group) provides a simple example of a sample-level descriptor. Other descriptors might only vary across studies (e.g., whether the population being assessed were college students). The coder must clarify the distinction between such sample-level descriptors and study-level descriptors that differ across, but not within, studies. This information is essential to inform selection of the analytic technique that best matches the data's structure.

One last piece of information about coding must also be provided. Unfortunately, the information sought by meta-analysts is not always presented in the primary studies. It is important for meta-analysts to clarify how they attempted to gather this kind of missing data as well as to detail the methods used to handle the missingness (see Desideratum 10).

## 6. Inter-Rater Reliability

Given the amount of information that needs to be gathered and coded in a meta-analysis, it is typical to involve at least a couple of researchers as coders. It is thus important to provide a description of the reliability of the coding that was conducted. If data indicate that coding is not reliable, further coding training should be conducted and consensus about each study's codes must be reached. At the very least, the average (median) percentage agreement for each variable should be reported in the meta-analysis. Use of kappa, weighted kappa, or the intraclass correlation to provide additional measures of inter-rater agreement is also encouraged (see Orwin, 1994, for additional details; see also Chapter 11, this volume). While it would be optimal for at least two coders to code every study in the meta-analysis, that sometimes is not feasible. If this is the case, then at least a reasonable proportion of studies should be coded by at least two raters with sufficient justification provided for not having two raters code all studies. Given a lack of complete agreement in the coding that is done by the two raters, the meta-analyst must describe how differences were resolved and consensus reached through discussion and possible respecification of codes used.

## 7. Study Quality

Since the introduction of the term *meta-analysis* in the 1970s (Glass, 1976), researchers have argued about how to handle differences in research designs' quality when synthesizing studies' results. Researchers agree that, at the outset, meta-analysts must select and justify a research design quality criterion for study inclusion. In addition, meta-analysts are encouraged to gather and code information (see Desideratum 5) on a study's design that might differentiate studies' ES results.

All sorts of factors might impact the quality of a study's design and thus also affect the ES results. Those factors include group selection and assignment, experimenter expectations (e.g., whether a study is blinded), psychometric properties of measures, and many more. It is up to the researcher to select the pool of possible design quality variables of relevance to the meta-analysis. Meta-analysts can use the resulting variables descriptively or use them as moderating variables in ensuing analyses (see Desiderata 13 and 14).

## 8. Weights

As with any consistent estimator, the precision of an ES estimate is greater when it is based on larger sample sizes. Thus, when pooling ES estimates, meta-analysts typically weight ESs by some function of their associated sample sizes (see Desideratum 13). When testing models (see Desideratum 14) designed to explore the variability in ESs using study and sample descriptors as moderators, meta-analysts frequently estimate models involving these same $N$-based weights. In either scenario, more weight is assigned to estimates based on larger sample sizes. The most commonly used weights are either the inverse of $N$ or the inverse of the variance of the ES of interest (which will also be a function of $N$). The weight entailing the inverse of the conditional variance results in the most efficient pooled estimate of the population ES and thus is recommended here. However, the meta-analyst should clarify the function of $N$ that is being used as the weight.

## 9. Handling Dependent ESs

Studies can frequently contribute multiple ESs to a meta-analysis. These multiple ESs can be considered dependent if they are based on the same sample. For example, in meta-analyses designed to assess intervention effectiveness (i.e., comparing two groups on an outcome), a study can provide results from comparing the two groups on each of multiple outcomes. Given that sufficient data are provided in the study for each outcome that corresponds to the construct of meta-analytic focus (e.g., depression and anxiety might both qualify as internalizing outcomes), an ES can be calculated. The resulting two standardized mean difference ESs are assumed dependent because the ESs describe a common sample. As important, the ESs are based on measures that are themselves correlated (e.g., depression and anxiety).

Alternatively, in a meta-analysis of the correlation between two variables, multiple dependent ESs would result from a study that provided correlation estimates between pairs of variables both of which matched the constructs of interest. This study would qualify as a *multiple-endpoint* study. For example, the meta-analyst might be interested in the correlation between internalizing disorders and academic achievement. If a study reports the correlation between, say, depression and SAT scores and the correlation between anxiety and SAT scores, then both correlations could be used to calculate ESs for later analysis. The dependence would again originate in the use of a common sample for estimation of the two ESs.

Another example of the source of possible dependence commonly found in meta-analyses of intervention research might originate in a study reporting results from comparing three groups on an outcome. This study would be an example of a *multiple-treatment* study. For example, a meta-analyst might be interested in summarizing the effectiveness of parental involvement interventions for improving internalizing disorders. A primary study might evaluate the internalizing disorders of three groups, two of which involve differing implementations of a parental involvement treatment and a control group. Two effect sizes could be calculated with one comparing the internalizing disorder scores of the first intervention group with the control group. The second ES would describe the difference in internalizing disorders between the second intervention group and control group. Given the involvement of the same control group in the calculation of the two ESs, the ESs would be considered dependent.

Meta-analysts have a choice of methods they can use to handle this dependence. Some choose to ignore the dependence which will negatively impact the validity of the associated statistical conclusions. Others might choose a single effect size to represent each study. For example, this "best" ES might be based on the measure with the best psychometric functioning in each study. Still others might calculate a weighted or simple average of each study's multiple ESs and use the result as the single ES for each study. While use of a single ES per study (selected via aggregation or deletion of the study's multiple ESs) does result in an analysis of independent ESs, it overly reduces the available database thereby reducing ensuing statistical power. It also unnecessarily reduces the possible heterogeneity in the ESs.

Another option available for handling dependent ESs involves modeling the multivariate nature of the dataset. Several options are available with use of generalized least squares (GLS) estimation procedures being the most commonly used method. The primary problem with the use of multivariate modeling to handle possible dependencies is that additional data must be gathered from the primary studies. For example, to use GLS for synthesizing results from multiple-endpoint studies, meta-analysts must use values for the correlation among scores on the multiple endpoints. However, it is possible to impute reasonable values for this correlation and, despite their complexity, GLS methods have been found to work very well for handling meta-analytic dependence. Meta-analysts are encouraged to consider using GLS methods and are referred to Gleser and Olkin (1994) for further details. Regardless of the method used, the meta-analyst must note the types of dependence that they encountered in their dataset. They must also describe and justify their choice of method used to handle this dependence.

## 10. Methods for Handling Missing Data

As with most social science datasets, analysis of meta-analytic datasets is also hampered by missingness. This can result from primary studies not reporting sufficient statistical information permitting calculation of an effect size. Alternatively, primary studies might not have gathered or not reported all information of interest to the meta-analyst. For example, a meta-analyst might be interested in explaining heterogeneity in an ES using a variable representing the percentage of participants who were female. Not every study will necessarily provide the percentage of participants who are female. Meta-analysts need to detail and justify how they handled missing data.

As with primary study analyses, there are a host of options that meta-analysts can use to handle missingness. There are similar caveats associated with these techniques when used in a meta-analytic context. For example, use of listwise or pairwise deletion still requires the assumption that data are missing completely at random and frequently results in large reductions in data available for a meta-analysis. These methods are not strongly recommended for use with meta-analytic data. Use of single-value imputation is not uncommon in meta-analysis (e.g., using a mean of reported values' information, or using a value that is reasonable based on patterns of values reported in other studies with similar participants). Single-value imputation can be recommended although its use inappropriately reduces the associated variability. Use of multiple imputation (MI) is still rare in meta-analysis but if it is used, then the missingness is assumed to be missing at random. Further methodological research is needed to assess the functioning of MI, however, it would seem likely to function best as a method for handling missing meta-analytic data.

Meta-analysis is also criticized for another form of missingness peculiar to this technique, namely, missingness due to *publication bias*. Publication bias is a term that refers to the scenario where only studies with statistically significant results are reported ("published") and only studies that are published (i.e., available) can provide data that can be synthesized in a meta-analysis. Clearly this kind of missingness will bias resulting ESs. There are many different ways researchers use to assess whether publication bias might exist. Graphical displays, such as the funnel plot, are sometimes used. ES estimates based on smaller sample sizes would be expected to vary more than for studies based on larger sample sizes although the average ES should not depend on sample size. Funnel plots involve graphing ES estimates against their associated sample sizes and provide a graphical way of assessing whether this pattern holds. If the plots are skewed, then this can be inferred as evidence of publication bias.

Indices are also available to assess potential publication bias. The fail-safe number, modifications thereof and trim-and-fill estimates can also be used to evaluate the potential for publication bias. Last, some meta-analysts use inferential tests of publication bias (e.g., Begg's rank correlation test, Egger's regression, and funnel plot regression). The reader is strongly encouraged to refer to any of the meta-analytic texts (especially Cooper & Hedges, 1994, and Rothstein et al., 2005) to find out further details about these different procedures. Meta-analysts are encouraged to use multiple methods for assessing publication bias including at least the trim-and-fill method and one of the regression methods despite their limited statistical power.

Meta-analysts should try to contact the primary study authors to obtain information that may not have been reported. In the absence of this information and if evidence supports the possibility of publication bias, meta-analysts are encouraged to use any of the variety of methods available for correcting for publication bias. In particular, the trim-and-fill correction and the use of weighted distribution theory-based approaches are recommended.

## 11. Correction for Artifacts

Some meta-analysts use artifact correction procedures to correct for artifactual errors resulting from imperfect research scenarios. These correction procedures are designed to correct resulting ES estimates so that they represent results under ideal research scenarios (for example, they can be used to correct an ES estimate so that it represents the ES estimate based on perfectly reliable and valid test scores). The most commonly used correction is the correction for attenuation that can result from the lack of perfect reliability of scores on social science measures. Other corrections include correction for dichotomization of continuous variables and for restriction of range. Use of these procedures involves obtaining additional information (e.g., internal consistency reliability estimates for the relevant outcomes) to correct the relevant ES as well as its associated variance estimate. Use of artifact correction can also affect the sampling distributions assumed for the resulting ESs. Meta-analysts must specify which artifacts they might be correcting for and how. There is no consensus in the field about the use of these artifact correction procedures. Given the difficulties encountered in terms of gathering realistic values to calculate the corrections and their effects on the ESs' sampling distributions, the validity of the resulting corrections and of analyses conducted using the corrected ESs seems questionable.

## 12. Homogeneity of ESs

Meta-analysis is used to synthesize results from a multitude of studies designed to assess the same research question. While replication is encouraged in research, most studies do not exactly mimic each other. Studies tend to involve some subtle (or not so subtle) variation on a previous but similar study. Samples from different populations might be used (e.g., adults versus adolescents or college students, clinical versus non-clinical respondents, populations with different demographic information). Different implementations of an intervention might be tested. Different measures of a related but distinct construct might be investigated. This means that the resulting effect sizes might not come from a single population (sampling distribution of effect size estimates) with a single true effect size. Instead, it is more likely that while some of the variability in effect size estimates is due to sampling error, some of the variability is also attributable to random effects. In other words, the estimates do not come from a single population.

   Meta-analysts should test the heterogeneity of the effect size estimates they gather. Methodological researchers have consistently supported use of the $Q$-test statistic designed to test the null hypothesis of homogeneous ESs. If the variability in the effect sizes is found to be more than could be solely attributed to sampling error, then this affects the model that should be assumed when conducting ensuing statistical analyses. Excess heterogeneity means that a random effects model should be assumed. If the effect sizes can be assumed homogeneous, then a fixed-effects model can be assumed. A meta-analytic researcher should clearly identify which model was assumed for all analyses including estimation of both pooled estimates as well as for analyses designed to investigate sources of variability in effect size estimates using the moderating variables detailed in Desideratum 5.

## 13. Descriptive Statistics

Meta-analysts should describe the resulting data that were gathered. This includes the availability of sample and study descriptors as well as information that could be used to calculate ESs. Some meta-analysts provide a table listing each study and associated descriptive information (such as the sample size underlying an ES as well as other study and sample descriptors as noted in Desideratum (5). This table usually also provides every ES or an overall ES for each study (see Desideratum 9). All meta-analysts present ES estimates pooled across studies for each outcome of interest and usually for levels of categorical moderating variables of interest. Along with all pooled estimates, associated standard error estimates (and/or confidence interval) should be provided. The (random- or fixed-effects) model that is assumed for the synthesis of estimates should already have been noted (see Desideratum 12).

## 14. Inferential Statistics

Results summarizing the tests of relation between moderators (see Desideratum 5) and the ES should be presented. Meta-analysts testing a number of moderating variables should consider use of (weighted) multiple regression as a model for testing the concurrent inter-relations. Conducting a multitude of statistical tests can lead to inflated Type I error for meta-analytic data as with any other kind of data. Controls such as the use of Bonferroni's correction to the nominal alpha level should be considered. Last, meta-analysts should appropriately model the meta-analytic data's structure. For example, in a meta-analysis involving multiple ES estimates per study, some of the moderators might be sample-level descriptors while others might be at the study level. Multilevel modeling suitable for use with meta-analytic data should be considered with this kind of clustered meta-analytic data.

## 15. Practical Significance

As with any empirical study, detection of statistical significance (or non-significance) should be interpreted within a context. While some researchers might cite rules of thumb for cutoffs representing small, moderate, and large effect sizes, interpretation of an effect size's magnitude should be made in the explicit context in which the effect size is calculated. For example, an ES estimate of 0.01 would qualify to most researchers as minuscule. However, in a test of aspirin for reducing heart attacks an ES estimate ($R^2$) of 0.011 was deemed sufficiently large that the trial was prematurely halted to stop "harming" placebo recipients who were not being given the aspirin (cited in Rosenthal, 1994). Thus, rules of thumb for describing an effect's size should be used with caution. Instead, the researcher should consider the magnitude and direction of the ES estimates in the context in which they are being assessed. Similarly, the strength (and direction) of the relation between the moderating variables and the ES should be interpreted at a practical rather than solely a statistical significance level.

## References

Cooper, H., & Hedges, L. V. (Eds.). (1994). *The handbook of research synthesis.* New York: Russell Sage Foundation.

Glass, G. V. (1976). Primary, secondary, and meta-analysis. *Educational Researcher, 5,* 3–8.

Gleser, L. J., & Olkin, I. (1994). Stochastically dependent effect sizes. In H. Cooper & L. V. Hedges (Eds.), *The handbook of research synthesis* (pp. 339–355). New York: Russell Sage Foundation.

Hedges, L. V., & Olkin, I. (1985). *Statistical methods for meta-analysis.* Orlando, FL: Academic Press.

Hunter, J. E., & Schmidt, F. L. (1990). *Methods of meta-analysis: Correcting error and bias in research findings.* Newbury Park, CA: Sage.

Lipsey, M. W., & Wilson, D. B. (2001). *Practical meta-analysis.* Thousand Oaks, CA: Sage.

Orwin, R. G. (1994). Evaluating coding decisions. In H. Cooper & L. V. Hedges (Eds.), *The handbook of research synthesis* (pp. 139–162). New York: Russell Sage Foundation.

Rosenthal, R. (1991). *Meta-analytic procedures for social research.* Newbury Park, CA: Sage.

Rosenthal, R. (1994). Parametric measures of effect size. In H. Cooper & L. V. Hedges (Eds.), *The handbook of research synthesis* (pp. 231–244). New York: Russell Sage Foundation.

Rothstein, H. R., Sutton, A. J., & Borenstein, M. (Eds.). (2005). *Publication bias in meta-analysis: Prevention, assessment and adjustments.* Hoboken, NJ: John Wiley and Sons, Inc.

Smith, M. L., & Glass, G. V. (1977). Meta-analysis of psychotherapy outcome studies. *American Psychologist, 32,* 752–760.

# 20
# Multidimensional Scaling

Mark L. Davison,[1] Cody S. Ding, and Se-Kang Kim

Multidimensional scaling (MDS) is a multivariate statistical method for estimating the scale values along one or more continuous dimensions such that those dimensions account for proximity measures defined over pairs of objects. It has been used to study such things as dimensions underlying perceptions of human speech, patterns of vocational/academic interests, and growth over time in reading and math achievement. We will limit ourselves to discussion of analyses based on Euclidean distance models. While this list is not exhaustive, there are three major applications of MDS which differ in the nature of the objects, the proximity measure defined over those objects, and the purpose. In *perception studies*, perhaps the most typical form of MDS, the objects are stimuli (e.g., speech samples), the proximity measures are judgments about the similarity of stimulus pairs (e.g., a rating of similarity), and the purpose is to identify the attributes (dimensions) along which stimuli are perceived to vary and that account for the similarity judgments. In *cross-sectional studies*, the objects are (typically continuous) variables measured at a single time point (e.g., score on a vocational interest scale). The proximity measure is an index of association defined over pairs of variables, such as a squared Euclidean distance or correlation coefficient. The purpose of the analysis of these proximity measures is to identify dimensions that point toward one or more within-person patterns needed to account for the associations among the variables. In *longitudinal studies*, which are a relatively recent extension of cross-sectional studies, the objects are occasions, and thus the data consist of a single variable (e.g., math achievement) measured at several time points. The proximity measure is an index of association among the occasions. The goal is to find patterns of growth, decay, or change that account for the associations among the occasions.

Most applications of MDS are exploratory; that is, they are designed to uncover dimensions accounting for the proximity data rather than test a priori hypotheses about dimensions that account for the proximity data. There are constrained versions of MDS designed for fitting a priori hypotheses, but not all of these constrained methods have well developed fit measures with which to evaluate the dimensional hypotheses. The cross-sectional and longitudinal applications of MDS can serve to generate hypotheses that will later be assessed for confirmation through methods discussed in other chapters (e.g., structural equation modeling, hierarchical linear modeling).

Kim, Frisby, and Davison (2004) described the application of MDS to cross-sectional data; Ding, Davison, and Petersen (2005) described the application to longitudinal data; and Kim, Davison, and Frisby (2007) discussed the translation of hypotheses generated by MDS into structural equation

Table 20.1 Desiderata for Multidimensional Scaling

| Desideratum | Manuscript Section(s)* |
| --- | --- |
| *General* | |
| 1. Describe the competing theories and prior research results leading to the study, including predictions about the dimensions or spatial configuration of stimuli needed to account for the data. | I |
| 2. State the purpose to which MDS will be put; for example, investigate perceptual dimensions of stimuli, recover within-person patterns in cross-sectional data, or explore patterns of growth and change in longitudinal data. | I, M, R |
| 3. Describe the sample of respondents; describe and justify the population to which results may be generalized. | M |
| 4. Describe the model of the proximity data. | I, M, R |
| 5. Describe the measures of fit that will be used to compare models or to decide on the dimensionality of the final solution. | R |
| 6. Describe and, if possible, justify the rotation of the reported solution. | M, R |
| *Perception Studies* | |
| 7. Describe the sample of stimuli; describe and justify the population of stimuli to which results can be generalized. | M |
| 8. Describe the judgment task by which proximity judgments were obtained. | M |
| 9. Explain how missing data, if any, were handled. | M, R |
| 10. Describe stimulus measurements collected for purposes of empirically confirming or disconfirming potential interpretations of dimensions or spatial configurations. | M |
| 11. Justify the final model selected, including the dimensionality of that model. | R |
| 12. Report the scale values of the final solution in tabular and/or graphical form. Explain and justify the interpretation of the dimensions or the spatial configuration of stimuli using additional analyses as needed. | R |
| *Within-Person Patterns in Cross-Sectional Studies* | |
| 13. Describe and justify the variables under study and the population of variables to which results can be generalized. | M |
| 14. Explain how missing data, if any, were handled. | M, R |
| 15. Describe and justify the proximity measure (measure of association between pairs of variables) selected for the study. Report descriptive statistics on the variables and report the proximity matrix. | M, R |
| 16. State and justify the model (if any) for the observed variables, the model for the proximity measures, the method of estimating the parameters in the model for the proximity measures, and the fit measure(s). | M, R |
| 17. Justify the proximity model selected, including the dimensionality of that model. | R |
| 18. Report the scale values of the final solution in tables or graphs. Explain and justify the interpretation of the dimensions or the spatial configuration of stimuli on which the final conclusions are based. | R |
| 19. Justify MDS over alternative analyses. Report or discuss parallel results from methods related to MDS (e.g., Q-factor analysis, cluster analysis.). | M, R |
| *Growth Patterns in Longitudinal Studies* | |
| 20. Describe and justify the variable, the time points under study, and the population of time points to which results can be generalized. | M |
| 21. Explain how missing data, if any, were handled. | M, R |
| 22. Describe and justify the proximity measure (measure of association between pairs of time points) selected for the study. | M, R |

23. State and justify the model for the variable at each of the several time points, the model for the      R
    proximity measures, the method of estimating parameters in the model for the proximity
    measures, and the fit measure(s).
24. Explain and justify the interpretation of the dimensions or the spatial configuration of stimuli on      R
    which the final conclusions are based.

---

\* *Note*: I = Introduction, M = Methods, R = Results, D = Discussion

models. More thorough treatments can be found in Borg and Groenen (2005), Cox and Cox (2001), and Davison (1991). SAS and SPSS contain MDS programs (ALSCAL, Takane, Young, & deLeeuw, 1977; PROXSCAL, Commandeur & Heiser, 1993). Other programs include MULTISCALE (Ramsay, 1977), and SMALLEST SPACE ANALYSIS (Lingoes, 1989).

In describing key methodological issues in MDS studies, we begin the table of desiderata and subsequent elaborations by describing those common to all three application types (perception, cross-sectional, longitudinal), followed by a discussion of issues specific to perceptual, cross-sectional, and longitudinal applications.

## 1. Theory and Prior Research

While MDS is usually exploratory, a solid grounding of the study in theory and prior research is still necessary. Existing theory is often insufficiently precise for purposes of specifying hypotheses to the degree of precision required by confirmatory analyses, hence the need for exploration.

The theory and prior research leading to the current study need to be explained. Competing theories, if any, should be included in the explanation. The explanation should include any dimensions or any spatial configuration suggested by that prior literature. The prior literature might be used to guide the number of dimensions included and the substantive interpretation of the final solution. It might also guide the selection of a respondent population or a variable population to be sampled. Further, it might suggest additional data to be collected for the purpose of confirming interpretations of the dimensions or the spatial configuration.

## 2. Purpose

In their write-up, authors need to explain the intended purpose for utilizing MDS. The stated purpose will differ depending on whether MDS is being used to study perceptions of stimuli, variables collected in a cross-sectional study, occasions in a longitudinal study, or some other entities. In perception studies, the researcher often intends to recover dimensions accounting for the perceptual judgments as well as a description of individual differences in the use of those dimensions. In cross-sectional studies, the researcher is often interested in describing dimensions that account for within-person variation, in which case, the researcher should so indicate. If the cross-sectional data are being collected for some other purpose, that other purpose should be described. Likewise, longitudinal data are often collected for the purpose of uncovering patterns of growth, decay, or change, and if so, that intent should be stated. If the data are being collected for purposes of hypothesis generation, any subsequently planned confirmatory analysis might also be described.

## 3. Sampling of Respondents

The Methods section must include a description of the respondent population sampled and the sampling method used. This will determine the limits of generalizability with respect to respondents.

The description should include a discussion of important subsamples that might vary in their perceptions of stimuli, their patterns of scores on cross-sectional variables, or their growth trajectories in longitudinal studies. In perception studies for which individual differences in perception are a focus, the subpopulations and the reasons for their inclusion may need particular attention. Selection of the respondents may indirectly influence the selection of stimuli, variables, or occasions and any such effects should be mentioned.

## 4. Model of Proximity Data

The researcher needs to describe the model, or models if several were tried, for the proximity data. This description should indicate whether the model is nonmetric and assumes that the proximity data form an ordinal scale, metric and assumes that the data constitute an interval scale, or metric and assumes that the data constitute a ratio scale. Individual differences parameters, if any, should be described. The description should indicate whether the model assumes that the data are related to Euclidean or non-Euclidean distances, and if Euclidean, whether Euclidean or squared Euclidean distances. Any constraints on parameter estimates should also be described. If models with varying numbers of dimensions were fitted to the data, the range of dimensionalities should be reported.

Consider the two models below:

$$\delta_{ii'} \approx f\left[\sqrt{\sum_k (x_{ik} - x_{i'k})^2}\right] \tag{1}$$

$$\delta_{ii'p} \approx \sqrt{\sum_k w_{pk}(x_{ik} - x_{i'k})^2} \tag{2}$$

The first is a nonmetric model which assumes that the proximity datum (judged a dissimilarity) for stimuli $i$ and $i'$ is a monotonically increasing function $f$ of Euclidian distances between points for stimuli $i$ and $i'$ whose locations in a $K$-dimensional space are given by the coordinates $x_{ik}$ and $x_{i'k}$ respectively. The second assumes that the proximity datum for person $p$ is a weighted Euclidean distance function with individual differences weights $w_{pk}$ representing individual differences in the salience of the perceived dimensions. Inclusion of the monotone function in Equation (1) implies weaker, ordinal assumptions about the proximity data, but the model does not accommodate individual differences in the stimulus perceptual process. The model of Equation (2), however, makes stronger, ratio scale assumptions about the data but allows for individual differences in the perceptual process through inclusion of the weight parameter $w_{pk}$. Models vary in several respects, such as their assumptions about the measurement scale of the proximity data $\delta_{ii'}$ and their assumptions about individual differences.

In cross-sectional or longitudinal applications, there may be two models, one for the variables from which measures of association were computed and a second model for the measures of association serving as proximity data. Ideally, the second model will be derived from and fully consistent with the first, and both the assumptions embodied in the model of the raw data and the model of the proximity measures will be explained.

## 5. Fit and Dimensionality

The various fit measures used to evaluate and compare solutions should be described. Criteria used to evaluate those fit measures should also be described. For instance, Kruskal (1964a, 1964b) provided guidelines for the least squares fit measure STRESS. A scree plot of the fit measure may also be used in deciding the number of dimensions to retain. In most multidimensional scaling applications in which global fit measures are employed, the number of dimensions to retain is the number at or above the elbow, not the number above the elbow as in factor analytic plots of eigenvalues. A distinctive "C" or "U" shape of the two dimensional configuration can be an indication that only one dimension is really needed to account for the data.

Because most authors have suggested that at least five stimuli are needed per dimension in order to estimate scale values with satisfactory precision, the number of stimuli will place an upper limit on the number of dimensions in the solution. Despite this guideline, some seemingly meaningful solutions have been obtained with as few as three stimuli per dimension in cross-sectional and longitudinal applications. Some computer programs print a warning if there are few stimuli per dimension and will not compute a solution if there are too few.

Like most iterative algorithms, MDS algorithms may fail to reach an optimal solution for several reasons, including local minima, solution degeneracy, or an inadequate number of iterations. Researchers should familiarize themselves with these problems, methods for detecting such problems, and methods for avoiding them. Obviously, non-optimal solutions should not be reported and steps taken to avoid such problems should be described.

## 6. Rotation of the Solution

Except in certain special cases, MDS solutions are subject to the same rotational indeterminacy as are exploratory factor solutions. Unfortunately, there are no widely accepted algorithms for optimizing the interpretability of the solution through rotation as in factor analysis. When interpreting the spatial configuration of stimuli, the rotation may not matter. For instance, if the stimuli have a circular configuration in two dimensions, that circular configuration is invariant with respect to rotation and, consequently, the rotational indeterminacy poses no limitation. On the other hand, when interpreting dimensions, the dimensional interpretation applies only to a particular rotation.

In some MDS models, most notably those based on the weighted Euclidean model, the rotation is determinate except in certain special cases. When the solution is based on a model in which rotation cannot be performed without loss of fit, this rotational determinacy should be noted.

When interpreting dimensions of the solution, rather than the more general spatial configuration of the stimuli, the researcher needs to justify the chosen rotation or acknowledge its indeterminacy as a limitation. Alternative dimensional interpretations corresponding to rotations of the solution need to be recognized. In perceptual studies, the researcher may wish to employ an analysis based on an individual differences model in which the rotation is generally determined by fit to the data in order to avoid the rotational problem.

## 7. Sampling of Stimuli in Perception Studies

The Methods section must describe and justify the sample of stimuli included in the study. MDS dimensions are those along which stimuli vary. In a study of occupational perceptions, for instance, the resulting dimensions are likely to include a dimension of occupational safety only if the sample of occupations includes both safe and dangerous occupations. The description should indicate whether the sample of stimuli is considered *fixed* (i.e., all stimuli of interest) or *random* (i.e., only a sample of stimuli of interest). Because the selection of stimuli can seriously influence the nature of dimensions, possible effects of stimulus selection may need consideration in the Discussion section.

## 8. Proximity Judgment Task in Perception Studies

In the most common form of perception studies, the respondent is shown a pair of stimuli and asked to judge the similarity of the two stimuli. For instance, the respondent might be shown two stimuli and asked to rate them on 7-point scale ranging from *highly similar* to *highly dissimilar*. Rating scales, however, are not the only possible task. For instance, the respondent can be shown three stimuli and then, from among the three stimuli, asked to select the two which are most alike and the two which are least alike. Indirect judgment tasks can also be used. For instance, the researcher can briefly present pairs of stimuli and then ask the respondent to indicate whether they were the same or different. The number of times two stimuli are confused, that is, incorrectly identified as being the same, can be considered a measure of their similarity. Whatever task is chosen, the researcher must describe the task and how responses were scored to obtain a proximity measure for each possible pair. This description should include the directions given to respondents or, at least, a summary of those directions. Because the order of stimulus presentation, both within and across pairs, can potentially have an effect on judgments, any steps taken to control order effects should be described. If the MDS analysis algorithm assumes that, within the limits of random error, similarity is symmetric (i.e., the similarity of pair (A, B) is the same as the similarity of (B, A) and that the order of presentation does not matter), the researcher would need to describe any asymmetries in the data and the method of handling such asymmetries.

## 9. Missing Data in Perception Studies

As the number of stimuli increases, the number of stimulus pairs increases rapidly. If $n$ is the number of stimuli, the number of stimulus pairs is $n(n-1)/2$. If the number of pairs is large, an incomplete data collection design can be employed to reduce the number of judgments by any one respondent. For any one respondent then, some judgments will be missing by design. Such missingness needs to be described, and its potential impact on the results should be discussed. Data may also be missing not by design. This missingness should also be described along with any steps to handle the missingness.

## 10. Additional Data in Perception Studies

Researchers often collect additional data to help confirm or disconfirm interpretations of dimensions. For instance, if the researcher suspects that perceived salary may be a dimension considered by respondents in making similarity judgments about occupations, the researcher may collect data on the salaries of the various occupations. These additional data may be objective (e.g., mean salaries of job incumbents) or subjective (e.g., perceived salary as rated by a sample of respondents). The Methods section should include a description of any such variables collected for the purpose of potentially confirming the interpretation of one or more MDS dimensions.

## 11. Final Model in Perception Studies

Often, the analysis will include a comparison of several models. Almost all research includes a comparison of models of varying dimensionalities. It may include models with and without constraints on scale values, models that do and do not include individual differences parameters, or models that are metric and nonmetric. Generally, the models are compared in terms of parsimony, fit to the data, interpretability of the dimensions, and replicability of dimensions across samples. All other attributes being equal, a model is preferred if it contains fewer dimensions or freely estimated parameters (parsimony), better fit to the data (fit), dimensions all of which are substantively interpretable or an interpretable spatial configuration of stimuli (interpretability), and dimensions or configurations that appear in the solutions of several samples (replicability).

## 12. Final Solution in Perception Studies

The scale values of the final solution should be reported in tabular form, graphical form, or both. When the solution is interpreted in terms of the stimulus configuration (e.g., a circular formation), graphical presentation is essential. Reported results should also include one or more fit measures. If an individual differences algorithm has been employed, estimated individual differences parameters should be reported in graphical or tabular form. If the sample size is large, however, these individual differences results may be reported in summary form (e.g., means and standard deviations of estimated individual differences parameters).

Following the presentation of the solution itself, the Results section should include results of any analyses that aid in the interpretation of the solution. Often these include various correlational analyses. For instance, in our hypothetical study of job perceptions, the scale values along Dimension 1 might be correlated with the median salary associated with each job in an attempt to determine whether Dimension 1 can reasonably be interpreted as reflecting salary. A cluster analysis of estimated dimension scale values might reveal distinct groupings of occupations. While the interpretation of the solution is ultimately subjective, it can be aided by additional data and analyses.

## 13. Variables in Cross-Sectional Studies

MDS can be used to study stimulus perceptions as described above, but it can also be used to study patterns of scores in cross-sectional data. A pattern is defined over a set of variables. Changing the variable set can alter the score patterns in the data. Therefore, the Methods section must describe the variables included in the set of measures analyzed and explain the rationale for their selection. Often the variables are scales which constitute an assessment inventory or test battery. The scales are commonly administered together and are often reported as a profile of scores in clinical, counseling, educational, or industrial/organizational psychology (e.g., the scales in an interest inventory or the scales in an intelligence test battery).

In some cases, the possible effects of changing the composition of the variable set will need to be discussed. Where the composition of the variable set is an issue, the researcher may want to empirically study the effect of adding or deleting variables. For instance, a researcher might examine the stability of Big Five personality scale patterns across two personality inventories, one of which included only Big Five personality scales and one of which included Big Five scales embedded in a larger set. If the variables have been sampled from a larger domain, the researcher will need to describe the larger domain from which the variables originate and discuss the generalizability of patterns to variable sets in the larger domain.

## 14. Missing Data in Cross-Sectional Studies

For any given person, data on some variables might be missing. The MDS proximity measures can often be computed within existing computer packages (e.g., SAS, SPSS), and these packages often include listwise and pairwise options for handling missing data. Such options can be justified if the data are missing completely at random (MCAR). When the MCAR assumption is satisfied, pairwise deletion uses larger samples for computing the results and therefore yields sample estimates of proximity measures with smaller standard errors. When the sample size is large, however, listwise deletion should yield proximity measures with sufficiently small standard errors and will ensure that every proximity measure is computed on the same sample of data.

If the MCAR assumption is not defensible, however, the researcher may want to employ some form of data imputation (usually multiple imputation), before the proximity measures are computed.

Existing statistical packages typically offer several imputation options. The option chosen should be described and justified.

## 15. Proximity Measure in Cross-Sectional Studies

Even casual inspection of the proximity module in any of the common statistical packages will reveal that there are many statistical measures of association that might be employed. The correlation and covariance statistics are probably the two most widely known in the social, behavioral, and education sciences.

The choice of proximity measure must be justified. In our opinion, the strongest basis for justification begins with a model of the raw data from which a proximity measure can be derived. An example of such a model is given in Equation (5).

In MDS, one plausible proximity measure is the squared Euclidean distance, which can be computed from the raw data for variable pair $(v, v')$ as follows:

$$\delta^2_{vv'} = \frac{\sum_p (y_{pv} - y_{pv'})^2}{P} .$$ (3)

In words, the squared Euclidean distance measure of proximity for variables $v$ and $v'$ is the squared difference between the score of person $p$ $(p = 1, \ldots, P)$ on variable $v$ and $v'$ averaged across all persons (however, SPSS computes the sum over all persons, not the average).

The most well-known measure of association, the Pearson product moment correlation coefficient, is closely related to the squared Euclidean distance. If the variables are in $z$-score form (i.e., variables $z_{pv}$ and $z_{pv'}$), then the squared Euclidean distance proximity measure has the following form:

$$\delta^2_{vv'} = 2 - 2r_{vv'}.$$ (4)

Equation (4) says that, when the squared Euclidean distance proximity measure is computed from standardized variables, the squared Euclidean distance proximity measure is linearly but inversely related to the correlation coefficient. In MDS, this means that an MDS of correlations among variables will yield exactly the same solution as an analysis of squared Euclidean distances computed from variables in standardized form if two conditions hold: (1) the correlations are treated as similarities whereas the squared Euclidean distances are treated as dissimilarities, and (2) both proximity measures are treated as ordinal or both are treated as interval level data points.

In our opinion, unless there is very good reason to do otherwise, researchers should base analyses of cross-sectional data on squared Euclidean distances or correlation coefficients, at least if the researcher intends to analyze a variables-by-variables matrix rather than a persons-by-persons matrix. Either of these measures can be justified from a plausible, explicit model of the original variables, $Y_v$. That model is described in the next section. An explicit model of the original variables can not only serve as the basis for justification of a proximity measure, but also it can enrich the interpretation of the resulting MDS dimensions. A table of proximity measures should be reported in the results section to facilitate later re-analysis of the data and meta-analysis.

## 16. Model of the Observed Variable in Cross-Sectional Studies

In cross-sectional applications of MDS, the phrase "model of the data" can mean one of two things: a model for the original variables $Y_{pv}$ or a model of the proximity measure. In this section, we are primarily concerned about a model of the original variables $Y_{pv}$, but in some cases the model of the original variables can be used to derive a model for the proximity measure. If the researcher has a model for the original data, the model should be stated. The model constitutes a statement of the assumptions on which the analysis is based. If there is no such model, and frequently there has not been one in cross-sectional applications, then the conditions under which the analysis is appropriate are unstated, seemingly unknown, and impossible to evaluate. Without such a model, there is no formal connection between the resulting MDS scale values and the original data, thus precluding formal explanations of the original variables in terms of the MDS solution.

One possible model from which a proximity measure can be derived is the Profile Analysis via Multidimensional Scaling (PAMS) model:

$$y_{pv} = c_p + \sum_k w_{pk} x_{vk} + e_{pv} \tag{5}$$

where $y_{pv}$ is the observed score of person $p$ ($p = 1, \ldots, P$) on variable $v$ ($v = 1, \ldots, V$), which represents the element in row $p$ column $v$ of the data matrix; $c_p$ is a level parameter which indexes the overall height of person $p$'s profile, $c_p = \dfrac{\sum_v y_{pv}}{V}$; the scale values $x_{vk}$ along dimension $k$ constitute a row vector $\mathbf{x}_k$ of contrast coefficients that depict a pattern of scores; $w_{pk}$ is a weight for person $p$ on dimension $k$ that indexes the degree of match between the pattern of person $p$'s observed scores and the pattern of the scale values in vector $\mathbf{x}_k$; and $e_{pv}$ is an error term. In essence, the PAMS model in Equation (5) represents each person's row vector of data as a linear combination of the patterns $\mathbf{x}_k$ represented as vectors of MDS scale values:

$$\mathbf{y}_p = c_p \mathbf{1} + \sum_k + w_{pk} \mathbf{x}_k + \mathbf{e}_p \tag{6}$$

where $\mathbf{1}$ is a row vector of 1's and $\mathbf{e}_p$ is a vector of error terms for person $p$. Readers familiar with factor analysis will recognize this as a linear model similar to that in factor analysis except that it includes an intercept term. In words, Equation (6) states that each person's row of data is a linear combination of pattern vectors $\mathbf{x}_k$. While multidimensional models including a person-specific intercept have long existed in the scaling literature, factor models with a random coefficient intercept are a more recent development.

Given appropriate assumptions (Kim et al., 2007) and if computed from the raw data according to Equation (3), the squared Euclidean distance proximity measure for each variable pair will have the following form:

$$\delta_{vv'}^2 = \sum_k (x_{vk} - x_{v'k})^2 + 2\sigma^2 = d_{vv'}^2 + 2\sigma^2 \tag{7}$$

where $\sigma^2$ is the variance of the errors in Equation (5). Hence, the proximity measures are a squared Euclidean distance function of parameters $x_{vk}$ in the model; a MDS of such proximity measures can be used to estimate the parameters; and the scale values in the MDS will constitute estimates of those parameters. Having estimated the parameters $x_{vk}$ through MDS, one can use the scale value estimates and regression to estimate the individual differences parameters $w_{pk}$ and $c_p$.

## 17. Proximity Model and Dimensionality in Cross-Sectional Studies

One must decide whether to consider the proximity data as ordinal, interval, or ratio in order to select an appropriate metric or nonmetric analysis. If there is a formal model of the original data from which the proximity measure has been derived, then the derived form of the proximity measure may determine the appropriate analysis. For instance, consider the model of the proximities in Equation (7) derived from the model for the raw data in Equation (5). According to Equation (7), the proximity data are not proportional to distances and therefore should not be treated as ratio data. It does, however, suggest that the proximity data are linearly related to squared distances and therefore could be treated as interval (or ordinal) level data for purposes of any analysis, such as that of ALSCAL, for which metric analyses include those based on the assumption of proximity data linearly related to squared distances. In most MDS analyses, however, the proximities would have to be treated as ordinal because most analytic models assume the data are monotonically but nonlinearly related to distances (rather than squared distances).

Final selection of a model also means deciding on the number of dimensions to retain. As described earlier, the decision can be based on the number of data points, parsimony, dimension interpretability, fit to the data, and dimension replicability across samples.

## 18. Final Solution in Cross-Sectional Studies

The Results section should contain a table, such as Table 20.2, showing the scale values for the final solution, and preferably with estimates of standard errors for those scale values (Kim et al., 2004). Table 20.2 shows a two-dimensional solution from an analysis of squared Euclidean distances for all possible pairs of the Woodcock-Johnson Psychoeducational Battery—Revised (Woodcock & Johnson, 1989) cognitive ability cluster scales in a sample of 357 respondents. In Table 21.2, scale values that are significantly different from zero are indicated by asterisks. Various plots can visually aid understanding of the dimensions.

When the analysis is based on the PAMS model (see Desideratum 16) and dimensions are interpreted in terms of score patterns, plots of dimension scale values against variables (Figure 20.1) can be used to portray the dimension patterns. The top figure, Dimension 1 scale values, shows a pattern marked by relative strengths in Speed of Processing (SPR) and Comprehension Knowledge (CKW), coupled with relative weaknesses in Long Term Retrieval (LTR), Auditory Processing (APR), and Visual Processing (VPR). The second dimension shows a pattern with relative strength in Speed of Processing (SPR) coupled with a relative weakness in Short Term Memory (STM). Note that only variables with scale values significantly different from zero in Table 20.2 were used to identify relative strengths and weaknesses along dimensions. Plots of scale values against variables (e.g., Figure 20.1) may only be useful when dimensions can be interpreted as patterns of relative strength and weakness. In other situations, other graphical forms may prove more informative.

The scale values can be interpreted either in terms of each dimension separately, as in Figure 20.1, or in terms of the overall configuration. For instance, theories positing a *circumplex* structure of

Table 20.2  Woodcock-Johnson Revised Ability Cluster Coordinates and Standard Errors: Standard Errors Estimated from 200 Bootstrap Replicated Samples

| Observed Variables | Dimension 1 | Dimension 2 |
| --- | --- | --- |
| LTR | −1.44* (.35) | .17 (.09) |
| STM | .01 (.14) | −1.43* (.37) |
| SPR | 1.04* (.27) | .51* (.19) |
| APR | −1.10* (.27) | .06 (.09) |
| VPR | −1.01* (.25) | .46* (.13) |
| CKW | 2.46* (.60) | .08 (.09) |
| FRE | .03 (.09) | .15 (.12) |

*Note:* Statistically significant scale value estimates are indicated by *. LTR = Long-term Retrieval; STM = Short-term Memory; SPR = Speed of Processing; APR = Auditory Processing; VPR = Visual Processing; CKW = Comprehension-Knowledge; FRE = Fluid Reasoning.

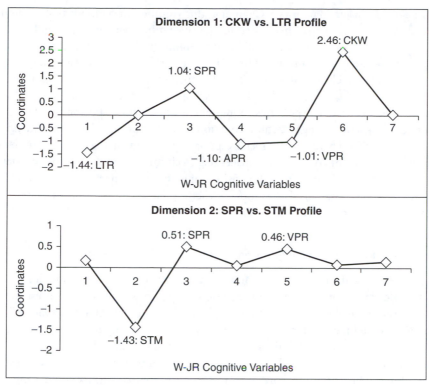

**Figure 20.1** Woodcock-Johnson Revised Latent Dimension Patterns.

variables lead to the prediction that the stimuli will fall in a circular two-dimensional arrangement. In such circumstances, Dimension 1 and Dimension 2 scale values are graphed against each other in a scatter plot and the overall configuration in the resulting plot is visually examined to evaluate whether the variables fall in a circular arrangement and whether they fall along the circle in the order predicted by theory. Whether in terms of separate dimensions or the overall stimulus configuration, the MDS scale values in the final solution should to be interpreted and related to theory.

## 19. Other Related Methods in Cross-Sectional Studies

In thinking about alternative analyses of cross-sectional data, to what analyses should MDS be compared? Because MDS and factor analysis (see Chapter 8, this volume) both yield representations of variables in a continuous space, it is rather natural to compare them. Davison (1983) described a seemingly common, if not universal, relation between unrotated components and MDS solutions in which the first (or general) component has no counterpart among the MDS dimensions but remaining unrotated components do have counterparts among the MDS dimensions. It can be argued, however, that MDS dimensions primarily describe within-person variation. This is consistent with Davison's finding that MDS dimensions contain nothing resembling a general component as the general factor primarily reflects between-persons, not within-person, variation. Because of its focus on within person variation, MDS and typical factor analysis would seem to serve somewhat different purposes.

Q-factor analysis, cluster analysis (see Chapter 4, this volume), or methods that combine the two (modal profile analysis) have often been used to describe within-person variation. For comparison purposes, researchers may want to present alternative solutions derived with clustering or Q-factoring either to corroborate or to complement the MDS solution. Alternatively, the researcher may wish to describe why MDS was chosen over other possible analyses of within-person variation. Kim et al. (2004) contrasted cluster, Q-factor, and MDS analyses of cross-sectional data.

MDS can be used to generate hypotheses about dimensions of within-person variation, hypotheses that are subsequently tested using structural equation modeling or mixed effects modeling in a later sample (Kim et al., 2007). This approach seems especially promising in the study of variables that display patterned covariance or correlation matrices (e.g., a circumplex matrix, a simplex matrix) given that such patterning arises from factors/dimensions of within-person variation. Repeated measurements of a single variable often display a simplex structure. The analysis of such repeated measures is the topic of our next section.

## 20. Sampling of Time Points in Longitudinal Studies

Whereas cross-sectional data consist of $V$ variables measured at a single occasion, longitudinal data consist of a single variable measured at $V$ time points. Equations (5) and (6) express the model on which the longitudinal analysis is based. When applied to longitudinal data, however, $y_{pv}$ is the measurement of person $p$ at time $v$; $c_p$ is an intercept for person $p$; $w_{pk}$ and $e_{pv}$ are interpreted as before; and $x_{vk}$ is a dimension scale value which is the score at time $v$ in the $k$th pattern of change. The scale values along each dimension are interpreted as a vector describing a pattern of growth, change, or decay. Each person's longitudinal vector of scores $\mathbf{y}_p = \{y_{pv}\}$ is represented as a linear combination of $K$ change patterns $\mathbf{x}_k = \{x_{vk}\}$. When the scale values for a given dimension are plotted against time, the plot visually displays one of the $K$ change patterns.

The Methods section should describe and justify the sample of time points selected for study. Typically, at least four time points or measurement occasions should be present per dimension. Because the selection of time points can seriously influence the form of the growth or change patterns, possible effects of time point selection may need to be considered in the Discussion section. Time points need not be equally spaced. For instance, in a study of gains in reading achievement, there could be a two-month interval between times 1 and 2, but a four month interval between adjacent time points thereafter. In any graphical representation of a change pattern plotted against time, the unequal spacing of time points should be displayed along the horizontal axis. Failing to accurately represent the unequal intervals between time points will distort the graphical representation of change rates over intervals.

## 21. Missing Data in Longitudinal Studies

Because longitudinal data are often incomplete, the researcher must describe and justify the method of handling missing observations. Some programs offer the researcher the option of listwise or pairwise deletion in the computation of a proximity measure, such as the squared Euclidean distance. Either pairwise or listwise is readily justified when data are missing completely at random (MCAR). In our opinion, however, data are seldom missing completely at random, and the MCAR assumption is even more difficult to justify for longitudinal data than for cross-sectional data. When the intervals between time points are long, there tends to be more missing data in longitudinal studies, often systematically related to variables inside or outside the study. For instance, in longitudinal studies of school achievement, at time 2 and beyond, data are more likely to be missing for low income students and low scoring students at time 1 because such students tend to change schools more frequently.

Therefore, researchers might want to apply some model-based, multiple-imputation technique before computation of the proximity measure. Model-based imputation may be necessary to account for the systematic nature of the missingness. The method of handling missing data needs to be described and justified.

## 22. Proximity Measure in Longitudinal Studies

Researchers must also describe their choice of proximity measure and explain why that particular proximity measure was chosen over the numerous other possibilities. As with cross-sectional data, the best justification, in our view, is one that starts with a model of the raw data, such as Equation (5), from which one can derive a proximity measure linearly or monotonically related to a distance function of the parameters to be estimated with MDS, the $x_{vk}$ in the case of Equation (7). Other forms of justification are presumably possible, however.

The model in Equation (5) leads to the choice of the squared Euclidean distance proximity measure defined over all possible pairs of time points and computed according to Equation (3). Under plausible assumptions, the squared Euclidean distance proximity measure computed from the raw data will be linearly related to the squared distance function of the parameters to be estimated, the $x_{vk}$, leading to the conclusion that MDS will provide plausible estimates of those parameters. Because the parameters have a natural interpretation in terms of change patterns, and the MDS scale values are estimates of those parameters, the MDS dimensions can be interpreted as estimates of change pattern vectors.

## 23. Model in Longitudinal Studies

Model justification can have several aspects. First, if the proximity measure is derived from a model of the raw data, then the plausibility of the raw data model needs to be considered. Second, model justification involves an explanation of the assumed form for the proximity measure. If, however, the researchers have no model of the raw data or no way of deriving a proximity measure from it, some other form of justification must be provided. In either case, the explanation must include a justification for the assumed measurement level of the proximity data. That is, does the researcher assume the proximity measures to be ordinal (i.e., monotonically related to a distance function of the scale values), interval (i.e., linearly related to a distance function of scale values), or ratio (i.e., proportional to a distance function of scale values)? This assumption will determine whether the MDS analysis will be of the nonmetric or metric form.

Finally, in explaining the model, the researcher must justify the number of dimensions retained in that MDS solution. As explained earlier, the justification may be based on the number of time points, model fit, parsimony, dimension interpretability, and dimension replicability across samples of respondents.

## 24. Interpretation of Dimensions in Longitudinal Studies

Interpretability depends primarily on whether what is known about the change process is consistent with the change patterns represented by scale values. For instance, if change is thought to be monotonically increasing with time, then a dimension along which scale values do not increase with time would be implausible and uninterpretable.

In longitudinal applications, the zero point along a dimension can be set in different ways without loss of fit to the data, and the different ways of setting the zero point lead to different interpretations of the intercept term. For instance, the zero point can be set so that the scale value at time 1 equals 0 for every dimension, in which case, $c_p$ becomes the model-based estimate of initial (time 1) status for person $p$. This alternative seems most useful in studies of growth or studies of decay, that is, studies in which change increases/decreases monotonically over time. If the measured variable oscillates non-monotonically over time, the zero point can be set so that the mean scale value equals zero along each dimension. In such cases, each dimension is interpreted as a pattern of oscillating change about person $p$'s "typical" level of performance represented by the intercept, $c_p$. In longitudinal applications, researchers need to explain and justify how the zero point along each dimension was set and the resulting interpretation of the intercept parameter.

While we have not discussed the correspondence weight parameters $w_{pk}$ in the model, they may sometimes enhance the interpretability and plausibility of a dimension. That is, if they enter into relations with external variables (e.g., individual differences in weights $w_{vk}$ are correlated with individual differences in ability, personality, or interests), then the correspondence weights may help understand the relation between individual differences in growth patterns and individual differences in the external variable(s). This explanatory power of the weights enhances the interpretability of the dimensions. The interpretability of a dimension depends on the explanatory power of both the dimensions' scale values and their corresponding weights.

As in cross-sectional applications, MDS can be used as an exploratory analysis by itself or as a way of generating hypotheses about longitudinal patterns needed to account for change in a particular variable, hypotheses that will be subsequently tested in a second sample using structural equation modeling or mixed effects growth modeling techniques (see Chapters 28 and 14, this volume). Theory may be too imprecise to generate the detailed growth curve hypotheses required by confirmatory methods, and therefore a combination of theory and exploratory analyses may be necessary for hypothesis generation. In comparison to other methods, advantages of MDS include: (1) no need for a priori growth trend specifications (e.g., linear or quadratic), (2) simultaneous estimation of multiple growth curves, (3) estimation of a growth rate for each time interval in each change pattern (dimension), and (4) ready accommodation of unequally spaced time points. Whether used to study perceptions, cross-sectional variables, or repeated measures, interpretation of the dimensions and/or the configuration need not rely solely on subjective judgment. Associations of dimension scale values with external variables and cluster analytically defined groupings of stimuli (or variables) in the solution space are just two examples of procedures that can be used to more objectively confirm or disconfirm interpretations of the solution. Purely subjective interpretations of solutions are to be avoided.

## Note

1  During the preparation of this chapter, the first author was partially supported by a grant from the U.S. Department of Education, grant number R305C050059.

## References

Borg, I., & Groenen, P. (2005). *Modern multidimensional scaling: Theory and applications*. New York: Springer.

Commandeur, J. J. F., & Heiser, W. J. (1993). *Mathematical derivations in the proximity scaling (PROXSCAL) of symmetric data*. (Tech. Rep. No. RR-93-03). Leiden, The Netherlands: Department of Data Theory, Leiden University.

Cox, T. F., & Cox, M. A. A. (2001). *Multidimensional scaling* (2nd ed.). New York: Chapman Hall.

Davison, M. L. (1983, 1991). *Multidimensional scaling*. New York: Wiley. (Reprinted by Krueger, 1991.)

Ding, C. S., Davison, M. L., & Petersen, A. C. (2005). Multidimensional scaling analysis of growth and change. *Journal of Educational Measurement, 42*, 171–191.

Kim, S.-K., Frisby, C. L., Davison, M. L. (2004). Estimating cognitive profiles using profile analysis via multidimensional scaling (PAMS). *Multivariate Behavioral Research, 39*, 595–624.

Kim, S.K., Davison, M. L., & Frisby, C. L. (2007). Confirmatory factor analysis and profile analysis via multidimensional scaling (PAMS). *Multivariate Behavioral Research, 42*, 1–32.

Kruskal, J. B. (1964a). Multidimensional scaling by optimizing goodness of fit to a nonmetric hypothesis. *Psychometrika, 29*, 1–27.

Kruskal, J. B. (1964b). Nonmetric multidimensional scaling: A numerical method. *Psychometrika, 29*, 115–129.

Lingoes, J. C. (1989). *Guttman-Lingoes nonmetric PC series manual.* Ann Arbor, MI: Mathesis Press.

Ramsay, J. O. (1977). Maximum likelihood estimation in multidimensional scaling. *Psychometrika, 42*, 241–266.

Takane, Y., Young, F. W., & de Leeuw, J. (1977). Nonmetric individual differences multidimensional scaling: An alternating least squares method with optimal scaling features. *Psychometrika, 42*, 7–67.

Woodcock, R. W. & Johnson, M. B. (1989). *Woodcock-Johnson Psychoeducational Battery—Revised.* Allen, TX: DLM.

# 21
# Multiple Regression

Ken Kelley and Scott E. Maxwell

Multiple regression (MR) has been described as a general data analytic system (e.g., Cohen, 1968) because many commonly used statistical models can be regarded as its special cases. Furthermore, many advanced models have MR as a special case. The ubiquity of MR makes this model one of the most important and widely used statistical methods in social science research. In general, the idea of the MR model is to relate a set of regressor (independent or predictor) variables to a criterion (dependent or outcome) variable, for purposes of explanation and/or prediction, with an equation linear in its parameters. More formally, the MR model is given as

$$Y_i = \beta_0 + \beta_1 X_{1i} + \ldots + \beta_k X_{ki} + \varepsilon_i \tag{1}$$

where $\beta_0$ is the population intercept, $\beta_k$ is the population regression coefficient for the $k$th regressor ($k = 1, \ldots, K$), $X_{ki}$ is the $k$th regressor for the $i$th individual ($i = 1, \ldots, N$), and $\varepsilon_i$ is the error for the $i$th individual, generally assumed to be normally distributed with mean 0 and population variance $\sigma_\varepsilon^2$. For contemporary treatments of MR applied to a wide variety of examples, we recommend Cohen, Cohen, West, and Aiken (2003), Pedhazur (1997), Harrell (2001), Fox (2008), Rencher and Schaalje (2008), and Muller and Fetterman (2002). Specific desiderata for applied studies that utilize MR are presented in Table 21.1 and explicated subsequently.

## 1. Research Goals

Standard textbook treatments of MR often emphasize that MR can be used for prediction or explanation. Depending on the goals of the researcher, prediction, explanation, or both might be desired. Although the MR model itself is exactly the same in both cases (i.e., Equation (1) does not change based on the goal), the distinction is nevertheless important because different statistical considerations arise for the two purposes. To clearly communicate the purpose of the study, it is important for authors to be clear about whether their purpose in using MR is prediction, explanation, or both.

The ultimate goal of explanation is to identify the causes of the outcome variable $Y$. Under ideal conditions, MR can identify such causes as having non-zero regression coefficients. To understand how a regression coefficient can potentially reflect a causal effect, we need to say what a regression

Table 21.1 Desiderata for Multiple Regression

| Desideratum | Manuscript Section |
|---|---|
| 1. The goals of the research and how multiple regression (MR) can be useful are explicitly addressed. | I |
| 2. The inclusion of each of the independent variables should be justified on theoretical and/or practical grounds. | I |
| 3. Each criterion and regressor variable should be described in detail, including scales of measurement, coding scheme, reliability etc., to convey how the MR model should be interpreted. | M |
| 4. Specific procedures for the computation and interpretation of effect sizes are delineated. | M |
| 5. Assumptions underlying the MR analyses and resulting inference are explicitly addressed. | M |
| 6. Variable selection techniques are justified. | M |
| 7. Sample sizes for all analyses are justified in terms of power, accuracy, and reproducibility of results. | M |
| 8. Methods of dealing with missing data are addressed. | M |
| 9. For models examining moderation, issues of interpretation, role of centering, and visualization are addressed. | R |
| 10. For models examining mediation, issues of interpretation and limitations due to cross sectional designs are addressed. | R |
| 11. Visual examination of data is addressed in order to assess model appropriateness and assumptions. | R |
| 12. Measurement error in predictor and/or outcome variables is addressed. | D |
| 13. Potential limitations of multiple regression in the current applied research context are explicitly stated. | D |
| 14. Alternatives to the MR model are given. | D |

\* *Note*: I = Introduction, M = Methods, R = Results, D = Discussion

coefficient represents. For example, when the model is correctly specified, the coefficient $\beta_k$ for $X_k$ reflects the relation between $Y$ and $X_k$ at a fixed value of all other regressors included in the model. In this sense the regression coefficient for $X_k$ is a measure of the extent to which $X_k$ and $Y$ are related when all other regressors in the model are held constant. Because the other regressors are held constant, any association between $X_k$ and $Y$ cannot be attributed to the other regressors. Thus, it is tempting to conclude that $\beta_k$ reflects the extent to which $X_k$ *causes* $Y$, in which case we have at least partly succeeded in explaining variation in $Y$. In fact, this reasoning is sometimes correct, but only under a set of restrictive conditions (e.g., Kenny, 1979). Unfortunately, it is difficult to justify these conditions unequivocally except in randomized experiments.

Predicting the value of a criterion variable given one or more regressors is another reason why MR is commonly used, especially in applied research. For example, a psychologist might use MR to predict how well pre-kindergarten children will be able to read at the end of first grade. The psychologist would use historical data (often called *training data*) containing scores on reading at the end of first grade as well as scores on a number of possible regressors. MR is then used to create a model where the value of the criterion is predicted based on one or more of the regressors. One of the real benefits of prediction is that the parameter estimates (i.e., the regression coefficients) obtained from the training data can be used to predict the value of an unknown (or yet to occur) criterion variable $Y$ based on the complete set of regressors used in the training data. There are many cases in which it is desirable to predict a criterion variable when it is as yet unknown (e.g., college grade point average or reading ability at the end of first grade) from a set of known regressors (e.g., high school grade point average, SAT scores, or pre-kindergarten measures of cognitive functioning). The ultimate goal is often selection, as in the college example, but can also be identifying at-risk individuals who might benefit from a relevant intervention.

Although we believe that recognizing the difference between explanation and prediction is critically important, there need not be such a rigid dichotomy between the two goals. In studies seeking to explain relations there can be prediction, and in studies that seek a way to predict there can be attempts at explanation. Pedhazur (1997) described predictive research having as its main emphasis "practical applications," whereas in explanatory research the main emphasis is "understanding phenomena" (p. 196). Huberty (2003) provided a discussion of the similarities and differences in research goals and reporting strategies when interest is primarily in prediction or explanation.

Statistical inference is important when a desire exists to generalize information obtained in a sample to the population from which the sample was drawn. Inference can be of two forms, confidence interval formation for the population effect sizes of interest and/or hypothesis testing for effect sizes. For purely predictive purposes, inferential procedures are not strictly necessary, but nevertheless provide information about the population of interest.

## 2. Justification of Regressors

MR can be applied along a continuum of research approaches anchored by *confirmatory* and *exploratory* research. The confirmatory anchor corresponds to a well-defined research question with a few theoretically justified variables, whereas the exploratory anchor corresponds to a diffuse research question with many variables included in one or more different analyses without explicit theoretical justification. Both confirmatory and exploratory analyses are beneficial, but care must be taken so that an exploratory analysis is not presented as if it were a confirmatory analysis. Provided the assumptions of the model are satisfied in the context of confirmatory studies, the probability values (i.e., the $p$-values) from null hypothesis significance testing and confidence interval coverages associated with the different effect sizes are meaningful. However, because exploratory analyses generally consist of systematic testing and retesting until settling on a satisfactory model, the process of test-retest renders the probability values and confidence interval coverages associated with the effect sizes as approximate at best, with such values being the starting point for future confirmatory research.

The reason probability values and confidence interval coverages are not correct in exploratory analyses where multiple models are evaluated is because of what is known as the *multiplicity problem*. The multiplicity problem describes the problem of multiple statistical tests being performed, where the effect sizes with small $p$-values are selected for inclusion in the presented statistical model. An outcrop of the multiplicity problem is that the obtained $p$-values are suspect, due to the sheer number of null hypothesis significance tests conducted. When many null hypothesis significance tests are conducted, even when all the null hypotheses are true, there is a high probability of finding some small $p$-values by chance. Thus, because of the suspect $p$-values and the associated confidence interval coverages associated with statistical inference in exploratory studies, it should be made clear if the study was confirmatory in nature or exploratory. In particular, exploratory approaches sometimes effectively are based on an informal variation of a formal variable selection method (such as stepwise regression, to be discussed in Desideratum 6), which may be fine for prediction but raises serious concerns about the meaningfulness of any claims regarding explanation. That is, some researchers reject the idea of stepwise regression, but themselves perform a more intuitive version of stepwise regression where many models are fitted, even when their purpose is explanation.

## 3. Descriptions of Criterion and Regressor Variables

A statistical model in and of itself is not very useful unless the variables in the model are understood in their appropriate context and have been discussed in enough detail to convey an understanding of the information they contribute to the research question. At a minimum, means and the covariance matrix or the correlation matrix (with accompanying standard deviations) should be provided for all variables used in the analysis. Furthermore, the type of variable (e.g., categorical or continuous) and the range over which values of the scale can vary (i.e., the limits of the scale) should be discussed. When categorical variables (e.g., grouping variables) are used, the coding scheme should be explicitly discussed. Without an explanation of the coding scheme, the estimated model parameters cannot be readily interpreted by others (e.g., for the "Sex" variable are females coded as 0 and males 1, females as 1 and males 0, or females −1 and males 1, etc.). Continuous variables should almost never be dichotomized (or polytomized more generally) but should instead be left in their continuous form. Examples of situations where it may sometimes be reasonable to polytomize continuous variables is when there are clear types or taxons of individuals or when the distribution of a count variable is highly skewed (MacCallum, Zhang, Preacher, & Rucker, 2002). It is clear, however, that median splits, a commonly used procedure for dichotomizing continuous data, is essentially never statistically justified. Where appropriate, the reliabilities of the variables should be given (see Desideratum 12).

## 4. Effect Sizes

As has been discussed a great deal in the methodological literature, effect sizes and their corresponding confidence intervals are widely recommended and should almost always be reported (e.g., Wilkinson & the American Psychological Association Task Force on Statistical Inference, 1999; see also Chapter 7, this volume). In MR, like many other statistical models, there are two types of effect sizes: *omnibus* and *targeted*.

The most widely used omnibus effect size in MR, and one of the most common in social science research in general, is the squared multiple correlation coefficient, whose population value is denoted $P^2$ (rho squared). The value of $P^2$ quantifies the proportion of variance in $Y$ that can be accounted for by the $K$ regressor variables. The typical estimate of $P^2$, $R^2$, is positively biased. Although confidence intervals and significance tests for $P^2$ are based on $R^2$, the adjusted value of $R^2$, denoted $R_A^2$, should also be reported and used as the best estimate of $P^2$. The typical adjusted estimate (e.g., Cohen et al., 2003; Harrell, 2001) is given as

$$R_A^2 = \max\left\{0, \left[1 - (1 - R^2)\left(\frac{N-1}{N-K-1}\right)\right]\right\},\tag{2}$$

where $\max\{\cdot,\cdot\}$ implies that the larger of the two values is taken.

Darlington (1968) explained that the adjustment shown in Equation (2) (developed by Ezekiel, 1930) will tend to overestimate the population validity of the sample regression equation. The idea here is that the adjustment estimates the population validity of the population regression equation. In other words, if the population regression coefficients were known, what proportion of the variance in $Y$ would this equation explain in the population? This makes sense when the goal is explanation, because one purpose here is to estimate the extent to which the regressors completely explain the variance in $Y$. However, this makes less sense when the goal is prediction, because in this context the sample regression equation derived in the training sample will be used to make predictions in a new sample. The key point is that the regression coefficients to be used for prediction are the values obtained in the training sample. However, these values will not be exactly the same as the optimal population values, thus lowering the resultant $R^2$ to some extent. For this reason, in prediction the population parameter of most interest is sometimes referred to as the population cross-validity, $P_C$, or the squared population cross-validity, $P_C^2$. Raju, Bilgic, Edwards, and Fleer (1999) described a variety of estimators of the population cross-validity and recommended an adjustment developed by Burket (1964):

$$R_C = \frac{NR^2 - K}{R(N-K)}.$$ (3)

Although effect size estimates are beneficial, an observed effect size is simply a point estimate that might differ considerably from the population value it estimates. Confidence intervals should be reported for any estimate that is itself deemed important enough to report. Confidence intervals for $P^2$ are not straightforward to construct and the appropriate confidence interval depends on whether or not regressors are regarded as fixed or random. Steiger (2004; see also Steiger & Fouladi, 1992), Algina and Olejnik (2000), and Kelley (2007), discussed methods of confidence interval construction and provided software solutions to implement such intervals.

Researchers should consider the squared semi-partial correlation coefficient, which is a targeted effect that describes the change in $R^2$ when the $k$th regressor is added to the MR model that already contains the other $K-1$ regressors. Thus, the squared semi-partial correlation coefficient quantifies the proportion of variance of $Y$ that is accounted for uniquely by a particular regressor in a model with other regressors. Such an effect size is useful when conveying the contribution of a regressor in a model with $K-1$ other regressors. Squared semi-partial correlation coefficients can also be used to quantify the proportion of variance of $Y$ that is accounted for by a particular set of regressors instead of just a single regressor.

Regression coefficients come in two forms: *unstandardized* and *standardized*, both of which represent targeted effects, which may or may not be causal in nature. Unstandardized regression coefficients can be transformed into standardized regression coefficients by multiplying the unstandardized regression coefficient by the quantity $\frac{s_{X_k}}{s_Y}$, which removes the scale of $X_k$ and $Y$, where $s$ denotes the standard deviation of the subscripted quantity. The process can be reversed by multiplying a standardized regressioncoefficient by $\frac{s_Y}{s_{X_k}}$. In general, either unstandardized or both unstandardized and standardized regression coefficients should be given, along with their corresponding confidence intervals. The $k$th regression coefficient quantifies the degree of linear relation between $Y$ and $X_k$, while holding constant the remaining $K-1$ regressors. Standardized regression coefficients are often an effective way of describing the effect of a regressor on the criterion variable when the scale of the measurements are not inherently meaningful. When standardized solutions are used in place of or in addition to their unstandardized counterparts, the measure of association is in terms of standard deviation units of the particular sample. For example, a standardized regression coefficient of .25 for $X_k$ in a standardized solution implies that a 1 standard deviation unit difference in $X_k$ is associated with a .25 standard deviation difference in $Y$ in the same direction, holding constant all other regressors.

Confidence intervals for unstandardized regression coefficients are easy to obtain and formulas are available in essentially all modern regression books and can also be obtained with popular statistical software. However, confidence intervals for standardized regression coefficients require the use of noncentral $t$-distributions and are more difficult to obtain (e.g., see Kelley & Maxwell, 2008, or Kelley, 2007, for a review and software solutions). In general, standardized regression coefficients are provided when there is a desire to remove the scaling of the measurement instrument so that each variable (regressors and criterion) has a mean of 0 and a standard deviation of 1. Doing so allows for relations to be framed in standard deviation units (as previously noted) and regression coefficients to be more directly comparable within an equation. That being said, there is no guarantee that the regressor with the largest regression coefficient is the "most important" independent variable in the equation (even when all variables are standardized). The meaning of "most important" might be different depending on the particular situation and goals of the study (Azen & Budescu, 2003).

## 5. Addressing Assumptions

Standard approaches to regression rely on ordinary least squares (OLS) to estimate model parameters. The OLS regression coefficients in MR minimize the sum of squared deviations between the model implied scores, denoted $\hat{Y}_i$ for the $i$th individual, and the observed scores (i.e., regression coefficients are chosen that minimize $\sum_{i=1}^{N}(Y_i - \hat{Y}_i)^2$). Estimation of the regression coefficients themselves does not require any parametric assumptions (although optimality properties of OLS depend on the validity of underlying assumptions). However, inferences from coefficient estimates do depend on assumptions. In particular, $p$-values and confidence intervals for regression coefficients from the regression model as specified in Equation (1) depend on four statistical assumptions: (a) errors (i.e., $e_i = Y_i - \hat{Y}_i$) follow a normal distribution; (b) error variance is homogeneous across all values of the regressors (*homoscedasticity*); (c) observations are independent of one another; and (d) the relation between $Y$ and the $K$ regressors is linear. It is important to note that no distributional assumptions are made about the regressors, meaning that, for example, skewness in a predictor is not by itself a problem. Also, the model does not assume that regressors are measured without error, but as we will discuss later, results obtained using regressors measured with error may differ substantially from results obtained when regressors are measured perfectly, so measurement error in the regressors often becomes an important consideration.

Although the linearity assumption (assumption d above) is fundamental, it is often overlooked in discussions and applications of MR. We agree with Gelman and Hill (2007) that, "The most important mathematical assumption of the regression model is that its deterministic component is a linear function of the separate predictors" (p. 46). This assumption is especially important in explanation because if this assumption is not true, then the regression coefficients in the model do not generally accurately reflect the relation between $Y$ and $X_k$ at a fixed value of the other regressors. As a result, the regression model might fail to hold the other regressors constant in attempting to estimate the relation between $Y$ and a specific $X$ variable. If linearity does not hold, then the model as specified in Equation (1) may not be appropriate for inferences, as Equation (1) is necessarily linear in form. When linearity does not hold, there are essentially three strategies: (a) transform one or more variables (one or more $X_k$ and/or $Y$) so that linearity in an additive model is a good approximation (e.g., $\sqrt{X_k}$ or $X_k^2$); (b) include an additional theoretically justified variable (e.g., $X_k^2$ in addition to $X_k$) that correlates with the outcome variable, in an attempt to explain some of the unaccounted for variability and/or (c) fit a nonlinear regression model (e.g., a negative exponential, Gompertz, logistic) instead of the traditional linear MR model (Seber & Wild, 1989).

## 6. Variable Selection Techniques Are Justified

In many situations, more regressor variables are initially included in the model than are ultimately desirable in the final model to be presented for interpretation. The way in which the researcher arrives at the final model should be made explicit. There are four common ways of selecting variables to be included in the analysis: (a) all analyses are theory driven, (b) model comparisons are performed, (c) stepwise methods are used, or (d) a variety of exploratory models and methods are fitted.

In many ways the ideal variable selection method is entirely theory driven and the regressors included are based on a priori theoretical arguments and/or previous literature. This method is ideal because a one-to-one mapping exists between the targeted nature of the research question and the targeted statistical analyses.

A model comparison approach (e.g., Maxwell & Delaney, 2004), where the inclusion of one or more variables is evaluated against a more basic model, is often the most straightforward way to evaluate competing nested models. The idea of the model comparison approach is to statistically compare nested models, where the models are compared most commonly in terms of $R_K^2$ and $R_{K+M}^2$, where $R_K^2$ is the model based on the $K$ regressors and $R_{K+M}^2$ is based on a richer model with an additional $M$ regressors.

A special type of model comparison is implemented through what is often termed *hierarchical regression* (not to be confused with hierarchical linear modeling, HLM; see Chapter 10, this volume). In hierarchical regression, not only are the variables selected by the researcher, so too is the order in which they enter the model. At each step of the procedure, the variables previously included remain in the analysis. When hierarchical regressions are performed, a series of fitted models should be provided as part of the reported results that shows the estimated model improvement when comparing the richer models to the simpler models. The improvement is generally gauged in terms of the change in $R^2$ when a single regressor variable is added, which again is the squared semipartial correlation coefficient, along with confidence intervals and significance tests for the change. It is also common to add a block of regressors in a hierarchical fashion. In such situations the change in $R^2$ is still of interest, but there the additional variability accounted for is due to the block of regressors.

When a large number of possible regressors exist, possibly for more than one criterion variable, data-driven selection methods are sometimes used. Whenever data-driven selection methods are used, a clear indication should be made that the study is not attempting to explain phenomena in a confirmatory fashion, but rather that the study is exploratory in nature. The type of data-driven selection procedure performed (e.g., forward selection, backward elimination, all possible subsets), and the selection criteria (e.g., a statistically significant change in $R^2$, or a change in $R^2$ of some specified magnitude, say .05) should be given. Also the particular computer program/package and its version should be provided, because different programs/packages and versions may implement data-driven selection procedures in different ways.

In short, there are many methodological problems that can arise when implementing a data-driven selection procedure. As Rencher and Pun (1980) illustrated, values of $R^2$ can be highly inflated and thus the obtained probability values can differ substantially from those reported as output in statistical software. When a large number of possible regressors exist in the context of a data-driven selection procedure, a model that accounts for a statistically significant proportion of variance in $Y$ can often be obtained even if the null hypothesis is true that all of the regression coefficients, less the intercept, are zero.

Vittinghoff, Glidden, Shiboski, and McCulloch (2005) provided an especially interesting perspective on model building by distinguishing three different purposes for selecting predictors: (1) evaluating a regressor of primary interest in the context of other possibly relevant regressors, (2) identifying the important regressors of an outcome, and (3) prediction. They emphasized that issues involved in predictor selection differ according to the purpose of the analysis. For example, suppose that two regressors $X_1$ and $X_2$ are highly correlated with one another. When the goal is prediction, it will generally be desirable to include only one of these two regressors in the model, and it may make little difference in the accuracy of prediction which of the two is included. Ironically, however, including both of the regressors will often worsen prediction because any gain in bias reduction is more than offset by an increase in the variance of predicted values. On the other hand, suppose the goal is to explain the relation between $X_1$ and $Y$. Should $X_2$ be controlled for and thus included in the model? We agree with Vittinghoff et al. (2005) that this question cannot be answered simply from knowing that $X_1$ and $X_2$ are highly correlated. Instead, for explanatory models it becomes necessary to consider a theoretical causal model for how the various regressors and $Y$ relate to one another. In particular, $X_2$ should be included in the model if it is a confounder, but not all variables highly correlated with the regressor of primary interest (i.e., $X_1$) are necessarily confounders. Vittinghoff et al. (2005), Jaccard, Guilamo-Ramos, Johansson, and Bouris (2006), and Hernan, Hernandez-Diaz, Werler, and Mitchell (2002) discussed various approaches for identifying whether a variable is a confounder and thus should be included in the regression model.

## 7. Sample Sizes Are Justified

Sample size is an important component to any research study. "Rules of thumb" that were once widely recommended for planning sample size are not generally appropriate and should not be used as justification (see Green, 1991, for a review). Instead, researchers should justify their sample size from either or both of two general approaches: power analysis and accuracy in parameter estimation (AIPE). The goal of the power analytic approach is to plan sample size so that a false null hypothesis can be rejected with some desired probability (i.e., power), whereas the goal of the AIPE approach is to obtain an accurate estimate of the population value, which is operationalized by a sufficiently narrow confidence interval with some desired degree of assurance (i.e., probability). In addition to deciding on whether power or AIPE is most appropriate, researchers also need to state whether primary interest is in an omnibus effect (i.e., the squared multiple correlation coefficient) or one or more targeted effects (i.e., regression coefficients), which is necessarily based on the question(s) of interest. In particular, questions of prediction are more likely to involve omnibus effects, whereas questions of explanation are more likely to involve targeted effects. Additional details are provided in Kelley and Maxwell (2008), who discussed sample size planning methods in a MR context in a $2 \times 2$ (power or AIPE × omnibus or targeted effect) framework.

In some cases existing/archival data become available to a researcher. Because the data have already been collected, sample size planning cannot be done as previously discussed, as it is implemented a priori in the design phase of the study. In general, power and AIPE are not often discussed for existing/archival data. However, power and AIPE can still be addressed, albeit in a different manner. In particular, for a specified value of an effect size at the size of the sample in the existing data, power and expected confidence interval width can be given. An appropriate value for the effect size to use is what can be termed the parameter of minimal importance (POMI), which is the parameter of smallest magnitude that is deemed to have scientific, clinical managerial, or practical importance/interest.

## 8. Missing Data

Missing data is a perplexing issue. There are three broad categories of missingness: (a) missing completely at random (MCAR), (b) missing at random (MAR), and (c) missing not at random (MNAR). MCAR is where missingness does not depend on either observed or missing values, whereas MAR is where missingness does not depend on the missing values but may depend on observed values. MNAR implies that missingness depends on an outside variable not in the model or depends on the variable itself (see Little & Rubin, 2002, for a summary of types of missing data and appropriate methods for dealing with the different types of missing data).

Although the specifics of the situation will differ, researchers should do their best to ensure the amount of missing data is minimized (e.g., remind participants about follow-up visits, check evaluations for blank responses before the participants leave, clearly state that sensitive data will remain confidential if appropriate). Generally, whenever missing data arises in a research study, it opens the possibility for criticism in the way it was (or was not) dealt with. Whenever there is a nontrivial amount of missingness, the data should be interrogated for patterns of missingness (Harrell, 2001). When apparent patterns are found, they should be reported and, if possible, a plausible explanation provided with a cautionary reminder given that exploratory methods were used to uncover any apparent patterns in the data. Regardless of the way in which missing data is dealt with, the method and the rationale for choosing the method should be discussed. That being said, some methods, in particular mean substitution and/or pairwise deletion, should not be used unless there is a good reason to do so with a clear explanation of why. We will briefly discuss three methods of dealing with missing data (see Schafer & Graham, 2002, for a thorough review).

When missing data does occur, casewise deletion is often recommended; however, casewise deletion can be problematic. Casewise deletion is when a participant is completely excluded, regardless of the amount of data available for the participant, if any data are missing for the analysis of a particular model. Casewise deletion generally yields unbiased estimates only under the very strong assumption that data are MCAR. At best, estimates obtained using casewise deletion are inefficient, implying less statistical power and estimation accuracy than would otherwise be the case. The reason casewise deletion is inefficient is because the sample size is reduced to only those with complete data sets, which tends to increase the sample standard error(s) and necessarily does so in the population. More important, however, is that estimates obtained using casewise deletion will often be biased, unless plausible arguments can be advanced for why missingness is likely to be MCAR.

Imputation or multiple imputation provides a reasonable way to deal with missing data in many situations. Imputation is when a plausible value is substituted for a missing value and multiple imputation is when this process is performed multiple times. The "plausible values" come from an imputation model that uses other data that are available to estimate the data that are not available. At first the idea of estimating data might seem problematic, but it is often better to estimate what is usually a small amount of data than to disregard valuable data with deletion (e.g., casewise) strategies (Harrell, 2001, section 3.4).

Full information maximum likelihood (FIML) and restricted maximum likelihood (REML) estimation are the most popular methods for dealing with missing data in multilevel models and structural equation models, likely because main-stream multilevel model and structural equation modeling programs can easily implement them (and usually do so by default). These maximum likelihood methods for dealing with missing data require that data are MCAR or MAR. Because FIML does not consider the degrees of freedom and uses the standard normal distribution instead of the $t$-distribution, sample size should not be small with this approach. Small sample sizes being used with the FIML approach to missing data will tend to yield differences in the empirical and nominal Type I error rates. REML, however, does consider the issue of degrees of freedom and is more appropriate in smaller samples. Another issue is that maximum likelihood estimation assumes multivariate normality, which might not always be reasonable (recall that the standard MR assumption is only that the errors are normally distributed). Enders (2001) provided a review and evaluation of maximum likelihood estimation when missing data exists in the context of MR. Our recommendation is to use either multiple imputation or maximum likelihood estimation when faced with missing data.

## 9. Models Examining Moderation

The regression model shown in Equation (1) assumes that the effects of each $X_k$ on $Y$ are additive. For example, with two regressors, this model assumes that the relation between $X_1$ and $Y$ is the same for every value of $X_2$ and similarly the relation between $X_2$ and $Y$ is the same for every value of $X_1$. In reality, however, the strength of the relation (or even the direction of the relation) between $X_1$ and $Y$ might depend on $X_2$, in which case $X_1$ and $X_2$ are said to *interact*. As a consequence, the regression model shown in Equation (1) might seem very restrictive, because it does not seem to allow for the possibility of an interaction between $X_1$ and $X_2$. Fortunately, this restriction is illusory, because modifications to the model allow $X_1$ and $X_2$ to interact. The ability to modify this model is critical because many theories in the social and behavioral sciences stipulate that the relation between a pair of values (e.g., $Y$ and $X_1$) depends on a third variable (e.g., $X_2$), which corresponds to an interaction effect.

The standard way of modifying the model in Equation (1) so as to allow for the possibility of an interaction (or equivalently, a moderator) is to add cross-product terms. For example, with two regressors, the model becomes

$$Y_i = \beta_0 + \beta_1 X_{1i} + \beta_2 X_{2i} + \beta_3 X_{1i} X_{2i} + \varepsilon_i \tag{4}$$

The inclusion the product term allows the relation between either $X$ and $Y$ to depend on the value of the other $X$. In particular, this model stipulates that the slope relating $X_1$ to $Y$ is given by

$$\frac{dY}{dX_1} = \beta_1 + \beta_3 X_2, \tag{5}$$

where $dY/dX_1$ is the derivative (instantaneous slope) of $Y$ with respect to $X_1$. If $\beta_3$ is non-zero, the relation between $X_1$ and $Y$ depends on $X_2$, so $X_2$ moderates the effect of $X_1$ on $Y$, or equivalently, $X_1$ and $X_2$ interact. However, many researchers might not realize that the product term represents a very specific type of interaction, namely a *bilinear* effect. In particular, Equation (5) shows that if $\beta_3$ is positive, the slope becomes increasingly higher for larger values of $X_2$. Similarly, if $\beta_3$ is negative, the slope becomes increasingly lower for larger values of $X_2$. Thus, researchers should consider whether this is the type of interaction they truly desire to detect. If not, more complicated models can be constructed, such as including quadratic terms for some or all regressors. Interested readers can consult Cohen et al. (2003) for additional details.

The best way to begin to interpret effects in moderator models is generally to plot the interaction. For example, suppose the primary interest involves the extent to which $X_2$ moderates the relation between $X_1$ and $Y$. Cohen et al. (2003) recommended plotting regression lines relating $Y$ and $X_1$ at three values of $X_2$ (typically at the mean of $X_2$ and also at scores one standard deviation below the mean and one standard deviation above the mean). We recommend that such a plot be included in a published paper involving moderator effects. Alternatively, what can be helpful is a three-dimensional representation of the relations, where $Y$ is plotted as a function of all possible scores on $X_1$ and $X_2$ within an appropriate range.

A point of some confusion historically has been how to interpret the $\beta_1$ and $\beta_2$ coefficients in the model in Equation (4). Some researchers have interpreted these coefficients as if they corresponded to main effects, but this is not generally true. Instead, they are conditional (i.e., simple) effects. For example, Equation (5) shows that $\beta_1$ is the slope of $Y$ on $X_1$ when $X_2$ equals 0. Unless the range of values of $X_2$ happens to include 0, the conditional effect in the interaction model will be meaningless. For this reason, it is often recommended that $X_1$ and $X_2$ be recoded so that a value of 0 takes on some meaning. Most commonly, both variables are centered by subtracting the sample mean from all scores (*mean-centering*), yielding a new coding with a mean of 0. In any event, it is critical that authors explain how regressors in interaction models have been coded, in order to facilitate interpretation of the corresponding regression coefficients.

Because of perceived complications of interpreting interactions between continuous regressors, some researchers decide to simplify analyses by categorizing either or both regressors. We strongly recommend that researchers avoid the temptation to categorize continuous variables. One reason to leave variables as continuous is that categorization can decrease power. Interestingly, Maxwell and Delaney (1993) have also shown that in some situations categorization can have the opposite effect of producing spurious effects, thus inflating the Type I error rate. Thus, statistically significant interaction effects based on artificially categorized variables cannot necessarily be trusted, strengthening the argument for leaving continuous variables as continuous.

Researchers should also be aware that several other factors affect the ability to detect interactions in regression models. First, when $X_1$ and $X_2$ are measured with error, the product term $X_1 X_2$ will generally be much less reliable than either $X_1$ or $X_2$, thus lowering power to detect an interaction. Researchers who use regression to investigate interactions need to consider carefully the reliability of regressors. Second, McClelland and Judd (1993) showed that the distribution of regressors in observational studies will often reduce power, especially when regressors correlate substantially with one another. Third, Lubinski and Humphreys (1990) showed that when regressors correlate substantially with one another, the Type I error for testing an interaction can be badly inflated if curvilinear effects exist but are not included in the regression model. Including higher order effects such as $X_1^2$ and $X_2^2$ can guard against spurious interaction effects, but also runs the risk of greatly lowering power to detect true interaction effects. There is no consensus among methodologists at this point about how best to resolve this dilemma. At the very least authors who want to investigate interactions in regression models should be clear about the extent to which their regressors correlate with one another as well as the extent to which theoretical considerations either do or do not rule out possible curvilinear effects. Given the scope of the topic of interactions, we recommend that readers consult such sources as Aiken and West (1991) and Jaccard and Turrisi (2003) for further information.

## 10. Models Examining Mediation

Baron and Kenny (1986) clarified the distinction between moderation and mediation. Both involve a role that $X_2$ (for example) may play in the relation between $X_1$ and $Y$, leading some researchers to confuse moderation and mediation. Thus, it is incumbent on authors of papers reporting either moderation or mediation to provide a clear theoretical rationale for their study.

The variable $X_2$ mediates the relation between $X_1$ and $Y$ when $X_1$ causes $X_2$ and $X_2$ in turn causes $Y$. Thus, mediation can be represented by a pair of regression models:

$$X_{2i} = \beta_0^* + \beta_1^* X_{1i} + \varepsilon_i^* \tag{6}$$

$$Y_i = \beta_0 + \beta_2 X_{1i} + \beta_3 X_{2i} + \varepsilon_i, \tag{7}$$

where the asterisk represents values from the model where $X_2$ is the dependent variable with $X_1$ as its regressor. From this perspective, $X_2$ is a mediator when both $\beta_1^*$ and $\beta_3$ are non-zero. In the special case where $\beta_2$ equals 0, $X_2$ is said to completely (or fully) mediate the relation between $X_1$ and $Y$; otherwise, $X_2$ partially mediates the relation.

Baron and Kenny (1986) suggested a four-step procedure for establishing mediation. Subsequent research has studied their approach as well as a variety of alternatives. This is an area of continuing methodological research, and at this point either of two different approaches seems advisable for establishing mediation. One approach involves bootstrap methods (Shrout & Bolger, 2002). The other involves the distribution of the product variable $\beta_1^* \beta_3$ (MacKinnon, Lockwood, & Williams, 2004). We recommend that authors use either of these two methods to test mediation. Authors should also report coefficients and corresponding confidence intervals for relevant parameters as shown in Equations (6) and (7).

Several other factors should be considered in a mediation analysis. First, it is well known that error of measurement in the mediator causes biased estimates of regression coefficients. In three-variable models such as those in Equations (6) and (7), random measurement error will tend to result in an underestimate of the mediated effect and an overestimate of the direct effect of $X_1$ on $Y$. Researchers should address this likely bias in any interpretation of their results unless the mediator is measured without error. Alternatively, a latent variable model might be used in order to address measurement error and its biasing effects. Second, Maxwell and Cole (2007) have shown that cross-sectional estimates of mediation can be seriously biased when mediation occurs over time. Researchers who rely on cross-sectional analyses need to interpret their results with appropriate caution, and should be encouraged to consider longitudinal designs instead of cross-sectional designs. Third, researchers should carefully consider necessary sample size to obtain adequate power. Fritz and MacKinnon (2007) provided useful guidelines. Fourth, further information about mediation, especially for more complicated models with more than three variables, may want to consult MacKinnon, Fairchild, and Fritz (2007) and MacKinnon (2008).

## 11. Checking Assumptions Visually

The assumptions of the MR model should be considered and evaluated whenever the model is used. As Anscombe (1973) noted, graphs can help researchers appreciate broad features of data *and* look beyond broad features to literally see potentially unexpected relationships, outliers, and violations of assumptions, et cetera. Anscombe went on to show four very different figures, three of which have gross violations of MR assumptions, yet where the results from the regression model were the same (i.e., estimates, *p*-values, confidence intervals, etc.). Recall that the linearity assumption is that the expected value of *Y* given the *K* regressors is a linear function of the *K* variables. We recommend a *conditioning plot* (also referred to as a *coplot*) for examining the critical assumption of linearity. Another useful set of plots for this purpose are *residual versus predictor* (RVP) and *component plus residual* (CPR) plots. One way to evaluate violations of this assumption for "obvious" violations is by plotting the residuals as a function of the model implied values. An obvious nonlinear relationship is evidence that the linearity assumption does not likely hold. When such is the case, there might be an important variable not included in the model, an interaction term might be appropriate, or the relation between the *K* regressors and the criterion might be nonlinear in nature. As previously noted, the latter, in our opinion, is not considered frequently enough, and correspondingly nonlinear models are not applied in many areas as often as it seems they should be, based on theory and empirical evidence. For example, sigmoidal forms or asymptotic values cannot adequately be modeled with linear models. We suggest readers consult Seber and Wild (1989) for a discussion of nonlinear regression models.

Recall that the errors in a MR model are assumed to be normally distributed for the validity of the significance test and confidence intervals. A normal quantile—empirical quantile plot (generally termed a *qq-plot*) is a two-dimensional plot where theoretical quantiles from the normal distribution are compared to the empirical quantiles of the observed errors. The qq-plot allows a visual evaluation of the assumption of normality of the errors. Gross violations of the normality assumption of the errors can often easily be seen with the use of a qq-plot. Although there are formal statistical tests to evaluate normality, visual displays are often extremely effective at identifying potential problems and are often easier to implement and interpret.

*Matrix scatterplots* (sometimes called *pairs plots*) are helpful to examine the bivariate relations among the $K + 1$ variables. These plots can also reveal observations that might be miscoded or identify potential outliers. Further, those cases that might not be considered outliers on either of two variables individually might be an outlier in a bivariate sense (which could heavily influence estimation and inference). For example, if there is a strong positive relation between $X_1$ and $Y$, yet one observation has a very low $X_1$ value and a very high $Y$ value, that point would disproportionally affect the estimate of the line of best fit (e.g., Cohen et al., 2003, for a review). Such a case would not be readily identified without visualization (or more formal outlier/influential data point checks), which could allow the possibility of further investigating such a unique case. Cases in such situations are said to be *leveraging points*. In general, formally operationalizing what constitutes an outlier and appropriately dealing with them can be difficult, but it is nevertheless important. Cohen et al. (2003, chapter 10) provided a detailed discussion of possible causes and possible remediations when outliers are believed to exist. Regardless of the exact way in which outliers are dealt with, transparency to the reader is key. Transparency is especially important because two researchers analyzing the same data might come to different conclusions when fitting the same model based only on how outliers are addressed.

In published work, space is often at a premium, which has the effect that figures that evaluate the model assumptions (e.g., RVP, CPR, qq-plots) are often unable to be printed. Nevertheless, even if such figures are not part of the published version of a work, there is little question that they can be very beneficial for authors, as well as satisfying reviewer curiosity on model fit and appropriateness, and can help to convey relationships to the reader more easily in visual form rather than in verbal form. We think it is generally wise for authors to include a brief discussion of the (published or unpublished) figures and the reasons why the model may seem either appropriate or inappropriate. Of course, if the figures help to identify weakness in the appropriateness of the model, other models should be considered and such a finding noted in the work. In short, visualization techniques should help justify the model chosen and this information should be conveyed to readers.

We are sensitive to the amount of journal space that such plots can consume. Due to limited journal space, editors may be reluctant to allow several pages of figures, even if they are informative. We believe a reasonable solution is for authors to produce supplemental material that can be referenced in the article but stored on a journal's supplemental materials web page, which many journals now make available.

## 12. Measurement Error

Measurement error can be conceptualized in a $2 \times 2 \times 2$ array, where depending on the specific conditions the effect of measurement error has different implications. The dimensions of the array are (a) type of measurement error (random or nonrandom), (b) type of variable (regressor or criterion), and (c) type of coefficient (unstandardized or standardized). We will briefly describe each dimension of the array below.

Random measurement error, which is omnipresent in research, is uncontrolled error that tends to have a mean of zero. Nonrandom measurement error will tend to have a mean that is not zero and/or to be correlated with errors. In short, nonrandom measurement error in the criterion and/or the regressor is problematic and can lead to biased estimates of model parameters. Because nonrandom measurement errors often represent a flaw in the measurement procedure, instrument, or design, we will simply say that MR is not generally appropriate in circumstances of nonrandom measurement error, with the exception being when the nonrandom error is so small that it has essentially no effect on the mean and covariance structures of the variables.

We will assume the random measurement errors have a mean of zero and are uncorrelated with measured variables, with their corresponding true scores, and with all other errors. Provided the regressors are unstandardized, any measurement error in $Y$ is absorbed into the model error term, $\varepsilon$ from Equation (1), and has no effect on the expected value of the regression coefficients. Thus, under the standard MR assumptions, the regression coefficients remain unbiased. However, because the model error variance increases, the estimate of the squared multiple correlation coefficient is systematically lowered. Because $R^2$ decreases—it is attenuated due to a larger error variance—the standard errors of the regression coefficients will also be larger, implying that statistical power and the accuracy of parameter estimates are reduced via a decrease in precision. However, in the situation where the regression model is standardized, the regression coefficients will be attenuated when the criterion is measured with error (Kenny, 1979). The attenuation occurs when the criterion is measured with error because for standardized regression coefficients the multiplier (i.e., $s_{xk}/s_Y$ for the $k$th regressor) of the unstandardized regression coefficient that yields the standardized regression coefficient has a denominator whose expected value is larger than the true value. The expected value of $s_Y$ is larger than $\sigma_{Y_T}$, the population standard deviation of the true scores of $Y$. From a classical test theory context for random measurement errors, the variance of $Y$ is the sum of the true score variance ($\sigma_{Y_T}^2$) and the error variance ($\sigma_{Y_E}^2$). Thus, $s_Y$ will tend to be larger than $\sigma_{Y_T}$, which leads to observed standardized regression coefficients smaller than their corresponding true values (Kenny, 1979, chapter 5).

In observational research the case of random measurement error in one or more regressors will generally lead to biased regression coefficients, regardless of whether or not the regressors are standardized. As Fox (2008) showed, in simple regression (i.e., when $K = 1$) when measurement error occurs in the (only) regressor, its regression coefficient is generally attenuated. However, with one exception, no general statement can be given for the effect of measurement error in one regressor on the regression coefficient for the other regressors in a MR model (i.e., when $K > 1$). As Kenny (1979) pointed out, measurement error in one regressor can attenuate regression coefficients, make the estimate of a regression coefficient that is zero be nonzero, and can change the sign of a regression coefficient (p. 104). The exception noted is for designed experiments, where the randomly assigned variable is uncorrelated with other regressors in the model. When the randomly assigned variable has measurement error, the regression coefficient is less accurate—unbiased, but less precise. Because the regression coefficient is less precise, the corresponding confidence interval tends to be wider and the test of the null hypothesis will not be as powerful.

In general, the difficulty in saying what happens when measurement error occurs in an observational application of MR lies in the multivariate nature of MR, as the properties of one regressor influence the regression coefficients of all other regressors. In short, when a regressor is measured with error in an observational application, its effects are not partialled out as fully as when it is measured without error. This concept is easiest to understand when one regressor is perfectly unreliable, and thus the effects of the true regressor have not been partialled in any way (Kenny, 1979). As a result, the coefficients for other regressors in the model are generally biased because the perfectly unreliable regressor has not been controlled for at all. The important point is that whenever a regressor is measured with error, not only is the coefficient associated with that regressor biased, but typically so are all of the other coefficients in the model, including even coefficients for any regressors that happen to be measured without error. Because the value of the regression coefficient for the variable that is measured with error is biased, being smaller in magnitude than it otherwise would have been if the variable were perfectly reliable, the bias will generally lead to an error variance larger than it would have been, which then leads to a negatively biased estimate of $P^2$ (i.e., $R^2$ is, on average, smaller than it should be), ultimately leading to larger standard errors for all of the regression coefficients in the model.

It is generally desirable to minimize measurement error in all uses of multiple regression. However, measurement error is especially problematic when the primary goal is explanation, because theoretical explanations virtually always relate to constructs, not to variables measured with error. When confronted with nontrivial measurement error, it is often advisable to obtain multiple measures of each construct and use structural equation modeling (see Chapter 28, this volume) instead of multiple regression. Measurement error can be less problematic when the goal is prediction, because the practical goal is often to determine how well regressors as measured can predict the criterion as measured. When the goal is explanation and nontrivial measurement error is likely to occur, we generally recommend obtaining multiple measures of each construct so that structural equation modeling can be used.

## 13. Statement of Limitations

MR is a flexible system for linking $K$ regressor variables to a criterion variable of interest. In many cases, MR is an appropriate statistical model for addressing common research questions, whether they be for purposes of explanation, prediction, or both. Nevertheless, MR has limitations that are defined in part by the model and its assumptions as well as by the research design. The limitations of MR in the specific context should be discussed.

MR has limitations, like other statistical models, when attempting to infer causality from a research design that was not experimental in nature (i.e., when random assignment of levels of the regressors to the participants was not part of the design). Although including additional regressors that are thought to be correlated with the regressor of interest adds a form of statistical control, with regard to causality there is no way to "control" all possible confounders unless randomization is an explicit part of the design. In purely observational designs, claims of causality should generally be avoided. The benefits of randomization cannot be overemphasized, even if for only some of the variables in the design, because randomization implies that the participants have equal population properties (e.g., mean and covariance structures) on all outside variables.

Variables termed "control" variables are often included in MR, as previously noted. However, including a control variable in the model in no way implies that the variable can literally be "controlled"—use of such a term is based on a precise statistical meaning and is not literal in the sense of everyday language. When something is "controlled for" it allows for the linear effect of each regressor on the criterion variable to be evaluated (i.e., a regression coefficient estimated), while holding constant the value of the other regressor variables. In practice, however, many variables cannot be controlled by the researcher, even in the most carefully designed studies. Thus, there is not literally any control by the researcher in an observational design over the variables said to be "controlled for." Rather, an effect can be examined while holding constant the other variables.

The reasonableness of temporal ordering of variables needs to be considered, as MR can be applied in ways such that an explanatory variable is nonsensically used to model a criterion variable. Although the MR model may account for a large proportion of variance, it might not make theoretical sense. For example, MR could be used to model "Time Spent Studying" as a function of "Test Score." However, such a model is nonsensical in the sense that "Time Spent Studying" would be an explanatory variable of "Test Score." This is a simple example of a causality problem, in the sense that the model itself does not make a distinction between what causes what. Theory, of course, should be the guiding principle of the specification and direction of causal relationships. Inferring causality can be difficult, especially because there technically needs to be some passage of time that occurs in order for a regressor to literally cause some change in a criterion.

## 14. Alternatives to Multiple Regression

When the assumption of normality of errors is violated, nonparametric approaches to inference for MR should be considered (e.g., Efron & Tibshirani, 1993; Györfi, Kohler, Krzyzak, & Walk, 2002). MR assumes that outcome variables are continuous and observed. However, when the criterion variable is censored, truncated, binary/dichotomous, ordinal, nominal, or count, an extension of the general linear model termed the generalized linear model, where a link function (e.g., exponential, Poisson, binomial, logit) relates the linear regression equation (analogous to the right hand side of Equation (1)) to a function of the criterion variable (e.g., probability of an affirmative response) can be used (e.g., Agresti, 2002; Long, 1997; McCullagh & Nelder, 1989).

Linearity is an assumption that is not reasonable in some situations, either based on theoretical or empirical evidence. *Spline* regression models allow different slopes over ranges of one or more regressors, in what has appropriately been termed a piecewise model (e.g., Fox, 2008; Ruppert, Wand, & Carroll, 2003). In spline regression multiple "knots" exists, where the slope of the regression line (potentially) changes over specified ranges (note that the slopes can be discontinuous in that they need not overlap at a knot). Another nonparametric regression procedure is known as *lowess* (locally weighted scatterplot smoothing) (also denoted *loess*; e.g., Cleveland, 1979; Fox, 2008), where MR models are fitted to areas/regions of the regressor(s) with "local" points receiving more weight than more distant points. The definition of "local" changes as a function of the width of the span selected, which is a parameter in the control of the analyst. For short spans the line of best fit can differ dramatically over a small range of a predictor, whereas a wide span tends to have a relatively smooth relationship between the regressor(s) and the criterion. Lowess techniques are most often used when $K = 1$. More general than lowess models are generalized additive models that allow some regressors to enter the model linearly and some to enter as splines (Ruppert et al., 2003, p. 215).

Applications of the general linear model are not robust to violations of the assumption of independent observations. Even for the simple case of the two independent group *t*-test, which can be considered a special case of MR, it is known that the nominal and empirical Type I error rate can be drastically different when the assumption of independence is violated (e.g., Lissitz & Chardos, 1975). When observations are not independent (e.g., students nested within classrooms, clients nested within therapists, observations nested within person), appropriate methods to explicitly control for the lack of independence should be used. A general approach to handling such nonindependence is multilevel models (also termed *hierarchical linear models*, *mixed effects models*, or *random coefficient models*; see Chapter 10, this volume).

When measurement error is serious, MR may not be an appropriate technique and latent variable models should be considered, especially when the primary goal is explanation instead of prediction. In particular, confirmatory factor analysis (see Chapter 8, this volume) and structural equation modeling (see Chapter 28, this volume) allow for explicitly incorporating error into the model of interest, which has the effect of separating the "true" part of the model from the "error" part.

# References

Agresti, A. (2002). *Categorical data analysis* (2nd ed.). Hoboken, NJ: John Wiley & Sons.
Aiken, L. S., & West, S. G. (1991). *Multiple regression: Testing and interpreting interactions*. Newbury Park, CA: Sage.
Algina, J., & Olejnik, S. (2000). Determining sample size for accurate estimation of the squared multiple correlation coefficient. *Multivariate Behavioral Research, 35*, 119–136.
Anscombe, F. J. (1973). Graphs in statistical analysis. *The American Statistician, 27*, 7–21.
Azen, R., & Budescu, D. V. (2003). The dominance analysis approach for comparing predictors in multiple regression, *Psychological Methods, 8*, 129–148.
Barron, R. M., & Kenny, D. A. 1986. The moderator-mediator variable distinction in social psychological research: Conceptual, strategic, and statistical considerations. *Journal of Personality and Social Psychology, 51*, 1173–1182.
Burket, G. R. (1964). A study of reduced rank models for multiple prediction. *Psychometric Monograph* (No. 12).
Cleveland, W. S. (1979). Robust locally weighted regression and smoothing scatterplots. *Journal of the American Statistical Association, 74*, 829–836.
Cohen, J. (1968). Multiple regression as a general data-analytic system. *Psychological Bulletin, 70*, 426–443.
Cohen, J., Cohen, P., West, S. G., & Aiken, L. S. (2003). *Applied multiple regression/correlation analysis for the behavioral sciences* (3rd ed). Mahwah, NJ: Erlbaum.
Darlington, R. B. (1968). Multiple regression in psychological research and practice. *Psychological Bulletin, 69*, 161–182.
Efron, B., & Tibshirani, R. J. (1993). *An introduction to the bootstrap*. New York: Chapman & Hall.
Enders, C. K. (2001). The performance of the full information maximum likelihood estimator in multiple regression models with missing data. *Educational and Psychological Measurement, 61*, 713–740.
Ezekiel, M. (1930). *Methods of correlational analysis*. New York: Wiley.
Fox, J. (2008). *Applied regression analysis and generalized linear models* (2nd ed.). Thousand Oaks, CA: Sage.
Fritz, M. S., & MacKinnon, D. P. (2007). Required sample size to detect the mediated effect. *Psychological Science, 18*, 233–239.
Gelman, A., & Hill, J. (2007). *Data analysis using regression and multilevel/hierarchical models*. New York: Cambridge University Press.
Green, S. B. (1991). How many subjects does it take to do a regression analysis? *Multivariate Behavioral Research, 26*, 499–510.
Györfi, L. Kohler, M., Krzyzak, A., & Walk, H. (2002). *A distribution-free theory of nonparametric regression*. New York: Springer.
Harrell, Jr., F. E. (2001). *Regression modeling strategies with applications to linear models, logistic regression, and survival analysis*. New York: Springer.
Hernan, M. A., Hernandez-Diaz, S., Werler, M. M., & Mitchell, A. A. (2002). Causal knowledge as a prerequisite for confounding evaluation: An application to birth defects epidemiology. *American Journal of Epidemiology, 155*, 176–184.
Huberty, C. (2003). Multiple correlation versus multiple regression. *Educational and Psychological Measurement, 63*, No. 2, 271–278.
Jaccard, J., Guilamo-Ramos, V., Johansson, M., & Bouris, A. (2006) Multiple regression analyses in clinical child and adolescent psychology. *Journal of Clinical Child and Adolescent Psychology, 35*, 446–479.
Jaccard, J., & Turrisi, R. (2003). Interaction effects in multiple regression (2nd ed.). Thousand Oaks, CA: Sage.
Kelley, K. (2007). Confidence intervals for standardized effect sizes: Theory, application, and implementation. *Journal of Statistical Software, 20*, 1–24.
Kelley, K., & Maxwell, S. E. (2008). Power and accuracy for omnibus and targeted effects: Issues of sample size planning with applications to multiple regression. In P. Alasuuta, J. Brannen, & L. Bickman (Eds.), *Handbook of social research methods* (pp. 166–192). Newbury Park, CA: Sage.
Kenny, D. A. (1979). *Correlation and causality*. New York: Wiley.
Little, R. J. A. & Rubin, D. A. (2002). *Statistical analysis with missing data* (2nd ed.). New York: John Wiley and Sons.

Lissitz, R. W., & Chardos, S. (1975). A study of the effect of the violation of the assumption of independent sampling upon the Type I error rate of the two-group *t*-test. *Educational and Psychological Measurement, 35*, 353–359.

Long, J. S. (1997). *Regression models for categorical and limited dependent variables.* Thousand Oaks, CA: Sage.

Lubinski, D., & Humphreys, L. G. (1990). Assessing spurious "moderator effects": Illustrated substantively with the hypothesized ("synergistic") relation between spatial and mathematical ability. *Psychological Bulletin, 107*, 385–393.

MacCallum, R. C., Zhang, S., Preacher, K. J., & Rucker, D. D. (2002). On the practice of dichotomization of quantitative variables. *Psychological Methods, 7*, 19–40.

MacKinnon, D. P. (2008). *Introduction to statistical mediation analysis.* Mahwah, NJ: Erlbaum.

MacKinnon, D. P., Fairchild, A. J., & Fritz, M. S. (2007). Mediation analysis. *Annual Review of Psychology, 58*, 593–614.

MacKinnon D. P., Lockwood C. M., & Williams, J. (2004). Confidence limits for the indirect effect: Distribution of the product and resampling methods. *Multivariate Behavioral Research, 39*, 99–128.

Maxwell, S. E., & Cole, D. A. (2007). Bias in cross-sectional analyses of longitudinal mediation. *Psychological Methods, 12*, 23–44.

Maxwell, S. E., & Delaney, H. D. (1993). Bivariate median splits and spurious statistical significance. *Psychological Bulletin, 113*, 181–190.

Maxwell, S. E., & Delaney, H. (2004). *Designing experiments and analyzing data: A model comparison perspective* (2nd ed.). Mahwah, NJ: Erlbaum.

McClelland, G. H., & Judd, C. M. (1993). Statistical difficulties of detecting interactions and moderator effects. *Psychological Bulletin, 114*, 376–390.

McCullagh, P., and Nelder, J. A. (1989). *Generalized linear models.* London: Chapman and Hall.

Muller, K. E., & Fetterman, B. A. (2002). *Regression and ANOVA: An integrated approach using SAS® Software.* Cary, NC: SAS Institute.

Pedhazur, E. J. (1997). *Multiple regression in behavioral research* (3rd ed.). Orlando, FL: Harcourt Brace.

Raju, N. S., Bilgic, R., Edwards, J. E., & Fleer, P. F. (1999). Accuracy of population validity and cross-validity estimation: An empirical comparison of formula-based, traditional empirical, and equal weights procedures. *Applied Psychological Measurement, 23*, 99–115.

Rencher, A. C., & Pun, F. C. (1980) Inflation of $R^2$ in best subset regression. *Technometrics, 22*, 49–54.

Rencher, A. C., & Schaalje, G. B. (2008). *Linear models in statistics* (2nd ed.). Hoboken, NJ: Wiley.

Ruppert, D., Wand, M. P., & Carroll, R. J. (2003). *Semiparametric regression.* New York: Cambridge University Press.

Schafer, J. L. & Graham, J. W. (2002). Missing data: Our view of the state of the art. *Psychological Methods, 7*, 147–177.

Seber, G. A. F., & Wild, C. J. (1989). *Nonlinear regression.* New York: Wiley.

Shrout, P. E., & Bolger, N. (2002). Mediation in experimental and nonexperimental studies: New procedures and recommendations. *Psychological Methods, 7*, 422–445.

Steiger, J. H. (2004). Beyond the *F* test: Effect size confidence intervals and tests of close fit in the analysis of variance and contrast analysis. *Psychological Methods, 9*, 164–182.

Steiger, J. H., & Fouladi, R. T. (1992). R2: A computer program for interval estimation, power calculation, and hypothesis testing for the squared multiple correlation. *Behavior Research Methods, Instruments, and Computers, 4*, 581–582.

Vittinghoff, E., Glidden, D. V., Shiboski, S. C., & McCulloch, C. E. (2005). *Regression methods in biostatistics: Linear, logistic, survival, and repeated measures models.* New York: Springer.

Wilkinson, L., & APA Task Force on Statistical Inference. (1999). Statistical methods in psychology journals: Guidelines and explanations. *American Psychologist, 54*, 594–604.

# 22
# Multitrait-Multimethod Analysis

Keith F. Widaman[1]

As Campbell and Fiske (1959) emphasized, every measurement we ever obtain in psychology is a trait-method composite—a measure purportedly of a particular trait construct obtained using a certain method of measurement. Campbell and Fiske introduced the *multitrait-multimethod* (MTMM) matrix as a tool for evaluating systematically the correlations among a set of measures. The primary utility of the MTMM matrix approach is the opportunity such a study affords to determine the preponderance of trait-related and method-related variance in measures in a battery. To aid in this evaluation, Campbell and Fiske argued that researchers should measure each of $t$ traits (e.g., Extraversion, Neuroticism, Fluid Intelligence) using each of $m$ methods (e.g., self-report, objective tests, observer ratings), so that each trait is measured using each method. By arranging trait measures in the same order within methods, the MTMM matrix should exhibit clear patterns to satisfy the dictates of convergent and discriminant validation. *Convergent validation* is satisfied if the researcher finds high correlations among measures of putatively the same construct, and *discriminant validation* is satisfied if low correlations are found among measures of presumably different constructs. Campbell and Fiske described several rules of thumb for evaluating patterns of correlations in a multitrait-multimethod (MTMM) matrix. Specifically, (a) correlations between measures of the same construct obtained using different methods of measurement should be large; (b) correlations between measures of the same construct obtained using different methods of measurement should be larger than correlations of those measures with measures of different constructs obtained using the same or different methods; and (c) the same pattern of trait correlations should hold for all methods and all combinations of methods.

Among others, Jöreskog (1971) pioneered the fitting of confirmatory factor (CFA) models to multitrait-multimethod data. The CFA approach circumvented several problems associated with the Campbell and Fiske (1959) rules of thumb. In particular, the CFA approach (a) yielded clear significance tests of differences between alternative models and of specific parameter estimates, whereas the ordinal comparisons involved in the Campbell–Fiske rules of thumb relied on dependent comparisons that compromised statistical tests; (b) allowed for tests of the amount of trait-related and method-related variance in the MTMM matrix; and (c) led to estimates of the amount of trait-related and method-related variance in each measure. Widaman (1985) systematized earlier work on CFA models and provided an informative taxonomy of models for MTMM data by cross-classifying available trait factor structures and method factor structures. In addition, Widaman discussed alternate

Table 22.1 Desiderata for Multitrait-Multimethod Analysis

| Desideratum | Manuscript Section(s)* |
|---|---|
| 1. Authors identify and justify nature of trait constructs under consideration. | I |
| 2. The ways in which the methods of measuring traits might influence measurements are discussed. | I |
| 3. The usefulness of trait and method variance estimates for understanding the underlying nature of measurements included in study is discussed. | I |
| 4. Path diagrams are presented for alternative models to be considered or for a general model within which alternative models are nested. | I |
| 5. The general strategy to be followed when comparing competing models is described. | I |
| 6. A description of the nature and size of participant sample, along with basic descriptive data, is presented. | M |
| 7. For each manifest variable a description is provided of the number of items, response scale, etc. | M |
| 8. The manifest variables associated with each trait and method factor are noted, verifying that each factor has a sufficient number of indicators and note method of identification. | M |
| 9. The name and version of program used is given, the method of estimation used is stated explicitly, and assumptions are addressed. | M |
| 10. The origins of missing data or outliers on manifest variables are described, and how they are handled is addressed explicitly. | M, R |
| 11. A description is provided of how problems (lack of convergence, improper estimates) are to be handled. | M, R |
| 12. The MTMM matrix is provided, along with means and standard deviations of manifest variables. | R |
| 13. Analytic strategy should dictate that certain (not all) models are evaluated. | R |
| 14. Likelihood ratio chi-square and practical fit indices are used, justifying any needed "fixes" to circumvent problems in the model fitting. | R |
| 15. Any needed post hoc modifications are described and justified theoretically and statistically. | R |
| 16. Parameter estimates are evaluated for statistical significance and with regard to interval estimates (i.e., considering standard errors for each estimate). | R |
| 17. The proportion of variance is computed due to trait, method, and unique factors. | R |
| 18. The quality of the different manifest variables for representing trait and/or method influence is discussed. | D |
| 19. A discussion is provided of how current results might impact future research on the traits and the substantive domain. | D |
| 20. Any limitations of the current study are addressed. | D |

* *Note*: I = Introduction, M = Methods, R = Results, D = Discussion

analytic strategies for exploring the magnitude of effects of trait and method constructs underlying manifest variables in an MTMM matrix. Building on earlier work by Kenny (1976), Marsh (1989) discussed an additional CFA method specification, using correlated uniquenesses to represent method effects. At about this time, Browne (1984) described a multiplicative model for fitting MTMM data. Reichardt and Coleman (1995) provided an advanced discussion of the relative fit of linear and multiplicative models. Recently, Eid, Lischetzke, Nussbeck, and Trierweiler (2003) discussed an approach leaving out one method factor to improve identification of model parameters. Lance, Noble, and Scullen (2002) and Eid and Diener (2006) provided overviews of the strengths and weakness of different analytic models. Stacy, Widaman, Hays, and DiMatteo (1985) and Widaman, Stacy, and Borthwick-Duffy (1993) conducted empirical studies with non-standard method factor specifications. Specific desiderata for studies that utilize CFA models for MTMM data are presented in Table 22.1 and then discussed. This chapter deals only with the structural equation modeling approach to evaluating MTMM matrices. Readers interested in other approaches can consult other references for more traditional or historical approaches. For the original approach involving comparing correlations, readers should refer to Campbell and Fiske (1959) or Ostrom (1969). Hubert and Baker (1978) presented a nonparametric approach for testing the patterns of differences among correlations in an MTMM matrix that improved statistically on prior methods. Analysis of variance (ANOVA) approaches have also been proposed, and the relative strengths and weaknesses of ANOVA relative to CFA modeling were discussed by Millsap (1995). An overview of alternative methods for analyzing MTMM data was provided by Eid (2006).

## 1. Substantive Context and Measurement Implications

One initial and important goal of any investigation that uses the multitrait-multimethod (MTMM) matrix is to identify and justify carefully the nature of the trait constructs that are the focus of the study. Trait constructs employed in psychology come in many different forms. In the mental ability domain, constructs such as fluid and crystallized intelligence or spatial ability can be studied, and researchers typically assume that these dimensions represent characteristics of persons that are rather stable across time. In the personality domain, researchers have recently emphasized the Big Five constructs of Extroversion, Agreeableness, Conscientiousness, Neuroticism, and Openness, which are also hypothesized to be stable characteristics. However, theory associated with certain other personality dimensions, such as State Anxiety, presumes that individual differences on these dimensions will exhibit notable fluctuation across time. In still other areas of study, the trait constructs may represent internal psychological processes, behavioral interaction styles, and so forth. Regardless of the constructs under investigation, the nature and definition of the constructs must be considered carefully, especially whether assessing each construct with each method of measurement is appropriate and theoretically justified.

Most state-of-the-art approaches to analyzing MTMM data involve confirmatory factor analysis (CFA; see Chapter 8, this volume) or other sophisticated approaches to analyses, which fall within the general family of methods of structural equation modeling (SEM; see Chapter 28, this volume). As will be discussed later (see Desideratum 9), the most common method of estimating parameters in CFA or SEM is maximum likelihood estimation, which requires that data satisfy certain assumptions. As a result, the researcher should verify that the measures of the trait constructs under study have been previously subjected to exploratory or confirmatory factor analyses or other forms of psychometric analysis to verify their basic psychometric properties. Chief among these properties are normal distributions of manifest variables and linearity and bivariate normality of relations among manifest variables. If data fail to satisfy these assumptions, other methods of estimation (e.g., robust weighted least squares) can be used that require less stringent assumptions. In addition, as discussed below (see Desideratum 9), the study should include at least three trait constructs in the MTMM matrix, to ensure identification of latent variables.

## 2. Methods of Measurement and Their Potential Impacts

To conduct an MTMM study, the investigator must select a set of methods, preferably three or more, to ensure identification of latent variables (see Desideratum 9). The researcher should also consider and discuss the ways in which the methods of measurement included in the study might influence those measurements of individual differences on trait dimensions. This is a difficult task, as a general theory of method effects, specifically in the context of MTMM studies, has never been developed.

Despite the lack of a general theory of method effects, researchers have used many different ways of operationalizing measurement methods, and the brief summary below is offered to assist in the challenging but essential task of understanding and interpreting method effects.

Many different types of measurement methods have been used in MTMM studies. In early studies of multiple dimensions of attitude (e.g., Ostrom, 1969), the methods included different ways of formatting items and/or developing scales (e.g., Guttman scaling, Thurstone scaling, rating scales). Some studies utilize different reporters as representing the different methods, and these can take any of several forms: (a) self, parent, and teacher report methods; (b) self, friend, and observer report methods; (c) mother, father, and teacher report; and so on. Still other studies survey methods more broadly, using life history (L), observer ratings (O), self-report (S), and objective test (T) indicators to serve as methods. These LOST indicators clearly represent a much broader selection of methods, as most researchers select multiple methods within a single LOST category to serve as the methods.

Methods may have many impacts on the trait measurements, an underemphasized aspect of most studies using the MTMM matrix. Self reports may be contaminated by notable amounts of response biases such as acquiescence, social desirability, or extremity bias. Observer reports are likely to reflect halo bias, or the tendency to rate a particular target similarly across trait ratings, particularly if the trait constructs have close connections (e.g., effort and performance). Objective tests are highly influenced by motivation to perform well at the time of the assessment, and such motivation may wax and wane over time. Life history data are more indicative of typical levels of performance than of maximal performance, which may lead to lower levels of convergence with objective test scores. The list of potential influences of methods on measurements is large, and research studies will be improved in the future if they are designed to shed light on the alternative sources of method effects.

## 3. Utility of Trait and Method Variance Estimates

Any findings regarding the trait and method variance in each of a set of measures is, of itself, an important contribution to the literature, especially if such estimates are not widely available from prior research. High levels of trait-related variance in measures support conclusions that the measures reflect the processes or constructs hypothesized, whereas low levels of trait-related variance should signal the need to revise measures to capture more adequately the underlying constructs. Either way, the use of the MTMM matrix can provide crucial information for the interpretation of research beyond the scope of the current study.

The researcher should discuss the usefulness of trait and method variance estimates for understanding the nature of measurements included in study. The history of every area of psychology is a history strewn with examples of theories built upon measures that were presumably reflective of specified underlying processes or theoretical constructs, measures later shown to be only weakly related to the underlying processes hypothesized. Much wasted effort might have been avoided if researchers had utilized MTMM matrix studies to investigate the properties of their measures.

Most research in psychology has a built-in confirmation bias, as researchers tend to search out and highlight positive correlations among measures of similar constructs. Some of these positive correlations are statistically significant, but fall in the range of .30 to .40, and correlations of this magnitude are not strong evidence that the measures are indicators of the same construct. Furthermore, finding low correlations between disparate measures can be just as important as, or more important than, finding high correlations between similar constructs. If a researcher found a .40 correlation between two measures of Construct 1, but found that both of these measures also correlate .40 with a measure of Construct 2, the researcher should be wary of the strength and importance of the former correlations. Use of the MTMM matrix approach forces researchers to confront research outcomes of this sort.

## 4. MTMM Path Diagrams

A path diagram or structural modeling diagram is a graphical presentation of the form of a statistical model. Most common linear models can be formulated as diagrams, and the statistical model represented by the diagram is often isomorphic with the diagram. In a path diagram, we typically denote manifest or measured variables as squares or rectangles and latent (or unmeasured) variables as circles or ellipses. Straight, single-headed arrows are used to indicate dependence or directed relations between variables, with the variable at the tail of the arrow a predictor of the variable at the head of the arrow. Finally, double-headed (and often curved) arrows are used to denote undirected relations, such as the covariance between two variables or the variance of a variable. In particular, a curved, doubled-headed arrows from a variable to itself represents a residual variance, with effects of all unidirectional influences on the variable controlled statistically. For example, in Figure 22.1, the curved, double-headed arrow from Trait Factor 1 to itself represents the entire variance of this trait factor because no unidirectional arrows are drawn toward the trait factor. In contrast, the curved, double-headed arrow from the manifest variable "Trait 1 Method A" to itself represents the unique variance of this indicator, which reflects variance in the indicator remaining after accounting for variance due to Trait Factor 1 and Method Factor A.

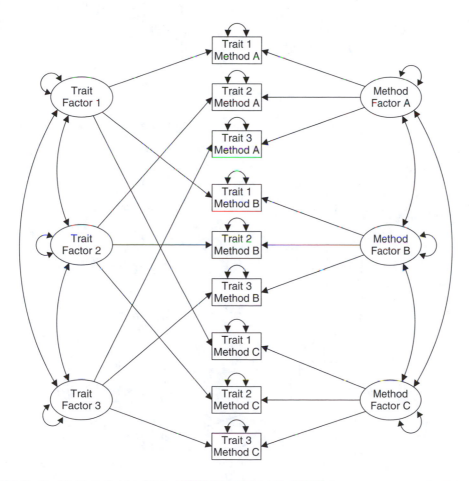

**Figure 22.1** The Correlated Trait–Correlated Method (CTCM) Linear CFA Model for MTMM Data.

In a study using the MTMM matrix, the researcher can and usually should present one or more path diagrams for alternative structural models to be investigated in the study. Two alternative path diagrams are shown in Figures 22.1 and 22.2. In Figure 22.1, nine manifest variables are shown in the rectangles in the middle of the figure: Traits 1, 2, and 3 each assessed using Methods A, B, and C. In the standard correlated-trait, correlated-method (CTCM) model, the researcher can hypothesize the presence of three trait latent variables, each associated with the manifest variables aligned with the trait construct and shown in the circles on the left side of the figure. Thus, Trait Factor 1 is shown having direct linear relations on the three manifest indicators of Trait 1, Trait Factor 2 has direct linear relations on the three manifest indicators of Trait 2, and so forth. The potential influence of three method factors is also represented by the three ellipses at the right side of Figure 22.1. Method Factor A is presumed to have direct linear effects on all manifest variables measured using Method A, and similar direct relations hold for Method Factors B and C. The double-headed arrows among the three trait factors reflect covariances (or correlations) among these latent trait factors, and the double-headed

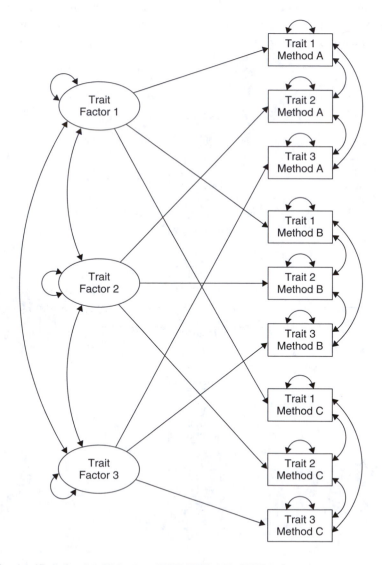

**Figure 22.2** The Correlated Trait–Correlated Uniqueness (CTCU) CFA Model for MTMM Data.

arrows among the three method factors reflect covariances among the latent method factors. The absence of double-headed arrows between trait and method factors indicates that these covariances are presumed to be nil and are typically forced to be zero.

The model in Figure 22.2 is often termed the correlated-trait, correlated-uniqueness (CTCU) model. As with the first model, nine manifest variables are shown in the rectangles in the figure. The trait factor specification in Figure 22.2 is identical to that in Figure 22.1, with trait factors having direct effects on their respective manifest variables and covariances posited among the trait factors. The major difference between the two figures is that method factors have been deleted in Figure 22.2 and have been replaced by covariances among unique factors or uniquenesses. That is, covariances have been specified among all measures obtained using Method A, among all measures obtained using Method B, and among all measures obtained using Method C. Note that the absence of double-headed arrows between measures obtained from different methods (e.g., no double-headed arrows between measures under Method A with measures under Method B) means that method effects are hypothesized to be statistically uncorrelated under the CTCU model and such effects are therefore fixed at zero.

## 5. Analytic Strategy for Comparing Models

Next, the researcher could describe a general analytic strategy that will be followed when comparing competing models. Analytic strategies have long been discussed for multiple regression analysis, and readers typically expect to read about the analytic strategy a researcher used in a complicated study employing regression analysis. In small studies, simultaneous regression—with all predictors included in a single regression model—is frequently used. But, if sample size is rather large and the number of potential predictors is also large, some form of theoretically or empirically driven hierarchical strategy is often used to understand the unique impacts of various sets of predictor variables.

Analogous analytic strategies should be used when evaluating the fit of a structural model to an MTMM matrix. Widaman (1985) outlined three alternative strategies, each based on a common strategy under multiple regression analysis. One strategy was akin to simultaneous regression analysis; under this approach, one might fit all of the models in the Widaman (1985) taxonomy and compare the importance of particular estimates whenever and wherever possible. A second strategy had similarity to forward selection in regression analysis. Under this strategy, one would start with the simplest model that might reasonably account for the data (i.e., a model with correlated traits and no method factors) and then add method factors and finally correlations among method factors if these were required to explain patterns in the MTMM matrix. The third strategy was similar to methods in regression analysis in which one partials out the influence of nuisance variables before testing the primary effects of interest. Here, the researcher could begin with a model that contains only correlated method factors and then adds trait factors and finally correlations among trait factors if needed to represent the relations in the MTMM matrix. Regardless of which analytic strategy is selected, the researcher should describe and justify the analytic approach taken in the study.

## 6. Sampling Method and Sample

As in any empirical study, the researcher should carefully and fully describe the nature and size of the sample of participants. Studies that utilize an MTMM matrix are oriented toward evaluating measurement properties, and measurement properties (e.g., reliability) are known to vary across samples, especially if a sample exhibits any restriction of range. Therefore, the researcher is advised to consider carefully the sample selection methods used, to arrive at as representative a sample from a population as possible. Samples of convenience are to be avoided in MTMM matrix studies, because the researcher intends to give general conclusions regarding the trait- and method-related variance in measures. Further, if any variables might be confounded in the correlations in the MTMM matrix, the effects of these should be estimated and, if large, controlled statistically. Potential confounding variables include nesting of participants within groups (e.g., students within classrooms), which can lead to failure to meet the assumption of independence of observations. With such effects, multilevel approaches to model fitting can be utilized. Or, if socioeconomic status, sex, or other background variables are related to the measures in an MTMM matrix, these variables could be included as control variables or covariates within the CFA model for the MTMM data.

The size of the sample of participants is of particular interest when structural equation modeling techniques are used. Monte Carlo studies of SEM have generally found that sample sizes of 150–200 or more should be used to estimate parameters and their standard errors (SEs) accurately if maximum likelihood (ML) estimation is used. If manifest variables fail to exhibit univariate and multivariate normality, then some more advanced methods of estimation (e.g., asymptotically distribution free [ADF], weighted least squares [WLS]) or robust adjustments to ML statistics (Satorra-Bentler corrections) might have to be used to ensure accurate estimation of parameter estimates and their SEs, and these methods typically require more substantial sample sizes (e.g., 1,000 for ADF). If a researcher has reasonable a priori estimates of trait and method factor loadings, along with correlations among the trait and among the method factors, the investigator should explore power analyses to ensure that s/he has sufficient power to detect the parameter estimates expected and/or to detect meaningful differences between competing MTMM models.

## 7. Identify and Describe Manifest Variables

Every empirical study in the social and behavioral sciences should have a clear description of the manifest variables included in the study. However, because the focus of any study featuring the use of the MTMM matrix is a full evaluation of the psychometric properties, particularly the trait and method decomposition, of manifest variables, the need to provide a careful description of all manifest variables is heightened. Most commonly, the manifest variables in an MTMM matrix study are not individual items, but are scale scores composed as the sum (or average) of multiple items. In a typical study, the researcher should separate out the description of each manifest variable in its own brief paragraph, so readers can easily identify the part of the Methods section that pertains to each manifest variable.

For each manifest variable, the researcher should provide a complete description of the way in which the manifest variable score was derived. This description should include the number of items in the scale, the response scale used for each items (i.e., the number of scale points and the scale endpoints), and the way in which the score on the manifest variable was derived. The scale or manifest variable score is usually obtained as the simple, equally weighted sum of all items that comprise the scale. However, if any differential weighting or other unusual method of arriving at a total scale score for the manifest variable is used, this must be clearly described and justified.

## 8. Latent Variables and Their Indicators

The manuscript should have a clear description of the manifest variables associated with each trait and method factor. As noted earlier, any MTMM matrix study should strive to have measures of at least three traits obtained using at least three methods to ensure adequate identification of all trait and method latent variables. Because the focus of an MTMM matrix study is on the measurement properties of manifest variables, the delineation of which variables load on which latent variables is usually a straightforward matter. Regardless of its often pro-forma nature, these relations should be clearly stated. Note that some researchers have formulated second-order MTMM designs (e.g., Marsh &

Hocevar, 1988). To employ a second-order MTMM design, one must have multiple indicators for each of the trait-method combinations in a study. Then, first-order factors are specified as having direct effects on appropriate manifest variables, and the MTMM matrix would be found in the matrix of correlations among first-order factors. Second-order factors would then be specified as having direct effects on appropriate first-order factors. The complexity of a second-order MTMM design underscores the importance of outlining carefully the relation of manifest variables to their associated latent variables and the place of measurements within the MTMM matrix.

The researcher should also clarify and justify how the latent variables in the model were identified. Many experts on SEM recommend fixing to 1.0 the factor loading of one indicator for each latent variable. This specification is usually sufficient to identify the scale of each latent variable and allow the estimation of all additional parameters of the model. However, this choice of identification constraint usually leads to latent variables with variances that depart from unity, so covariances among trait factors and among method factors are in a metric that can be difficult to interpret. Thus, it is often advisable to fix to 1.0 the variance of all latent variables in the model, which ensures that covariances among latent variables will be scaled in the metric of correlations, making them much easier to interpret. In addition, this choice for model identification leads to the estimation of all trait factor and method factor loadings and their associated SEs, enabling a more informative set of estimates for these key model parameters.

## 9. Analytic Methods

Many different SEM programs can be used to fit the standard structural models for MTMM data. These programs are listed in Chapter 28 in this volume. Most structural models for MTMM data are relatively simple to specify, so virtually all SEM programs will give identical parameter estimates and SEs for these models. Still, the researcher should provide the name and version number of the program used for analyses, as SEM programs are in a continual state of revision and readers should be informed of the specific program used for analyses. The researcher should also clearly state the method of estimation used. ML estimation is the default in virtually all SEM programs and is often the method used for analyses of MTMM data. However, ML estimation rests on stringent assumptions, including the assumption of multivariate normality of manifest variables. Thus, the researcher should address whether the MTMM data meet these assumptions; for example, the researcher should report univariate and multivariate indices of kurtosis, as departures from optimal kurtosis can lead to bias in model fit statistics. If the data exhibit either platykurtosis or leptokurtosis, then some method other than ML may be more appropriate and should be explored. Readers are referred to the chapter on SEM in the current volume (Chapter 28) for more details on this point.

In addition to other technical details on analytic methods, the researcher should note clearly whether the structural models were fit to the MTMM matrix in a correlational or covariance metric. Most structural models are covariance structure models, not correlation structure models. As a result, fitting covariance structure models to correlational data can lead to inaccurate estimation of many statistics, including the overall chi-square index of fit, factor loadings, factor intercorrelations, and SEs of all parameter estimates. Even when all other statistics are estimated accurately, the SEs of parameter estimates will almost always be biased if covariance structure models are fit to a correlation matrix. Unfortunately, this matter received too little attention in the analysis of MTMM data; indeed, most studies in which structural models have been fit to such data apply covariance structure models to the MTMM matrix in correlational metric. To ensure proper estimation of parameter estimates and SEs, it is strongly recommended that models be fit to MTMM data in covariance metric, and standardized estimates can still be reported to allow less complicated interpretation.

## 10. Missing Data and Outliers

In many types of study (e.g., longitudinal studies), missing data are more the norm than the exception. In contrast, given the fact that studies that use the MTMM matrix approach usually require only a single time of measurement, many such studies do not have missing data. Regardless, the researcher should describe the extent and the origins of any missing data in a study. Researchers often avoid missing data by using listwise deletion of participants, which leads to the dropping of participants with missing values on any manifest variable. This is not a generally recommended practice, because it leads to bias in the estimation of correlations among manifest variables. Instead, researchers should explore the use of full information maximum likelihood (FIML) estimation, in which models are fit directly to raw data matrices with missing data, or multiple imputation (MI), which analyzes and aggregates results from several datasets containing randomly varying imputed values. These approaches should lead to less bias in the estimation of the initial relations among manifest variables and then less bias in estimates of model parameters and standard errors.

The researcher should also discuss the presence of any univariate or multivariate outliers on manifest variables and how such outliers are to be handled. Outlier detection is rarely discussed in studies that use the MTMM matrix approach, yet outliers can distort relations in the MTMM matrix, so are not a trivial issue. Greater detail on outlier detection and how to handle outliers is provided in the SEM chapter (Chapter 28) so will not be discussed in further detail here. Suffice it to say that optimal results in MTMM matrix studies will be obtained only if the influence of outliers is minimized, so researchers are strongly encouraged to use state-of-the-art methods of detecting and handling outliers.

## 11. Problems with Model Fitting

Studies that use SEM frequently encounter problems in the fitting of models to data. These problems in model fitting can be of several forms, including lack of convergence and the presence of improper parameter estimates. The investigator should describe how any problems in model fitting will be handled. Lack of convergence arises when the iterative estimation routines in an SEM program fail to meet the convergence criterion and therefore fail to arrive at the ML solution within a specified number of iterations. Each SEM program has a default number of estimation iterations, which are often a function of the size of the problem or of the number of estimates in the model. The output from the SEM program should indicate in some fashion why the program failed to converge, although the user might have to be especially attentive to note whether convergence was achieved or whether iteration halted because the default number of iterations was exceeded. If the default was exceeded, the user can increase the number of iterations. If this does not solve the convergence problem, the model may requirerespecification because the model may not be sufficiently well identified empirically.

The presence of improper parameter estimates is the second major class of problems that often arise when fitting models to MTMM data. The most common types of improper estimates are (a) estimated correlation coefficients that fall outside the range from −1.0 to +1.0 and (b) negative variances. At times, the estimated correlation between a pair of trait or method factors falls outside the mathematically acceptable range from −1.0 to +1.0, because latent variable variances and covariances are estimated separately and no joint constraints are automatically imposed to ensure that correlations remain within mathematically acceptable bounds. If unacceptable estimates of this nature occur, the factors with unacceptably high correlations cannot be distinguished empirically, and the model must be respecified so that all model estimates are acceptable. For example, if Trait Factor 1 and Trait Factor 2 are estimated to correlate +1.1, one could either respecify the model so that all indicators for both Trait Factors 1 and 2 load on a single trait factor that subsumes the two domains of content or constrain the estimate of the correlation between Trait Factors 1 and 2 to fall on or within the boundary of the acceptable parameter space (i.e., fall between −1.0 and +1.0).

With regard to the second class of problematic estimates, the estimate of a variance parameter can be negative, and a negative variance is unacceptable because a variance is the square of the corresponding SD so must be greater than or equal to zero. Various possible bases for negative variances have been discussed (van Driel, 1978); regardless of the basis, a negative variance is unacceptable. In responding to this problem, one can either (a) fix the variance estimate to a value that is on or within the boundary (e.g., fix the variance to zero or to some small positive value) or (b) constrain the variance estimate to be greater than or equal to zero or to some small positive value. If the negative variance is a unique variance, then an appropriate model constraint may be employed. Suppose the researcher has an estimate of the reliability of the manifest variable, where $r_{yy}$ is used to denote the reliability of variable $Y$. The researcher could then constrain the unique variance for the manifest variable to be greater than or equal to $s^2_y(1 - r_{yy})$, where $s^2_y$ is the variance of variable $Y$. The quantity $s^2_y(1 - r_{yy})$ for a manifest variable represents its estimated error variance, which is a lower bound estimate of unique variance for the manifest variable. Regardless of how the problem is handled, a negative error variance is often a symptom of a larger problem with the model, so some form of model respecification is typically needed.

## 12. Descriptive or Summary Statistics

The key set of descriptive statistics for any MTMM study is the matrix of correlations among the manifest variables, along with the means and SDs of the manifest variables. Following the original presentation by Campbell and Fiske (1959), researchers usually array the measures of the $t$ trait factors in the same order within each of the $m$ methods. That is, the rows and columns of the correlation matrix are arranged so that the first $t$ measures are the trait measures obtained using the first method of measurement, the next set of $t$ measures are the trait measures obtained using the second method, and so forth. Arrayed in this fashion, the several key parts of the MTMM matrix—including the validity diagonals that contain the convergent validities—will be in the expected places that will ensure easy interpretation of the MTMM matrix by readers.

In addition to the simple presentation of a table with the MTMM matrix of correlations, the investigator should offer some summary observations on the correlations contained in the matrix. For example, Campbell and Fiske (1959) argued that convergent validities should be statistically significant and sufficiently large to encourage further research with the manifest variables. Campbell and Fiske then discussed how one should compare the validity diagonal elements to other elements in the matrix to evaluate the discriminant validity of the measures. As a result, prior to fitting structural models to the data, the investigator should describe the general levels of convergent and discriminant validity exhibited by manifest variables in the MTMM matrix. This would include noting both the general magnitude of the convergent validities and the degree to which the convergent validities tend to exceed the magnitudes of other relevant correlations and thereby exhibit discriminant validity.

## 13. Fit Relevant MTMM Models

The researcher should next select an analytic strategy for fitting relevant structural models to the MTMM data. Widaman (1985) discussed three general analytic strategies, the first of which is a simultaneous strategy, which involves the fitting of the entire set of models in the taxonomy of models he proposed. With four trait structures and four method structures, this would entail the fitting of 16 different structural models to the MTMM data. The investigator could supplement this set of models with additional models proposed by Marsh (1989), who incorporated a fifth method structure—the correlated uniqueness method specification—into the taxonomy proposed by Widaman (1985).

Rather than fitting all possible models discussed by Widaman (1985), research in a given area may be sufficiently advanced that only a subset of models should be considered. If a more restricted set of models should be of interest, one could implement one of two stepwise analytic strategies. The first of these is a forward selection strategy, in which one would first fit a model with correlated trait factors, then add to this model orthogonal method factors, and finally allow the method factors to correlate. The second approach is based on first partialing irrelevant sources of variance. Under this approach, one might first fit a model with correlated method factors (because these represent construct-irrelevant variance) and then subsequently add trait factors and correlations among the trait factors to determine whether the addition of trait factors leads to improved representation of the data over and above the estimation of method factors. The primary factor guiding the choice of an analytic strategy is this: Prior research and theory should dictate which sources of variance—trait factors and/or method factors—might be expected to explain correlations among the manifest variables, and these considerations should, in turn, dictate the set of models to be fit to the data.

## 14. Evaluate Relative Overall Fit of Competing Models

Once the set of MTMM structural models are fit to the data, the researcher should evaluate the relative fit of alternative models. The Widaman (1985) taxonomy, supplemented with the Marsh (1989) method structure, can be employed to determine which model comparisons are legitimate when testing statistical differences in fit between models. Statistical differences in fit between two models can be tested if one model is nested within the second model. The latter, more highly parameterized, model must have all parameter estimates in the first, more restricted model; the more restricted model can be obtained from the more highly parameterized model by fixing one or more parameter estimates in the latter model to zero. For example, a model that contains correlated trait and correlated method factors is a highly parameterized model, which can be designated Model 3. One could obtain one more restricted model, Model 2, by fixing correlations among method factors to be zero; a researcher could obtain a still more restricted model, Model 1, by fixing all method factor loadings to zero and thereby eliminating method factors. With this set of models, Model 1 would be nested within Model 2, and Model 2 would be nested within Model 3. Given this set of nesting relations, differences in fit between the various models can be studied.

Differences in the fit of nested models can be evaluated in several ways. The most common tool for investigating the differences in fit of nested models is to use the likelihood ratio (chi-square difference) test. Basically, if one model is nested within another, the difference in chi-square values is distributed as a chi-square variate with degrees of freedom equal to the difference in degrees of freedom for the two models. The likelihood ratio chi-square test of a model, and any related chi-square difference tests comparing nested models, are heavily influenced by sample size. In such situations, researchers often rely on practical fit indices to evaluate the difference in fit between models. Certain practical fit indices are termed measures of parsimony-corrected absolute fit because they contain a correction or adjustment for model complexity, yet index fit for a given model without regard to other, more restricted models; among these measures, the root mean square error of approximation (RMSEA) is perhaps the most useful. Other practical fit indices are termed measures of relative or incremental fit because they involve the comparison of fit of a given model to that of a more restricted, null model. Of the relative fit indices, the comparative fit index (CFI) and the Tucker-Lewis index (TLI) appear to be optimal indices of fit. The interested reader is referred to Widaman and Thompson (2003) for a more in-depth consideration of these measures.

Finally, model comparisons can be made among models that have no nesting relations. The most common measures used for such comparisons are information indices, such as the Akaike Information Index (AIC) and the Schwarz Bayesian Information Criterion (BIC). For the AIC and

BIC, the model with the smaller value of the index is the preferred model. These information criteria are also useful if two trait factors or two method factors are merged because the correlation between the two factors nears 1.0. Although the merging of two factors does lead to nesting of models under typical definitions of nesting, the resulting likelihood ratio difference test statistic does not follow a chi-square distribution because the correlation between factors is fixed at a boundary value (i.e., 1.0).

## 15. Post Hoc Model Modifications

The investigator should clearly describe and justify any post hoc modifications to the a priori specification of models to fit the MTMM matrix. The specification of trait factor and of method factors is a relatively simple and straightforward enterprise. However, the correlated trait—correlated method (CTCM) is often poorly identified empirically, so is prone to problems of lack of convergence and improper estimates. To counter identification problems, modifications to the a priori specification of a model can lead to a more well-conditioned model that converges and has acceptable estimates of all parameters. One example of a post hoc modification is the constraining to equality of all factor loadings on each method factor (presuming all variables to be in the same metric). Many times, method factor loadings are not large, and estimation problems can arise in such situations. These estimation problems can be avoided by constraining to equality all factor loadings on a given method factor.

A second class of post hoc modifications is the respecification of the nature of method factors. The studies by Stacy et al. (1985) and Widaman et al. (1993) exemplify principled respecifications of method factors. In the latter study, the authors used three methods—standardized instrument, day shift ratings, and evening shift ratings—of four trait dimensions of adaptive behavior—cognitive competence, social competence, social maladaption, and personal maladaption. The original method factor specification had all four measures from a given method of measurement loading on a single factor for that method, but this method failed to achieve proper estimates. Based on examination of model fit, Widaman et al. respecified the method structure with two method factors for each method of measurement—one factor with loadings from the two dimensions of competence (cognitive and social) and a second factor with loadings from the two dimensions of maladaption (social and personal)—and allowing these method factors to correlate within methods, but not between methods. This respecified model fit the data very well. Regardless of the form of post hoc respecification, any altered specification of a model must be justified on both theoretical and empirical grounds.

## 16. Evaluate Fit of Optimal Model to Manifest Variables

Once a final, acceptable model is selected to represent the MTMM matrix, the researcher should carefully interpret and evaluate all parameter estimates in the model. Evaluation of estimates is usually done in several ways. First, the point estimates of the factor loadings should be noted, at least with regard to their mean (or median) value and range. Thus, the investigator might say that the trait factor loadings were moderate to strong, with a mean loading of .65 and a range from .55 to .80. Similar comments can be made regarding the method factor loadings and the correlations among the trait and/or method factors.

The researcher should also evaluate each parameter estimate for statistical significance and likely population value. The critical ratio of the parameter estimate divided by its standard error yields a large-sample $z$ statistic, and the common practice in the field is to require a $z$ statistic to be 2.0 or greater to declare the associated parameter estimate significant at the $p < .05$ level. The investigator could also provide an interval estimate of each parameter estimate or of certain, key selected parameter estimates. An approximate 95% confidence interval can be constructed as the parameter estimate plus or minus twice its standard error, and this interval estimate is often more useful as an indication of likely population values of a given parameter than is the simple $z$ statistic.

As discussed by Widaman (1985), certain parameter estimates in the CTCM model are directly related to the stipulations regarding convergent and discriminant validity discussed by Campbell and Fiske (1959). The following estimates are of particular interest: (a) trait factor loadings are the primary indicator of convergent validity, because they represent the relation of the trait latent variable to its indicators, so higher trait factor loadings indicate higher levels of convergent validation; (b) correlations among trait factors are the principal basis for representing discriminant validity, as the correlations among trait factors indicate how discriminable the latent factors are empirically, so correlations among trait factors that tend toward zero indicate better discriminant validity than do high correlations among trait factors; and (c) method factor loadings are the primary indicator of the how strongly methods affect or influence manifest variables, so lower method factor loadings are preferred. The investigator should discuss these different sets of parameter estimates to provide the interested reader with a concise description of the degree to which the manifest variables in the MTMM matrix exhibit optimal patterns of convergent and discriminant validity.

## 17. Report Estimates of Trait and Method Variance

The investigator should compute and report the proportion of variance in each manifest variable due to trait, method, and unique factors. This is a relatively simple task if the manifest variables are in standardized, or correlational, metric. Provided that no correlations are allowed between trait and method factors and that each manifest variable loads on only its appropriate trait and method factors, then the proportions of variance explained by trait and method factors are simply the squares of the standardized trait and method factor loadings, and the sum of trait, method, and unique variance should be unity.

By default, structural equation modeling programs typically fit covariance structure models to data, and these models should be fit to covariance matrices. Unfortunately, most analyses of MTMM data involve the fitting of models to correlation matrices, as noted in Desideratum 9 above. If a researcher performs analyses in an optimal fashion and fits MTMM structural models to the MTMM matrix of covariances among manifest variables, the computation of variance due to trait, method, and unique factors is somewhat, but only slightly more complicated. That is, the variance of each manifest variable is represented in such a model as the additive sum of the square of the trait factor loading, the square of the method factor loading, and the unique variance, a total variance that usually departs from unity. Dividing each source of variance (e.g., the squared trait factor loading) by the total variance will provide the desired estimate of variance due to that source. Alternatively, the researcher can fit an MTMM CFA model to the covariance matrix and then present parameter estimates from the standardized solution. The researcher could additionally report a confidence interval for each variance estimate to supplement the point estimate of variance that is usually reported.

## 18. Implications of Modeling for Manifest Variables

The investigator should discuss the results of the MTMM analyses for the manifest variables included in the matrix. One of the key points made by Campbell and Fiske (1959) was that evidence for convergent validation of measures should be sufficiently strong to encourage further research in a given domain and, presumably, with the particular manifest variables in the analysis. The results of fitting structural models to MTMM data provide estimates of trait-related and method-related variance. Although no hard-and-fast rules for adequacy of trait-related variance in measures, some guidelines can be suggested. In much Monte Carlo work on factor analysis, standardized factor loadings of .4, .6, and .8 have been used to represent low, medium, and high levels of communality. After squaring these loadings, this means that researchers generally consider explained variance figures of .16, .36, and .64 to reflect low, medium, and high levels of saturation of the manifest variable with the factor. Whether these guidelines are acceptable in any particular application of structural modeling to MTMM data is a subject matter concern. Thus, in some domains, saturation of .16 with a trait factor for one or more manifest variables may be considered adequate; in other domains, minimal levels of trait factor saturation might be .25 or .30. Prior research in the area—and prior research using the manifest variables used in the current study—would be very valuable information when interpreting results.

The investigator should keep in mind the fact that, in the typical MTMM structural model, each manifest variable is influenced by two factors—one trait factor and one method factor. As a result, higher levels of factor saturation (e.g., above .60) are unlikely to be achieved routinely for saturation with a trait factor because the trait and method factors are each attempting to explain variance of the manifest variables. Still, the guidelines listed above may prove useful when interpreting the magnitude of effects of trait factors and method factors on the manifest variables.

## 19. Implications of Modeling for Future Research

The researcher should also discuss the implications or impacts of the results of the current study for future research on the traits and the substantive domain. Provided that very strong convergent validation of measures was obtained (see, e.g., Stacy et al., 1985), the researcher might suggest that little is gained by inclusion of many methods of measurement in the future. Still, to avoid bias associated with any particular method of measurement, one might reasonably recommend the use of several methods of measurement in any future study in order to triangulate in the assessment of key constructs of interest.

In many areas of research in the social and behavioral sciences, researchers continue to use favorite ways of assessing constructs because these approaches have been widely used in the past. But, if MTMM studies have not been performed in a particular domain, questions may linger regarding whether the results obtained are closely related to the trait construct presumably assessed by a measure or whether method influences on the measure might represent a major contaminant on scores. Because of this, research in any domain of investigation should be supplemented with MTMM data collection to verify the amounts of trait- and method-related variance in manifest variables. Successful MTMM studies will buttress traditional measures of constructs, adding important measurement-related information to the existing literature. On the other hand, MTMM studies that yield rather low levels of trait-related variance for key manifest variables can lead to important reorientations in fields of inquiry. Regardless of the degree of success in confirming high amounts of trait-factor saturation in measures, the results of MTMM studies have important implications for research on the trait constructs and the methods of assessing these constructs, implications that should be drawn out with clarity.

## 20. Limitations

As a final note, the researcher should discuss any limitations of the current study. Limitations may arise at many levels. For example, use of a sample of participants from a university subject pool may result in a rather restricted sample, and the results obtained may not generalize to the population. Or, the precise manifest variables included in the study may be problematic in some ways. Because MTMM studies often include more variables than typical studies, researchers may use shortened forms of measures due to the need to assess a large number of constructs. If these shortened forms fail to exhibit high amounts of trait-related variance, the problem may be due more to the use of shortened forms with lowered levels of reliability than to the problematic nature of assessing the constructs under study.

Sample size is always a consideration when using structural equation modeling, and studies using MTMM data and models are no exception. Preferably, sample size should be rather large, generally with a minimum sample size of 150 or 200 participants, and larger samples should be sought as the size of the model increases. MTMM structural models tend to be somewhat less stable and therefore more prone to lack of convergence than do typical structural models. Therefore, a researcher should be encouraged to obtain as large a sample of participants as possible. Regardless of any limitations, the use of the MTMM approach to construct validation is still considered one of the strongest approaches one can take, and the results of MTMM studies will always represent important contributions to the research literature.

## Note

1 This research was supported by grants from the National Institute of Child Health and Human Development, the National Institute on Drug Abuse, and the National Institute of Mental Health (HD047573, HD051746, MH051361, and DA017092) (Rand Conger, PI).

## References

Browne, M. W. (1984). The decomposition of multitrait-multimethod matrices. *British Journal of Mathematical and Statistical Psychology, 37,* 1–21.

Campbell, D. T., & Fiske, D. W. (1959). Convergent and discriminant validation by the multitrait-multimethod matrix. *Psychological Bulletin, 56,* 81–105.

Eid, M. (2006). Methodological approaches for analyzing multimethod data. In M. Eid & E. Diener (Eds.), *Handbook of multimethod measurement in psychology* (pp. 223–230). Washington, DC: American Psychological Association.

Eid, M., & Diener, E. (Eds.). (2006). *Handbook of multimethod measurement in psychology.* Washington, DC: American Psychological Association.

Eid, M., Lischetzke, T., Nussbeck, F. W., & Trierweiler, L. I. (2003). Separating trait effects from trait-specific method effects in multitrait-multimethod models: A multiple-indicator CT-C(M-1) model. *Psychological Methods, 8,* 38–60.

Hubert, L. J., & Baker, F. B. (1978). Analyzing the multitrait-multimethod matrix. *Multivariate Behavioral Research, 13,* 163–179.

Jöreskog, K. G. (1971). Statistical analysis of sets of congeneric tests. *Psychometrika, 36,* 109–133.

Kenny, D. A. (1976). An empirical application of confirmatory factor analysis to the multitrait-multimethod matrix. *Journal of Experimental Social Psychology, 12,* 247–252.

Lance, C. E., Noble, C. L., & Scullen, S. E. (2002). A critique of the correlated trait-correlated method and correlated uniqueness models for multitrait-multimethod data. *Psychological Methods, 7,* 228–244.

Marsh, H. W. (1989). Confirmatory factor analyses of multitrait-multimethod data: Many problems and a few solutions. *Applied Psychological Measurement, 13,* 335–361.

Marsh, H. W., & Hocevar, D. (1988). A new, more powerful approach to multitrait-multimethod analyses: Application of second-order confirmatory factor analysis. *Journal of Applied Psychology, 73,* 107–117.

Millsap, R. E. (1995). The statistical analysis of method effects in multitrait-multimethod data: A review. In P. E. Shrout & S. T. Fiske (Eds.), *Personality research, methods, and theory: A Festschrift honoring Donald W. Fiske* (pp. 93–109). Hillsdale, NJ: Erlbaum.

Ostrom, T. M. (1969). The relationship between the affective, behavioral, and cognitive components of attitude. *Journal of Experimental Social Psychology, 5,* 12–30.

Reichardt, C. S., & Coleman, S. C. (1995). The criteria for convergent and discriminant validity in a multitrait-multimethod matrix. *Multivariate Behavioral Research, 30,* 513–538.

Stacy, A. W., Widaman, K. F., Hays, R., & DiMatteo, M. R. (1985). Validity of self-reports of alcohol and other drug use: A multitrait-multimethod assessment. *Journal of Personality and Social Psychology, 49,* 219–232.

van Driel, O. P. (1978). On various causes of improper solutions of maximum likelihood factor analysis. *Psychometrika, 43,* 225–243.

Widaman, K. F. (1985). Hierarchically nested covariance structure models for multitrait-multimethod data. *Applied Psychological Measurement, 9,* 1–26.

Widaman, K. F., Stacy, A. W., & Borthwick-Duffy, S. A. (1993). Construct validity of dimensions of adaptive behavior: A multitrait-multimethod evaluation. *American Journal on Mental Retardation, 98,* 219–234.

Widaman, K. F., & Thompson, J. S. (2003). On specifying the null model for incremental fit indices in structural equation modeling. *Psychological Methods, 8,* 16–37.

# 23

# Multivariate Analysis of Variance

## Stephen Olejnik

Multivariate analysis of variance (MANOVA) is a statistical model that is appropriate for both experimental and non-experimental research contexts where the relations between one or more explanatory (independent) variables and multiple outcome (dependent, response) variables are of interest. While in general explanatory variables may be quantitative or qualitative (see Chapter 3, this volume, on canonical correlation), this chapter focuses on the analysis and interpretation of statistical models involving only *qualitative* explanatory variables, that is, variables that are used to group the available units, typically human participants. As presented here, MANOVA is viewed as an extension of the univariate general linear model (see Chapter 1, this volume, on between-groups ANOVA) to examine population differences on one or more linear composites of correlated outcome variables. The correlations among the outcome variables suggest that one or more constructs might underlie the observed measures (here, "constructs" are conceptualized somewhat differently than in a structural equation modeling context; see Chapter 28, this volume). Composites are weighted combinations of the observed variable scores with the estimated weights chosen to maximize group separation. These composites are called *linear discriminant functions* and each function defines an independent construct. It is the difference between populations on these constructs that is of primary interest to the researcher. The purpose of a multivariate analysis of variance therefore is to identify, define, and interpret the constructs determined by the linear composites separating the populations being compared. The careful selection of the outcome variables to study is essential for a meaningful analysis. A researcher may begin a study having some idea regarding the underling constructs, but often unanticipated constructs are identified. MANOVA can be used to support the researcher's beliefs regarding the assessed constructs as well as to reveal hidden constructs underlying the observed outcome measures.

Additional discussions of the application and interpretation of MANOVA that are less technical can be found in Hair, Anderson, Tatham, and Black (2005), Huberty and Olejnik (2006), and Stevens (2009). More mathematical discussions of MANOVA can be found in Anderson (2003), Johnson and Wichern (2007), and Rencher (2002). Guidelines for preparing or evaluating studies using MANOVA are presented in Table 23.1. These guidelines are elaborated upon in subsequent sections.

## 1. Research Questions and Design

The substantive questions that motivated the research study should be explicitly stated and justified based on theory, previous research findings, and/or the researcher's experiences. The research

Table 23.1 Desiderata for Multivariate Analysis of Variance

| Desideratum | Manuscript Section(s)* |
|---|---|
| 1. Specific research questions and the research design are explicitly stated. | I |
| 2. A rationale for the use of a multivariate analysis is provided. | M |
| 3. Constructs of interest are discussed and the selection of the outcome variables as indicators of those constructs is presented. | M |
| 4. In a series of preliminary analyses, data are screened and the appropriateness of the statistical model is verified. Any modifications of the original data are reported. | M |
| 5. The versions of the statistical software packages and the specific programs used in the analyses are stated. | M, R |
| 6. Hypothesis tests on the relations between the grouping and outcome variables implied by the research questions are reported. | R |
| 7. The strength of the observed relations between the grouping and the outcome variables are reported. | R |
| 8. Outcome variables are ordered on the basis of their importance for establishing a relation between the grouping and the outcome variables. | R |
| 9. The number of meaningful constructs that underlie the observed set of outcome measures is determined and reported. | R |
| 10. The linear discriminant function (LDF) coefficients are presented along with both the group-mean centroids in LDF space and a plot of these centroids. | R |
| 11. The identified constructs are defined on the basis of the relation between the LDFs and the outcome variables. | R |
| 12. Focus tests examining specific relations or group differences specified by the research questions are reported. | R |
| 13. Relations between grouping variables and the linear composites of the observed outcome variables which define the constructs that underlie the observed measures are discussed. Unanticipated constructs identified are highlighted. | D |
| 14. The current findings are generalized and related to previous findings. Limitations are recognized and additional questions are raised. | D |

* Note: I= Introduction, M=Methods, R=Results, D=Discussion

questions of interest provide the first indication as to whether a multivariate analysis is appropriate. Research questions answered through MANOVA examine the relations between one or more grouping variables (e.g., participation in an Exercise Program emphasizing Strength, Aerobics, or a Combination of Strength and Aerobics Training) and two or more outcome variables (e.g., Heart Rate, Blood Pressure, Oxygen Uptake, Mood, Attitude). It is essential then, as part of the explication of the research questions, that researchers explicitly define both sets of variables.

A multivariate analysis might be justified if the questions raised involve examining sets of outcome variables which in some sense "hang" together to measure a construct that cannot be adequately represented by a single outcome variable. If the research questions are stated in terms of individual outcome variables, a multivariate analysis will be inappropriate and multiple univariate analyses may be justified (see Chapter 1, this volume). But when multiple outcome variables are considered together through the formation of one or more composites, a multivariate analysis is justified. A composite is a weighted combination of outcome variable scores that represents a theoretical construct. The determination of those weights and their interpretation are critical steps in MANOVA.

The research questions under investigation should involve a construct (in MANOVA, "constructs" are often operationalized differently than in structural equation modeling; see Chapter 28, this volume). An example of a construct that cannot be easily measured by a single outcome variable is Wellness. This construct may be defined many ways and might include both physical and psychological indicators. No single indicator can be expected to adequately define Wellness, but taken together the construct can be better assessed. The appropriate weighting of these indicators is, in part, the role of a multivariate analysis.

The number, definition, and formation of the grouping variables determine the research design and must be explicitly stated for an appropriate interpretation of the relations stated in the research questions. Research questions stating or implying causal relations require the formation of groups through the random assignment of units (e.g., college students) to the levels of the grouping variable. For example a group of 60 college student volunteers may be recruited to participate in a study comparing the benefits of three types of exercise programs. From this group, 20 students might be assigned at random to each program. But *random assignment* alone does not guarantee causal inference. Causal inference requires that no confounding variables are present. That is, no other variable(s) can be identified that offer a convincing alternative explanation for an observed relation or difference between populations. If all confounding variables are eliminated or controlled, the design may be described as experimental. On the other hand, when groups are formed by selecting units from existing populations, (e.g., students are identified who have self selected an exercise program that emphasizes strength, aerobics, or a combination of strength and aerobics) the research design is non-experimental or correlational and relations identified are functional rather than causal. *Random selection* of units from the target populations, while desirable, is rarely achievable and is not required for appropriate application of MANOVA. The selection process used in the study does not determine the nature of the relation found but does determine the generalizability of the relation. It is important for researchers to describe in some detail the units involved in the study in order to provide some guidance as to the appropriateness of generalizing the findings.

Research questions may reflect conditional, omnibus (main effect), or specific focused relations. Conditional relations are examined through the interaction of two or more grouping variables. Only explanatory (e.g., grouping) variables may interact. It is inappropriate to refer to an interaction between a grouping variable and a set of outcome variables. An omnibus relation is a main effect, examining the relation between a grouping variable and the outcome variables generalized across all levels of the additional grouping variables. And specific relations are examined through focused tests or contrasts, comparing specific levels or combination of levels of a single grouping variable.

The research questions and design therefore lay the foundation for the appropriate analysis and presentation of the study's findings. The need for an explicit statement of the research questions cannot be over-emphasized.

## 2. Rationale for a Multivariate Analysis

Researchers, when describing the statistical models used, should provide a rationale for choosing MANOVA. The inclusion of multiple outcome variables is a necessary but insufficient justification for choosing a multivariate analysis. An appropriate justification for MANOVA is an interest in the study of one or more constructs defined by linear composites of the observed outcome measures. When multiple outcome variables are examined individually, chances increase that one or more relations will be judged to be real when in fact they are simply functions of sampling error (i.e., Type I error). That is, when testing group differences on each outcome variable individually, the risk of making one or more Type I errors ($\alpha$) increases as the number of outcome variables studied increases. Some researchers believe that the use of MANOVA will avoid making such errors. But this "protection" will be provided *only* for the unusual situation where none of the outcome variables are related to the grouping variables. That is, MANOVA controls the experimentwise Type I error rate to its nominal $\alpha$ level only under the complete null condition. However, if there are $p$ outcome variables and $m$ of these variables are in fact related to a grouping variable, then the overall MANOVA test will generally be statistically significant. Following the significant MANOVA test with $p$ separate univariate ANOVA tests will offer no protection from the experimentwise Type I errors over the $p - m$ true null hypotheses, whose risk of at least one false rejection will be $1 - (1 - \alpha_v)^{p-m}$ (where $\alpha_v$ is the statistical criterion used for each of the univariate tests on the $p$ outcome variables). So for example, if the relationship between a grouping variable and each of eight outcome measures ($p = 8$) are tested with $\alpha_v = .05$ but only three outcomes are related to the grouping variable ($m = 3$), the familywise error rate or probability of at least one Type I error among the five non-related outcome measures is .226 [i.e., $1 - (1 - .05)^5$].

If multiple univariate hypothesis tests are truly of interest and there is a concern regarding a problem with the experimentwise Type I error rate, MANOVA does *not* provide an adequate solution and the use of this model is not justified. To control the experimentwise Type I error rate a Bonferroni or modified Bonferroni approach may be used for the multiple univariate tests (see, e.g., Shaffer, 2002).

## 3. Selecting Multiple Outcomes

Because the purpose of conducting a MANOVA is to examine the construct(s) that separate the populations, the selection of outcome variables to be included in the analysis is a critical step in the planning of the inquiry. A rationale for including the selected variables should be provided. Too frequently researchers include outcome variables simply because they are easily obtainable or happen to be available at the time of data collection. Inclusion of such variables can reduce statistical power, add little to the understanding of population differences, and make the interpretation of the results much more difficult. Outcome variables selected for analysis should cluster into one or more groupings. These groupings should reflect the constructs that the researcher believes to be relevant for the determination of group separation with a single grouping variable or in the case of multiple grouping variables, population differences. The final analysis of the selected variables may or may not support the researcher's belief. Discovering unanticipated constructs supported by the data may be among the most interesting findings of the study.

## 4. Preliminary Analyses

While a researcher might be tempted to proceed directly to answering the stated research questions by testing specific hypotheses and examining the linear composites or identified constructs, such temptation must be avoided. Before the statistical model can be examined in depth, it is important for researchers to examine basic characteristics of their data set and report their findings. Before comparing groups, the data within groups must be examined first. Data should be examined to determine if they are complete (i.e., there are no missing observations or scores) and whether they include unusual or outlying observations. Within each group, score reliability for each measure (see Chapter 25, this volume) and the shape of the data distributions should also be examined. The results of these analyses should be reported and any action taken to correct or alter the original data must be reported. The results of these analyses might justify deleting some measures, or perhaps when sample sizes are relatively small (e.g., the number of outcome measures is similar to or greater than the error degrees of freedom) or if after examining the within-group correlation matrices it might be judged that too many highly related outcomes are included in the data set, it may be desirable to combine measures through a principal components analysis (see Chapter 8, this volume).

Once the researcher is satisfied that the data adequately represent the variables of interest within the groups, some preliminary comparisons among groups is appropriate. For each group the sample size, outcome means, and standard deviations must be reported. Initial insight on the constructs underling outcome variables and support for justifying the use of a multivariate analysis can be gained by reporting and interpreting the pooled within-group correlation matrix. Note that this matrix is *not* the total sample correlation matrix among outcome variables ignoring group membership; ignoring group membership can result in spuriously high or low correlations among variables because of differences among group means. If the outcome variables are uncorrelated, then multiple univariate analyses would likely be more appropriate than a multivariate analysis.

Univariate analysis of variance comparing outcome means may also be useful in providing an initial understanding on how the groups differ. As a preliminary analysis, the univariate *F*-tests can provide an indication of the relationship between individual outcome variables and the grouping variable(s). This analysis is for description only and provides insights similar to those gained by examining of the univariate pooled-within-group correlation matrix. Further if there is no indication that an individual outcome variable is related to the grouping variable(s) the researcher might consider dropping that measure from further analyses. Because this is a preliminary analysis, an increased risk of a Type I error might be tolerable and a Bonferroni-type adjustment would not be necessary.

The within-group covariance matrices should be examined and compared across groups. Such a comparison is particularly important when group sample sizes are substantially different from each other (say, by a factor of 2). A statistical test for covariance equality (e.g., Box, 1949) may be used to formally compare the matrices, but these tests tend to be overly sensitive to small departures from covariance homogeneity because: (1) they are sensitive to distributional non-normality, (2) they involve a large number of degrees of freedom, and (3) the number of variances and covariances being compared is generally large. An alternative strategy to a formal statistical test is a comparison of the log-determinants for the separate group-covariance matrices. The determinant of a covariance matrix is referred to as the *generalized variance*, a measure of total variance in the outcome variables, and taking the logarithm helps to put that value on a useful metric for comparison. Comparing the individual group log-determinants along with the log-determinant of the pooled covariance matrix provides an indication as to whether the assumption of equal covariance matrices is reasonable. If the determinants are in the same "ballpark" the researcher may be justified in pursuing the multivariate analysis. The correlation between group size and the log-determinant of the group-covariance matrix may also be examined. A strong negative correlation would indicate a liberal MANOVA test for the relation between the grouping variable and the linear composites while a positive relation would indicate a conservative hypothesis test. While a violation of the covariance homogeneity assumption is less serious for MANOVA hypothesis tests when sample sizes are approximately equal, covariance inequality is still a potentially serious problem when examining the constructs underlying the observed outcome measures even when sample sizes are equal. If the researcher judges that covariance matrices are problematically unequal, several options exist. A statistical test for comparing the populations on the outcome measures that does not assume equal covariance matrices are available (e.g., Johansen, 1980; Yao, 1965). Alternatively, the researcher might examine the covariance matrices within groups and identify one or two variables that contribute disproportionately to the covariance inequality. It might be reasonable to analyze these variables separately and continue the multivariate analyses with a slightly smaller set of outcome variables, if the analysis of the remaining variables is still meaningful. Still another alternative might be to reduce the number of groups being compared. Perhaps there are several subsets of the grouping variable that do have similar covariance matrices but the subsets' covariance matrices deviate greatly from each other. Separate multivariate tests might make sense to compare those groups that have similar covariance matrices. If only one group differs from the others, focused tests involving that group using the Yao procedure may be used to address the research questions.

## 5. Computer Software

Both SPSS and SAS computing software packages include several analytic procedures that are useful for carrying out a MANOVA analysis. Huberty and Olejnik (2006, p. 123) summarized the output provided by five of the analysis procedures in these packages. It is highly unlikely that any single procedure will provide all of the analyses needed for a complete MANOVA analysis. Further, because these procedures are revised on occasion, the version of the software and the specific programs used by the data analyst should always be reported.

## 6. Examining Group Differences

Once the researcher is satisfied that the necessary data conditions for MANOVA have been reasonably well satisfied, the analyses to answer the researcher's stated questions may proceed. Answers to the research questions are stated as hypotheses and these hypotheses are tested statistically using the MANOVA model. For example, a hypothesis in the null form might be stated as: There is no relation between participation in an Exercise Program (Strength Training, Aerobic Training, Combined Strength and Aerobics Training) and Wellness as measured by five outcome measure (Heart Rate, Blood Pressure, Oxygen Uptake, Mood, Attitude). While a univariate analysis compares the populations represented by the available groups on each individual outcome measure separately, the MANOVA tests the hypothesis on the construct (e.g., Wellness) by comparing the group means on the set of outcome measures simultaneously. The set or vector of outcome means is called a *mean centroid* and the null hypothesis could be stated as: There are no differences in the mean centroids among individuals participating in Strength, Aerobics or Combined Strength and Aerobic training program.

*Conditional tests.* If multiple grouping variables are included in a factorial research design, it is likely that the research questions include inquiries into group differences among levels of one grouping variable conditioned or depending on the level of a second grouping variable. These inquiries should be addressed first. For example, consider a $3 \times 3$ factorial design to investigate the relation between three Exercise Programs (e.g., Strength, Aerobics, Combined Strength and Aerobics), three levels of Exercise Intensity (e.g., Twice a Week, Three Times a Week, or Daily), and Wellness as measured by several physiological and psychological measures (e.g., Heart Rate, Blood Pressure, Oxygen Uptake, Mood, Attitude). One research question may ask whether the relation between Exercise Program and Wellness varies for different levels of Exercise Intensity. This question is one of an *interaction* between the two grouping variables. Alternatively several research questions might ask about the relation between one of the grouping variables and the outcome variables at each (or at a specific) level of a second grouping variable. For example, a researcher may be interested specifically in the relation between Exercise Program and Wellness for each level of Exercise Intensity. These alternative questions are referred to as *simple effects*. Both interaction and simple effect questions are conditional, meaning that a relation between a grouping variable and the outcome variables may vary with specific levels of a second grouping variable.

Generally, when multiple grouping variables are included in a factorial design the interaction tests precede tests of simple effects but that need not be the case. The number of interactions that may be tested is determined by the number of grouping variables being considered and the meaningfulness of such analyses. All interactions need not be tested however. Interactions involving more than three grouping variables are often difficult to interpret and often lack statistical power. In those cases the sum-of-squares associated with higher order interactions may be pooled with the within-group or error sum-of-squares.

If an interaction is statistically significant, the factorial design may be simplified into a series of simpler designs. For example, given a statistically significant interaction between Exercise Program and Exercise Intensity, the research questions may focus on the relationship between Exercise Program and Wellness for each level of Intensity. Or the comparison of two specific Exercise Programs at each level of Exercise Intensity might be examined. These are questions that are answered with tests of simple effects. Three points might be made regarding simple effects. First, simple effects are simplifications of a factorial design. They are alternative conceptualization of the linear model that combines the effect of a grouping variable with an interaction effect. A three factor design can be simplified to three two-factor designs; a $3 \times 3$ factorial factor design can be simplified to two sets of three one-factor designs. An interaction test need not precede the simplification of a factorial design. The research questions raised may call for the simplification of the design.

Second, a simple effect resulting from the simplification of a two-factor design to a number of one factor designs is not equivalent to a MANOVA with one grouping variable. A simple-effect hypothesis test uses the pooled error sum-of-squares across all groups in the factorial study when computing the test statistic of interest and determining the constructs underlying the outcome variables, whereas a MANOVA for a single grouping variable pools the error sum-of-squares for only those groups being compared. When the MANOVA assumptions are met, the difference between a simple effect and a MANOVA with one grouping variable is statistical power. The simple effect will have greater error degrees of freedom and therefore greater sensitivity to identify population differences.

And third, when simple effects are considered the researcher may want to adjust the criterion used to judged statistical significance to limit the overall Type I error rate. While a good argument may be made for not making any adjustments, the problem of increasing the risk of a Type I error across all of the tests conducted should at least be addressed.

*Omnibus or main-effect tests.* An omnibus or main-effect test examines the relation between the outcome measures and one grouping variable generalized across all levels of additional grouping variables. If an interaction has been identified, such tests although valid are not likely to be meaningful. However, if Exercise Program and Intensity do not interact, a more general question on the relationship between Program and Wellness can be addressed. That is, across all levels of Exercise Intensity is there a relation between Exercise Program and Wellness. The answer to this question requires the simultaneous comparison of all three identified Exercise Programs. If the relation is found to be statistically significant the results do not identify which Program(s) best promotes Wellness. Specific comparisons through focused tests would be needed. Focused tests are discussed in some detail in Desideratum 12.

In a factorial study when the number of observations per group is unequal and disproportional, the research design is *nonorthogonal.* In this case, the hypothesis tests on the grouping variables in the model are not independent of one another. For the results to be interpreted appropriately it is essential that the method used to compute group means on the outcome variables be explicitly stated. For example, in a two-factor design a marginal column mean may be computed as the average of all observations in a column or as the average of the cell means in the column. The first approach is a *weighted* (also referred to as *type I* or *hierarchical* sum-of-squares) approach, where each cell mean in the column is weighted by the ratio of the number of units in the cell to the number of units in the column. Thus, the contribution a cell mean makes to the calculation of the column mean is in proportion to the number of observations in the cell. The second approach is an *unweighted* (also referred to as *type III* or *regression* sum-of-squares) approach where cell means within a column contribute equally to the calculation of the column mean, regardless of group size. The weighted and unweighted approaches for computing marginal group means test somewhat different hypotheses but both are valid. Unless the method used to compute the marginal means is explicitly stated, it is not clear what relations are tested (Carlson & Timm, 1974; Pendleton, Von Tress, & Bremer, 1986). Generally, in a multi-factor nonorthogonal research design the unweighted approach is preferred when the inequality in sample sizes is unintended and do not represent true differences in population sizes but rather reflect a random loss of participants. This may occur in experiments when an equal number of participants are randomly assigned to each condition but participants are lost due to reasons unrelated to the conditions studied (e.g., illness). The unweighted solution is also preferred because the effect of each explanatory variable can be tested after controlling or considering the effect of the additional factors and interactions in the model. Occasionally the weighted solution might be used if the differences group sample sizes are proportional to differences population sizes they represent.

Four different multivariate test criteria have been developed and are reported on computer printouts to evaluate interaction, simple effects, main effects, and focused hypothesis tests. The SPSS MANOVA procedure reports these criteria as: Wilks, Pillais, Hotellings, and Roys while the SAS and SPSS GLM procedures reports these criterion as Wilks's Lambda ($\Lambda$), Pillai's Trace, Hotelling-Lawley Trace, and Roy's Greatest Root. These four criterion provide exact tests of hypotheses with identical $F$ and significance values when the hypothesis degrees of freedom equals one or when only two outcomes are measured. For other contexts the four criteria are approximate tests that provide slightly different results but generally the same conclusion. While there is no consensus as to which criterion is best, Rencher (2002) recommended Roy's criterion if the outcome variables represent a single construct and he recommends Pillai's criterion when multiple constructs are assessed. Olson (1974) examined the robustness of the four multivariate criteria and recommended Pillai's criterion because it is less sensitive to covariance inequality and provides adequate statistical power. Researchers should be explicit regarding the MANOVA criterion used by stating its numerical value, $F(df_{num}, df_{den})$-value, and $p$-value (e.g., Pillai's = .164, $F(4,126) = 2.840$, $p = .027$).

## 7. Strength of Relations

Using one of the four multivariate test criteria mentioned in the previous section researchers might infer whether an observed relation between a grouping variable and the outcome variables appears to reflect a true relation or whether the relation can be explained by sampling error. The identification of true relations, however, is generally not sufficient for interpretation purposes. In addition to meeting statistical criteria, the strength of the association is needed to judge the meaningfulness of the results. A number of multivariate effect-size measures have been suggested. For contexts when the grouping variable has only two levels and situations involving a multi-level grouping variable where two or more of the levels are compared using a pairwise or complex contrast, the Mahalanobis $D$ statistic can be useful. Mahalanobis $D$ is a measure of distance between two mean centroids that is analogous to Cohen's standardized mean difference statistic, $d$.

When three or more levels of a grouping variable are examined simultaneously a measure of association between the grouping variable and the outcome variables can be useful. Several multivariate measures of association have been proposed with the most popular ones being associated with each of the multivariate test criteria discussed in the previous section (see Huberty & Olejnik, 2006, pp. 62–65). While the SAS computing package does not report these multivariate measures of association, the SPSS MANOVA procedure computes them and refers to them using the same label as the test criterion (i.e., Pillais, Hotellings, Wilks). The GLM procedure in SPSS refers to these measures of association as partial eta squared. While the four measures of association are identical for multivariate comparisons of two groups or for complex one-$df$ contrasts involving multiple groups, they provide slightly different estimates of relation when multiple levels of a grouping variable are assessed simultaneously. Kim and Olejnik (2005) discussed the different definitions of effect used by these measures of association. In addition, these measures of association overestimate the strength of relation. An adjustment for the overestimation provided by the effect estimated by using Pillai's test criterion was suggested by Serlin (1982) based on an adjustment for the squared multiple correlation in a multiple regression context by Ezekiel (1930). Kim and Olejnik (2005) showed that Serlin's adjustment also works well for measures of association based on Wilks's and Hotelling's test criteria. The multivariate measure of association reported should be consistent with the researcher's multivariate test criterion choice. However from an interpretation perspective the effect-size measure associated with Pillai's criterion may be the most useful. Pillai's criterion is based on the sum of squared canonical correlations. A canonical correlation is a measure of association between the grouping variable and an estimated construct measured by the outcome measures composite. Generally two or more constructs are estimated (not all are necessarily meaningful) in any set of outcome measures. Pillai's effect-size measure reports the average squared canonical correlation. That is, it reports an average association between the constructs measured and the grouping variable. In a sense it is similar to eta-squared often reported in ANOVA, but in MANOVA the focus is on the estimated constructs while ANOVA the focuses on an individual outcome variable.

One difficulty with the multivariate measures of association lies in the interpretation of the computed values. In the case of univariate analyses, guidelines have been suggested by Cohen for interpreting $d$, eta-squared, omega-squared, and $R^2$. Further, over the past ten years these statistics have been increasingly reported in the empirical literature and so they have become somewhat familiar to the applied researcher. In the case of MANOVA, guidelines have not been proposed nor have these effect-size measures been consistently reported along with multivariate hypothesis tests. Another problem with multivariate effect-size measures is that they are influenced by the number and choice of outcome variables included in the study. Consequently, it is not clear how to characterize the strength of relations in MANOVA. It is unlikely that Cohen's univariate guidelines would be useful in multivariate contexts. Because several outcome variables generally reflect a construct in a multivariate analysis, one might expect a stronger relation between the grouping variable and a construct than might be expected when the relation is based on an individual variable.

Up to this point the analyses of multiple outcome variables with MANOVA parallels what is typically done in a univariate ANOVA context. However, while similar in procedure, MANOVA's focus is uniquely multivariate. Researchers should generally *not* follow the MANOVA by examining the univariate relations between each outcome variable and the grouping variable after identifying a significant multivariate relation. This is an error because such a strategy ignores the relations among the outcome variables, it ignores the relations between the identified constructs and the grouping variables, and the questions answered by the univariate analyses are inconsistent with the multivariate analyses just conducted.

## 8. Variable Importance

Frequently, when there is some evidence to indicate that there is a relation between a grouping variable and the outcome variables studied, researchers are interested in determining which of the outcome variables are primarily responsible for determining the relation. Several options have been considered for determining variable importance. One option that should *not* be used to identify variable importance is the order in which variables are entered as a result of stepwise discriminant analysis (see Chapter 6, this volume). Like stepwise multiple regression, stepwise discriminant analysis has several serious limitations, including (1) inflated Type I errors, (2) inappropriate use of the $F$ distribution, and (3) lack of reproducibility. Alternatively, some methodologists prefer to examine the standardized discriminant function weights while others prefer to examine structure $r$ values (Huberty & Wisenbaker, 1992); both of these statistics are discussed under Desideratum 11. But for variable importance, Huberty and Olejnik (2006) preferred to rank variables based on the numerical value of Wilks's $\Lambda$ for group differences when the $i^{th}$ variable is deleted. Wilks's $\Lambda$ is one of the criteria used to test for statistical relation between the outcome variables and a grouping variable. Small values for Wilks's $\Lambda$ indicate a strong relation between the grouping variable and the construct. Thus, a relatively large Wilks's $\Lambda$ value upon deletion of a variable indicates a relatively important variable;

as such, variables with the largest deleted Wilks's $\Lambda$ values would be judged at the most important variables.

It should be recognized that the approached outlined here only provides one strategy to guide researchers in the identification of relatively "important" outcome measures. It is not intended to be prescriptive. Concluding that one variable is more important and another based solely on the numerical value of Wilks's $\Lambda$ or the $F$ statistic is not recommended. Some judgment on the part of the researcher is needed to determine whether the difference between two or more Wilks's $\Lambda$ values is meaningful. It might be that several deleted variables result in "similar" values for Wilks's $\Lambda$, in which case such variables would be judged to be of comparable importance. It must also be remembered that a statistic like Wilks's $\Lambda$ is a random variable subject to sampling error. A replication of the study could very well alter the ordering of variables and the identification of the "most important" outcome measure.

## 9. Determining Constructs

A statistically significant relation between a grouping and the outcome variables suggests that there is a relation between the grouping variable and at least one construct. It is not clear on the basis of the omnibus test, however, how many constructs or what variables define the construct(s). Here, the determination of the number of constructs is considered while the next desideratum considers the process for defining them. The univariate analysis of each outcome variable is not useful in determining the number of constructs underlying the variable space, but univariate analyses are related to a procedure considered in the next section for defining the constructs.

The number of constructs or linear discriminant functions (LDF) that can be estimated is determined by the degrees of freedom $(J-1)$ for the grouping variable with $J$ levels or the number of outcome variables $(p)$, whichever is smaller. Generally, there are fewer between-group degrees of freedom than outcome variables. For example, when there are only two levels of a grouping variable or when focused tests of pairwise or complex contrasts are of interest, the between-group degrees of freedom equal one. Regardless of the number of outcome variables studied, only one LDF is formed. For a two factor $(J \times K)$ interaction, the number of degrees of freedom is the product of the degrees of freedom for each factor $[df_{J \times K} = (J-1)(K-1)]$, yielding as many LDFs (if sufficient outcome variables exist). Each LDF represents a separate, independent construct that underlies the observed outcome variables. Typically, not all of the estimated constructs are meaningful or statistically significant.

The number of meaningful constructs that underlie the observed variables can be determined using a two step process. In the first step a series of statistical tests are conducted. Wilks's $\Lambda$ was mentioned as one of the four criteria useful for testing the overall relation between a grouping and outcome variables. Statistical significance here suggests that at least one construct underlies the observed variables. To determine whether additional constructs are represented by the observed variables, the variation associated with the first construct is removed and Wilks's criterion is modified. The modified $\Lambda$ is then used to test whether the remaining variation among the outcomes is statistically significant. If this test is statistically significant, it suggests that group separation exists on at least two constructs. Testing continues by further partitioning or adjusting Wilks's criterion and testing stops when the remaining variation in the variables is no longer statistically significant.

The second step in determining the number of meaningful constructs represented by the outcome variables is to examine the proportion of variation in the outcome variables that is explained by each statistically significant construct or LDF identified in step one. Not all statistically significant LDFs explain a meaningful amount of variation in the outcome variables. Guidelines for defining meaningful proportions of variations do not exist but the original research questions and the earlier justification given for variable selection should be influential in the decision process.

## 10. Linear Discriminant Functions

Linear discriminant functions (LDFs) are weighted combinations of the outcome variables to represent underlying constructs. These LDFs are useful to *describe* how groups differ. They are *not* used, however, for predictive or classification purposes. Because some computer software programs do not clearly distinguish between descriptive and predictive discriminant functions researchers often confuse them and sometimes refer to them interchangeably. Chapter 6 of this volume discusses in detail differences between these two types of discriminant functions, but briefly predictive discriminant (or classification) functions (PDF) differ from LDFs in (1) purpose, (2) number, (3) calculation, (4) equal covariance matrices requirement (DDF requires it, PDF does not), and (5) application. Researchers should be clear that when presenting MANOVA results, LDFs are interpreted as descriptive and not predictive (PDFs).

## 11. Defining Constructs

Once the number of constructs underlying the observed variables has been determined, attention must then be given to defining the constructs. Constructs are believed to be real, yet are unobservable. The outcome variables, on the other hand, are observable and interpretable. The definition of the latent construct relies on the relation between the construct and the outcome variables (here, "latent constructs" are conceptualized differently than in structural equation modeling; see Chapter 28, this volume). The relation between the each identified construct and the outcome variables is obtained by correlating each outcome variable with the discriminant function composite score (computed by summing the products of the raw discriminant function weights and the outcome variable scores). The correlation between an outcome variable and the composite score is called a *structure r*. While the sign for *r* is arbitrary, it must reflect the theoretical direction of the relation between the construct and each outcome variable. For example, in measuring Wellness, three variables—Heart Rate, Blood Pressure, and Oxygen Uptake—might be expected to define the physical aspect of Wellness. The signs of the structure *r* values should reflect a theoretically reasonable pattern. One might expect Heart Rate and Blood Pressure to be related to a common construct in the same direction while Oxygen Uptake should be related in an opposite direction (e.g., low heart rate and low blood pressure are associated with Wellness while high oxygen uptake is associated with Wellness). If all structure *r* values were positive or all negative, it might be difficult to explain and could raise some question regarding the construct being measured. A relatively high value for |r| indicates that the latent construct shares some meaning with the outcome variable. When several outcome variables have relatively high |r| values, the composite is measuring the latent variable believed to be common with the outcome variables. The construct definition therefore is determined by whatever is common to the outcome variables that are correlated with the composite score. A table of structure *r* values therefore is very useful when interpreting the nature of group differences and should be reported as part of the study's findings.

While structure *r* values provide a useful approach to define latent constructs, one problem with using structure *r* values is that they are highly related to the univariate *F* statistics for testing group differences on each of the outcome variables. The absolute value of the structure *r* will be large for an outcome variable that has a large computed univariate *F* statistic. This relation between *r* and *F* can be interpreted to mean that the structure *r* is not a true multivariate statistic and consequently should not be used to define the constructs (see Rencher, 2002, p. 317).

An alternative approach advocated by some is to examine standardized discriminant function weights. Similar to multiple regression contexts, the numerical value of a raw (unstandardized) discriminant weight is determined to a great extent by the scale or variance of the outcome variable: generally, the greater the variance the smaller the raw discriminant function weight. Standardizing each of the outcome variables creates a common scale for all outcome variables and consequently makes it possible to compare the weights associated with the outcome variables. Constructs are sometimes interpreted based on the variables that have the greatest standardized weights. One problem with using standardized weights to define constructs is that their numerical values are influenced to some extent by the degree of correlation among the outcome variables. This problem is analogous to the multicollinearity issue in multiple regression analysis. The standardized discriminant function weights provide an index of the unique contribution individual outcome variables make to the composite score. If some outcome variables are highly related, the unique contribution those variables make will be small and consequently the weights will be small. Variables that contribute to construct formation may thus be overlooked.

Both structure *r* values and standardized discriminant function weights should be reported when interpreting the results of MANOVA. They address different issues regarding the identified constructs underlying the variable space created by the outcome variables. A structure *r* provides an index for relating an outcome to a composite score and a standardized weight reflects the unique contributions an individual variable makes to the composite score. Both are useful for interpreting the discriminant functions separating the groups.

## 12. Focused Tests

Examining conditional relations and omnibus relations are generally the starting point for most MANOVA analyses. But ultimately in most research studies the questions asked will focus on pairwise or complex contrasts between specific levels of a grouping variable. As was stated earlier but is worth repeating, in many research contexts after determining that the data meet the necessary model conditions, it may very well be the case that the only analyses needed are those accomplished through contrast analyses. The results of a specific contrast can be substantially different than the results from omnibus analyses. First, while the omnibus analysis may result in the identification of more than one construct, a contrast has only one degree of freedom and consequently only one construct may be identified. Second, with omnibus tests the separation among all levels of the grouping variable is considered simultaneously, thus all levels of the grouping variable are compared on constructs defined by the same outcome variables. With contrasts, on the other hand, the outcome variables that define construct which separates the groups being compared can differ depending on which groups are being compared. For example the difference between Strength and Aerobics Training may be identified as Physical but the difference between Strength and the Combined Strength and Aerobics Training might be identified as Psychological. And third, if variable importance is of interest, the "most important" variables that separate all levels of the grouping variable can be different from the "most important" variables that separate the groups identified in the contrast.

Both the simultaneous comparisons of multiple levels of the grouping variable and specific comparisons through contrasts provide useful information regarding differences among the populations studied. The results of MANOVA will typically include both sets of analyses.

## 13. Answering Research Questions

Following the analyses of the data, the results should be summarized in relation to the research questions stated at the beginning of the article. The original research questions were introduced in the context of theory, practice, and/or previous research so the current findings must again be presented within those contexts. Outcome variables were chosen to reflect anticipated constructs and the extent to which the results support those constructs should be discussed. Equally important unanticipated constructs, if identified, must be discussed. Differences in construct formation obtained in omnibus and focused tests between specific populations might also be highlighted.

## 14. Generalizing Findings

It is important that the researcher relate the current results to those previously reviewed. Both consistencies and inconsistencies with theory and previous research should be highlighted. Inconsistencies merit additional discussion with possible explanations. All research studies impose a number of limitations on the scope and execution of the inquiry. These limitations should be made explicit. While care must be given not to overstate the implications of the current findings, it is also important not to minimize the contribution of the current findings as well. Being overly cautious with the interpretation can also be a serious mistake. Finally, directions for future research should be explicitly stated. It may be a trite statement but it is nevertheless true: good research introduces at least as many questions as it answers.

## References

Anderson, T. W. (2003). *An introduction to multivariate statistical analysis* (3rd ed.). New York: Wiley.

Box, G. E. P. (1949). A general distribution theory for a class of likelihood criteria. *Biometrika, 36*, 317–346.

Carlson, J., & Timm, N. (1974). Analysis of nonorthogonal fixed effects design. *Psychological Bulletin, 81*, 563–570.

Ezekiel, M. (1930). *Methods of correlational analysis.* New York: Wiley.

Hair, J. F., Anderson, R. E., Tatham, R. L., & Black, W. C. (2005). *Multivariate data analysis* (6th ed.). Upper Saddle River, NJ: Prentice Hall.

Huberty, C. J., & Olejnik, S. (2006). *Applied MANOVA and discriminant analysis* (2nd ed.). New York: Wiley.

Huberty, C. J., & Wisenbaker, J. M. (1992). Variable importance in multivariate group comparisons. *Journal of Educational Statistics, 17*, 75–91.

Johansen, S. (1980). The Welch-James approximation to the distribution of the residual sum of squares in a weighted linear regression. *Biometrika, 67*, 85–92.

Johnson, R. A., & Wichern, D. W. (2007). *Applied multivariate statistical analysis* (6th ed.). Englewood Cliffs, NJ: Prentice Hall.

Kim, S., & Olejnik, S. (2005). Bias and precision of measures of association for a fixed-effect multivariate analysis of variance model. *Multivariate Behavioral Research, 40*, 401–421.

Olson, C. L. (1974). Comparative robustness of six tests in multivariate analysis of variance. *Journal of the American Statistical Association, 69*, 894–908.

Pendleton, O. J., Von Tress, M., & Bremer, R. (1986). Interpretation of the four types of analysis of variance tables in SAS. *Communications in Statistics – Theory and Methods, 15*, 2785–2808.

Rencher, A. C. (2002). *Methods of multivariate analysis* (2nd ed.). New York: Wiley.

Serlin, R. C. (1982). A multivariate measure of association based on the Pillai-Bartlett procedure. *Psychological Bulletin, 91*, 413–417.

Shaffer, J. P. (2002). Multiplicity, directional (Type III) errors, and the null hypothesis. *Psychological Methods, 7*, 356–369.

Stevens, J. P. (2009). *Applied multivariate statistics for the social sciences* (5th ed.). Hillsdale, NJ: Erlbaum.

Yao, Y. (1965). An approximate degrees of freedom solution to the multivariate Behrens-Fisher problem. *Biometrika, 52*, 139–147.

# 24
# Power Analysis

**Kevin R. Murphy**

One of the most common applications of statistics in the social and behavioral science is in testing null hypotheses. For example, a researcher wanting to compare the outcomes of two treatments will usually do so by testing the hypothesis that in the population there is no difference in the outcomes of the two treatments. If this null hypothesis can be rejected, the researcher is likely to conclude that there is a real difference in treatment outcomes. The power of a statistical test is defined as the likelihood that a researcher will be able to reject a specific null hypothesis when it is in fact false. One of the key determinants of power is the degree to which the null hypothesis is false; if treatments in fact have a very small effect, it may be difficult to reject the hypothesis that they have no effect whatsoever. However, effect size is not the only determinant of power. The power of statistical tests is a complex function of the sensitivity of the test, the nature of the treatment effect, and the decision rules used to define statistical significance.

There are several statistical models that have been used in defining and estimating the power of statistical effects. Kramer and Thiemann (1987) derived a general model for statistical power analysis based on the intraclass correlation coefficient, and developed methods for evaluating the power of a wide range of test statistics using a single general table based on the intraclass correlation. Lipsey (1990) used the $t$-test as a basis for estimating the statistical power of several statistical tests. Murphy and Myors (2003) developed a model based on the noncentral $F$ distribution and showed how it could be used with virtually all applications of the general linear model.

Cohen (1988), Lipsey (1990), and Kraemer and Thiemann (1987) provided excellent overviews of the methods, assumptions, and applications of power analysis. Murphy and Myors (2003) extended traditional methods of power analysis to tests of hypotheses about the size of treatment effects, not merely tests of whether or not such treatment effects exist. All of these sources describe the two main applications of power analysis, in designing studies (e.g., determining sample sizes, setting criteria for significance) and in evaluating research (e.g., understanding why particular studies rejected or failed to reject the null hypothesis)

Desiderata for studies that apply power analysis are described in Table 24.1, and are explained in the sections that follow.

Table 24.1 Desiderata for Power Analysis

| Desideratum | Manuscript Section(s)* |
|---|---|
| 1. The hypotheses being tested are defined, alternative hypotheses are laid out and analytic methods are chosen. | I, M |
| 2. Three factors that determine statistical power—effect size, sensitivity, and decision criteria—are examined. | I, M |
| 3. Statistical power is estimated, sample size requirements are determined, and decision criteria are evaluated. | M |
| 4. The results of power analysis are reported. | M, R |
| 5. Statistical power is considered in evaluating existing research and in planning future studies. | D |

* *Note*: I = Introduction, M = Methods, R = Results, D = Discussion

## 1. Defining the Null Hypothesis

The first step in power analysis is to define the specific hypothesis that is tested (the null hypothesis) and to define the alternative hypothesis that will be accepted if a statistical test leads the researcher to reject the null. The term *null hypothesis* is typically used to refer to a specific point hypothesis (e.g., that the population difference between experimental and control conditions is zero) that can be tested and potentially rejected on the basis of data collected in a sample. There is considerable debate in behavioral and social sciences about the value and relevance of null hypothesis tests (Cohen, 1994; Cortina & Dunlap, 1997; Hagen, 1997; Harlow, Mulaik, & Steiger, 1997; Morrison & Henkel, 1970); one of the principal objections to this type of test is that the null hypothesis that is most likely to be tested (e.g., that the population correlation between two variables is zero, or that the difference between two treatments is zero) is often one that is very unlikely to be true (Murphy, 1990).

Numerous alternatives to traditional null hypothesis testing have been suggested. Serlin and Lapsley (1985) laid out procedures for creating and testing hypotheses that the effects of treatments or interventions are sufficiently close to those predicted on the basis of substantive theory to justify the conclusion that the theory is supported. Rouanet (1996) described Bayesian models for hypothesis testing. Murphy and Myors (1999, 2003) described methods for forming and testing hypotheses that the effects of treatments or interventions exceed some minimum value. They also examined in detail the power of tests of the hypothesis that the effects of treatments are either trivially small or are large enough to be of substantive interest.

Statistical analysis should not normally be limited to tests of the traditional null hypothesis, in part because of the very low likelihood that this hypothesis is correct (Meehl, 1978; Murphy & Myors, 2003). At a minimum, studies that test traditional null hypotheses should also report information about the importance and accuracy of results (e.g., effect size estimates, confidence intervals). Alternatives to traditional null hypothesis tests should be carefully considered and used where applicable.

## 2. Factors That Affect Power

The power of a statistical test is a function of its sensitivity, the size of the effect in the population, and the standards or criteria used to test statistical hypotheses. Tests have higher levels of statistical power when studies are highly sensitive, when effect sizes (*ES*) are large, and/or when the criteria used to define statistical significance are relatively lenient. Studies should, if possible, be designed so that they achieve power levels of .80 or greater (i.e., so that they have at least an 80% chance of rejecting a false null hypothesis; Cohen, 1988; Murphy & Myors, 1993). When power is less than .50, it is not always clear whether statistical tests should be conducted at all, because of the substantial probability that they will lead to Type II errors (Murphy & Myors, 2003).

Sensitivity refers to the ability of a study to consistently detect relatively small deviations from the null hypothesis. Researchers can increase sensitivity by using better measures, or using study designs that allow them to control for unwanted sources of variability in their data. The simplest and most common method of increasing the sensitivity of a study is to increase its sample size ($n$). Large samples should be used wherever possible; as $n$ increases, statistical estimates become more precise and the power of statistical tests increases.

The formula for the standard error of the mean helps to illustrate how and why sample size affects sensitivity. The standard of the mean ($SE_{\bar{X}}$) is given by:

$$SE_{\bar{X}} = \frac{\sigma}{\sqrt{n}} \tag{1}$$

where: $\sigma$ is population standard deviation of scores on this test. As this formula shows, as sample size ($n$) increases, the size of the standard error decreases, indicating smaller and smaller differences between population values and sample values (i.e., more precision). This formula also illustrates one of the most challenging barriers to achieving high levels of precision and sensitivity in a study, the curvilinear relationship between sample size and precision. For example, doubling the sample size does not usually double the precision of sample estimates. Because the size of the standard error is usually a function of the square root of $n$, one implication is that in order to double the precision of estimates, one must often increase the $n$ by a factor of four. The nature of the relationship between $n$ and precision reinforces the recommendation that the largest possible samples be used.

The second factor that influences power is the size of the effect of treatments or interventions. If treatments have large effects, it is relatively easy to reject the null hypothesis that the effect of treatments is zero, whereas small treatment effects might be difficult to detect reliably. Effect sizes are often measured in terms of statistics such as the standardized mean difference ($d$, the difference between treatment means divided by the pooled standard deviation) or the percentage of variance explained by treatments, interventions, and so forth (see Chapter 7, this volume). Lipsey and Wilson (1993) reviewed effect sizes typically reported in the social and behavioral sciences. Effect size estimates may be obtained from meta-analyses of research literature (see Chapter 19, this volume), or may be derived from a substantive theory about the phenomenon being studied, but most power analyses depend on conventions of particular research communities to define the size of the effect that is expected or that is used in designing studies.

On the basis of surveys of research literature, Cohen (1988) suggested a number of conventions for describing treatment effects as "small," "medium," or "large." Table 24.2 presents conventional val-

Table 24.2 Some Conventions for Defining Effect Sizes

| | PV | $r$ | $d$ | $f^2$ | Probability of a higher score in treatment group |
|---|---|---|---|---|---|
| Small effects | .01 | .10 | .20 | .02 | .56 |
| Medium effects | .10 | .30 | .50 | .15 | .64 |
| Large effects | .25 | .50 | .80 | .35 | .71 |

From: Cohen (1988), Grissom (1994)

*Note:* Cohen's $f = R^2/(1-R^2) = \eta^2/(1-\eta^2) = PV/(1-PV)$, where $\eta^2 = SS_{treatments}/SS_{total}$

ues for describing large, medium, and small effects, expressing these effects in terms of a number of widely used statistics.

For example, a small effect might be described as one that accounts for about 1% of the variance in outcomes, or one where the treatment mean is about one fifth of a standard deviation higher in the treatment group than in the control group, or as one where the probability that a randomly selected member of the treatment group will have a higher score than a randomly selected member of the control group is about .56. In the absence of an acceptable estimate of the effect size expected in a particular study, it is common practice to assume that the effect will be small (e.g., $d = .20$, 1% of the variance accounted for by treatments) and to plan studies accordingly. Unless you have a good reason to believe that the effects of treatments are moderately large or larger, it is always best to design studies so that they have sufficient power for detecting small effects. Studies designed to detect small effects will also have sufficient power for detecting larger effects.

Very large samples are often needed to yield adequate power for detecting small effects. For example, if the expected difference between control and treatment means is about one fifth of a standard deviation (i.e., $d = .20$), a sample of $n = 777$ will be needed to achieve power of .80 for rejecting the null hypothesis. In a study where subjects are randomly assigned to one of five treatments and where treatment differences are expected to account for 1% of the variability in outcomes, a sample of $n = 1170$ will be needed to achieve this same level of power. The smaller the expected effect of treatments or interventions, the more important it is to consider power in the design of studies.

Third, the power of statistical tests is affected by the standards or criteria used to define statistical significance, usually defined in terms of the alpha ($\alpha$) level. Alpha is defined as the conditional probability that a statistical procedure will reject a null hypothesis, given that this null hypothesis is in fact true; conventional approaches to null hypothesis testing define statistical significance criteria in such a way that the maximum value of $\alpha$ will be small. In the behavioral and social sciences, differences in treatment outcomes are usually regarded as statistically significant if the results obtained in a sample are outside of the range of results that would have been obtained in 95% of all samples drawn from a population in which the null hypothesis is true (i.e., $\alpha = .05$ is the most common threshold for "statistical significance" in the behavioral and social sciences). Some researchers use more stringent criteria when defining statistical significance, for example demanding that the alpha level should be set at .01 or smaller before sample results are declared statistically significant. The use of stringent criteria (e.g., $\alpha = .01$ or lower) for defining statistical significance is not recommended. Unless there are good reasons to believe that the null hypothesis might be true (because the null hypothesis is a point hypothesis, this is rarely the case; Cohen, 1994; Meehl, 1978), use of stringent criteria for defining statistical significance will normally lead to reductions in power without providing substantial benefits.

## 3. Power, Sample Size, and Criteria for Significance

The power of a null hypothesis test is a function of $n$, ES, and $\alpha$, and the equations that define this relation can be easily rearranged to solve for any of four quantities (i.e., power, $n$, ES, and $\alpha$), given the other three. The two most common applications of statistical power analysis are in: (1) determining the power of a study, given $n$, ES, and $\alpha$, and (2) determining how many observations will be needed (i.e., $n$ required), given a desired level of power, an ES estimate, and an $\alpha$ value. Both of these analyses are extremely useful in planning research, and are usually so easy to do that they should be a routine part of designing a study.

Both of these applications of power analysis assume that a decision has been made about the significance criterion to be used (e.g., $\alpha = .05$) and that there is some basis for estimating the size of treatment effects, or the degree to which the null hypothesis is likely to be wrong. It is best to be conservative in estimating effect size; as noted earlier, if you have no credible a priori basis for making

Table 24.3  Two Ways of Displaying the Outcomes of a Power Analysis

| Power Levels | | | | n Required for Power of .80 | |
| --- | --- | --- | --- | --- | --- |
| n | PV = .02 | PV = .05 | PV = .10 | PV | n |
| 100 | .27 | .61 | .92 | .02 | 387 |
| 200 | .50 | .91 | .999 | .05 | 153 |
| 500 | .90 | .999 | .999 | .10 | 75 |

this estimate, it is best to base power analyses on the assumption that treatment effects will be small. In some studies, observed treatment effects are used to make an estimate of population treatment effects, but such post hoc power analyses are usually discouraged (Hoening & Heisey, 1971), in part because they tend to provide overestimates of power.

Once ES has been estimated and $\alpha$ has been chosen, it is easy to either determine the power a particular study will provide or to determine the sample sizes needed to reach specific levels of statistical power. The equations that define the power of various statistical tests are not complex, but the calculation of power is somewhat tedious, and it is common practice to use power tables or power analysis software to perform the necessary calculations. Given a particular sample size ($n$) and alpha level, the power of a statistical test is a nonlinear monotonic function of the effect size (ES) that asymptotes at or near 1.0. That is, the likelihood that a study will reject any particular null hypothesis approaches unity as the gap increases between the null hypothesis (e.g., that treatments have no effect) and the reality that they might have large effects. Cohen (1988) provided among the most complete sets of power tables readily available, based on calculations of power for a variety of different statistical tests, whereas Murphy and Myors (2003) provided a smaller set of tables that can readily be adapted to most of the statistical tests discussed by Cohen.

An example may clarify the two main applications of power analysis. Suppose a researcher is comparing two different methods of instruction. Table 24.3 displays the results of a power analysis in two different ways, first showing the level of power for statistical comparisons ($\alpha$ = .05) of two groups given various values of $n$ and ES (expressed in terms of the percentage of variance in the dependent variable explained by treatments), then showing the values of $n$ that would be needed to yield power of .80 given different ES values.

If the percentage of variance in outcomes explained by treatments is relatively small (e.g., PV = .02), a relatively large sample ($n$ = 387) will be needed to attain power of .80. This power analysis shows that when $n$ = 200 and PV = .02, the probability that the null hypothesis will be (correctly) rejected is .50, suggesting that with a sample of 200 cases, null hypothesis tests will essentially be a coin flip. On the other hand, when the effect of treatments is large (e.g., PV = .10), samples with $n$ = 75 will have an 80% chance of correctly rejecting the null hypothesis.

Table 24.3 can also be used to estimate the types of effects that could be detected with a fixed level of power, given $n$ and $\alpha$. For example, assume that 100 subjects are available for a study and that the researcher desires a power of .80 or greater for statistical tests that employ a .05 $\alpha$ level. Table 24.3 makes it clear that this level of power will only be achieved if the ES is a bit greater than PV = .05, but less than PV = .10 (a PV of approximately .066 is required to achieve power of .80).

Finally, power analyses may be used to aid in making rational decisions about the criteria used to define statistical significance (Cascio & Zedeck, 1983; Nagel & Neff, 1977). For example, suppose researchers are comparing two treatments with 200 subjects assigned to each treatment. The researchers expect a relatively small treatment effect of, say, PV = .02. Using $\alpha$ = .05, power would be 0.64. If $\alpha$ = .01 is used, power drops to 0.37 (Cohen, 1988). The trade-off between Type I error protection and power suggests that a researcher must balance risk and consequences of a Type I error with risk and consequences of a Type II error.

In general, the choice of a more stringent $\alpha$ level (e.g., choosing .01 rather than .05) will lead to reductions in power. This choice might make sense if researchers are more concerned with the possibility of falsely rejecting the null hypothesis (Type I error) than with the possibility of failing to reject the null hypothesis when it is indeed false (Type II error). Cascio and Zedeck (1983) presented equations for estimating the relative weight given to Type I vs. Type II errors in various research designs, which can help researchers evaluate these tradeoffs. They showed that the apparent relative seriousness (ARS) of these errors implied by a study design can be estimated using:

$$ARS = \frac{p(H_1)(1-\text{power})}{(1-p(H_1))\alpha} \tag{2}$$

where: $p(H_1)$ = probability that $H_0$ is false.

For example, if the researcher believes that the probability that treatments have *some* effect is .7, and $\alpha = .05$ and the power is .80, the choice of the $\alpha = .05$ significance criterion implies that a mistaken rejection of the null hypothesis (i.e., a Type I error) is 9.33 times as serious [i.e., $(.7^*.2)/(.3^*.05) = 9.33$] as the failure to reject the null when it is wrong (i.e., a Type II error). In contrast, setting $\alpha = .10$ leads to a ratio of 4.66 [i.e., $(.7^*2)/(.3^*.10) = 4.66$], or to the conclusion that Type I errors are treated as if they are 4.66 times as serious as a Type II error (see also Lipsey, 1990).

The first advantage of Equation (2) is that it makes explicit the values and preferences that are usually not well understood, either by researchers or by the consumers of social science research. In the scenario described above, choice of an $\alpha$ level of .05 makes sense only if the researcher thinks that Type I errors *are* over nine times as serious as Type II errors. If the researcher believes that Type I errors are only four or five times as serious as Type II errors, he or she should set the significance level at .10, not at .05.

## 4. Reporting Results

There is no standard format for reporting the outcomes of a power analysis, but it is relatively simple, on the basis of the known determinants of power, to determine what information should be reported. Because power is a function of three variables (i.e., $n$, ES, and $\alpha$), it is best to include information about all three in discussions of statistical power. For example, one might report,

> On the basis of our review of the literature, we expected that the difference between two treatments would correspond to a medium-sized effect (i.e., $d = .50$). Our study included 120 subjects who were randomly assigned to treatment and control conditions. We used a two-tailed test ($\alpha = .05$) to compare group means. The power of this test for detecting medium effects, under assumed conditions (normality, independence of observations, homogeneity of variance), is .77.

It is also important to report the method or the statistical software used to estimate power. Power analyses are included as part of several statistical analysis packages (e.g., SPSS provides Sample Power, a flexible and powerful program) and it is possible to use numerous websites to perform simple power analyses. Some power analysis textbooks (e.g., Murphy & Myors, 2003) include software for performing these analyses. Three notable software packages designed for power analysis are:

- *G\*Power* (Faul, Erdfelder, Lang, & Buchner, 2007; http://www.psycho. uni-duesseldorf.de/abteilungen/aap/gpower3/) is distributed as a freeware program that is available for both Macintosh and Windows environments. It is simple, fast, and flexible.

- *Power and Precision*, distributed by Biostat, was developed by leading researchers in the field (e.g., J. Cohen). This program is very flexible, covers a large array of statistical tests, and provides power analyses and confidence intervals for most tests.
- *PASS*, distributed by NCSS, this program covers a wide of range of tests and provides particularly useful graphical output.

## 5. Evaluating Existing Research and Designing Future Studies

Power analyses are extremely valuable for understanding the outcomes of significance tests in the published literature. For example, suppose a researcher reports a statistically significant correlation between pre-employment drug tests and subsequent job performance. If the study uses very large samples, a statistically significant finding might not be very meaningful. If $n = 5000$, the power for detecting a correlation as low as .04 exceeds .80. Similarly, a researcher who reports that there is no statistically significant correlation between mothers' health and the health of children, based on a sample of $n = 30$, might lead readers seriously astray. If $n = 30$, power is less than .80 for detecting correlations as large as .45, and it is possible that a sample this small could miss a substantial correlation between these two variables. Both of the examples illustrate a key point about null hypothesis testing. Describing the correlation between two variables as "statistically significant" does not necessarily mean that it is large or important, and describing this same correlation as "statistically nonsignificant" does not necessarily mean that it is small and unimportant. It is recommended that whenever interpreting the results of statistical significance tests, power should be considered. When tests or the null hypothesis are carried out with either very high levels of power or very low levels of power, the outcomes of these tests are virtually a foregone conclusion, and the power of these tests should be routinely considered when evaluating their results.

Power analyses can be useful in understanding the likelihood that the results in a particular study will replicate well in future studies. For example if a power analysis indicates that power is quite low (e.g., .60), it is still possible that a study will reject the null hypothesis. However, if the estimates of effect size are reasonably accurate, one should not expect that replications of that study will consistently reject the null hypothesis. On the contrary, if effect size estimates are accurate and there are ten replications of a study, the best guess is that only six of these will reject the null hypothesis. The alpha level is often misinterpreted as an indication of the likelihood that test results will replicate. Assessments of power are recommended as a much more accurate barometer of whether future tests conducted in similarly designed studies are likely to lead to the same outcomes.

Power analyses can be quite useful in settings where one wants to argue that the null hypothesis is at least close to being correct. For example, if one wanted to argue that a new type of training has no real effects, one way to make this argument is to design a study that has a high level of power. If a powerful study is conducted and it still fails to reject the null hypothesis, one might conclude that this null is at least reasonably close as an estimate of the true state of affairs. Whenever researchers want to use the failure to reject the null hypothesis as evidence that this hypothesis is at least approximately true, they should first demonstrate that their studies have sufficient power to reject the null when it is meaningfully wrong.

Finally, power analysis should be carefully considered when designing future studies, particularly when making choices about sample size. There are times when practical constraints make it impossible to obtain the large samples needed to reliably detect small but potentially important effects. If there are constraints on the maximum sample size that can be attained, power analysis can be used to determine the type of effect that can be detected, given a fixed level of power, or the level of power that can be attained given a fixed effect size. For example, suppose a research team is interested in comparing the effects of two drugs and uses a two-tailed $t$ test (e.g., $\alpha = .05$) to determine whether there are differences in the drug effects. They expect a relatively small difference in the effectiveness of the drugs (e.g., $d = .15$), and a power analysis shows that a sample of at least $n = 1398$ will be needed to reach power of .80 for tests of the null hypothesis that the drugs have identical effects. They can afford to sample only 800 subjects. With a sample this large, they will have power of .80 or above for detecting somewhat larger effects ($d = .20$), and will have to make a decision about whether it is realistic to expect effects this large. If they are confident that the effect will be approximately $d = .15$ and they are limited to a sample of 800 participants, power for testing the null hypothesis will be only .56, suggesting that they almost as likely to conclude that there is no detectable difference between the drugs than they are to conclude that there is a small, but potentially important difference in the effects of the drugs.

In sum, power analysis is extremely useful as a tool for planning and evaluating research studies. Studies in the behavioral and social sciences are often conducted with low levels of power (Cohen, 1988; Murphy & Myors, 2003). Researchers who pay careful attention to power analysis are less likely to make Type II errors or to misinterpret the outcomes of null hypothesis tests.

## References

Cascio, W. F., & Zedeck, S. (1983). Open a new window in rational research planning: Adjust alpha to maximize statistical power. *Personnel Psychology, 36,* 517–526.
Cohen. J. (1988). *Statistical power analysis for the behavioral sciences* (2nd ed.). Hillsdale, NJ: Erlbaum.
Cohen. J. (1994). The earth is round ($p < .05$). *American Psychologist, 49,* 997–1003.
Cortina, J. M., & Dunlap, W. P. (1997). On the logic and purpose of significance testing. *Psychological Methods, 2,* 161–173.
Faul, F., Erdfelder, E., Lang, A.-G., & Buchner, A. (2007). G*Power 3: A flexible statistical power analysis program for the social, behavioral, and biomedical sciences. *Behavior Research Methods, 39,* 175–191.
Hagen, R. L. (1997). In praise of the null hypothesis statistical test. *American Psychologist, 52,* 15–24.
Harlow, L. L., Mulaik, S. A., & Steiger, J. H. (Eds.) (1997). *What if there were no significance tests?* Mahwah, NJ: Erlbaum.
Hoening, J. M., & Heisey, D. M. (1971) The abuse of power: The pervasive fallacy of power calculations for data analysis. *The American Statistician, 55,* 19–24.
Kraemer, H. C., & Thiemann, S. (1987). *How many subjects?* Newbury Park. CA: Sage.
Lipsey, M. W. (1990). *Design sensitivity.* Newbury Park, CA: Sage.
Lipsey, M. W., & Wilson, D. B. (1993). The efficacy of psychological, educational, and behavioral treatment. *American Psychologist, 48,* 1181–1209.
Meehl, P. (1978). Theoretical risks and tabular asterisks: Sir Karl, Sir Ronald, and the slow progress of psychology. *Journal of Consulting and Clinical Psychology, 46,* 806–834.
Morrison, D. E., & Henkel, R. E. (1970). *The significance test controversy: A reader.* Chicago: Aldine.
Murphy, K. R. (1990). If the null hypothesis is impossible, why test it? *American Psychologist, 45,* 403–404.
Murphy, K. R., & Myors. B. (1999). Testing the hypothesis that treatments have negligible effects: Minimum-effect tests in the general linear model. *Journal of Applied Psychology, 84,* 234–248.
Murphy, K., & Myors, B. (2003). *Statistical power analysis: A simple and general model for traditional and modern hypothesis tests* (2nd Ed). Mahwah, NJ: Erlbaum.
Nagel, S. S., & Neff, M. (1977). Determining an optimal level of statistical significance. *Evaluation Studies Review Annual, 2,* 146–158.
Rouanet, H. (1996). Bayesian methods for assessing the importance of effects. *Psychological Bulletin, 119,* 149–158.
Serlin, R. A., & Lapsley, D. K. (1985). Rationality in psychological research: The good-enough principle. *American Psychologist, 40,* 73–83.

# 25
# Reliability and Validity of Instruments

Thomas R. Knapp and Ralph O. Mueller

Both reliability and validity are essential parts of the psychometric properties of a measuring instrument.[1] The *reliability* of an instrument is concerned with the consistency of measurements: from time to time, from form to form, from item to item, or from one rater to another. On the other hand, the *validity* of an instrument is usually defined as the extent to which the instrument actually measures "what it is designed to measure" or "what it purports to measure." Validity is therefore concerned with the relevance of an instrument for addressing a study's purpose(s) and research question(s). Both reliability and validity are context-specific characteristics: for example, researchers are often interested in assessing if a measure remains reliable and valid for a specific culture, situation, or circumstance (e.g., a psychological test might be highly reliable and valid in a population of Caucasian adults but not in one of African American children). The conceptualization and specific definitions of reliability and validity have changed over time, as reflected in the various editions of *Educational Measurement* (Cronbach, 1971; Cureton, 1951; Feldt & Brennan, 1989; Haertel, 2006; Kane, 2006; Messick, 1989; Stanley, 1971; Thorndike, 1951). Table 25.1 contains a list of desiderata regarding reliability and validity of measuring instruments that should be followed in any empirical research report.

## 1. Instrument Description and Justification

Empirical data for analysis during research studies are collected with the aid of measuring instruments, be they laboratory equipment or, more common in the social and behavioral sciences, surveys, achievement batteries, or psychological tests. Because study results should only be trusted when investigators collect good data, authors should ensure that readers can judge the "goodness" of the data for themselves. Thus, at a minimum, a full description of the instrument(s) used is necessary (and should be followed by an assessment of reliability and validity; see Desiderata 2 and 3, respectively), including the purpose(s) and intended use(s), item format(s), and scales of measurement (i.e., nominal, ordinal, interval, or ratio). Obviously, a specific instrument is appropriate for use in some contexts, but not necessarily in others (e.g., a high school reading test is likely to be inappropriate to measure a middle schooler's intelligence). Authors must take care in justifying the choice of instrument(s) and making explicit the link to the study's purpose(s) and research question(s). Often, the description and justification for a particular instrument is presented in a manuscript's Instrumentation subsection of the Methods section but could also be accomplished in the Introduction.

Table 25.1 Desiderata for Reliability and Validity of Instruments

| Desideratum | Manuscript Section(s)* |
|---|---|
| 1. Each instrument used in the study is described in sufficient detail. The appropriateness of the instrument to address the study's purpose(s) and research question(s) is made explicit. | I, M |
| 2. Appropriate reliability indices are considered: The study's purpose(s) guide the choice of indices calculated from current data and/or examined from previous research. | M, R |
| 3. Suitable validity evidence is gathered: The study's purpose(s) determine the type of validity support gathered from current data and/or consulted from related literature. | M, R |
| 4. Applicable reliability and validity evidence is reported and interpreted. The study's conclusions are placed within the context of such evidence (or lack thereof). | D |

\* *Note*: I = Introduction, M = Methods, R = Results, D = Discussion

## 2. Reliability Indices

Several approaches exist to assess an instrument's reliability, whose appropriateness is dependent on a study's specific purpose. Four traditional strategies often found in the literature are briefly discussed below: test-retest, parallel forms, internal consistency, and rater-to-rater. All four are based on classical test theory, but alternative conceptualizations of reliability exist that are based on other analytical frameworks: generalizability theory (see Chapter 9, this volume), item response theory (see Chapter 12, this volume), and structural equation modeling (see Chapter 28, this volume).

*Test-Retest Reliability.* If a study's purpose is to assess measurement consistency of one instrument from one time point to another, a straightforward way to collect reliability evidence is to measure and then re-measure individuals and determine how closely the two sets of measurements are related (i.e., the *test-retest* method). In studies assessing psychological constructs such as attitude, a question with often serious ramifications is how much time should be allowed between the first and second testing. If the interval is too short, measurement consistency might only be due to the fact that individuals being tested "parrot back" the same responses at Time 2 that they gave at Time 1. If the interval is too long, some items might no longer be developmentally appropriate (e.g., academic achievement items on a middle school test administered to students entering high school) which could impact the validity of the instrument as well. Even in studies with physical variables, the length of time between measurements might be crucial for some (e.g., repeated weight measurements during a health-awareness program might fluctuate widely, depending on weight loss/gain), but not for other variables (e.g., repeated height measurements of adult participants are likely to remain consistent, irrespective of the time-lag between measurements). In general, authors should defend their choice of time intervals between measurements as the "correct" amount of time is situation specific and somewhat subjective. An assessment of test-retest reliability can be accomplished in either an absolute manner (e.g., the median difference between corresponding measurements) or relative manner (e.g., the correlation between the two sets of measurements) with the latter approach being more common than the former.

*Parallel Forms Reliability.* If a measuring instrument is available in two *parallel* (i.e., psychometrically equivalent and interchangeable) *forms*, say Form A and Form B, with measurements having been taken on both forms, reliability evidence can be obtained by comparing the scores on Form A with the scores on Form B, again either absolutely or relatively. The time between the administrations of the two forms is still an important consideration, but because the forms are not identical there is no longer the concern for "parroting back" if the time interval is short.

*Internal Consistency Reliability.* Given the disadvantage of multiple test administrations for test-retest and parallel forms reliability (e.g., increased costs, time lag between measurements, and missing data due to non-participation during the second testing), a commonly used alternative is the estimation of the *internal consistency* of an instrument. Here, an instrument consisting of multiple items measuring the same construct is administered only once, but now treating the items as forming two parallel halves of the instrument. The two half-forms are created after the actual measurement, traditionally by considering the odd-numbered items as one form and the even-numbered items as the other form (though other ways to split an instrument are certainly possible, e.g., random assignments of items to halves). The scores on the two forms are then compared, usually relatively by computing the correlation between the scores on the odd numbered items with the scores on the even numbered items. But the correlation must be "stepped up" by using the Spearman-Brown formula (Brown, 1910; Spearman, 1910), in order to estimate what the correlation might be between two full-forms as opposed to two half-forms. That estimate is obtained by multiplying the correlation coefficient by two and then dividing that product by one plus the correlation. The type of reliability evidence thus produced is strictly concerned with internal consistency (from half-form to half-form) since time has not passed between obtaining the first set of measurements and obtaining the second set of measurements.

Another type of internal consistency reliability is from item to item within a form. Such an approach was first advocated by Kuder and Richardson (1937) for dichotomously scored test items, and was subsequently extended to the more general interval measurement case by Cronbach (1951). Their formulae involve only the number of items, the mean and variance of each, and the covariances between all of the possible pairs of items. Cronbach called his reliability coefficient *alpha*. It is still known by that name and is by far the most commonly employed indicator of the reliability of a measuring instrument in the social sciences.[2]

*Rater-to-Rater Reliability.* When the data for a study take the form of ratings from scales, the type of reliability evidence that must be obtained before such a study is undertaken is an indication of the extent to which a rater agrees with him(her) self (*intra-rater reliability*) and/or the extent to which one rater agrees with another (*inter-rater reliability*). Several options exist to assess rater-to-rater consistency, with the *intraclass coefficient* and Cohen's *kappa* (1960) being among the most popular (see Chapter 11, this volume).

*Norm- vs. Criterion-Referenced Settings.* Literature devoted to reliability assessment within norm-referenced versus criterion-referenced frameworks is plentiful. Most users of *norm-referenced* tests—where scores are primarily interpreted in relation to those from an appropriate norm or comparison group—have adopted approaches to reliability assessment similar to those summarized thus far, with particular emphasis on correlations that are indicative of relative agreement between variables. *Criterion-referenced* (or *domain-referenced*) *measurement* is concerned with what proportion of a domain of items has been answered successfully and whether or not that portion constitutes a "passing" performance (e.g., "John spelled 82% of the words on a spelling test correctly, which was below the cut point for progressing to the next lesson."). Here, reliability assessment concentrates on measurement errors in the vicinity of the cut point with a particular interest on the reliability of the pass-or-fail *decision*. In parallel-form situations, for example, the matter of whether a person passes both Form A and Form B or fails both Form A and Form B takes precedence over how high the correlation is between the two forms.

## 3. Validity Evidence

In physical science research the usual evidence for the validity of measuring instruments is expert judgment and/or validity-by-definition with respect to a manufacturer's specifications. For example, "For the purpose of this study, body temperature is the number of degrees Fahrenheit that the Smith thermometer reads when inserted in the mouths of the persons on whom the measurements are being taken." As evidence of validity, the researcher might go on to explain that the Smith thermometer is regarded as the 'gold standard' of temperature measurement.

In the social and behavioral sciences, investigators are often urged to provide evidence for *content validity* (expert judgments of the representativeness of items with respect to the skills, knowledge, etc. domain to be covered), *criterion-related validity* (degree of agreement with a "gold standard"), and/or *construct validity* (degree of agreement with theoretical expectations) of the measuring instruments used in their substantive studies. More recently, all three validity types have been subsumed under an expanded concept of construct validity, but not without controversy. Whatever conceptualization is used, researchers must be clear that instrument validity is *not* context free: a measure might be valid in one situation or for one population, but not in or for another (e.g., the Scholastic Aptitude Test, SAT, is often argued as being valid to assess high school seniors' potential for success in undergraduate higher education, but not for measuring their intelligence or potential to succeed in vocational training).

*Criterion-related Validity.* When a measure is designed to relate to an external criterion, its validity is judged by either *concurrent* or *predictive* assessments (i.e., degrees to which test scores estimate a specified present or future performance). For example, a passing score on a driver's permit test with acceptable concurrent validity will allow the test taker to immediately drive motor vehicles, assuming an associated road test has been passed. On the other hand, evidence of predictive criterion-related validity is often helpful for judging instruments that are designed to measure aptitude, with passing achievement scores serving as the standards for whether or not the aptitude tests are predictive of achievement. But, herein also lies an interesting dilemma: How does one know that the achievement tests themselves are valid? Do the standards need to be validated against an even higher standard? Or, if the standards' validity is established by expert judgment, why not appeal to experts directly for validity assessments of the aptitude measure? Furthermore, if expert judgment is to be the ultimate arbiter, who are the experts and who selects them?

*Construct Validity.* In order to judge the degree to which a theoretical construct accounts for test performance, a researcher must assess the test's construct validity. Supportive evidence usually comes from exploratory or confirmatory factor analyses (see Chapter 8, this volume) in which the dimensionality and the degree of correlation of the variables comprising the instruments are investigated. The most popular approach is the *convergent/discriminant* strategy first recommended by Campbell and Fiske (1959): researchers determine the extent to which measurements obtained with the instruments in question correlate with variables with which they are theoretically expected to correlate (convergent) and the extent to which those measurements correlate with other variables with which they are theoretically *not* expected to correlate (discriminant). See also Chapter 22, this volume, on multitrait-multimethod analysis.

## 4. Reporting and Interpreting Reliability and Validity Results

Before reporting a study's main findings, investigators should discuss evidence of the reliability and validity of the instrument(s) used. Ideally, such evidence should come from both a thorough search of the related literature and an assessment based on current study participants. A comparison of present reliability and validity information with that gleaned from related literature is helpful to readers, especially when such information might be contradictory, as, for example, when earlier reliability/validity evidence could not be reproduced based on the current sample's data.

Certainly when no previous reliability and validity information is available—as is the case when investigators construct their own instruments— authors must report psychometric properties of the instrument(s) based on an analysis of the current data. But even if reliability/validity evidence is identified from previous studies, it is often the case that it does not generalize to the current population under study. Thus, it is incumbent upon each investigator to provide a thorough justification for why the instruments used are appropriate for the current sample of participants.

In the social and behavioral sciences, reliability and validity coefficients in the .70 or .80 or above range are often considered acceptable with values below these cut-offs being acknowledged as study limitations. However, the acceptability of coefficients should be judged with caution as value adequacy certainly depends on the particular phenomenon under study. Nevertheless, the interpretation of main results should commence from within the context of reliability and validity results as unreliability and/or invalidity usually attenuate the magnitudes of expected findings and lead to wider confidence intervals and less likelihood of the detection of effects and relationships in the data.

## Notes

1 This chapter does not deal with the internal and external validity and reliability of a chosen research design (but see Chapter 26, this volume). Also, in fields outside the social/behavioral sciences, validity and reliability are sometimes known by different names. For example, in epidemiology, *reproducibility* is generally preferred over reliability. In engineering and related disciplines, equipment is said to be reliable if it does not, or is very unlikely to, break down. Also, the ambiguous term *accuracy* is sometimes used in lieu of either reliability or validity.
2 Kuder and Richardson actually derived several formulae for internal consistency by making successively relaxed assumptions, and they numbered them accordingly. The formula that is most frequently used to compute Cronbach's alpha is actually a direct extension of Kuder and Richardson's Formula #20 for dichotomous data.

## References

Brown, W. (1910). Some experimental results in the correlation of mental abilities. *British Journal of Psychology, 3*, 296–322.

Campbell, D. T., & Fiske, D. W. (1959). Convergent and discriminant validation by the multi-trait multi-method matrix. *Psychological Bulletin, 56*, 81–105.

Cohen, J. (1960). A coefficient of agreement for nominal scales. *Educational and Psychological Measurement, 20*, 37–46.

Cronbach, L. J. (1951). Coefficient alpha and the internal structure of tests. *Psychometrika, 16*, 297–334.

Cronbach, L. J. (1971). Test validation. In R. L. Thorndike (Ed.), *Educational measurement* (2nd ed.) (pp. 443–507). Washington, DC: American Council on Education.

Cureton, E. F. (1951). Validity. In E. F. Lindquist (Ed.), *Educational measurement* (pp. 621–694). Washington, DC: American Council on Education.

Feldt, L. S., & Brennan, R. L. (1989). Reliability. In R. L. Linn (Ed.), *Educational measurement* (3rd ed.) (pp. 105–146). New York: Macmillan.

Haertel, E. H. (2006). Reliability. In R. L. Brennan (Ed.), *Educational measurement* (4th ed.) (pp. 65–110). Westport, CT: American Council on Education/Praeger.

Kane, M. L. (2006). Validation. In R. L. Brennan (Ed.), *Educational measurement* (4th ed.) (pp. 17–64). Westport, CT: American Council on Education/Praeger.

Kuder, G. F., & Richardson, M. W. (1937). The theory of the estimation of test reliability. *Psychometrika, 2*, 151–160.

Messick, S. (1989). Validity. In R. L. Linn (Ed.), *Educational measurement* (3rd ed.) (pp. 13–103). Washington, DC: American Council on Education.

Spearman, C. (1910). Correlation calculated from faulty data. *British Journal of Psychology, 3*, 171–195.

Stanley, J. C. (1971). Reliability. In R. L. Thorndike (Ed.), *Educational measurement* (2nd ed.) (pp. 356–442). Washington, DC: American Council on Education.

Thorndike, R. L. (1951). Reliability. In E. F. Lindquist (Ed.), *Educational measurement* (pp. 560–620). Washington, DC: American Council on Education.

# 26
# Research Design

**Sharon Anderson Dannels**

The definition of research design is deceptively simple: it is a plan that provides the underlying structure to integrate all elements of a quantitative study so that the results are credible, free from bias, and maximally generalizable. "Research design provides the glue that holds the research project together" (Trochim, 2006, Design, ¶ 1). The research design determines how the participants are selected, how variables are manipulated, data are collected and analyzed, and how extraneous variability is controlled so that the overall research problem can be addressed. Regardless of the sophistication of the statistical analysis, the researcher's conclusions might be worthless if an inappropriate research design has been used. Thus, design decisions both constrain and support the ultimate conclusions (Miles & Huberman, 1994).

Research designs may be identified as a specific design (e.g., a pretest-posttest control group design or a nonequivalent control group design) or by the broader category of experimental or nonexperimental. *Experimental* designs are used in experiments to investigate cause and effect relationships. In contrast, *nonexperimental* designs are used in more naturalistic studies or in situations where the primary purpose is to describe the current status of the variables of interest. The latter designs are distinguished by the absence of manipulation by the researcher, with an emphasis on observation and measurement.

The adequacy of the research design to produce credible results, most notably to make causal inferences, is evaluated in terms of two primary types of validity: internal and external (Campbell & Stanley, 1963). *Internal validity* refers to the confidence that the specified causal agent is responsible for the observed effect on the dependent variable(s). *External validity* is the extent to which the causal conclusions can be generalized to different measures, populations, environments, and times. In addition, *statistical conclusion validity* is considered with internal validity and refers to the appropriate use of statistics. *Construct validity*, the ability to generalize the research operations to hypothetical constructs is a companion to external validity (Cook & Campbell, 1979).

Campbell and Stanley (1963) and Cook and Campbell (1979) produced the seminal works defining *quasi-experimental design* (see Desideratum 3) from which much of the literature on research design is extrapolated. Shadish, Cook, and Campbell (2002) revisited the initial works, providing greater attention to external validity, randomized designs, and specific design elements rather than prescribed research designs. Also, a contemporary white paper prepared for the American Educational Research Association (AERA) by Schneider, Carnoy, Kilpatrick, Schmidt, and Shavelson (2007) specifically addressed the issue of causal inferences using large-scale datasets, experimental, and nonexperimental

Table 26.1 Desiderata for Research Design

| Desideratum | Manuscript Section(s)* |
|---|---|
| 1. The research design is foreshadowed and follows logically from the general problem statement and associated research questions. | I |
| 2. The research problem is clearly articulated and researchable. | I |
| 3. The research design is appropriate to address the research problem and is clearly articulated. | M |
| 4. Variables are identified and operationalized; sampling, instrumentation, procedures, and data analysis are detailed. | M, R |
| 5. The research design is internally consistent (e.g., the data analysis is consistent with the sampling procedures). | M, R, D |
| 6. The design is faithfully executed or, if applicable, explanations of necessary deviations are provided. | I, M, R |
| 7. Extraneous variability is considered and appropriately controlled. | M, R |
| 8. Potential rival hypotheses are minimized. Threats to internal validity and statistical conclusion validity, and the adequacy of the counterfactual, are considered. | M, D |
| 9. Conclusions as to what occurred within the research condition are appropriate to the design. | M, D |
| 10. Generalizations, if any, are appropriate. External validity and construct validity are considered elements of the design. | M, D |
| 11. The limitations of the design are articulated and appropriately addressed. | D |

* *Note*: I = Introduction, M = Methods, R = Results, D = Discussion

designs. Texts by Keppel, Saufley, and Tokunaga (1998), Fraenkel and Wallen (2006), or Huck, Cormier, and Bounds (1974) provide more thorough introductory treatments, whereas the text by Keppel and Wickens (2004) presents more advanced coverage. Table 26.1 contains specific desiderata to guide reviewers and authors as they make decisions regarding quantitative research design.

## 1. The Research Design is Foreshadowed

From within a quantitative social and behavioral science research framework, the discussion of research design usually begins with the methods or procedures used to conduct the study (e.g., the selection and/or assignment of participants, the operationalization of the variables, the procedures for data collection and analysis) and the subsequent implications for conclusions. However, the research design and methods utilized should not come as a surprise at the end of the Introduction, but rather should be an extension of the foundation that has been developed therein.

In describing research design for qualitative research, Maxwell (2005) identified five components that comprise his model for research design, much of which is applicable, yet has remained only implicit, for quantitative research design. The five interacting components that Maxwell identified include the goals, conceptual framework (which includes the theoretical framework and research literature), research question(s), methods, and validity. Although Maxwell included elements within these that are not appropriate to quantitative research (e.g., the inclusion of personal experience within the conceptual framework), and he envisioned these elements dynamically interacting rather than the more sequential linear procedure of quantitative research, he did make explicit the need to evaluate the conclusions of a study within this larger context. It is in the Introduction that researchers should identify what variables will be attended to and which will be ignored. The congruence of the Introduction, including its review of the literature, with the research design is necessary to evaluate the overall contribution of the study.

## 2. The Research Problem

It is impossible to evaluate the adequacy of a particular research design if there is no clearly articulated statement of the research problem. The research problem may be expressed in the form of research question(s) and/or hypotheses, and serves to formalize the research topic into an operational guide for the study, connecting the conceptual framework to the methods (Fraenkel & Wallen, 2006; Maxwell, 2005). The description of the research problem should identify the target population, the variables, and the nature of any anticipated relation between the variables and thereby focus the data collection and presage the data analysis. Hypotheses are not necessary, but are often stated when a specific prediction to be tested is made.

Terms used in the research problem must be defined in such a way that the questions are focused and testable. For example, *What is the best treatment for anxiety?* is not a testable question. Without defining "best" there is insufficient information to guide the study. Does "best" mean the most economical, the most consistent, or possibly the most permanent? The question also does not identify the population (e.g., children, teens, adults), or what types of treatment will be evaluated (e.g., psychotherapeutic, pharmacological, social behavioral), or what type of anxiety (e.g., self-report, clinically diagnosed, theoretically defined, physiologically measured). Without these further clarifications, it is not possible to assess whether or not the research design is appropriate to address the research problem.

Not only does the research problem suggest the appropriate research design, it also clarifies the specific type of data to be collected and thereby influences the data collection procedures. Questions can be classified as instrumentalist or realist (Cook & Campbell, 1979; Maxwell, 2005). *Instrumentalist* questions rely on the utilization of observable measures and require direct observation or measurement. *Realist* questions are about feelings, attitudes, or values that cannot be directly observed. The type of question, instrumentalist or realist, should connect the purpose of the study with the type of data collected. For example, if the purpose of the study is to provide information about teaching effectiveness, an instrumentalist question would be posed. It then would not be appropriate to collect the data using a survey within a survey research design to garner information from teachers as to their *perceived* effectiveness.

Designs developed for use with instrumentalist questions require greater inference and therefore might be more susceptible to bias. Yet as Tukey (1986) stated, and is often quoted, "Far better an approximate answer to the right question which is often vague, than an exact answer to the wrong question, which can always be made precise" (p. 407). Within the quantitative paradigm authors should clearly state their efforts to minimize bias and/or include appropriate caveats urging caution when interpreting and generalizing the results.

## 3. Articulation of the Research Design

The type of question(s), realist or instrumentalist (see Desideratum 2), will determine the type of data collected (e.g., self-report or performance). However, more fundamentally the research question(s) will determine the appropriate type of research design. Questions about relations among variables or questions about the current status of variables can be answered with a nonexperimental design. To answer questions about cause and effect, an experimental research design provides the stronger evidence.

The types of research design are distinguished by the degree to which the researcher is able to control the research environment. Four types of control are evaluated: (a) the researcher's ability to control the selection and/or assignment of participants to groups, (b) the manipulation of the independent variable(s), (c) how any dependent variables are measured, and (d) the timing of the measurement(s). The types of experimental designs vary significantly with regard to the type of control that the researcher is able to exert. Nonexperimental designs offer very little control, and experimental designs require more control.

Questions of cause and effect should be addressed using experimental designs. They are *experiments* in the sense that the researcher is able to control or deliberately manipulate conditions in order to observe the varying outcomes. As Shadish et al. (2002) stated:

> Experiments require (1) variation in the treatment, (2) posttreatment measures of outcomes, (3) at least one unit on which observation is made, and (4) a mechanism for inferring what the outcome would have been without treatment—the so-called "counterfactual inference" against which we infer that the treatment produced an effect that otherwise would not have occurred. (p. xvii)

Within the category of experimental designs, *randomized* designs (sometimes referred to as *true experimental* designs), are distinguished by the researcher's ability to control the experimental conditions, most specifically the random assignment of participants to conditions. *Quasi-experimental* designs comprise a separate category because, although the researcher can manipulate the proposed causal variable and determine what, when, and who is measured, he/she lacks the freedom to randomly assign the experimental units or participants to the treatment conditions. Without this random assignment the researcher must be more circumspect when making causal inferences (Cook & Campbell, 1979; Shadish et al., 2002).

In addition to the randomized and quasi-experimental designs, methodologists have referred to *pre-experimental* or *pseudo-experimental* designs (Cook & Campbell, 1979; Huck et al., 1974) as forms of experimental designs. These designs are separated from quasi-experimental because of their lack of experimenter control and subsequent weaker claims of causality. It is imperative that researchers and reviewers attend to how the various types of control (and more specifically, the lack of control) can impact both the internal validity (see Desideratum 8) and external validity (see Desideratum 10).

Questions of cause and effect require a comparison. The ideal, but impossible, comparison is the *counterfactual* (Cook & Sinha, 2006). Whereas the experimenter is able to measure what occurs when a treatment is introduced, he/she cannot say what would have occurred to that individual had the treatment not been introduced—the counterfactual. Thus, any experiment requires an approximation of the true counterfactual. "The better the counterfactual's approximation to the true counterfactual, the more confident causal conclusion will be" (Cook & Sinha, 2006, p. 551).

Specific experimental research designs are distinguished by how this counterfactual is constructed. Some designs use a control group and/or an alternate treatment, whereas others use a pretest measure to compare to the outcome measure. Some designs combine more than one approach (e.g., pretest-posttest control group design) to improve the quality of the counterfactual. The logic of this approach is that the counterfactual represents what would be in the absence of the treatment. Unfortunately, this belief cannot always be justified. For example, the use of a control group presumes that this group is identical to the treatment group in all ways except for the existence of the treatment. Similarly, the use of a pretest presumes that all else remains the same, except the exposure to the treatment. Clearly, these assumptions cannot always be defended and the researcher should provide as much evidence as is reasonable to support his/her claims of the adequacy of counterfactual that serves as the comparative. Additional variables should be tested to further support the argument of equivalence of a control group to which the participants have not been randomly assigned.

Nonexperimental designs are usually restricted to descriptive or associational research, where the main purpose is at most to provide evidence of relations between two or more variables. However, there are two classes of research that use nonexperimental designs to explore causal relationships:

*ex post facto* or *causal-comparative* (Fraenkel & Wallen, 2006) and studies utilizing statistical modeling procedures. Due to the absence of researcher control over not only the assignment of participants, but also the manipulation of a hypothesized causal agent, casual conclusions can only be tenuously advanced from studies that use a causal comparative design. Studies that utilize some form of statistical modeling (e.g., structural equation modeling; see Chapter 28, this volume) rely upon an a priori theory and stochastic assumptions to make causal claims. Despite the statistical sophistication, methodologists remain divided on whether nonexperimental designs can provide convincing evidence that warrant claims of causality (see, for example, Shaffer, 1992).

Researchers who rely upon extant databases should be attentive to the quality of the original research design and how the design decisions impacted the data collected. For example, large datasets frequently result from sampling strategies that have implications for how the data should be evaluated (see Chapter 30, this volume). Researchers using existing data, including those performing statistical modeling, should (a) disclose information relevant to how the data were obtained, (b) provide sufficient detail of the a priori theory or theories, (c) faithfully execute the chosen statistical procedure after adequately addressing associated underlying assumptions, and (d) acknowledge the limitations of the study to make claims of causality.

The selection of the research design should consider the research problem within the larger context of the research topic. Careful consideration should be given to whether a longitudinal within-subjects design (see Chapter 2, this volume) or a cross-sectional between-subjects design (see Chapter 1, this volume) would be better suited to address the research problem. For example, either design can answer the question of whether or not there is a difference in performance on some defined measure of knowledge of teenagers and septuagenarians. However, if the hypothetical construct being measured is long-term memory, the longitudinal design will enable greater confidence that the difference in test performance is due to memory rather than learning. If however, the hypothetical construct represented by the test performance is learning, the less time and cost consuming, cross-sectional between subjects design would be adequate and consistent with research in this field.

Once determined, the research design helps authors to coordinate how participants are selected, how variables are manipulated, how data are collected and analyzed, and how extraneous variability is controlled. Discussion of specific designs can be found in Cook and Campbell (1979), Huck et al. (1974), Shadish et al. (2002), or Creswell (2005). Each element of the research design (sampling, instrumentation, experimental procedures and data collection, and data analysis) should be described with sufficient detail that the study can be replicated. All variables (i.e., independent, dependent, moderator, mediator, or control) should be defined, and the measurement should be congruent with the presentation in the Introduction. The type of design (i.e., experimental, or nonexperimental) or the specific design (e.g., groups × trials mixed between-within design, or nonequivalent control group design) should be stated. Adherence to a specific design is not required and the inclusion of additional procedures to control extraneous variability is encouraged (e.g., the inclusion of a pretest or a control group). In their follow-up text to Campbell and Stanley (1963) and Cook and Campbell (1979), Shadish et al. (2002) emphasized the value of design elements rather than designs per se. In essence the design is constructed rather than being limited to selection from a prescribed list. The inclusion of each design element should be evaluated in terms of the potential impact on both internal and external validity (see Desiderata 8 and 10).

It is not uncommon for the researcher to omit any explicit reference as to what research design or design elements are used. Yet as Maxwell (1996) noted, "Research design is like a philosophy of life; no one is without one, but some people are more aware of theirs and thus able to make more informed and consistent decisions" (p. 3). When design elements have not been explicated, the degree to which the researcher has made conscious design decisions is unknown. In this case, not only must the reviewer be vigilant in evaluating the credibility of what is reported, but he/she must also try to reconstruct the design that was used by what is reported in the Methods and Results sections. Without the aid of the author to define the counterfactuals used, the reviewer and reader are left to not only evaluate their effectiveness, but to also identify what they are. This is essential to determining whether the research design can support the stated conclusions.

## 4. Specific Design Elements

The first element of the research design is a description of the participants. The selection of participants should be consistent with the identified design. The type of design will determine first whether group assignment is necessary, and second, if so does assignment precede or follow selection. If a sample is used, the sampling frame and the population should be identified. The sampling procedure should be specified and there should be a justification of the sample size (see Chapter 24, this volume). An appropriate sample, in size and composition (i.e., representativeness), is foundational to the conclusions of the study (see Desideratum 9). In addition, how participants are assigned to treatment conditions (if appropriate) is important to the determination of the strength of any inference of causality (see Desideratum 8). Not only is it important to report what definition or instrument is used to select or assign participants, it is also important to report the reliability, validity, and cut scores of that instrument. This provides confidence that the participants met the criterion established and allows comparison to previous research. For example, "extraverts" as defined by the Eysenck Personality Inventory (EPI) are not the same as "extraverts" defined by the Myers-Briggs Type Indicator®. Defining extraverts as those scoring above the sample mean on the EPI may not be the same as extraverts defined as those scoring above a normed score. The method by which participants are placed into groups (i.e., no groups, randomly assigned, or pre-existing groups) is essential to the type of research design being used and therefore to the conclusions that can be drawn (see Desiderata 8 and 9).

An integral element to the integrity of any study is the reliability and validity of the instrument(s) used to collect the data (see Chapter 25, this volume, for specifics on validity and reliability assessments). Not only is it necessary to provide evidence of the appropriate types of reliability and validity that have been established, but to make the case for why the author would expect that this evidence would apply to his/her use of the instrument. Citing extensive previous use is not sufficient evidence of reliability or validity.

The experimental and/or data collection procedures comprise the next element of the research design. There should be a detailed description of any experimental conditions, including any control conditions, if a treatment is introduced. This should include precise details of time intervals—duration of exposure to the treatment as well as time lapse between exposure and data collection, dosages, equipment settings, and research personnel. How and when the data are collected should be clearly described, making special note if the timing or the mode of collection could affect the response. For survey research designs this should include the number of reminder contacts, the timing of the reminders, and the mode of contact.

The final element of the research design before the discussion of study results is the presentation of the data management and data analysis. Data reduction and transformations, including the treatment of missing values, should be articulated, highlighting any deviations from standard procedures. The data analysis should explicitly address demographic data that are useful for the discussion of appropriate generalizations (see Desideratum 10) or the equivalency of groups (see Desideratum 3). Data from instruments with total or scale scores should be analyzed for internal consistency reliability and compared to previous uses of the instrument. The specific test(s) used to address each research question and/or hypothesis should be named, including any information necessary for the reader to determine the appropriateness of the test or decision (e.g., degrees of freedom, alpha level, $p$ values). Post hoc tests for the interpretation of omnibus test results (e.g., post hoc comparisons following an ANOVA) must be included and should be identified by name (see Chapters 1 and 2, this volume). When using samples, there should be a test at every point of decision. Evidence should be presented to confirm that the assumptions of statistical tests were met or that appropriate adjustments were made.

## 5. Internal Consistency of Research Design

A research design that lacks internal coherence creates problems for the interpretation of the results. This problem emerges particularly when the research design has not been explicated. Beginning with the Introduction, which should establish the need for the study and what precisely will be studied, through the statement of the research problem and research question(s), the way the sample is selected, the independent variable(s) are manipulated, the data collected, and how the data are analyzed, each design element should logically follow. If the researcher claims that a randomized design is used, it then follows that participants must be randomly assigned to the experimental conditions. The most blatant example of inconsistency is when the statistical analysis is not appropriate for the type of research question or how the data were collected (e.g., using a test of correlation to answer questions of cause and effect, especially when no temporal order has been established). Similarly, if participants are selected because they represent the two extremes of a grouping variable, it would be inappropriate to use correlation to evaluate the relationship between the two variables.

## 6. Design Execution

Details that researchers present in the Methods section must be consistent with what they intended at the outset of the research, as expressed in the Introduction. The procedures detailed in the Methods section should be evaluated to verify that they were faithfully executed. Small departures from the original design are often unavoidable—even anticipated, however, they require explanation.

A common problem is that the number of anticipated observations is not equal to the sample size upon which the conclusions are based, likely reducing the desired power level (see Chapter 24, this volume). This issue is particularly prevalent in survey research. Even in studies where researchers over-sample in an effort to achieve the desired sample size, the total response rate is often less than desired; the response rate for an individual survey item (i.e., missing response) may be considerably lower (Jackson, 2002). Often, studies are designed with equal or proportional sample size in each cell, yet frequently when the results are reported the cells are uneven. This has implications for how missing data are treated, for statistical assumptions, for the power of the test (see Chapter 24, this volume), as well as for other design implications. It is therefore incumbent upon the researcher to account for missing values and evaluate the implications for the design. One consideration is whether the nonresponses adversely affect the representativeness of the sample. More specifically, consideration must be given to whether missing values represent a threat to internal validity (see Desideratum 8). The disproportionate loss of participants from one treatment condition might suggest a threat to internal validity (Cook & Campbell, 1979). This is not only true of quasi-experimental designs but also of randomized designs, which by virtue of random assignment of participants are protected from most other threats.

Reviewers of manuscripts should be vigilant for evidence suggesting that procedures were counter to stated claims. For example, if a researcher claims that participants were randomly assigned, but then later suggests that the treatment was assigned to pre-existing groups, this changes the design from a randomized design to a quasi-experimental design with all the attendant issues that must be addressed.

Valid conclusions about a causal relation between treatment and outcome are dependent upon the treatment (and control) condition(s) being faithfully delivered and the dependent variable reliably measured. There should be evidence that the researcher (or whoever is providing the treatment) has been trained and dependably delivered the specified treatment. Evidence in the form of manipulation checks should be provided to verify that experimental manipulations were effective. For example, experiments that rely on deception require that the participants are indeed deceived, and a well-designed study will provide evidence to this effect. In addition to ensuring that the experimental conditions are consistent with what is reported there should be evidence that the researcher has sufficient training to collect the data (e.g., training for interviewers, inter-rater reliability).

In addition to considering whether the researcher has delivered treatment successfully, the researcher and the reviewer should consider the plausibility of participant noncompliance with treatment. Drug trials are dependent upon participants actually consuming the prescribed dose; training is dependent upon participant attendance. Without evidence of participant compliance there is insufficient evidence that the research design has been implemented.

## 7. Control of Extraneous Variability

Without appropriate control of extraneous variability it can be difficult to isolate and observe the effect(s) of the hypothesized causal variable(s) on the dependent variable(s). Control of extraneous variability is therefore one of the primary functions of a research design. Rigorous adherence to carefully designed research procedures can help to minimize the effect of unintended influences. However, the design element that has the greatest impact on the control of extraneous variability is the selection and/or assignment of participants. Random assignment to treatment conditions is the principal means by which a research design avoids the systematic influence of unintended variables. The advantage of this method is that it controls for the influence of a number of variables, even those unidentified. However, it alone might be insufficient if an extraneous variable has a stronger effect than the causal variable that is being considered. Manuscript reviewers should note any mention of one or more variables in the Introduction (or from previous content knowledge) that is known to have a strong relation to any of the dependent variables, and ensure that its influence is considered in the research design chosen by the study's authors. In fact, an alternative research design might have been more appropriate. For example, the effect of an extraneous variable might be controlled by including it as an additional variable in the design (i.e., randomized block design) or by restricting the population of the study to only one level of the extraneous variable (e.g., only include women in the study).

*Matching* is a procedure whereby participants are paired on their scores for a specific variable(s) and then each member of the pair is assigned to a different treatment condition. The intent of using this procedure is to equate the groups in terms of this specific variable, a variable that is believed to influence the dependent variable. This procedure should be used judiciously. Although it might equate the groups on that one specific variable, matching interferes with the ability to randomly assign participants, and thereby forfeits the benefit of randomization. The implications for the data analysis also must be considered. The matched pairs cannot be treated as independent observations and the data analysis must reflect this. The use of these procedures must be considered within the larger context of the overall design to ensure that their use is reflected in other design decisions (e.g., data analysis) and conclusions.

Sometimes, statistical procedures can also be used to control extraneous variability. The use of covariates can help adjust the scores on the dependent variable before testing for group differences if the extraneous variable is measured as a continuous variable. Propensity scores from a logistic regression (see Chapter 17, this volume) might also be used to evaluate the equivalence of treatment groups, improve matching, or be used as a covariate (Pasta, 2000). Although the use of propensity scores is a means of controlling for the effect of more than one extraneous variable, it is still limited to the control of only those variables that are identified and quantitatively measured. Rather than attempt to improve the equivalence of groups, Rosenbaum (1991) suggested a procedure (hidden bias sensitivity analysis) to assess how much bias would be necessary between the treatment and control groups for bias to be a viable alternative explanation for the treatment outcome. Shadish et al. (2002) warned that the use of these, or other advanced statistical procedures, is not a substitute for good design. Where possible, extraneous variability should be controlled by the research design, and then if appropriate augmented by available statistical procedures.

## 8. Internal Validity and Statistical Conclusion Validity

The careful construction and faithful execution of the research design provides the foundation for the research conclusions. Each element of the design relates to the validity of the study. Research conducted to test causal relations relies on the adequacy of the constructed counterfactual to represent the true counterfactual (see Desideratum 3). The adequacy is evaluated in terms of the ability to rule out rival hypotheses or alternative explanations for the outcome. In 1957, Campbell first coined the term *internal validity*, which was further elaborated by Campbell and Stanley (1963) as the confidence that the identified causal variable is responsible for the observed effect on the dependent variable and not due to other factors. They identified a list of threats to internal validity, which should be considered when constructing the design as well as when evaluating the conclusions. The list of threats to internal validity, with some modifications, can be found in most research design textbooks (also see Shadish et al., 2002; Shadish & Luellen, 2006; for discussions on threats relevant to specific designs see Cook & Campbell, 1979; Huck et al., 1974).

Threats to internal validity are usually discussed in terms of quasi- experimental designs because they result from the inability to randomly assign participants to treatment conditions. That is, random assignment reasonably protects the study from most threats to internal validity; however, such threats should be considered for any study that seeks to make causal inferences, with or without random assignment. Some threats (e.g., mortality or attrition, the disproportional loss of participants from one condition) occur after the assignment to experimental condition or as a result of something that occurs during treatment delivery, which thereby jeopardize causal interpretations of even a randomized design.

The potential of a threat to internal validity in and of itself is insufficient to dismiss a researcher's claims of causality. When evaluating the potential threats, Shadish and Luellen (2006) advocated the consideration of three questions:

(a) How would the threat apply in this case? (b) Is the threat plausible rather than just possible? and (c) Does the threat operate in the same direction as the observed effect so that it could partially or totally explain that effect? (p. 541)

If it can be conceived how a specific threat would offer a rival hypothesis, which is probable—not just possible, and explains the direction of the outcome, only then would the internal validity be challenged. The careful researcher will consider these threats in the design of the study, anticipating those with potential relevance. If considered prior to the execution of the study, it may be possible to alter a design element(s) to avoid a potential threat, or additional data may be collected to provide evidence to argue against a threat's explanatory ability (see Desideratum 3).

Cook and Campbell (1979) further refined the discussion of internal validity by introducing *statistical conclusion validity* as a distinct form of validity related to internal validity. Statistical conclusion validity refers to the "appropriate use of statistics to infer whether the presumed independent and dependent variables covary. Internal validity referred to whether their covariation resulted from a causal relationship" (Shadish et al., 2002, p. 37). Threats to statistical conclusion validity provide reasons why the researcher might be wrong about (a) whether a relationship exists and (b) the strength of the relationship. A list of threats to statistical conclusion validity can be found in Shadish et al. (2002, p. 45). Attention to statistical power, assumptions of the statistical tests, inflated Type I error, and effect size, as well as issues related to the measurement and sampling, fall within the purview of statistical conclusion validity.

Generally, it is not appropriate to refer to the internal validity of nonexperimental designs, with one exception: specific designs that are being used to make causal inferences (e.g., causal comparative). However, the validity of conclusions reached still requires evaluation. Each design decision affects the validity, with decisions regarding the appropriate sampling, instrumentation, and statistical analysis of particular importance for the nonexperimental design. The sample size and representativeness of the population, the reliability and validity of measurement, and the appropriate statistical analysis are key to the conclusions of a nonexperimental study.

Authors and reviewers must keep in mind that "Validity is a property of inferences. It is *not* a property of designs or methods, for the same design may contribute to more or less valid inferences under different circumstances" (Shadish et al., 2002, p. 34). Executing a prescribed design does not guarantee valid inferences, nor does the rigid adherence to a checklist of potential threats to validity. Neither are adequate substitutes for the researchers' sound logic.

## 9. Conclusions Are Appropriate

Miles and Huberman (1994) noted that design decisions both support and constrain the conclusions of research. Just as the genesis of the research design is before the Methods section, its influence extends beyond the Results. Researchers are responsible for presenting conclusions that are consistent with and appropriate to the design. The adage "correlation is not causation" is just one example for the necessity to ensure that claims in the Discussion do not exceed what the research can support. Design decisions, such as the decision to control extraneous variability to only one level of an extraneous variable (e.g., women only) restrict the conclusions to only that group.

Careful articulation of the research design elements, with attention to potential threats to internal and statistical conclusion validity (see Desideratum 8), prepares the researcher to present the conclusions within the context of the existing literature. Causal claims should not be made without ruling out threats to internal validity. With a nonexperimental design utilizing only descriptive statistics to report the findings from a sample, it is inappropriate to make comparisons between groups (e.g., "women scored higher than men"). There must be a test at the point of decision.

Without appropriate supporting evidence it is inappropriate to draw conclusions from statistical nonsignificance. For example, when testing for mean differences between treatment populations, nonsignificance should not be interpreted to imply that there is no difference between the populations' means or that the population means are therefore equal. Statistical nonsignificance means that the researcher has failed to show a difference of sufficient magnitude that cannot be reasonably explained by chance alone. That the population means are equal is only one possible explanation. It is also possible that the sample size was insufficient or the measurement not sensitive enough to detect true differences.

The tendency to overstate findings is not limited to misrepresenting statistical conclusions or failing to recognize threats to internal validity, but also includes making claims beyond what was studied. For example, if a study using a survey to measure the level of teacher satisfaction shows that 65% of teachers report being *slightly dissatisfied* with teaching, it is inappropriate for the researcher to conclude that his/her study found that teachers will be leaving their schools, or that teachers should be paid higher salaries. The author should be diligent to ensure that recommendations from the study are not presented in such a manner that they can be construed as findings.

## 10. External Validity and Construct Validity

Technically, the *external validity* of a research design refers to the degree to which a study's observed *causal* relations are generalizable; that is, it helps characterize "to what populations, settings, treatment variables, and measurement variables can this effect be generalized" (Campbell & Stanley, 1963, p. 5). Internal and external validity are considered to be complementary: whereas the former addresses the question of what can be inferred about cause and effect from this instance, the latter assesses the degree to which the causal findings can be generalized. Frequently what will increase internal validity may decrease external validity and vice versa. In their 1979 work, Cook and Campbell extended their dichotomous discussion of validity into the typology that comprised internal, statistical conclusion, external, and construct validity. Whereas internal and statistical validity (see Desideratum 8) are relevant to the inferences that derive from the specifics of the study procedures, construct and external validity relate to whether the inferences can be extended beyond the current situation. *Construct validity* generalizations refer to "inferences about the constructs that research operations represent" and external validity generalizations are "inferences about whether the causal relationship holds over variation in persons (or more generally: units), settings, treatment, and measurement variables" (Shadish et al., 2002, p. 20). From these definitions it becomes apparent that with nonexperimental designs that are used to describe the current status or noncausal relations between variables, it is inappropriate to discuss external validity. Instead, construct validity is the more appropriate consideration. Thus, nonexperimental designs that are not used to test causality should be evaluated for construct validity, and nonexperimental designs that are used to evaluate causal relations and experimental designs should be evaluated for both construct and external validity.

Construct validity is inherent in social and behavioral research and as an issue is twofold: definition and measurement. Every construct has multiple facets or features, with some being more central than others. Thus, defining the construct requires identifying multiple components, with the core being those features to which there is the greatest agreement. Once defined, the question becomes one of how to represent the construct, and more specifically how to measure it. Determining how multifaceted constructs can be reduced to a manageable size, yet still represent the higher order construct, is the dilemma of construct validity. Each study uses a limited set of conditions in terms of the population, the treatment, the setting, and the outcome; from which the desire is to make statements about the higher order construct. Each element of the research design should be evaluated for the construct(s) it represents. Researchers tend to focus on only the treatment variable, if there is one, and the outcome measure. Clearly, discussions limited to the construct validity of the outcome measure are insufficient as they address only one of the constructs in the study. How will the sample selected reflect the larger construct that it represents? For example, how does a sample consisting of students two grades below reading level represent a population of "students at risk"? How does conducting the study in the laboratory represent the larger construct of the settings where the conclusions would apply? These questions need to be considered in addition to the more obvious examination of how the treatment and outcome constructs are operationalized. There is no one-to-one correspondence of the operationalization of the study and the constructs; the question is: *How great is the disparity?* A list and further discussion of potential threats to construct validity is presented in Shadish et al. (2002).

External validity refers specifically to whether or not observed *causal* relations can be extended across individuals, settings, treatments, and/or outcome measures. The use of probability sampling is the foundation for external validity. Probability sampling requires that each item in the domain has a nonzero chance of being randomly selected. This condition is infrequently met when sampling participants for a study, much less when sampling from the domains that describe the other elements of an experiment (i.e., the setting, treatment conditions, outcome measures). Shadish et al. (2002) explicated a more heuristic approach to determine causal generalizations. They proposed five principles for consideration: surface similarity, ruling out irrelevancies, making discriminations, interpolation and extrapolation, and causal explanation. Too frequently external validity is only discussed in terms of generalizing to populations, either those internal or external to the study, and the ability to generalize to the other elements receives scant attention. With a design seeking to establish evidence of a causal relationship, the researcher and reviewer should examine the degree to which the design elements that are included represent a random sampling of the construct domain, be it the population, setting, treatment, or outcome. In the absence of random sampling, the principles described by Shadish et al. provide a systematic means to evaluate external validity. Shadish et al. also present a list of common threats to external validity.

## 11. Design Limitations

The diligent researcher will acknowledge weaknesses in the research design and present the implications of the shortcomings. For example, by recognizing in advance that the use of pre-existing groups compromises the internal validity (see Desideratum 8), the researcher has the opportunity to offer explanations, possibly even statistical evidence (see Desideratum 7), to argue the equivalence of the groups prior to the introduction of the treatment. By ignoring any reference to this design decision, the reviewer and reader are left to decide whether the potential pre-existing group differences are sufficient to explain the outcome. More significantly, this can create a lack of confidence. The question becomes: *If the researcher does not know enough to discuss the implications of the use of pre-existing groups, what other relevant information might he/she not recognize the necessity to reveal?* Does the researcher understand enough about the research design to adequately convey the information necessary for the reader to make an independent decision as to the appropriateness of the conclusions?

In theory, many of the weaknesses can be avoided by assiduous attention to the research design, yet this is not always the case. Weaknesses result from a lack of feasible alternatives, unforeseen occurrences during the study, and/or from poor research design. The credibility of the researcher is enhanced if he/she is able to eliminate him/herself from the latter category by anticipating and addressing criticism from the knowledgeable reviewer or reader.

## References

Campbell, D. T., & Stanley, J. C. (1963). Experimental and quasi-experimental designs for research on teaching. In N. L. Gage (Ed.), *Handbook of research on teaching* (pp. 171–246). Chicago: Rand McNally.

Cook, T. D., & Campbell, D. T. (1979). *Quasi-experimentation: Design and analysis issues for field settings.* Boston: Houghton Mifflin.

Cook, T. D., & Sinha, V. (2006). Randomized experiments in educational research. In J. L. Green, G. Camilli, & P. B. Elmore (Eds.), *Handbook of complementary methods in education research* (pp. 551–565). Mahwah, NJ: Erlbaum.

Creswell, J. (2005). *Educational research: Planning, conducting, and evaluating quantitative and qualitative research.* Upper Saddle River, NJ: Prentice Hall.

Fraenkel, J. R., & Wallen, N. E. (2006). *How to design and evaluate research in education* (6th ed.). New York: McGraw Hill.

Huck, S., Cormier, W., & Bounds, W. G. (1974). *Reading statistics and research.* New York: Harper & Row.

Jackson, G. B. (2002). *Sampling for social science research and evaluations.* Retrieved January 26, 2009, from http://www.gwu.edu/~gjackson/281_Sampling.PDF

Keppel, G., Saufley, W. H., Jr., & Tokunaga, H. (1998). *Introduction to design and analysis* (2nd ed.). New York: Freeman.

Keppel, G., & Wickens, T. D. (2004). *Design and analysis: A researcher's handbook* (4th ed.). New York: Pearson Prentice Hall.

Maxwell, J. A. (1996). *Qualitative research design: An interactive approach.* Thousand Oaks, CA: Sage.

Maxwell, J. A. (2005). *Qualitative research design: An interactive approach* (2nd ed.). Thousand Oaks, CA: Sage.

Miles, M. B., & Huberman, A. M. (1994). *Qualitative data analysis: An expanded sourcebook.* Thousand Oaks, CA: Sage.

Pasta, D. J. (2000). Using propensity scores to adjust for group differences: Examples comparing alternative surgical methods. *Proceedings of the twenty-fifth annual SAS® Users Group International Conference.* Cary, NC: SAS Institute Inc.

Rosenbaum, P. R. (1991). Discussing hidden bias in observational studies. *Annals of Internal Medicine, 115,* 901–905.

Schneider, B., Carnoy, M., Kilpatrick, J., Schmidt, W. H., & Shavelson, R. J. (2007). *Estimating causal effects using experimental and observational designs: A think tank white paper.* Prepared under the auspices of the American Educational Research Association Grants Program.

Shadish, W. R., Cook, T. D., & Campbell, D. T. (2002). *Experimental and quasi-experimental designs for generalized causal inference.* Boston: Houghton Mifflin.

Shadish, W. R., & Luellen, J. K. (2006). Quasi-experimental design. In J. L. Green, G. Camilli, & P. B. Elmore (Eds.), *Handbook of complementary methods in education research* (pp. 539–550). Mahwah, NJ: Erlbaum.

Shaffer, J. P. (Ed.). (1992). *The role of models in nonexperimental social science: Two debates.* Washington, D.C. American Educational Research Association and American Statistical Association.

Trochim, W. M. K. (2006). Design. *Research Methods Knowledge Base.* Retrieved February 18, 2009, from http://www.socialresearchmethods.net/kb/design.php

Tukey, J. W. (1986). The collected works of John W. Tukey: 1949–1964. In L. V. Jones (Ed.) *Philosophy and principles of data analysis 1949–1964, Volume III.* Boca Raton, FL: CRC Press.

# 27
# Single Subject Design and Analysis[1]

**Andrew L. Egel and Christine H. Barthold**

Single subject research is a form of rigorous investigation in which the individual is the unit of analysis. Individual variability of each participant is measured as opposed to the mean performance of groups. This allows for the examination of individual responding of participants during and following an intervention using previous performance (baseline measures) as a control (Sidman, 1960). Variability and experimental control are evaluated through visual inspection of graphed data (Skinner, 1938) and confirmed through independent and systematic replication both within and across research studies. Therefore, it is a misnomer that single subject designs have an $N$ of 1; in fact, most studies have three or more participants.

Single subject designs can be used to study any construct defined so that it can be observed and measured over time, such as methods for effective teaching in higher education (Saville, Zinn, Neef, Van Norman, & Ferreri, 2006), teaching math skills to at-risk children (Mayfield & Vollmer, 2007), or psychological phenomena such as severe phobias (Jones & Friman, 1999). The results of single subject designs are often the identification of effective interventions for heterogeneous populations where random assignment can be compromised or when information is needed about variable performance within groups (e.g., students with learning disabilities or autism). Kennedy (2005) suggested that single subject designs can be used to demonstrate the effectiveness of interventions, to compare two or more interventions, and to complete parametric and component analyses. Some useful resources for those new to single subject designs are Alberto and Troutman (2003), Kennedy (2005), Bailey and Burch (2002), and Horner et al. (2005).

## 1. Review of the Literature

Those who do not learn from history are doomed to repeat it.

George Santayana

As with most studies, a review of the literature for a single subject design informs the reader of the theoretical importance of the study as well as what dimensions of the research question have been studied in the past. Almost all single subject studies are built upon previous applied, basic, or theoretical findings.

The literature review should end with a statement of the current research question(s) in light of the existing literature. Null hypotheses are rarely articulated in single subject designs; instead, studies are

Table 27.1 Desiderata for Single Subject Design and Analysis

| Desideratum | Manuscript Section(s)* |
|---|---|
| 1. A description of the purpose(s) of the present study is provided within the context of a review of the literature describing and evaluating research that has been conducted previously. | I |
| 2. For purposes of future replication, participants and setting(s) are described in detail. | M |
| 3. All independent and dependent variables are operationally defined. | M |
| 4. The experimental procedures utilized are described in enough detail to allow for replication and control for threats to internal validity. | M |
| 5. The experimental design is selected based on the specific research question(s) investigated. | M |
| 6. Data collection procedures are selected based upon the dimensions of behavior recorded as per the operational definition. | M |
| 7. Inter-observer agreement coefficients are presented, together with associated formulae used (dependent on the dimension of behavior recorded). The inter-observer agreement is sufficient to indicate reliable interpretation of operational definitions. | M, R |
| 8. Results are independently replicated at least twice across participants, settings, and/or behaviors. | R |
| 9. Data are presented graphically to allow for visual inspection and decisions on when to alter variables are based on level, trend, variability, and overlap. | R |
| 10. Data are interpreted relative to the research question(s) and available literature. Limitations of the research and directions for future investigations are proposed. | D |

* *Note*: I = Introduction, M = Methods, R = Results, D = Discussion

built around one or more research questions. Good research questions should be stated in observable and measurable terms (Johnston & Pennypacker, 1993). These questions are usually very specific to a problem that needs to be addressed, such as increasing appropriate behavior (Kennedy, 2005).

## 2. Participants and Setting

Careful description of the participants' characteristics and where the intervention is carried out is of high importance in any type of study. However, more emphasis is placed on participants and setting in single subject research because of the emphasis placed on understanding individual variability. Several factors have been identified in the literature that influence selection of settings and participants.

*Important factors in choice of setting.* Settings must have face validity relative to the settings in which the intervention is likely to be adopted. For example, research designed to increase social interactions between children with autism spectrum disorders and typical children should be conducted in schools if teachers and classroom staff are the intended audiences. Conducting the same study in a child's home would not have face validity.

Settings should also be stable and flexible throughout the study. Settings with rapidly changing schedules or ongoing issues (e.g., a classroom with substitute teachers) present confounding variables that are not easily controlled. Settings with little flexibility in scheduling and/or choices of intervention might make data collection difficult or impossible (Bailey & Burch, 2002).

*Factors that affect the selection of participants.* Care must be taken to provide a detailed description of the participant, including characteristics that might be related to the study's outcomes (e.g., communication ability, psychiatric diagnoses, age). Complete subject descriptions can, at times, provide possible explanations for failure in replication and lead to a greater understanding of the generality of results (Kazdin, 1981).

Bailey and Burch (2002) identified several characteristics that will affect selection of participants. They noted that participants must reflect the dimensions of the general population that are being addressed in the study. For example, if the investigation is designed to increase interview skills, then the participants should be individuals who are struggling with those behaviors. In addition, the authors noted that participants need to be readily available over the course of the study, and demonstrate stability with respect to health, cooperation, and attendance.

## 3. Independent and Dependent Variables

Researchers using single subject research designs must implement the independent or treatment variable (IV) repeatedly over time. Ensuring that the IV is delivered reliably requires that it be *operationally defined*. Operationally defining an IV requires researchers to describe all components in concrete, observable, and measurable terms. An IV that is not operationally defined increases the likelihood that it will not be implemented with consistency and, as a result, the ability to demonstrate functional control might be compromised. The quality of an operational definition can be determined by collecting data on the implementation of the IV. High treatment integrity would be evidence that the IV was well defined.

It is also critical that researchers using single subject designs provide operational definitions for dependent variables (DV). Describing DVs in concrete, observable terms allows observers to agree on whether the behavior is occurring or not occurring. Furthermore, operational definitions of the DV also facilitate replication by other investigators because they know exactly what was measured during the original study.

## 4. Experimental Procedures

Experimental procedures should be selected so that functional relations between the dependent and independent variable can be demonstrated, threats to internal validity controlled, and social validity (i.e., the extent to which the change in the target behavior improves the participant's quality of life) maximized. Most research studies using single subject designs will have, at a minimum, a baseline and intervention phase. *Baseline* refers to the experimental condition that precedes the intervention phase. Baselines allow for a contextual evaluation of the effects of the independent variable. Although baseline is often thought of as a control condition, another experimental condition might serve as baseline as well. For example, if a person with heart disease is participating in a study to determine the effects of an experimental medication, baseline might be taken on symptoms while the participant is on a more well-established medication. All aspects of both the baseline and intervention conditions, including the exact materials used, any instructions provided, and levels of feedback for correct/incorrect responding, must be described in enough detail to permit replication.

It should also be clear that the researchers designed the experiment to control for threats to internal validity. Like all research designs, the validity of single subject designs can be threatened by variables such as history and maturation. Conversely, threats such as regression to the mean, participant selection bias, and selective attrition are not considered threats to internal validity given the single subject nature of the designs (Kennedy, 2005). However, there are some threats to internal validity that are of particular concern to the single subject researcher, as described below.

*Testing*, or repeated exposure, is of concern because the frequent number of measurements characteristic of single subject designs could result in the participant learning the response(s) independent of intervention. Testing effects can be minimized by spacing out observations and/or choosing an experimental design in which the number of data points needed is minimized.

*Multiple treatment interference* occurs when participants receive more than one intervention in a condition and it is a serious threat to the internal validity of a single subject design. Multiple treatment interference does not allow a researcher to determine which of the interventions, alone or in combination, were responsible for changes in behavior.

*Sequence effects* can also influence interpretation of data collected within a single subject design. They occur when a variable is introduced or removed in a particular order and affect responding in subsequent conditions. Sequencing the introduction and removal of variables is crucial in single subject designs and a researcher must be sensitive to the possibility of their occurrence. Counterbalancing or randomizing the sequence of conditions across participants is one way to control for sequence effects.

## 5. Experimental Design / Research Question Correspondence

There are several single subject designs that can be used to analyze the effects of an intervention. The design a researcher selects will depend on the specific question asked as well as the resources that are available at the time. In each of the designs discussed below the individual serves as his or her own control and the experimenter replicates the effect(s) of the independent variable in order to establish experimental control.

The *reversal design* has, historically, been one of the most frequently used single subject design. This design requires that the experimenter implement both the baseline and treatment phases multiple times in order to demonstrate experimental control. An A-B-A-B reversal design is typically used because it allows researchers to replicate both the baseline ("A") and intervention ("B") conditions.[2] Experimental control is established when the data patterns in each condition co-vary with the introduction and removal of an independent variable. The effects must be replicated across each condition. An A-B-A-B reversal design is illustrated in Figure 27.1 as well as in the study by Ahearn, Clark, MacDonald, and Chung (2007). Ahearn et al. used a reversal design to evaluate the effects of response interruption and redirection (RIRD) on the occurrence of vocal stereotypy in four children with

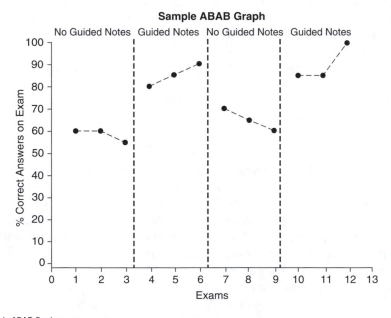

**Figure 27.1** Sample ABAB Design.

autism spectrum disorder (ASDs). The authors found substantial decreases in vocal stereotypy for each participant when RIRD was implemented.

Although reversal designs can be used to demonstrate the effects of an independent variable, there are circumstances where it would be inappropriate to use such a design, including cases where the behavior in question is irreversible or cases where reversing the specific behavior puts the participant at risk of injury. In these circumstances, other single subject designs would be more appropriate. A *multiple baseline design* is one of the design alternatives. Multiple baseline designs require the concurrent collection of two or more baselines (across participants, behaviors, or settings). When responding is consistent across baselines, the intervention is introduced systematically to one baseline at a time. Experimental control is demonstrated when the behavior changes only when the intervention is implemented and the effects of intervention are replicated across participants, behaviors, or settings. An example of a multiple baseline design across participants is presented in Figure 27.2 and can be seen in the investigation by Reeve, Reeve, Townsend, and Poulson (2007). Reeve et al. examined the extent to which a multicomponent intervention package could be used to teach generalized helping responses to four children with ASDs. The authors introduced the intervention successively across participants, in a multiple baseline fashion. The results showed that all participants acquired the helping responses. The responses also generalized to discriminative stimuli that were not used during training.

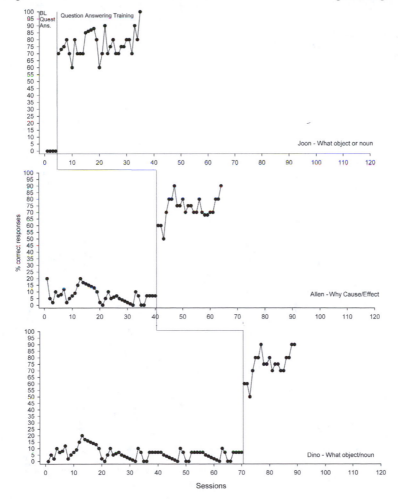

**Figure 27.2** Sample Multiple Baseline Graph.

Horner and Baer (1978) introduced a variation of the multiple baseline design for situations in which baseline responding will be zero or when extended baselines can result in high rates of problematic behavior (e.g., tantrums, noncompliance). The *multiple probe design* requires that baseline sessions be collected intermittently rather than continuously as in the multiple baseline design. According to Horner and Baer (1978), implementation of a multiple probe design requires that an initial baseline probe be collected across each behavior, participant, or setting. One to two additional probes are conducted on the first tier of the multiple baseline design. Intervention is subsequently implemented on the first tier, while no data are collected on any of the remaining tiers. Once the data in the first tier show an effect or reach criterion, an additional probe is implemented on each remaining tier. Additional baseline probes are collected on tier two until there is at least one more probe session than occurred in tier one. Horner and Baer (1978) referred to these consecutive probes as the "true" baseline sessions and they always preceded the implementation of intervention. Thus, each tier has at least one more "true" baseline session than the preceding tier. An example of a multiple probe design is presented in Figure 27.3 and can be seen in the study by Secan, Egel, and Tilley (1989). Secan and her colleagues used a multiple probe design to assess the impact of an intervention using pictures as a referent on the acquisition, generalization, and maintenance of wh-question answering skills in four students with ASDs. The results of the study showed that each student learned how to answer

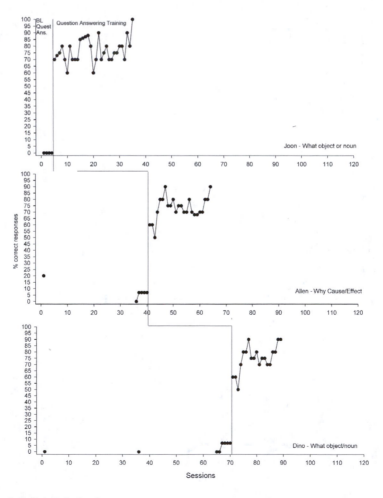

**Figure 27.3** Sample Multiple Probe Graph.

wh-questions, although generalization was affected by whether or not the relevant cue was visible in the generalization pictures.

Multiple probe designs are more efficient than a multiple baseline design because baseline data are collected intermittently. However, this also presents a problem because a researcher would not see abrupt changes that might occur during baseline.

A second variation of the multiple baseline design is a *changing criterion design* (Hartman & Hall, 1976; Tawney & Gast, 1984). The design is typically used to evaluate behaviors that will increase in a gradual, stepwise fashion. Implementation of the design requires a baseline phase during which data are collected to show both the pre-intervention level of the behavior and to determine an initial criterion level (e.g., the average level of responding). Intervention is implemented subsequently until the target behavior reaches the first criterion level. Once the first criterion level has been reached, a more stringent criterion is established and intervention continues until that second criterion has been met. This pattern continues until responding is at the terminal criterion. The data collected at each criterion level serve as a baseline for the subsequent phase. Experimental control is demonstrated when behavior changes to the new criterion level each time the criterion is changed. The change should occur rapidly and responding should stabilize at the specific criterion level before the criterion is changed. An example of a changing criterion design is found in Figure 27.4. An example from the literature can be found in the study by Warnes and Allen (2005). Warnes and Allen assessed whether electromyographic (EMG) biofeedback could affect paradoxical vocal fold motion (PVFM) in a 16-year-old participant. Baseline levels of muscle tension were recorded initially followed by intervention with EMG biofeedback. Once lower muscle tension had occurred and was maintained at the criterion level, the criterion was changed and treatment continued until the new criterion level was met. This continued in a changing criterion design format until typical levels of muscle tension were attained. The results demonstrated that the intervention was effective in reducing muscle tension.

*Alternating treatments designs* (ATDS), also known as *multi-element designs, multiple schedule designs,* or *randomized designs,* are also considered to be an extension of the reversal design. In an alternating treatments design, two or more independent variables are alternated rapidly so as to compare

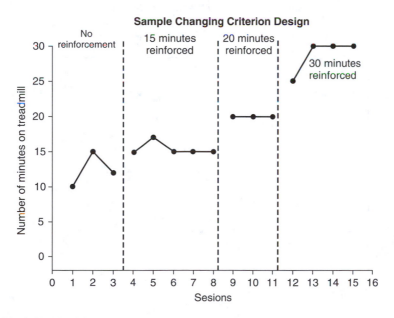

**Figure 27.4** Sample Changing Criterion Design.

their differential effects. A distinct stimulus is always associated with each condition to facilitate the participant's discrimination. This helps reduce the possibility of carryover effects and increases the likelihood that differences in responding are a function of different levels of effectiveness and not the participant's inability to discriminate the presence of the different conditions. Unlike the designs discussed above, an ATD does not require baseline data, a reversal, or stability in order to demonstrate experimental control. Fractionation, or salient changes in level between the two conditions, signifies effects (Barlow & Hayes, 1979; Tawney & Gast, 1984).

When alternating between conditions, it is best to randomize the presentation of the two conditions to avoid sequence effects. Ulman and Sulzer-Azaroff (1975) recommended that the presentation be random with no more than two of the same conditions presented in succession.

Other concerns for researchers include multiple treatment interference, generalization, and contrast effects. With generalization or carryover effects, responding is diffused among conditions by virtue of exposure. This is likely to occur when participants do not discriminate between the conditions. In contrast effects, the presence of one condition serves to suppress responding in another condition. Generalization and contrast effects can be avoided by making conditions salient. The researcher might assign certain uniforms or colors of stimuli to each condition. A classic article that utilizes an ATD is by Iwata, Dorsey, Slifer, Bauman, and Richman (1994). Iwata et al. sought to determine the relation between self-injurious behavior (SIB) and specific environmental events in an effort to identify the function of different forms of SIB prior to intervention. Participants were observed in four different conditions that were introduced in an ATD fashion: social disapproval, academic demand, unstructured play, and a final condition where the participants were placed in a room without access to any materials or persons. The results showed that, for the majority of participants, SIBs were consistently associated with one of the conditions listed above. These findings provided evidence that SIB may be a function of different sources of reinforcement.

Experimental designs are selected based upon the salient aspects of the research question(s). In addition, dimensions of the operational definition as well as the constraints of the setting should be considered when selecting a particular research design. Ethical factors, such as the feasibility of with-

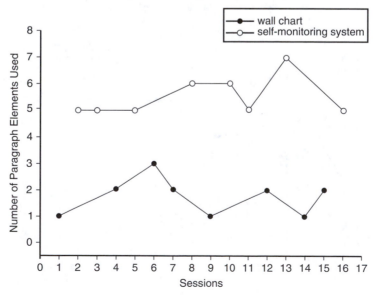

**Figure 27.5** Sample Alternating Treatments Graph.

drawing the independent variable, should be considered as well. A sample alternating treatments graph is presented in Figure 27.5.

## 6. Data Collection Procedures

There are several methods that have been used by single subject researchers to collect data. Selection of the method should be based upon the specific research question, the operational definition of the target behavior(s), and practical/environmental factors. If a data collection procedure is selected that does not reflect the operational definition and research question, is difficult to record (such as the number of times a child taps his or her pen), or there are restrictions on data collection (such as not having the ability to videotape a language sample), data taken might not be representative of the environment. Therefore, data collection procedures should be selected that maximize the probability that the data collected are representative of the actual environmental events.

The length of the observation will be dependent upon the probability that the behavior will be observed. If the response is occurring consistently throughout the day, then a shorter sample will more likely be representative of environmental events. Data collection can also be restricted to times when behavior is most likely to occur (e.g., if a child is most likely to have a tantrum after getting off of the bus, then data collection can be restricted to that particular time). If responding is more diffuse or less probable, then the researcher will need to lengthen the data collection period to insure a representative sampling of the dependent variable.

One form of data collection used by single subject researchers is *frequency*. Frequency data are collected on behaviors that have discrete onset and offset and occur at a distinct moment in time, in other words, behaviors that are "countable." For example, the number of times a person checks email during office hours is an example of a behavior that would be measured using frequency data. Sometimes a researcher will know in advance that the target behavior occurs more/less frequently during a specific time period within an observation. In this situation, the time period would be broken into equal intervals, and the frequency of the behavior would be measured within each interval.

*Duration data* can be used to measure how long a behavior occurs. Tawney and Gast (1984) suggested that duration data could be collected in two ways depending on the characteristics of the target behavior. *Total duration* refers to the total amount of time a participant is engaged in a defined behavior and can be used when the target behavior occurs continuously. For example, one might be interested in recording the total amount of time a student spent studying in an evening. *Duration per occurrence* is a measure of all contiguous occurrences of behavior. In the last example, a researcher would use duration per occurrence if studying was constantly interrupted, and s/he was interested in the duration of each discrete study period.

*Latency data* collection is used when an investigator is concerned with the length of time between the cue to respond and how long it takes for responding to begin, as opposed to how long it takes to complete a task. In the latter scenario, duration recording would be more appropriate. Latency recording would be appropriate, for example, if the researcher were interested in measuring the amount of time it took for a student to begin lining up at the door following a teacher's direction to engage in that behavior.

In some cases, the dependent variable might not have a clear onset or offset, or might occur too rapidly for reliable data collection using the methods mentioned previously. For example, if an experimenter is taking data on conversational skills with multiple components, it can be difficult to discern precisely where the interaction begins and where it ends. In this case, recording methods such as *partial interval*, *whole interval*, or *time sampling* might be appropriate. Interval measures require that observation periods are broken down into equal intervals (sometimes also known as *bins*). For partial and whole interval, time is split between observation intervals that are typically 10–20 seconds in length, and recording intervals that are typically 3–5 seconds in length. During observation intervals, the observer is monitoring for the occurrence of the dependent variable. No data are recorded until the end of the observation interval; conversely, no responses emitted during the recording interval are applied to the observation interval. If the observer is using a partial interval system, the dependent variable is recorded as occurring if it is emitted at any time during the observation interval. For whole interval data collection, responses are recorded *only* if the response is emitted for the entire observation interval.

*Time sampling* is used when continuous observation of the dependent variable is either impractical or unnecessary. In time sampling, time is also broken into equal intervals. Data are only recorded in the instant that the interval expires. The observer marks a "+" on the data sheet if the participant is engaged in the response at that moment. If the participant is not engaged, the observer marks a "–". Time sampling is often recommended when the observer is taking data simultaneously with other responsibilities, such as teaching. However, time sampling is not recommended in situations where the response occurs infrequently due to the lack of continuous observation.

When using any of the above interval measures, the length of observation per session will depend upon the topography of the behavior. For example, high frequency behaviors of short duration would require shorter time intervals (10–20 seconds) while behaviors that are of low frequency with long durations would require longer observation intervals. However, intervals larger than 2 minutes have been shown to decrease the validity of the data (Cooper, Heron, & Heward, 2007).

## 7. Inter-Observer Agreement Coefficients

Researchers employing single subject designs have historically used human observers to record the occurrence or nonoccurrence of behavior. Human observers, however, increase the likelihood of variability during observations (Kazdin, 1977). As a result, procedures were developed for measuring whether or not independent observers record data in a reliable manner (i.e., the operational definition is interpreted consistently). Inter-observer agreement scores are typically reported as an overall percentage of agreement together with the range across experimental sessions. The literature recommends that inter-observer agreement be assessed on at least 30% of sessions, although this figure did not evolve from any research investigations. An acceptable level of observer agreement is usually 80% or above; however, this figure has also not been established through systematic research.

Different formulas have been developed for calculating inter-observer agreement depending on the type of data collected (see Chapter 11, this volume). Occasionally, an agreement coefficient such as *kappa* (Cohen, 1960) is calculated in single subject research; however, the following formulae are seen most frequently in the single subject literature. *Total agreement* (Kennedy, 2005) is typically used for *frequency, duration,* and *latency* data. The formula used to calculate total agreement between two observers is:

(Smaller total of behavior recorded / Larger total of behavior recorded) × 100%

One major limitation of this formula is that it does not take into account whether or not two observers ever agreed on the occurrence of individual instances of behavior. As a result, high levels of agreement between observers may occur, even though they have never agreed on the occurrence of a single behavior. Bailey and Burch (2002) suggested that calculating a block by block percentage agreement and then averaging the scores would be one way to correct for the above problem.

Calculating percentage agreement scores for data collected using *interval* measures (e.g., partial interval, whole interval, and time sample) involves a different formula that compares observer recordings interval by interval. The formula used to calculate percentage agreement in this manner is:

(Agreements / Disagreements) × 100%

An agreement was scored if both observers recorded the occurrence or nonoccurrence of a behavior in the same interval; disagreements occurred when one observer recorded an occurrence of a behavior and the second observer did not.

A more stringent method for determining agreement for interval type data is to use the same formula to calculate reliabilities for the occurrences and nonoccurrences separately. For example, percentage agreement for occurrence data would be calculated using the formula:

[(Agreements of Occurrence) / (Agreements of Occurrence + Disagreements)] × 100%

Percentage agreement for nonoccurrence data would be calculated using the same formula except the focus would be on agreements of nonoccurrence.

## 8. Replication

Replication is a critical feature of single subject designs because it is through replication that both internal and external validity are demonstrated. Replication can take many forms. *Direct replication* is the replication of the procedures by the same experimenter using the same procedures either within or across participants. Sidman (1960) stated that there are two types of direct replication – *intrasubject*, where a participant is exposed to the intervention at multiple points in time (e.g., a reversal design), and *intersubject*, where a new participant is introduced to the study to replicate effects.

*Systematic replication* is a form of replication in which some part of the original study is altered to increase the generality of the results. For example, a conflict-resolution strategy that has been shown in the literature to be effective with adults might be applied to an adolescent population. Systematic replication typically occurs in a new study and with new researchers.

Replication in investigations using single subject designs is also used to establish the external validity of experimental findings. Birnbrauer (1981) argued that the external validity of data from single subject designs is established through systematic replication of effects by other researchers in settings different from the site in which the original study occurred. The focus in research using single subject designs is to not only demonstrate the replicability of findings but to also establish the parameters of the intervention.

## 9. Presentation and Summarization of Data

The visual analysis of graphic data both within and across conditions of a research study represents the most frequently used data analysis strategy employed by researchers using single subject designs. Each single subject design requires a specific pattern of data in order for an investigator to conclude that implementation of the independent variable is responsible for changes in the dependent variable. Researchers typically evaluate several factors when visually analyzing data: *Trend, level, variability, immediacy of effect*, and *percentage of overlap*. *Trend* refers to the slope of the data, and should not be confused with trend analysis typically used with inferential statistics. The trend of the data should be evaluated both within and between conditions.

A minimum of three data points must be collected in each condition in order to establish level and trend. This is especially true in baseline, which Birnbrauer (1981) equated to the descriptive data often seen in group designs. That is, baseline describes the participants' responding under naturalistic conditions before the application of the dependent variable. More data might be taken in cases where level and/or trend are not easily evaluated (such as with highly variable data) or in the case where extended baselines show the independence of baselines, such as in a multiple baseline or multiple probe design.

Fewer than three data points might be taken in cases where extending a baseline or treatment condition might result in physical or psychological harm, such as in the case of severe self-injury.

Data should be stable (that is, three data points with little change in slope) or trending in the opposite direction to the desired effect before moving to another condition. Whether the desired trend is ascending or descending depends upon the conditions in place and the operational definition. For example, if a reversal design is used to determine the effects of praise on homework completion, one would want a stable or descending trend in baseline before proceeding. If homework completion was ascending during baseline and continued to increase during intervention, it would be difficult to determine whether the treatment per se was responsible for changes in responding. However, if the dependent variable was calling out, the investigator would want to have a stable or ascending baseline before proceeding.

The investigator should also see changes in trend between conditions as well. For example, few conclusions about the effectiveness of the dependent variable can be made if an ascending trend is observed in baseline and continues in treatment. However, if baseline responding is stable or descending and treatment results in an increasing trend, the independent variable clearly had an effect on the dependent variable.

*Level* refers to the height of the data on the *y* axis. Kennedy (2005) referred to level as "the average of the data within a condition" (p. 197). Level of data is often used to make a comparison between two phases. Immediate changes in level upon changes in condition suggest that the application of the treatment variable(s) is responsible for changes in the dependent variable.

*Variability* refers to the patterns observed between individual data points. Most data will not have a clean ascending or descending trend; however, the less variable the data, the more clear the pattern. Occasionally, data will be variable/ stable, in which case a pattern of variability can be easily seen. For example, if two data points are ascending and one is descending, and this same pattern is observed at least once more, the data are considered to be *variable/stable*. It would be appropriate to move to the next phase of the investigation if this were the case. When data points are scattered in such a way that no pattern can be discerned, the researcher must attempt to control for the variability in the data. Unless the source of variability cannot be determined, the investigator should avoid moving to the next phase of the investigation.

The visual analysis of data requires the researcher to evaluate carefully each of these factors to determine whether any changes in behavior from baseline to intervention can be attributed to the independent variable. For example, one could conclude that the independent variable was effective only if a clear change in trend and/or level occurred from baseline to intervention. Conclusions may also be affected/limited the more time that passes between implementation of the intervention and changes in behavior. The percentage of overlap between baseline and intervention data also will affect conclusions. In general, stronger statements about the effectiveness of the intervention can be made if there is very little overlap between baseline and intervention data.

It is crucial that changes in treatment conditions (e.g., moving from baseline to treatment) are made based upon careful analysis of the trend, level, and variability of the data. Inferential statistics are rarely used in single subject designs; instead, visual inspection of data is used. Functional relationships are documented and analyzed by looking at patterns in the data. Statements such as "data were collected for three days for all participants" suggest that the schedule of the application of the independent variables were determined before the study began, and that changes in data patterns were not considered.

The description of the data in the narrative should match the graph, but be written with enough detail that it can stand by itself. Any changes in trend and level of the data should be described in detail. The mean and range of responding per condition is typically presented, and can help the researcher understand how responding changes over time.

## 10. Discussion of Data

The Discussion sections of single subject manuscripts do not differ significantly from those using other types of research designs. The results of data collection are discussed relative to the research question. The research question should be answered in this section; that is, the authors should address whether the data support that the intervention(s) was effective in producing behavior change. Previous literature should be cited as it relates to the interpretation of the research question; however, introducing new literature not previously cited in the Introduction should be avoided unless absolutely necessary. The researcher is free to make generalizations and suppositions based upon the data, but should do so with care, and on the basis of the data collected.

New data, aside from anecdotal evidence collected during the study, should not be introduced in this section. Relevant and/or idiosyncratic variables for replication that might not be addressed in the procedures should be presented in the discussion section. Limitations to the study should also be addressed. For example, if a participant withdrew from the study and their results were counter to the results collected from full participants, that information should be discussed. Possible uncontrolled threats to internal and external validity should be described in detail (see Chapter 26, this volume).

Most single subject research leads to information that informs some sort of systematic replication. New research based upon some dimension of the study (e.g., using basic findings to investigate an innovative treatment) is occasionally observed as well. Suggestions for future research and replication should be presented at the end of the Discussion.

## Notes

1  Preparation of this chapter was supported in part by USDE OSEP grant #H355A040025.
2  An A-B-A is the basic form of the reversal design. Although experimental control can be established using this form of the design, researchers prefer to use the A-B-A-B because it allows for additional replication and ends on an intervention phase.

## References

Ahearn, W. H., Clark, K. M., MacDonald, R. P. F., & Chung, B. I. (2007). Assessing and treating vocal stereotypy in children with autism. *Journal of Applied Behavior Analysis, 40*, 263–275.
Alberto, P. A., & Troutman, A. C. (2003). *Applied behavior analysis for teachers* (6th ed.). Upper Saddle River, NJ: Merrill-Prentice-Hall.
Bailey, J. S., & Burch, M. R. (2002). *Research methods in applied behavior analysis.* Thousand Oaks, CA: Sage.
Barlow, D. H., & Hayes, S. C. (1979). Alternating treatments design: One strategy for comparing the effects of two treatments in a single subject. *Journal of Applied Behavior Analysis, 12*, 199–210.
Birnbrauer, J. S. (1981). External validity and experimental investigation of individual behavior. *Analysis and Intervention in Developmental Disabilities, 1*, 117–132.
Cohen, J. (1960). A coefficient of agreement for nominal scales. *Educational and Psychological Measurement, 20*, 37–46.
Cooper, J. O., Heron, T. E., & Heward, W. L. (2007). *Applied behavior analysis* (2nd ed.). Upper Saddle River, NJ: Pearson-Merrill-Prentice-Hall.
Hartman, D. P., & Hall, R. V. (1976). The changing criterion design. *Journal of Applied Behavior Analysis, 9*, 527–532.
Horner, R. D., & Baer, D. M. (1978). Multiple-probe technique: A variation on the multiple baseline. *Journal of Applied Behavior Analysis, 11*, 189–196.
Horner, R. H., Carr, E. G., Halle, J. W., McGee, G., Odom, S., & Wolery, M. (2005). The use of single-subject research to identify evidence-based practice in special education. *Exceptional Children, 71*, 165–180.
Iwata, B. A., Dorsey, M. F., Slifer, K. J., Bauman, K. E., & Richman, G. S. (1994). Toward a functional analysis of self-injury. *Journal of Applied Behavior Analysis, 27*, 197–209.
Johnston, J. M., & Pennypacker, H. S. (1993). *Strategies and tactics of behavioral research* (2nd ed.). Hillsdale, NJ: Erlbaum.
Jones, K. M., & Friman, P. C. (1999). A case study of behavioral assessment and treatment of insect phobia. *Journal of Applied Behavior Analysis, 32*, 95–98.
Kazdin, A. E. (1977). Artifact, bias, and complexity of assessment: The ABCs of reliability. *Journal of Applied Behavior Analysis, 10*, 141–150.
Kazdin, A. E. (1981). External validity and single-case experimentation: Issues and limitations (A response to J. S. Birnbrauer). *Analysis and Intervention in Developmental Disabilities, 1*, 133–143.
Kennedy, C. H. (2005). *Single case designs for educational research.* New York: Allyn & Bacon.

Mayfield, K. H., & Vollmer, T. R. (2007). Teaching math skills to at-risk students using home-based peer tutoring *Journal of Applied Behavior Analysis, 40,* 223–237.

Reeve, S. A., Reeve, K. F., Townsend, D. B., & Poulson, C. L. (2007). Establishing a generalized repertoire of helping behavior in children with autism. *Journal of Applied Behavior Analysis, 40,* 123–136.

Saville, B. K., Zinn, T. E., Neef, N. A., Van Norman, R., & Ferreri, S. J. (2006). A comparison of interteaching and lecture in the college classroom. *Journal of Applied Behavior Analysis, 39,* 49–61.

Secan, K. E., Egel, A. L., & Tilley, C. S. (1989). Acquisition, generalization, and maintenance of question-answering skills in autistic children. *Journal of Applied Behavior Analysis, 22,* 181–196.

Sidman, M. (1960). *Tactics of scientific research: Evaluating experimental data in psychology.* Boston: Authors Cooperative.

Skinner, B. F. (1938). *The behavior of organisms.* New York: Appleton-Century-Crofts.

Tawney, J. W., & Gast, D. L. (1984). *Single subject research in special education.* Columbus, OH: Charles E. Merrill Publishing Company.

Ulman, J. D., & Sulzer-Azaroff, B. (1975). Multielement baseline design in educational research. In E. Ramp & G. Semb (Eds.), *Behavior analysis* (pp. 371–391). Engelwood Cliffs, NJ: Prentice-Hall.

Warnes, E. D., & Allen, K. D. (2005). Biofeedback treatment of paradoxical vocal fold dysfunction and respiratory distress in an adolescent female. *Journal of Applied Behavior Analysis, 38,* 529–532.

# 28
# Structural Equation Modeling

Ralph O. Mueller and Gregory R. Hancock

Structural equation modeling (SEM) represents a theory-driven data analytical approach for the evaluation of a priori specified hypotheses about causal relations among measured and/or latent variables. Such hypotheses may be expressed in a variety of forms, with the most common being *measured variable path analysis* (MVPA) models, *confirmatory factor analysis* (CFA) models, and *latent variable path analysis* (LVPA) models. For analyzing models of these as well as more complex types, SEM is not viewed as a mere statistical technique but rather as an analytical process involving model conceptualization, parameter identification and estimation, data-model fit assessment, and potential model respecification. Ultimately, this process allows for the assessment of fit between correlational data (obtained from experimental or non-experimental research) and one or more competing causal theories specified a priori; SEM is explicitly not designed for exploratory purposes. Software packages such as AMOS, EQS, LISREL, and Mplus are utilized to complete the computational, but not the substantive aspects of the overall SEM process. For contemporary treatments of SEM we recommend texts by Byrne (1998, 2001, 2006), Kline (2005), and Loehlin (2004), or, for more advanced readers, books by Bollen (1989), Hancock and Mueller (2006), and Kaplan (2000). Specific desiderata for applied studies that utilize SEM are presented in Table 28.1 and explicated subsequently.

## 1. Substantive Theories and Structural Equation Models

Early in a manuscript, each model under investigation must be thoroughly justified by a synthesis of the theory thought to underlie that model. In typical SEM applications, an operationalized theory assumes the form of a *measured variable path analysis* (MVPA), *confirmatory factor analysis* (CFA), or *latent variable path analysis* (LVPA), although the analysis of more complex models (e.g., latent means, latent growth, multilevel, or mixture models) is becoming more popular in the applied behavioral and social science literatures. Regardless of model type, there must be strong consonance between each model and the underlying theory, as a lack thereof can undermine the modeling process. SEM's main strength lies in its ability to help evaluate a priori theories, not to generate them post hoc (equally, not to evaluate theories derived through prior exploration of the same data, for example with an exploratory factor analysis). Often, the articulation, justification, and testing of competing alternative models strengthens a study as it provides a more complete picture of the current thinking in a particular field (also see Desideratum 14). Thus, authors must convey a firm overall sense of what

Table 28.1 Desiderata for Structural Equation Modeling

| Desideratum | Manuscript Section(s)* |
|---|---|
| 1. Substantive theories that led to the model(s) being investigated are synthesized; a set of a priori specified competing models is generally preferred. | I |
| 2. Path diagrams are presented to facilitate the understanding of the conceptual model(s) and the specification of the statistical model(s). | I |
| 3. If applicable, latent factors are defined and their status as latent (vs. emergent) is justified. | I, M |
| 4. Measured variables are defined and, if applicable, their appropriateness as indicator variables of associated factors is justified. | M |
| 5. Latent factors are indicated by a sufficient number of appropriately measured variables; how the latent factors are given scale within the model(s) is addressed. | M |
| 6. How theoretically relevant control variables are integrated into the model is explained. | M |
| 7. Sampling method(s) and sample size(s) are explicated and justified. | M |
| 8. The treatment of missing data and outliers is addressed. | M, R |
| 9. The name and version of the utilized software package is reported; the parameter estimation method is justified and its underlying assumptions are addressed. | M, R |
| 10. Problems with model convergence, offending estimates, and/or model identification are reported and discussed. | R |
| 11. Summary statistics of measured variables are presented; if raw data were analyzed, information on how to gain access to the data is provided. | R |
| 12. For models involving structural relations among latent variables, a two-phase (measurement, structural) analysis process is followed and summarized. | R |
| 13. Recommended data-model fit indices from multiple classes are presented and evaluated using literature-based criteria. | R |
| 14. For competing models, comparisons are made using statistical tests (for nested models) or information criteria (for non-nested models). | R |
| 15. For any post hoc model re-specification, theoretical and statistical justifications are provided. | R |
| 16. Latent factor quality is addressed in terms of validity and reliability. | R |
| 17. Standardized and unstandardized parameter estimates together with information regarding their statistical significance are provided; $R^2$ values for key structural outcomes are presented. | R, D |
| 18. Appropriate language regarding model tenability and structural relations is used. | D |

* Note: I = Introduction, M = Methods, R = Results, D = Discussion

theoretical and/or prior empirical evidence has led to the initial conceptualization of the model(s) under study.

## 2. Path Diagrams

A *path diagram* is a graphical depiction of a theory relating measured and possibly latent variables and is helpful not only in representing the conceptual links among those elements but also in the specification of the statistical model. By convention, *measured variables* are represented by rectangles/squares and *unobserved factors* are expressed by ellipses/circles (and, in the case of mean structure models, occasionally a triangle is used to facilitate the modeling of means and intercept terms). Directional (one-headed) arrows point from hypothesized *causes* to *effects*, while non-directional (two-headed) arrows represent non-causal covariation between elements of a model (or a variance of an element, when the non-directional arrow returns directly to the element of origin).

In investigations proposing models that involve structural relations among factors (whether those factors are *latent* or *emergent*; see Desideratum 3), a distinction should be made between the *measurement* and *structural* phases of the modeling process (see Desideratum 12). Often, the *structural portion* of such a model is the focus of the study as it represents the main theory to be tested; thus, a path diagram of the structural portion should be presented and justified early in a manuscript (see Figure 28.1a). The model may be depicted with the associated *measurement portion* which focuses on how latent variables are manifested in observable data, that is, it explicates the hypothesized relations between latent factors and their chosen measured indicators. A model with both structural and measurement portions is shown in Figure 28.1b (assuming three indicators per factor for simplicity), although details of the measurement model might be better presented and explained in the Methods section of a manuscript where specifics about the instrumentation are typically discussed. In sum, authors are encouraged to utilize appropriately detailed path diagrams to complement and illustrate their written explanations of the theory being tested.

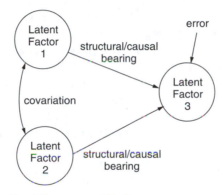

(a). Structural Model without Measurement Portion

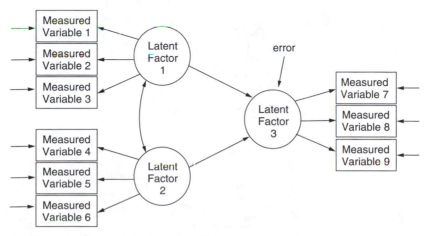

(b). Structural Model with Measurement Portion

**Figure 28.1 (a and b)** Path Diagrams.

## 3. Latent vs. Emergent Factors

Structural equation modeling (SEM), in its common forms of confirmatory factor analysis (CFA) and latent variable path analysis (LVPA), addresses theory-based relations among latent variables (factors). Here, the term *latent* adds meaning to its popular definition of being unobserved or unobservable: it connotes a factor hypothesized to have a causal bearing on one or more measured indicator variables. This causal relation implies that the latent factor explains variance in its measured indicator variables and induces covariance among them. That is, individuals vary on the measured indicators in part because they vary on the underlying factor and their indicator scores covary because they have a common latent cause. SEM analyses involving factors typically assume them to be latent, and researchers must explain and justify that the factors are indeed causal agents of their observable *effect* indicators (see Figure 28.2a).

On the other hand, theory and/or the nature of the available data occasionally dictate that measured variables serve as *cause* indicators of constructs therefore described as *emergent* (rather than effect indicators of constructs considered latent). Variation in such measured cause indicators (which themselves might or might not covary) is now hypothesized to cause and partially explain variation in the emergent factor (see Figure 28.2b). For example, the construct *socioeconomic status* (SES)—measured by indicators such as parental *income*, *education*, and *occupational prestige*—should be modeled as emergent, not latent: It seems to make little theoretical sense, for instance, to propose that SES is the cause of parents' education level. More sensible seems an emergent system, whereby changes in one's SES are caused (at least in part) by changes in parental income, education, and occupational prestige. The analysis of models involving emergent constructs is certainly possible but generally more challenging (for example, parameter estimation difficulties can occur) than models involving latent variables only. It is incumbent upon the researcher to justify *each* factor's status as latent (or emergent) rather than simply presume a latent status for all factors. The latter could lead to a misspecified measurement model and, in turn, to incorrect inferences regarding the relations between the construct in question and other variables (latent or measured) within the model. For a more detailed discussion on latent vs. emergent variable systems, consult Kline (2006).

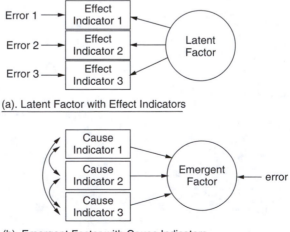

(a). Latent Factor with Effect Indicators

(b). Emergent Factor with Cause Indicators

**Figure 28.2 (a and b)** Latent vs. Emergent Variables.

## 4. Measured Variables

All measured variables in a model should be defined clearly, or if they are previously established measures, an accessible reference to their description should be provided. Measured variables may be *exogenous* (independent; having no causal inputs) or *endogenous* (dependent; having one or more causal inputs) within a model. Each may serve as (a) an effect indicator of a latent factor (where it is theoretically clear that the factor has a causal bearing on the measured variable); (b) more rarely, a cause indicator of an emergent factor (where it is theoretically clear that the variable has a causal bearing on the factor; see Desideratum 3); or (c) a stand-alone variable, not serving as an indicator of any factor in the model (e.g., sex of respondent).

Measured variables can be individual items from larger scales or scores obtained by summing across several response items. Exogenous variables may be dichotomous (two categories) or continuous without requiring any special model estimation procedures. The analysis of ordinal variables with fewer than five scale points (exogenous or endogenous, including effect indicators of latent factors) typically requires special estimation procedures and/or input data consisting of alternate measures of association (e.g., polychoric correlations, see Chapter 5, this volume). Other measured variable types that might warrant accommodation, or a defense as to why no such accommodation was made, include count variables and *censored* variables (e.g., an income variable with an upper category of $50,000 and up).

## 5. Indicator Variables of Latent Factors

Without context, latent factors may have any number of measured variable indicators, but in order to determine if an adequate number was utilized in a particular application, three key issues should be considered. First, overall model identification (the ability to estimate all model parameters) can be helped or hindered by the chosen number of indicators per latent factor. With three or four indicators, the factor does not require the estimation of more parameters (e.g., loadings, error variances) than the data supplied from its measured indicators. Having fewer than three indicators is certainly possible and occasionally even necessary due to limited availability. Using a single indicator variable of a factor may be accomplished by setting the error variance of the indicator variable to zero (effectively equating the variable with the factor) or to some portion of the measured indicator's variance, depending upon known reliability of the indicator variable (e.g., a reliability coefficient of .80 for a measured variable implies that 20% of its variance is residual, unexplained by its latent factor). Using two indicator variables for a latent factor typically requires no fixing of error variances to known constants as long as the factor relates to one or more other factors in the model. This case does run the risk, however, of creating an *empirical under-identification* problem should the relation of the factor to others within the model be estimated as zero or near zero.

Second, having three or more indicators tends to enhance the quality of the construct (its theoretical breadth as well as its ability to replicate across samples, that is, construct validity and reliability). Contrary to early methodological literature, having additional indicators does not appear to burden estimation even for relatively modest sample sizes. At some point, however, diminishing returns are expected. Our experience is that having four to six indicators of reasonable quality (e.g., standardized loadings exceeding .6 or .7 in absolute value) is practically ideal, although more indicators of slightly lesser quality can be feasible as well. Researchers with many indicators at their disposal might consider forming composites (*parcels*) of items to serve as indicators.

Third and finally, as factors are latent, they have no inherent *metric*. Hence, they must be assigned units within the model. For an exogenous factor, this can be accomplished by either fixing its variance (typically to 1, giving the factor standardized units) or by keeping the variance freely estimated and instead fixing a path to one of its indicator variables (typically to 1, giving the factor the metric of the indicator variable). For an endogenous factor, assigning units is typically accomplished by the latter method, although some software programs allow the specification of the former as well. Researchers should be explicit in how latent factors were scaled and if their software package handled the assignment automatically.

## 6. Control Variables

Researchers might examine parameter estimates in a model of theoretical interest after controlling for the effects of specific variables, thus ensuring that the estimates are above and beyond the linear effects of external elements (including background variables capturing individual characteristics, such as IQ, and/or environmental conditions, such as family income or parental education). Although some applied researchers have attempted such control by a priori partialing the variables from the data, the practice of analyzing such *residualized* data with structural equation modeling is questionable. Instead, incorporating the control variables directly into the model is the preferred strategy. Typically all control variables (measured and/or latent) are allowed to covary as exogenous predictors within the model. Each has a structural path to all exogenous and endogenous factors (and stand-alone variables) within the structural, but not the measurement portion of the model. Interpretation of the paths within the structural portion proceeds normally, now acknowledging that they have been purged of the linear effects of the control variables.

Two additional points of elaboration are worthwhile. First, control variables should not generally be modeled as indicators (cause or effect indicators) of a single factor (e.g., a demographic factor) unless indeed the factor makes sense as a continuum in its own right. Second, the type of control described here is *linear*, not nonlinear or multiplicative. Thus, for example, if a researcher believes the structural paths of interest actually differ as a function of the control variables (i.e., that the control variables *moderate* the structural relations), then a model with multiplicative predictor terms or, in some cases, a multisample analysis (see Chapter 29, this volume) should be considered. Either way, the researcher should clarify the expected nature of the control variables' relations with the structural elements of the theoretical portion of the model, and how the modeling strategy employed exacts the proper type of control.

## 7. Sampling Method and Sampling Size

Sampling methods (e.g., random, stratified, cluster) must be made explicit in the text. Stratified sampling techniques typically yield sampling weights (see Chapter 30, this volume) whose purpose, when applied to the sample data, is to weigh individual cases more or less in order to maximize the sample's representativeness of the target population. In such cases researchers should delineate how these weights were incorporated into their modeling. Cluster sampling approaches, where groups of individuals are sampled at a time, will yield a *hierarchical* (nested) structure to the data (e.g., students within classrooms within schools). This introduces some dependence among observations in the sample, thereby violating the assumption of independence of observations. In this case researchers should utilize a multilevel structural equation modeling approach or present evidence of sufficiently small effect of the dependence among clusters of cases (e.g., a small design effect) to justify not pursuing the more complex multilevel approach.

In order to determine the sample size required for a structural equation modeling (SEM) analysis, researchers should consider both adequacy for correct parameter estimation and for desired level of statistical power. A common guideline for obtaining trustworthy *maximum likelihood* (ML) estimates is to have at least five cases per model parameter (not per variable); when employing other estimation methods that require less stringent distributional assumptions and/or that are tailored specifically for ordinal data (e.g., *Asymptotically Distribution Free*; *Arbitrary Generalized Least Squares*; *Weighted Least Squares*; *Satorra-Bentler rescaling corrections*), more cases are generally needed. If the available sample size is substantially smaller, researchers must provide justification from the SEM methodological literature (e.g., sample size necessary for sound parameter estimation can be reduced by having latent factors of high quality; see Desideratum 16). Regarding power, sample size should be justified on two fronts. First, authors should explain how the available sample size provides adequate power for relevant tests of the data-model fit as a whole (e.g., using the confidence interval for the *Root Mean Square Error of Approximation*, RMSEA; see Desideratum 13). Second, the sample size should provide sufficient power to detect individual model relations of key theoretical interest (e.g., structural connections among latent factors). For a detailed discussion of statistical power in SEM, consult Hancock (2006).

## 8. Missing Data and Outliers

Within the context of structural equation modeling (SEM), classic missing data techniques (e.g., listwise or pairwise deletion, mean substitution) are generally considered inadequate unless the amount of missing data is so small as to be trivial. Methods currently considered acceptable include *full-information maximum likelihood* (FIML) estimation and *multiple imputation* (MI), but their availability varies across SEM software packages (for details, see Enders, 2006). The former implicitly allows all subgroups of individuals, defined by differing patterns of available data, to contribute to those parameters' estimation which their data are able to inform. The latter, on the other hand, imputes values for those missing but does so multiple times to determine average parameter estimates from across several possible sets of partially imputed data. Both methods assume that data are *missing at random*, that is, that the missing data mechanism for each variable is independent of that variable (e.g., people failing to provide income information has nothing to do with their income). Authors must identify and justify which missing data algorithm was utilized (FIML or MI) and report the proportions of cases missing for each applicable variable.

With regard to cases that were not missing but were otherwise aberrant, criteria for the detection and possible removal of such outliers from further analysis must be addressed. To start, if descriptive/background data led the researcher to conclude that specific cases were inappropriate for generalizing to the population of interest (e.g., some data came from foreign students in otherwise English-speaking classrooms), such preliminary (but post hoc) exclusion criteria should be made explicit. From a univariate statistical perspective, outliers can be determined by comparing cases' standard ($z$) scores against some reasonable threshold, such as ±3; cases exceeding that threshold could be removed. Multivariate outliers (where a case's scores on individual variables may not be unusual but its combination of scores sets it apart) may be diagnosed using Mahalanobis distance ($D$) or squared distance ($D^2$). Because $D^2$ values follow a $\chi^2$ distribution in large samples, each case has an associated $p$-value; cases with extreme values, say $p < .001$, could be removed as outliers. Finally, cases may be evaluated in terms of their influence on the overall multivariate kurtosis (to which normal-theory estimation methods such as maximum likelihood are particularly sensitive); criteria for cases' removal are relative to the metric of kurtosis employed by the software. Thus, with regard to outliers, authors are responsible for detailing the criteria used for outlier removal or for addressing other methods automatically employed by software packages (such as down-weighting extreme cases).

## 9. Software and Estimation Method

Typical structural equation models may be estimated using a wide variety of software packages, including, but not limited to, AMOS, EQS, LISREL, Mplus, Mx, and SAS PROC CALIS. These packages tend to produce very similar if not identical results for most common applications. As modeling needs become more demanding, however, either in terms of addressing assumption violations (e.g., nonnormality; categorical data) or model complexity (e.g., multilevel models), packages differ in their capabilities and/or the manner in which they meet such needs. Authors should report which software package was used and the specific version of that software, as these packages are constantly evolving. If a package was chosen to meet specific modeling needs, an explanation thereof is warranted.

Regarding estimation methods specifically, although maximum likelihood (ML) is a default in virtually all packages, it should not be chosen without an understanding of its limitations (e.g., inflated parameter $z$-values and model $\chi^2$ values under nonnormality), an explicit rationale for its selection, and data-based justification thereof. Summary statistics detailing a lack of skewness (e.g., < 2) and kurtosis (e.g., < 7) for measured variables can help assuage concerns regarding nonnormality at a univariate level. Addressing more directly the multivariate assumption underlying normal-theory methods (e.g., ML; *Generalized Least Squares*), Mardia's *normalized coefficient of multivariate kurtosis* should also be reported. Although no universal guideline exists, values around 3 or less can be reassuring. By extension, authors utilizing an alternate estimation procedure (e.g., Asymptotically Distribution Free; Weighted Least Squares) should fully justify the selection, including a discussion of its assumptions (or specific lack thereof) and its applicability to the data and model(s) at hand.

## 10. Problems with Convergence, Estimates, and Identification

In practice, structural equation modeling (SEM) analyses seldom proceed smoothly. Whether by programming error or uncooperative data, problems involving convergence, estimation, and/or identification inevitably occur. The estimation process may fail to converge within a default number of iterations; linear dependencies may occur that prevent some parameters' identification and hence estimation; offending estimates such as negative error variances (Heywood cases) may arise; matrices may be reported as nonpositive-definite; and so forth. As SEM is a process, documenting the steps followed is necessary just as in other scientific endeavors. To facilitate model convergence, perhaps it was necessary to specify start values other than the default (typically derived from *Two Stage Least Squares*), or some variables needed to be rescaled to make their variances less extreme relative to others. Perhaps an error variance needed to be constrained (e.g., to zero or slightly above) to eliminate a Heywood case, or a ridge regression correction was necessary to address a nonpositive-definite covariance matrix. Regardless of the issues faced, whether those described above or others, authors are responsible for detailing challenges that arose and the corrective actions taken, or for stating explicitly that no such problems were encountered.

## 11. Data Display and Accessibility

In order to facilitate verification of study results, and allow for exploration of competing explanations for those results, the American Psychological Association (APA) requires that in their manuscripts, authors present "informationally adequate statistics." Although, clearly, not all journals containing structural equation modeling analyses are governed by APA, the rationale for this requirement is sound and is herein endorsed. Authors should provide adequate summary information for readers to be able to verify the presented results (e.g., sample size, covariance matrices or correlation matrices with standard deviations, and possibly other relevant statistics such as means and reliabilities). In instances where the number of variables modeled prohibits tabling summary information economically, information as to how to acquire such information should be provided. Where advanced estimation methods are used that draw directly from raw data and cannot be accomplished using summary statistics, authors should provide information as to how readers may gain access to the data.

## 12. Two-Phase Modeling Approach

For the analysis of latent variable path models, a two-phase modeling process is generally recommended to facilitate the diagnosis and potential remediation of data-model misfit. In the first or *measurement* phase, the model is temporarily respecified such that all latent variables are allowed to freely covary (along with stand-alone and control variables). If the data satisfactorily fit this measurement model (see Desideratum 13), the second phase can commence; if not, respecification of the measurement model may be entertained. Such modifications are typically informed by theoretical considerations, *Lagrange multiplier tests* (*modification indices*), and/or relatively large residuals, and most often take the form of either error covariances or cross-loadings (paths from factors to secondary measured indicators) not already in the model; see Desideratum 15. If satisfactory data-model fit was not achieved even after such modification(s), further modeling should be terminated. If, however, reasonable fit is achieved, the second phase of modeling is entered.

In the second or *structural* phase, the originally hypothesized structural relations are re-inserted among the factors (and stand-alone and control variables) while preserving whatever measurement model modifications might have been made during the first phase. Poor data-model fit for this initial structural model can no longer be due to the measurement portion of the model but to the structural portion. At this point, both statistical and practical evaluation of the overall model should occur. Because this initial structural model is nested within the final measurement model, a $\chi^2$ difference test should be conducted to assess the statistical difference between the two (see Desideratum 14). Ideally, fit would not degrade statistically significantly, however a fit difference in the final measurement model and initial structural model is typically expected since structural perfection is unlikely. If the data do not fit the initial structural model to a satisfactory degree, it should be rejected and authors should explicitly acknowledge so. If, on the other hand, authors respecified the conceptual structure, strong statistical and theoretical justifications must be provided and the now exploratory nature of the analysis acknowledged (see Desideratum 15).

## 13. Data-Model Fit

A central issue addressed by any structural equation modeling analysis is the assessment of the fit between observed data and the hypothesized model (for a now classic overview, see Bollen & Long, 1993). While a $\chi^2$ test is commonly reported for this purpose, it is viewed by most as overly strict given its power to detect even trivial deviations of a data from the proposed model. Researchers should therefore report multiple fit indices, typically drawing from three broad classes (while many indices are available, only recommended ones are listed together with empirically derived target values to retain a model):

- *Absolute indices* evaluate the overall discrepancy between observed and implied covariance matrices (and possibly means); fit improves as more parameters are added to the model: the *Standardized Root Mean Square Residual* (SRMR) should fall below .08.
- *Parsimonious indices* evaluate the overall discrepancy between observed and implied covariance matrices (and possibly means) while taking into account a model's complexity; fit improves as more parameters are added to the model, as long as those parameters are making a useful contribution: the *Root Mean Square Error of Approximation* (RMSEA) and its associated 90% confidence interval should fall below .05.
- *Incremental indices* assess absolute or parsimonious fit relative to a baseline model, usually the null/independence model (which specifies no relations among observed variables): the *Normed Fit Index* (NFI), *Nonnormed Fit Index* (NNFI; also referred to as *Tucker-Lewis Index*, TLI), and/or *Comparative Fit Index* (CFI) have .95 as a minimum target value.

Because absolute, parsimonious, and incremental data-model fit indices can lead to inconsistent conclusions, reviewers should insist on fit results from different classes. If, after considering several
indices, data-model fit is deemed acceptable (and judged best compared to competing models, if
applicable), the model is retained as tenable and individual parameter estimates may be interpreted.
If, however, evidence suggests unacceptable data-model fit, it *might* be appropriate to respecify the
model (see Desideratum 15) to improve fit.

## 14. Model Comparisons

In as much as authors follow the recommendation to propose and test competing theories in their
investigations (see Desideratum 1), they are then obligated to offer comparative judgments regarding
the relative tenability of these alternative explanations of what gave rise to the observed data. These
judgments can be based on statistical and/or descriptive evidence, depending upon the *nested* (hierarchical) nature of the structural equation models. When two models, say Model 1 and Model 2, are
nested (such as when the estimated parameters in the former are a proper subset of those associated
with the latter), fit comparisons can be accomplished with a formal $\chi^2$ difference test (also referred to
as a *likelihood ratio test*). That is, if Model 1 (with $df_1$) is nested within Model 2 (with $df_2$), their $\chi^2$ fit
statistics may be statistically compared by $\Delta\chi^2_{(df_1-df_2)} = \chi^2_{(df_1)} - \chi^2_{(df_2)}$, which itself follows a $\chi^2$ distribution
with $df = df_1 - df_2$ (under conditions of multivariate normality and reasonable models). For example,
an orthogonal CFA model (factors are specified to be independent) is nested within an oblique one
(factors are allowed to covary), and data will fit the former less well than the latter as evidenced by a
larger model $\chi^2$. However, if the $\chi^2$ difference test indicates no statistically significant difference in fit,
a researcher focusing on model parsimony might prefer an orthogonal over an oblique explanation as
no statistical distinction has been established.

When Model 1 and Model 2 do not have a nested relation, researchers must then compare the
models descriptively, often using an information criterion such as the *Akaike Information Criterion*
(AIC), which expresses the expected ability of models to cross-validate: Models associated with
smaller AIC values are preferred over models with larger AIC values. Descriptive comparisons of
other data-model fit indices are also possible to discern practical differences in fit, but irrespective
of comparison approach, authors must be clear that these results are relative, not absolute:
while data might be judged to fit one model better than another, both models might exhibit unsatis-
factory data-model fit as gauged by indices designed to evaluate each model individually (see
Desideratum 13).

## 15. Model Respecification

In a strict sense, *any* hypothesized model is, at best, only an approximation to reality; the remaining
question is one of degree of that misspecification. With regard to *external* specification errors—when
irrelevant variables were included in the model or substantively important ones were left out—remediation can only occur by respecifying the model based on additional relevant theory. On the other
hand, *internal* specification errors—when unimportant paths among variables were included or when
important paths were omitted—can potentially be diagnosed and remedied using *Wald statistics* (predicted increase in $\chi^2$ if a previously estimated parameter were fixed to some known value, e.g., zero)
and *Lagrange multiplier statistics* (also referred to as *modification indices*; estimated decrease in $\chi^2$ if a
previously fixed parameter were to be estimated). As these tests' recommendations are directly motivated by the data and not by theoretical considerations, any resulting respecifications must be
acknowledged as data driven and exploratory in nature and might not lead to a model that resembles
reality any more closely than the one(s) initially conceptualized.

For example, a statistically nonsignificant path could merely be the result of insufficient statistical power, and its removal from a model could become theoretically misleading. We believe authors should leave nonsignificant paths in the model, as it preserves the originally theorized model while still communicating that, within the model's context, a hypothesized relation did not establish itself beyond chance. On the other hand, implementing suggestions from modification indices to add paths to the measurement or structural portion of a model could in fact merely be a capitalization on random covariation in the current sample and not be indicative of true population relations. In the case of measurement model modifications (e.g., cross-loadings, error covariances), statistical support (e.g., change in $\chi^2$ values) should be supplemented by a theoretical rationale for each parameter addition. For structural model modifications, such as adding paths among factors not originally hypothesized, statistical and theoretical support should be strengthened by an explicit discussion addressing the tension between the original hypotheses and the exploratory and apparently contrary respecifications.

## 16. Validity and Reliability of Latent Factors

Each latent factor should be evaluated in terms of its *validity*, that is, its ability to represent the construct it is hypothesized to represent, as well as its *reliability*, that is, the ability to replicate across new sets of data. With regard to the former, researchers should address the extent to which the factor relates to elements it should, and does not relate to elements it should not. Those elements first and foremost are the factor's own effect indicators; researchers should observe patterns of loadings that are relatively high for measured variables expected to reflect the factor, and relatively low (ideally zero) for variables intended to reflect other factors. One validity index commonly recommended is the *variance extracted* (the average squared standardized loading for a factor's indicators) with target values around .50 and above. Another gauge of a factor's validity worth addressing is the degree to which its relation to other factors matches theoretical expectations.

With regard to reliability, while in theory a factor is perfectly reliable as it is an error-free entity, in practice a factor would not be expected to replicate perfectly should the same individuals provide new scores on the factor's indicators (assuming no recollection of the prior measurements). Thus, reliability reflects the extent to which a factor is expected to replicate (i.e., correlate with itself), given that it is indicated by data containing error. Although some researchers report Cronbach's $\alpha$ based on the sum of a factor's indicators (typically after standardization), this index is inappropriate as it reflects reliability of a composite rather than the reliability of the factor as reflected in its measured indicators. Thus, we recommend instead that authors report maximal reliability (*Coefficient H;* e.g., Hancock & Mueller, 2001) for each factor, as it is an estimate of the correlation the factor is expected to have with itself over repeated administrations. Values above .70 would be considered desirable in order to establish the stability of each hypothesized factor.

## 17. Results for Specific Structural Equations

For either measured or latent variable path analysis models, once the overall data-model fit has been assessed and deemed satisfactory (see Desideratum 13) and, if applicable, evidence of the quality of latent variables has been presented (e.g., construct validity and reliability; see Desideratum 16), more detailed results regarding the structural relations of interest should be offered. Individual parameter estimates (i.e., the *direct* structural effects from one variable to another) should be listed in standardized and unstandardized form to facilitate comparisons to results obtained in subsequent studies (though technically, one set of estimates might be acceptable, given that it is derivable from the other if sufficient information is provided). Statistical significance information for key parameters is essential, in the form of symbolic designations (e.g., asterisks) or actual test statistics (e.g., $z$-values), and authors should present coefficients of determination ($R^2$ values) for each measured or latent outcome of theoretical interest. Depending on the study's purposes, a presentation of the *indirect* and *total* structural effects, along with their statistical significance information, might also prove useful in understanding and interpreting associations among structurally related measured and/or latent variables. Note, however, that structural equation modeling software packages do not indicate the statistical significance of individual indirect effects but only of the cumulative indirect effect via all potentially intervening mediators, and of the overall total structural effect of one variable on another.

## 18. Interpretative Language

Readers should be wary of authors' claims that acceptable data-model fit implies a model was "confirmed" or that a particular theory was proven to be "true," especially after post hoc respecifications (see Desideratum 15). Such statements are grossly misleading given the typically nonexperimental nature of the data's origins, and given that alternative, structurally different, yet mathematically equivalent models always exist that would produce identical fit results and thus would explain the data equally well. At most, a model with acceptable fit may be interpreted as *one* tenable explanation for the associations observed in the data.

Following satisfactory data-model fit, the interpretation of individual parameter estimates is permitted to involve explicit causal language, *as long as this is done from within the context of the particular causal theory proposed* and the possibility/probability of alternative explanations is raised unequivocally. Though some might disagree, we think that explicit causal statements are more honest than implicit ones and are more useful in articulating the study's practical implications within the guiding theoretical framework (for more on the role of causality in structural equation modeling and other statistical analyses, consult Mulaik, 1987, or Pearl, 2000). In the end, structural equation modeling is a powerful tool at the researcher's disposal for testing and interpreting theoretically derived causal hypotheses from within an a priori specified causal system of measured and/or latent variables. However, we urge reviewers to continually remind authors to resist the apparently still popular belief that the main goal of SEM is to achieve satisfactory data-model fit results; rather, it is just to get one step closer to the truth. If it is true that a proposed model does not reasonably approximate reality, then reaching a conclusion of *mis*fit between data and model should be a desirable goal, not one to be avoided by careless respecifications until satisfactory levels of fit are achieved.

## References

Bollen, K. A. (1989). *Structural equations with latent variables*. New York: Wiley.

Bollen, K. A., & Long, J. S. (1993). *Testing structural equation models*. Newbury Park, CA: Sage.

Byrne, B. M. (1998). *Structural equation modeling with LISREL, PRELIS, and SIMPLIS*. Mahwah, NJ: Erlbaum.

Byrne, B. M. (2001). *Structural equation modeling with AMOS*. Mahwah, NJ: Erlbaum.

Byrne, B. M. (2006). *Structural equation modeling with EQS: Basic concepts, applications, and programming*. Mahwah, NJ: Lawrence Erlbaum.

Enders, C. K. (2006). Analyzing structural models with missing data. In G. R. Hancock & R. O. Mueller (Eds.), *Structural equation modeling: A second course* (pp. 313–342). Greenwich, CT: Information Age Publishing.

Hancock, G. R. (2006). Power analysis in covariance structure modeling. In G. R. Hancock & R. O. Mueller (Eds.), *Structural equation modeling: A second course* (pp. 69–115). Greenwich, CT: Information Age Publishing.

Hancock, G. R., & Mueller, R. O. (2001). Rethinking construct reliability within latent variable systems. In R. Cudeck, S. du Toit, & D. Sörbom (Eds.), *Structural equation modeling: Present and future—A Festschrift in honor of Karl Jöreskog* (pp. 195–216). Lincolnwood, IL: Scientific Sofware International.

Hancock, G. R., & Mueller, R. O. (Eds.). (2006). *Structural equation modeling: A second course*. Greenwich, CT: Information Age Publishing.

Kaplan, D. (2000). *Structural equation modeling: Foundations and extensions*. Thousand Oaks, CA: Sage.

Kline, R. B. (2005). *Principles and practice of structural equation modeling* (2nd ed.). New York: Guilford Press.

Kline, R. B. (2006). Formative measurement and feedback loops. In G. R. Hancock & R. O. Mueller (Eds.), *Structural equation modeling: A second course* (pp. 43–68). Greenwich, CT: Information Age Publishing.

Loehlin, J. C. (2004). *Latent variable models* (4th ed.). Hillsdale, NJ: Lawrence Erlbaum.

Mulaik, S. A. (1987). Toward a conception of causality applicable to experimentation and causal modeling. *Child Development, 58,* 18–32.

Pearl, J. (2000). *Causality: Models, reasoning, and inference.* Cambridge, UK: Cambridge University Press.

# 29

# Structural Equation Modeling
*Multisample Covariance and Mean Structures*

**Richard G. Lomax**

This chapter represents an extension to Chapter 28 in this volume on basic structural equation modeling (SEM). Here we consider the development and testing of theoretical models in two more advanced contexts. First, multiple sample SEM (MS-SEM) considers the invariance (or equality) of parameters across populations (i.e., equivalence of covariance structures). For example, Jöreskog and Sörbom (1993, Example 10) tested whether the same factor structure of the SAT verbal and math sections is present in two groups of students (e.g., by examining invariance of factor loadings, the factor correlation, and the measurement error variances). Second, structured means SEM (SM-SEM) additionally assesses mean differences between populations (i.e., intercept parameters). For example, Jöreskog and Sörbom (1993, Example 13) examined mean differences between academic and non-academic boys on the latent variables of verbal ability at grades 5 and 7.

The hypothesized models can be tested utilizing SEM software such as AMOS (Arbuckle, 2007), EQS (Bentler, 2006), LISREL (Jöreskog & Sörbom, 2006), or Mplus (Muthén & Muthén, 2006) with data from experimental, quasi-experimental, cross-sectional, or longitudinal studies. For full-length descriptions of basic SEM analyses, including some discussion of MS-SEM and SM-SEM, we recommend the following textbooks: Byrne (1998, 2001, 2006), Kline (2004), Loehlin (2004), Maruyama (1998), and Schumacker and Lomax (2004). Examples of multiple sample and structured means models are described in the SEM software manuals as well as in articles such as Aiken, Stein, and Bentler (1994), Lomax (1983, 1985), and Shumow and Lomax (2002). Table 29.1 indicates the desiderata for MS-SEM and SM-SEM around which the remainder of this chapter is organized.

## 1. Justification for Theoretical Model

In reading an SEM application, it is often difficult to determine exactly where the theoretical model(s) tested come from. Here we are specifically talking about the structural model, that is, the model where the relations among the latent variables are hypothesized. We often come to the Results section of an SEM study and suddenly theoretical model(s) appear, seemingly out of thin air. Where exactly do these theoretical models come from? Does some bird or deity whisper them into the researcher's ear? Assumedly not, but the reader often has no knowledge of the basis of the models, just that they are presented as a path diagram in one or more figures of the Results section.

Table 29.1 Desiderata for Structural Equation Modeling: Multisample Covariance and Mean Structures

| Desideratum | Manuscript Section(s)* |
|---|---|
| 1. Justification for theoretical model(s) to be tested, including the latent factors therein, is made based on the available theory and research in the substantive area. | I |
| 2. Path diagram(s) are presented to display the theoretical model(s) to be tested. | I |
| 3. Populations and samples are described, including sampling method(s) and sample sizes. | M |
| 4. The measured variables that are used to indicate the latent factors are fully described (e.g., scales, reliability, validity). | M |
| 5. Name and version of the software utilized is reported. In addition, the method of parameter estimation is discussed. | M, R |
| 6. Methods of treating missing data and outliers are described. | M, R |
| 7. Table(s) of correlations, means, and standard deviations for each sample are presented, and possibly access to raw data is facilitated. | M, R |
| 8. Problems with identification, convergence, non-positive definite matrices, and inadmissible solutions are reported and resolved. | R |
| 9. Parameter estimates and statistical significance (including expected direction, such as positive or negative directionality for positive and negative relationships, respectively) are reported in tables, path diagrams, and/or text. | R |
| 10. Various global goodness-of-fit indices are reported for each model and chi-square difference tests are reported to compare nested models, if applicable. | R |
| 11. Model modifications (if any) are reported, including theoretical and statistical justifications. | R |
| 12. Series of models are tested to check for invariance of parameter estimates in multiple sample modeling (MS-SEM) (i.e., the covariance structure portion of the model). | R |
| 13. Mean structure parameter estimates are reported in structured means modeling (SM-SEM; i.e., the mean or intercept portion of the model), including effect size measures where relevant. | R |
| 14. Results are discussed in the context of assessing the (non)invariance of the theoretical model(s) tested. | D |

\* *Note*: I = Introduction, M = Methods, R = Results, D = Discussion

The theoretical model(s) should be developed from the available theory and research in the substantive area being studied. This is the major purpose of the literature review in research utilizing SEM. More specifically, the literature review should compile the research on each of the specified relations among the latent variables (i.e., the path or structure coefficients). Each path or structure coefficient is essentially a mini-hypothesis of the relation between two latent variables. Evidence could be drawn from whatever types of research are available (e.g., correlational, experimental, non-experimental, qualitative), as well as theoretical work. After reading the literature review, the reader should be presented with the theoretical models to be tested and knowledge about where the models came from. In short, the literature review should guide the reader from the available literature to the theoretical models to be tested.

For MS-SEM and SM-SEM, the literature review should also lead up to whether the researcher expects to find group differences in the theoretical model(s). For example, the research might suggest that male and female adolescents develop differently for specified constructs. This would then lead to hypotheses that certain parameters in the SEM model(s) will behave differently by gender. In MS-SEM we can test whether specific parameters appear to be invariant (the same) across populations or not invariant (different) across populations. In SM-SEM we can also test whether there appear to be latent mean differences across the populations. As an example, in the Lomax model of schooling (1985), a series of MS-SEM models determined that the factor loadings and structure coefficients appeared invariant for the public and private school populations. In subsequent analyses, an SM-SEM model inferred that there were several latent mean differences due to school type, such as for constructs relating to home background, academic orientation, and extra-curricular activity.

## 2. Path Diagrams

A path diagram should be included for each theoretical model tested. Such path diagrams can be presented in at least two different forms depending on their intended purpose. One purpose of a path diagram is to present the theoretical model in the literature review. This "conceptual diagram" will not include any results or estimates, but merely displays the latent variables and how they are theorized to relate to one another. In other words, this type of path diagram describes the theoretical structural model only. The measurement model is implied by the description and/or list of observed variables and which latent variables they assess (see Desideratum 4). The conceptual path diagram can either be depicted at the beginning of the literature review for framing purposes, or at the end of the literature review for summative purposes. Details of the observed variables are contained in the Methods section.

A second purpose of a path diagram is to display the results of the analysis. In other words, key parameter estimates (perhaps even including standard errors, or $z$ values, or statistical significance, in parentheses) are shown in the diagram. Thus, in one diagram the most important results of a particular model can be shown. Unless the model is quite small and simple in nature, only the structural model is usually depicted in this type of path diagram. In other words, the measurement model is not included unless there is sufficient space in the figure. Finally, in the situation where there are multiple samples, the estimates for each sample should be shown.

## 3. Sampling Issues

Sampling is an area where reporting in SEM studies tends to be less than adequate. More specifically, a typical SEM study states the sample size used in the analysis but little else. This is not a sufficient level of reporting for sampling related issues. There is additional information about three issues that the reader should be made aware of in terms of the sampling frame of the study. This information allows for judging the sampling adequacy of the sample as well as which populations the results can be generalized to.

First, information about the definition of the populations must be included (e.g., males and females aged 13–17 in the U.S. who are attending public schools and receiving regular education). Second, the type of sampling procedure utilized needs to be described (e.g., random, cluster, stratified, convenience). That is, authors should describe how these particular sample observations were selected from their respective populations (e.g., a strategy of stratified random sampling by ethnicity was followed to ensure adequate representation of different ethnic groups). And finally, precisely how the data were collected from these samples needs some attention (e.g., assessments were individually administered to subjects over two days by a trained specialist); that is, details need to be presented on how the data were actually collected from each of the samples. Information about these sampling issues will also be useful when discussing missing data (see Desideratum 6).

## 4. Latent Variables and Observed Variables

A key portion of the Methods section for any SEM study is the selection and preparation of the observed variables (for an example, see Shumow & Lomax, 2002). There are at least five measurement issues that need to be considered. First, for each latent variable or factor, one or more observed variables are utilized. One method for dealing with this is in the form of a table listing the latent variables and each of their associated observed measures. Alternatively, the measurement of the latent variables through the observed variables can be described in the text (e.g., a section for each latent variable, with a subsection for each observed variable). This information helps to define the measurement model in terms of latent variables and observed variables. Second, the types of measures actually being used should be described. Are the observed variables individual items, composites of items (i.e., unweighted or weighted sums of individual items in a subscale, such as item parcels), or factor scores? In other words, the type of measures being utilized should be discussed.

Third, what are the psychometric properties of the observed variables? Except for the most commonly used and well-known measures, information should be given on the reliability and validity of each observed variable. This information needs to include some basic evidence through reliability and/or validity coefficients, which often can be documented in a sentence or two for each observed variable.

Fourth, what is known about the measurement scales being used? More specifically, it would be useful to the reader to have some information about the possible scale values (e.g., minimum/maximum or range), type of measurement scale (i.e., nominal, ordinal, interval, ratio), the number of categories for ordinal measures, and whether any rescaling or recoding of the variable values has been undertaken. Oftentimes the scale of an original variable has been altered to fit the needs of the individual study (e.g., to rescale an observed variable when there is a negatively worded item; or to generate a more normally distributed observed variable through the use of a statistical transformation, see Desideratum 5).

Finally, how are the observed variables distributed? In other words, are the measures reasonably normally distributed, what do the distributions look like in terms of skewness and kurtosis (univariate and multivariate), and how is any non-normality taken into account (e.g., statistical transformations; or the use of an estimation procedure more robust to non-normality)? This information is also particularly relevant for the selection of an appropriate method of estimation, discussed next (see Desideratum 5).

## 5. Software and Method of Estimation

There are a number of software packages available for conducting SEM analyses; thus, it is useful to report the name and version of the software utilized. All of the packages are somewhat comparable, and thus no overarching recommendation can be made. However, each program does have certain features that distinguish it from the others. As well, because the field of SEM has changed so rapidly over the past two decades, utilizing a recent version of the software is recommended.

As in any quantitative research study involving the use of inferential statistics, applicable statistical assumptions must be evaluated and potential violations dealt with in an appropriate manner. In other words, statistical assumptions form part of the foundation for any inferential statistic. Without reasonably meeting those assumptions we cannot count on our results having much value.

In SEM studies, most methods of estimation make some assumption about the distribution of the observed variables. Maximum likelihood estimation is the most commonly used method of estimation in SEM and assumes that the variables are multivariate normal. Thus, it is crucial to determine whether the data meet this distributional assumption. If evidence (e.g., skewness, kurtosis, statistical test of normality) suggests that the data are not reasonably multivariate normally distributed, then maximum likelihood may not be appropriate. Here the offending observed variables could be (a) transformed into a new variable that is somewhat better behaved, (b) deleted if there are a sufficient number of observed variables to assess each latent variable, or (c) a method of estimation more robust to non-normality be utilized instead of maximum likelihood. Note that this only addresses univariate distributional issues; two or more variables would have to be considered simultaneously to address multivariate distributional issues.

Maximum likelihood is typically recommended unless the data deviate substantially from multivariate normality and/or include categorical variables. Thus, the choice of a method of estimation needs to be tied to the distributional properties of the data, as well as to the measurement scales of the variables, all of which need to be reported. In other words, an SEM study needs to describe the distribution of the data, the method of estimation, and the measurement scales of the observed variables. With multiple samples, this also means that each of these needs to be described for each sample. As described in the Shumow and Lomax (2001) model of school safety, the observed variables were fairly well normally distributed in a univariate sense (e.g., in terms of skewness and kurtosis) in each of the ethnic group samples, thus maximum likelihood was selected as the method of estimation.

### 6. Treatment of Missing Data and Outliers

While the final sample size is generally reported, in many studies little, if any, information is provided about the presence and treatment of missing data. We find this curious because rarely does a social or behavioral scientist have a dataset with absolutely no missing data. The inclusion of sampling framework information can easily lead into a discussion of how missing data were treated. Were only complete cases used (i.e., listwise deletion), were complete cases included for each pair of observed variables (i.e., pairwise deletion, where different pairs of variables could have different sample sizes), was some sort of missing data replacement method utilized (e.g., mean imputation, similar response pattern imputation, multiple imputation), or was a method used that works around the missing data (i.e., does not impute the missing data, but works with the available data, such as full-information maximum likelihood)? A description of how the missing data were treated and a justification for that method should be included, in addition to the initial and final sample sizes. For example, Arbuckle and Wothke (1999, Example 17) considered a dataset with a sample of 73 girls on six psychological tests. Approximately 27% of the data were missing, complete data were only available for seven cases, and thus full-information maximum likelihood was selected as the method for dealing with the missing data.

Another data treatment issue deals with outliers. An outlier is an observation that is quite different from the rest (e.g., often defined in a univariate sense as more than 2 or 3 standard deviations beyond the mean). Outliers can cause normality problems, improper solutions (see Desideratum 8), correlations being reduced in strength or magnitude (i.e., closer to zero), among other serious issues. Thus, it is always important to examine the data to detect outliers. Outliers can be a function of any of the following: (a) a malfunctioning instrument (e.g., a computerized testing or tape recorder problem); (b) a data recording error (e.g., recorded a 60 instead of a 6 when the data were gathered); (c) a data entry error (e.g., typed a 60 into the data file instead of a 6 when the data were entered); (d) an error in observation (e.g., observed and coded an incorrect behavior); (e) an inappropriate use of administration instructions (e.g., gave subjects more time than the directions called for); or (f) an accurate observation (i.e., a true outlier). Obviously errors should be corrected whenever possible. Otherwise, a rationale should be made for the treatment of outliers, and the number of outliers deleted should be reported.

### 7. Data Table for Samples

Once the data have been fully prepared, then the analysis can proceed accordingly. An initial piece of information to present to the reader is a summary table of correlations, means (for SM-SEM), and standard deviations (or variances) for the set of observed variables. This is typically presented through a table in matrix form. In the case of multiple samples, then either (a) separate tables can be shown for each sample, or (b) pairs of samples can be given in one table with one sample being shown above the main diagonal of the matrix and a second sample below the main diagonal (e.g., see Tables 2 and 3 in Shumow & Lomax, 2002).

The summary table allows readers to examine the data from a descriptive perspective and even to conduct their own analyses (i.e., either to verify the results or to test additional models). The complete matrix should be given in the data table for each sample, unless there are too many observed variables to include for practical purposes. In this case the table should be made available by some other means (e.g., website, e-mail). In summary, there should be sufficient information in any SEM application to allow the reader to be able to replicate the results (e.g., theoretical model, sampling information, measures, software and method of estimation, data tables, and results).

## 8. Problems in Obtaining a Proper Final Solution

Researchers who have utilized SEM have likely run across various and sundry error messages, such as "model not identified," "model did not converge after 100 iterations," or "matrix not positive definite." In other words, a proper final solution does not result 100% of the time even after a theoretically justifiable model has been specified. These error messages need to be taken quite seriously. For example, if a variance estimate is found to be negative (i.e., a Heywood case), then this is not a proper result (and this will generate the non-positive definite matrix message). Similarly, correlations beyond 1.00 (in either the positive or negative direction) are not admissible. Most methods of estimation in SEM are iterative, meaning they take many steps or iterations before reaching a final solution. When a final solution is not generated, the convergence error message will be shown, in which case the estimates cannot be taken seriously (e.g., obtaining a standardized factor loading of 500,000).

If any of these problems come up, then they must be dealt with rather than be ignored. Otherwise the results will be less than meaningful. Solutions to these problems might include the following: (a) allowing the SEM program to run through more iterations; (b) constraining (i.e., forcing) negative variances to be positive or eliminating the offending observed variable; and (c) solving identification problems (e.g., if an attempt is made to estimate more parameters than there are non-redundant elements in the covariance matrix of the observed variables). The use of multiple samples will certainly not alleviate these kinds of problems; in fact, they might even be more likely to occur. Overlooking these serious errors and not reporting them is not sound research, as the results of the analysis cannot be trusted. Thus these errors need to be reported as well as resolved.

## 9. Parameter Estimates and Statistical Significance

One of the most important parts of the Results section is a complete reporting of the estimates, standard errors, and statistical significance for every individual free parameter (e.g., through a $z$ value, and/or some sort of notation regarding statistical significance). With MS-SEM, this means that all of the results need to be reported for each sample. With multiple samples, the rows of the table can be the individual parameter estimates and the columns can be the samples (e.g., Tables 4 and 5 in Shumow & Lomax, 2002). For SM-SEM, the estimates of the mean difference parameters will also need to be reported (e.g., either in the bottom portion of the results table, such as Table 7 of Lomax, 1985, or alternatively in a separate table just of structured means results, such as Table 2 of Lomax, 1983).

Key results of the structural model can be reported in the form of a path diagram. A full reporting of the results from both the measurement and structural models usually requires one or more tables. For ease of reading these tables, we recommend that the results be presented using observed and latent variable names, rather than Greek symbols, and that the different types of parameters be clearly labeled (e.g., factor loadings, measurement error variances, path or structure coefficients). More specifically, it is recommended that the unstandardized results be reported because it is those estimates that are being tested for invariance in MS-SEM and SM-SEM.

## 10. Global Goodness-of-fit Indices and Model Comparisons

Another essential part of the SEM results that must be reported are the global goodness-of-fit indices selected and the criteria used to evaluate those indices. Of the issues described in this chapter, the reporting of global fit has seen the greatest research and development over the past twenty years. As a result, there are well over 50 fit indices currently available in SEM software. Numerous simulation studies of fit indices have been conducted over those years and lead us to make the following two recommendations: (a) the chi-square measure can be greatly influenced by several characteristics, including sample size, model complexity, and non-normality, and thus should never be used as the sole measure of model fit; and (b) no other single fit measure has been shown to be "best" in all contexts. In short, it is always recommended that multiple fit indices be reported in SEM studies, typically three to five measures.

Not only is it important to select several different fit indices that are in common usage, but it is also essential to evaluate those indices according to known criteria (e.g., Schumacker & Lomax, 2004). Although most criteria are subjective in nature (e.g., fit indices that are scaled from 0 to 1.00, requiring a criterion value of at least .90 or .95), some movement toward more empirically based criteria has occurred in recent years (e.g., Hu & Bentler, 1999). Most researchers tend to have a short list of favorite fit indices that they prefer for assessing the fit of a single theoretical model. Our particular favorites include chi-square (with the chi-square value, degrees of freedom, and $p$ value being reported), GFI (goodness-of-fit index), SRMR (standardized root mean square residual), and RMSEA (root mean square error of approximation). No blanket recommendation of any particular set of indices has yet been accepted in the SEM literature. See the Shumow and Lomax (2002) example for the fit indices and criteria that they utilized with the parental efficacy model.

When using MS-SEM and SM-SEM, the chi-square difference test is one fit index that is always recommended in order to examine a series of nested models (i.e., one model is nested within another model when the variables are the same, but one or more parameters are either included or deleted). For example, to compare an initial model with no parameters invariant across populations to a second model with only factor loadings invariant across populations, a difference in chi-square statistic can show the extent to which the fit of the second model has deteriorated. If the chi-square value does not decrease by a statistically significant amount, then the factor loadings have been shown to be statistically invariant across the samples tested; as such we would retain a hypothesis of loading invariance across populations. If the chi-square value does decrease by a statistically significant amount, then the factor loadings have been shown to be statistically different across the samples tested; we would thus infer some degree of loading noninvariance across populations. Thus, the chi-square difference test can help assess whether the factor loadings appear the same or different across populations. The invariance of other types of parameters (e.g., path or structure coefficients) can also be evaluated by comparing two nested models.

## 11. Model Modification

In contrast to model fit, model modification is probably one of the most poorly reported aspects in SEM studies. When the initial model does not have adequate model fit, then modifications to that model are often considered. There are four questions in the area of model modification that need to be addressed in all SEM studies. First, were relevant parameters statistically different from zero (typically evaluated through $z$ values) at some nominal $\alpha$ level and in the expected direction (see also Desideratum 9)? In other words, do the expected relations in the model have coefficients that are statistically different from zero and in the expected direction based on previous research and theory?

Of greatest interest are the path or structure coefficients and the factor loadings. For example, we may see an SEM application where the fit of the theoretical model is seen as acceptable, but no information is provided about the significance of the estimates. We have even seen several instances where the model fit was quite strong, but the estimates did not make sense and/or were not significant. Thus, an adequate model fit means little if the parameter estimates do not support the hypothesized model (i.e., when the estimates do not make sense and are not statistically significant).

Second, what portions of the resultant SEM output were considered for possible model modification? For example, SEM software generates considerable information useful for model modification, such as residual matrices (e.g., raw and standardized residual matrices), as well as various types of indices and statistics (e.g., $z$ values, modification indices, expected parameter change statistics, Lagrange multiplier statistics). Unfortunately, rarely is such modification information reported in applications. Even a description of which information was used to make model modification would be an improvement.

Third, was the initial theoretical model modified in some way and how? That is, were new relations added to the initial model, were non-significant relations trimmed from the original model, and what were those modifications? In most SEM applications, only the results of the final model are presented to the reader. Without further information, the implication is that only one model was ever tested. Experienced modelers know that this is rarely the case. In the interest of journal space, while only the final model results should commonly be fully reported, a brief overview of the models tested and the modifications made should be included.

Lastly, what is the basis for each of the model modifications made? Any modifications made need to be justified both statistically (based on modification information given in the SEM output) and substantively (based on the modifications making substantive theoretical sense). No modification should be made without a substantive rationale, that is, with only a statistical justification (which is simply number crunching). In addition, experience has taught us that only one modification should be made at a time in an SEM model. The rationale is that all of the parameters and potential parameters in an SEM model are related in some fashion. So it is difficult to predict what effect the first modification will have on subsequent modifications. Thus, making an initial single modification will better inform the modeler where to go for a second modification, instead of simultaneously making two modifications.

For the Shumow and Lomax (2002) example, all of the hypothesized structure coefficients and factor loadings were significantly different from zero ($p < .05$) and in the expected direction. In terms of model modification, the standardized residual matrix and the modification indices were examined. The original model was maintained, without any parameter trimming or deletion, but with the addition of two measurement error covariance terms (due to shared method variance, where the exact same method of measurement was utilized for those observed variables). Only three total models were tested, the original model and two modified models, each with one additional measurement error covariance term included due to shared method variance.

## 12. Multiple Sample Modeling

An initial question to pose is when multiple sample MS-SEM should be considered. It is often the case that the researcher has a dataset consisting of various subsamples of the overall sample (e.g., by age, ethnicity, SES, gender). The idea is to evaluate each subsample against each of the theoretical model(s) to be tested. Here the researcher wants to know whether a particular model holds for each of these sub-samples, or whether there are some differences among the subsamples for various parameters. For example, does the same model of learning apply across different age, or gender, or ethnic group sub-populations? In particular, we might want to know whether the factor loadings are invariant (the same) across the groups when developing a scale, and/or whether the path or structure coefficients are invariant across the groups when testing the generalizability of a theory. These are the types of research questions that MS-SEM can answer.

It should also be mentioned that in MS-SEM, one group serves as the reference group (e.g., females) and the unstandardized parameter estimate(s) in the other group(s) (e.g., males) are constrained or forced to be equal to the values obtained in the reference group. These constraints are part of the input to the SEM software. This also needs to be done in SM-SEM.

To be more specific about the conduct of MS-SEM, a prescribed series of nested models is typically evaluated. The initial model tested, Model 1, is one in which none of the parameters are invariant across populations. As an example using gender, this would mean that none of the parameters in the female population model are constrained to be equal to their respective parameters in the male population model. The second model tested, Model 2, constrains the factor loadings to be the same across the populations, but not other model parameters (i.e., parameters other than the loadings are free to vary across populations). If the fit according to the chi-square difference test substantially deteriorates from Model 1 to Model 2, then the factor loadings are statistically different for the samples and we infer some degree of noninvariance across populations. If the fit does not deteriorate substantially, then the factor loadings are not statistically different for the samples, and a hypothesis of population noninvariance is retained.

The following additional models can then be tested. Keep in mind though that once particular parameters have been deemed to be invariant (by virtue of no statistical differences across samples), they remain as such throughout the analyses. Model 3 considers whether the measurement error variances are invariant (which occurs only on rare occasions, in our experience). In Model 4 we assess whether the latent independent variable variances and covariances are invariant. Model 5 is used to determine whether the path or structure coefficients are invariant. Finally, Model 6 examines whether the structure or prediction error variances are invariant (also rare).

The idea is to test a series of nested models to determine which sets of parameters are invariant and which are not. Thus, MS-SEM is a method for determining whether the same covariance structure is operating across the populations. In addition, tests can be done at the individual parameter level, for example, if a researcher wants to assess whether or not *one* particular factor loading is invariant across populations as opposed to *all* factor loadings being invariant. A table of results should be provided that lists each model tested, including the chi-square value (with degrees of freedom and $p$ values reported), chi-square difference tests, and some indication of the statistical significance for each chi-square difference test (e.g., see Table 6 of Lomax, 1985).

Unfortunately, over the years we have not seen many applications of MS-SEM in the social and behavioral science literature, although the numbers are on the rise. One reason why MS-SEM results can be so valuable lies in their assistance in validating theoretical models across different populations or subgroups. This would indicate that a particular theoretical model applies rather broadly across those populations. Alternatively, the researcher might find that different theoretical models are in operation for different groups. For example, perhaps a model of learning should look different for different groups, that is, all individuals do not learn in the same fashion. Thus MS-SEM can be quite useful in exploring the generalizability of a particular theoretical model across different populations.

In the Shumow and Lomax (2002) model of parental efficacy, three major ethnic groups were represented in the original sample. As it was believed theoretically that parental efficacy operated differently by ethnicity, MS-SEM was conducted. As the overall sample was sufficiently large, such analyses were reasonable to consider (i.e., 319 European Americans, 187 African Americans, and 200 Hispanics). While the observed measures performed similarly across these groups (i.e., the factor loadings were deemed invariant across the ethnic groups), there were some relationships in the structural model that differed by ethnic group. The results of the structural model were depicted in a path diagram using abbreviations for each of the groups to indicate statistical significance (see Figure 1 of Shumow & Lomax, 2002).

## 13. Structured Means Modeling

In many multiple sample settings, the researcher is also interested in examining mean difference (or intercept) parameters. This requires the use of structured means structural equation modeling (SM-SEM). For example, we might want to know if there is a mean difference in an SES latent variable for public versus private school populations (as in Lomax, 1985). Or we might want to know whether there is a significant mean difference in a student achievement latent variable for these same two samples, controlling for SES (Lomax, 1985). These are the types of questions that SM-SEM can answer.

A typical single sample SEM model does not assume anything about the means of either the latent variables or the observed variables. In fact, all of the means are typically fixed at zero in both the structural and measurement models by the SEM software, mainly because there is no way to estimate mean parameters (for an exception, though, see latent growth curve models in Chapter 14, this volume). However, once we include multiple groups, this then allows the researcher the possibility of estimating mean parameters. Thus, if one is interested in the estimation of mean parameters, then there needs to be multiple groups where a particular mean parameter of a reference group (e.g., male achievement) is fixed to zero in order to allow the estimation of the corresponding mean parameter in another group (e.g., female achievement). This is basically what happens in SM-SEM where conceptually this latent mean parameter is estimating a gender difference.

Unfortunately, we have seen even fewer applications of SM-SEM in the social and behavioral sciences literature than MS-SEM. In part this is because these kinds of models are a little more challenging to consider, although they have been greatly simplified in the more recent versions of the SEM software. However, these types of models can be quite informative, as we see next.

In Lomax (1985), a series of MS-SEM models indicated that the same model of schooling appeared to apply equally well to both public and private schools (i.e., the same covariance structure existed both for the factor loadings and for the structure coefficients). Subsequently, the following results were found in a SM-SEM model: (a) private school students had a significantly better home background than their public school counterparts; (b) academic orientation and level of extra-curricular activity were both significantly higher for private school students when controlling for home background; and (c) aspirations (educational and occupational) and achievement were all lower for private school students when controlling for academic orientation, level of extra-curricular activity, and home background (all $p < .05$). While the first two findings supported previous research, the third finding was not something that had been previously investigated. Thus, the superiority of the private school students was not a consistent finding in this study.

The SM-SEM results can either be presented in the text, or in some sort of a table (see Desideratum 8) when there are a substantial number of mean difference parameters. In addition, to determine the strength of any effect in SM-SEM, an effect size measure can be computed. For example, a standardized effect size measure can be determined where the intercept of an equation for the second group is divided by the square root of the disturbance or error variance of that equation (see, for example, Hancock, 2001). These effect sizes can then be judged in accordance with guidelines such as those provided by Cohen (1988).

## 14. Discussion of Results in Context of Theoretical Model(s)

One of the main purposes of the Discussion section of an SEM study is to relate the results of the theoretical model(s) tested to the literature reviewed. The following questions could be considered in the Discussion. What evidence is there that a particular model has been supported for one or more populations? Are there some portions of the model that fit the data rather well and others that are not? Should additional latent variables be included and/or other latent variables eliminated based on the analysis? Do we need to refine our list of observed variables in terms of seeking out higher quality measures? Do some of the paths need to be eliminated? These are just some of the issues that the Discussion section can address in the examination of a single sample.

Additionally, the study of multiple samples allows us to perform a form of model validation as to whether a particular model fits well for different samples (e.g., for males and females), or in different contexts (e.g., for different countries or for different cohorts). The Discussion section could also describe future methods for validating a model or models in other samples or contexts. For example, does the same model of schooling apply in the United States, Estonia, and Japan?

It should also be noted that MS-SEM is only one way in which models can be validated. Other approaches available to researchers include the following methods: (a) traditional cross-validation where the researcher splits a single sample in half and cross-validates one half against the other half (meaning larger samples are necessary to start); (b) expected cross-validation on a single sample without splitting the sample in half (meaning smaller samples can be used); (c) simulation methods (where a large number of sample datasets is generated by the computer; thus the data are computer-generated rather than actually being collected); (d) the bootstrap method (where the resampling of observations is done from a single sample); and (e) the jackknife method (where resampling of observations is done by leaving one observation out of each sample, sometimes known as the leave-one-out method of sampling). For more details of these methods, the reader is referred to, for example, Schumacker and Lomax (2004).

## References

Aiken, L. S., Stein, J. A., & Bentler, P. M. (1994). Structural equation analyses of clinical subpopulation differences and comparative treatment outcomes: Characterizing the daily lives of drug addicts. *Journal of Consulting and Clinical Psychology, 62,* 488–499.

Arbuckle, J. L. (2007). AMOS (Version 7). Chicago: SPSS.

Arbuckle, J. L. & Wothke, W. (1999). *Amos 4.0 user's guide.* Chicago: SPSS.

Bentler, P. M. (2006). EQS (Version 6.1). Encino, CA: Multivariate Software.

Byrne. B. M. (1998). *Structural equation modeling with LISREL, PRELIS, and SIMPLIS.* Mahwah, NJ: Erlbaum.

Byrne, B. M. (2001). *Structural equation modeling with AMOS.* Mahwah, NJ: Erlbaum.

Byrne, B. M. (2006). *Structural equation modeling with EQS: Basic concepts, applications, and programming.* Mahwah, NJ: Erlbaum.

Cohen, J. (1988). *Statistical power analysis for the behavioral sciences* (2nd ed.). Hillsdale, NJ: Erlbaum.

Hancock, G. R. (2001). Effect size, power, and sample size determination for structured means modeling and MIMIC approaches to between-groups hypothesis testing of means on a single latent construct. *Psychometrika, 66,* 373–388.

Hu, L., & Bentler, P. M. (1999). Cutoff criteria for fit indices in covariance structure analysis: Conventional criteria versus new alternatives. *Structural Equation Modeling: A Multidisciplinary Journal, 6,* 1–55.

Jöreskog, K. G., & Sörbom, D. (1993). *LISREL 8: Structural equation modeling with the SIMPLIS command language.* Chicago: Scientific Software International.

Jöreskog, K. G., & Sörbom, D. (2006). *LISREL (Version 8.8).* Lincolnwood, IL: Scientific Software International.

Kline, R. B. (2004). *Principles and practice of structural equation modeling* (2nd ed.). New York: Guilford Press.

Lochlin, J. C. (2004). *Latent variable models* (4th ed.). Mahwah, NJ: Erlbaum.

Lomax, R. G. (1983). A guide to multiple sample structural equation modeling. *Behavior Research Methods and Instrumentation, 15,* 580–584.

Lomax, R. G. (1985). A structural model of public and private schools. *Journal of Experimental Education, 53,* 216–226.

Maruyama, G. M. (1998). *Basics of structural equation modeling.* Thousand Oaks, CA: Sage.

Muthén, B. O., & Muthen, L. K. (2006). Mplus (Version 4). Los Angeles: Muthén & Muthén.

Schumacker, R. E., & Lomax, R. G. (2004). *A beginner's guide to structural equation modeling* (2nd ed.). Mahwah, NJ: Erlbaum.

Shumow, L., & Lomax, R. G. (2001). Predicting perceptions of school safety. *The School Community Journal, 11,* 93–112.

Shumow, L., & Lomax, R. G. (2002). Parental efficacy: Predictor of parenting behavior and adolescent outcomes. *Parenting: Science and Practice, 2,* 127–150.

# 30
# Survey Sampling, Administration, and Analysis

## Laura M. Stapleton

The use of surveys in social science research is abundant and may appear straightforward, but can involve a complex set of procedures. *Total survey error* refers to all of the errors that could occur in researchers' attempts to gain valid information about a population with the use of surveys. These errors have been placed into five categories: coverage error, nonresponse error, editing and processing errors, measurement error, and sampling errors (Groves, 1989). *Coverage error* refers to the failure to provide some members of the population the chance to be selected into the sample, *nonresponse error* refers to the failure to obtain responses from all members of the selected sample, *editing and process errors* refer to the failure to capture data accurately from the respondents, *measurement error* represents the failure of the observed response to reflect the true opinion of the sample member, and *sampling error* refers to the fact that sample statistics are not expected to exactly reflect population parameters.

Methods to acknowledge and address each of these types of errors are covered in this chapter and should be, to the extent possible, addressed in research manuscripts. Specifically, this chapter addresses three main areas of the survey process: appropriate methods of sampling from a specified population, developing survey items and administering the survey, and undertaking appropriate evaluations of data and analyses based on the sampling design. The sampling design defines how one obtains a sample intended to be representative of the population to which researchers would like their results to generalize. Sampling is discussed in Lohr (1999, 2008), and for the more advanced reader, Kalton (1983) and Kish (1965). Development of survey items, and the appropriate methods of administering the survey in order to obtain high response rates and quality responses, is multifaceted and development strategies depend on the topic to be measured and the target population of the survey. An excellent, comprehensive resource with practical guidelines is provided by Dillman (2000) and Forsyth and Lessler (1991). More theoretical treatment regarding the cognitive response process is provided by Tourangeau, Rips, and Lapinski (2000). Finally, analysis of data is straightforward if a simple random sample has been taken (analysis covered in most behavioral and social science textbooks assume such a sampling strategy). However many surveys are not conducted with a simple random sampling technique and thus special analytic procedures must be undertaken and these procedures might include sampling weights, and strata and/or cluster indicators. Analyses undertaken using such sampling design elements are

**Table 30.1** Desiderata for Survey Sampling, Administration, and Analysis

| *Desideratum* | *Manuscript Section(s)** |
|---|---|
| 1. The population for generalization is described and justified. | I, D |
| 2. The survey or questionnaire development process is described and includes a discussion of the evaluation of item validity. | M |
| 3. The sampling frame is defined and the procedures for obtaining it are described. | M |
| 4. The type of sampling process utilized is explained (such as multistage, random, systematic) and defended as to why it is appropriate for use with the target population. | M |
| 5. The survey administration method is outlined including the mode (such as face-to-face, phone interview, web survey, or mailed survey), use of incentives, and number of contacts. The administration process is defended as to why it is appropriate for use with the particular sample and questionnaire content. | M |
| 6. The response rate is provided and discussed. | R |
| 7. An analysis of possible sources of non-response is conducted and the possibility of the creation of post-stratification weights is addressed. | M, R |
| 8. The data analysis includes components to address any disproportionate selection probabilities (or non-response adjustments). | M, R |
| 9. If probability sampling other than simple random sampling is used, the estimated design effects of the means of key variables are provided. | R |
| 10. The analysis includes an appropriate method to adjust for any dependencies resulting from multistage sampling or efficiencies gained from stratified sampling. | M, R |
| 11. A discussion of the limitations with regard to questionnaire item validity, sampling strategy, response rate and analysis decisions is provided. | D |

* *Note:* I = Introduction, M = Methods, R = Results, D = Discussion

detailed at a basic level in Lohr (1999) and Stapleton (2008), and at a more advanced level in Lee, Forthofer, and Lorimor (1989), Kalton (1983) and Kish (1965). A simple list of best practices for the entire survey research process is provided by the American Association of Public Opinion Research (2008) and a new text edited by de Leeuw, Hox, and Dillman (2008) contains accessible chapters that relate to every step of the survey process. Specific desiderata for studies that incorporate the use of surveys for collection of behavioral and social science data are presented in Table 30.1 and are described in detail in the following sections.

## 1. Population for Generalization

Although the term *survey* has been used to connote different things across people and disciplines, at its heart, the term should imply that a questionnaire has been administered to a sample of members from a given population to which researchers would like to generalize; a questionnaire that is administered to all members of a given population is considered to be a census and not a survey. Therefore it is imperative that in an applied study involving the administration of a survey, authors specify the population to which results are intended to generalize. It is from this target population that the sampling frame is developed (see Desideratum 3). There are times when researchers would like to generalize to a large population, such as all U.S. adults aged 18 to 65, but the sampling procedures do not allow such a broad generalization; for example, if a telephone survey is administered based on random digit dialing, the actual population from which the sample is drawn only includes adults with a cellular or land-line telephone. This narrowed population, then, is referred to as the sampling frame population. If there is a discrepancy between the target population and sampling frame, authors should carefully describe the members of the target population who will be excluded by sampling from only the narrower sampling frame (in terms of both numbers and characteristics). To the extent that the members of the target population and sampling frame differ in their behaviors and attitudes measured on the survey, this discrepancy is termed coverage error.

## 2. Questionnaire Development Process

The questionnaire development process is one that is poorly defined in many research manuscripts, if not totally ignored. It can be argued, however, that it is one of the most important aspects of survey research; if validity of the measures obtained from the survey cannot be assured then any subsequent data analysis and interpretation of parameter estimates is questionable. Given research findings with regard to the effect of item wording on response quality, it is essential that authors describe the process undertaken to develop and evaluate the questionnaire. If using a previously developed questionnaire, authors should cite a reference to the development and validation information for that questionnaire. If using a newly developed questionnaire, authors should provide a description of how the items were generated and whether items were reviewed by outside parties. Additionally, a discussion should be provided of the pilot testing process and the results shared from reliability assessments (both internal consistency for scales and test-retest reliability of specific items and scale scores, as appropriate) and validity assessments (such as cognitive interviewing results or statistical estimates of relations with external criteria).

Specific issues in item development that have been found to affect response quality and thus induce measurement error and that should be considered in questionnaire development include, for rating scale measures, the *valence* of the item stem (whether the stem is phrased negatively or positively), the number of response options, and whether anchors are provided for those options. Although surveys often include the "strongly agree" to "strongly disagree" Likert format, this type of measurement has been found to be problematic for two reasons: without extreme statements there usually is limited variability in the obtained responses, and the responses are difficult to interpret across individuals due to the vague quantification of response options.

Not only the wording and structure of the actual items are of concern but the format of the questionnaire has been found to affect response rate and response quality. Dillman (2000) provided excellent suggestions for the actual page layout and construction. Authors should provide a copy of the questionnaire used or provide an internet link to allow readers to evaluate the survey for possible order or questionnaire construction effects, as well as item wording effects.

## 3. Sampling Frame

To obtain a representative sample of a given population of interest, a sampling frame needs to be determined. As specified in Desideratum 1, given a target population of interest, the researcher must then attempt to identify a list of all members of that population from which to sample. If a list is indeed available (e.g., all students at a particular university of interest) then the concept of a sampling frame is fairly simple (although the definition of "enrolled at the university" on a specific date would need to be determined and reported). Often, however, no list of members of the target population exists; in this case a sampling frame might have to be pieced together from separate sources (and duplicate and missing units identified and addressed), or the sampling frame might have to be built as part of the sampling process. For example, if researchers were interested in a target population of all undergraduate students at 4-year U.S. colleges and universities, they would be quick to note that no such list currently exists. However, each of the 2,000+ colleges and universities in the population of interest does have a list of its own students. Therefore, the sampling frame might exist in theory and can only be practically accessed once specific higher level units (the institutions in this example) are contacted. Whether the list already exists or must be constructed as sampling takes place, authors should delineate the process of building the sampling frame from which the final sample was determined.

## 4. Sampling Design

There are several types of sampling designs that are appropriate for generalization to a target population. A few sampling strategies exist that might be less appropriate and are discussed briefly at the end of this section. For generalization, especially with correlational research designs, it is best to have what is referred to as a *probability sample* where each member of the sampling frame has a known probability of being selected into the sample.

One probability sampling scheme is *simple random sampling* (SRS) in which each unit in the sampling frame has an equal chance of selection. SRS can only be achieved if the researcher has a list of the entire sampling frame (either physically or theoretically as in random digit dialing whereby a computer can generate all possible combinations of numbers). Another SRS-equivalent sampling strategy is referred to as *systematic sampling*. In this process, a sampling interval is determined to arrive at a final sample number. For example, if there are 5,000 students at a university and a sample of 200 students is desired, then a list of the students sorted in random order can be created and every 5,000/200 = 25th student on this list is selected for the sample. Note that each population member's probability of selection ($\pi$) is the inverse of this interval (e.g., $\pi = 1/25 = .04$). Systematic sampling can also be used when a list is not available, by selecting, for example, every 10th shopper to enter a store or every 20th voter to arrive at a polling place. In order to use this systematic sampling strategy, an estimated size of the population coming to the location would be needed, along with a desired sample size; these two values would provide the appropriate sampling interval.

Another probability sampling design is termed *stratified sampling*. In this sampling method, members of the sampling frame are split into mutually exclusive categories (strata) and then elements are sampled from each category to ensure specific representation of members of each stratum; the selection rate might or might not be proportional to the population size within each stratum depending on the researchers' needs. The resulting sample will be more representative based on the strata as compared to the use of SRS and systematic sampling, but this result has implications for assumptions underlying sampling error estimates as will be discussed in Desideratum 10.

*Multistage sampling* designs refer to the selection of *primary sampling units* (PSUs) as a first stage of sampling (such as the random selection of U.S. colleges and universities) and then sampling of one or more lower level units within each of the selected PSUs (such as faculty or students within those selected universities). Multistage sampling is usually undertaken when a list of units in the population is not available or it is more efficient to collect data within clusters. For example, it would be easier to collect opinions of 10 students at each of 20 college and universities than to go to possibly 200 different campuses to reach 200 randomly selected students.

As part of multistage sampling, a technique called *probability proportionate to size* sampling can be used where the probability of selection of the PSU depends on the size of the PSU. If an equal number of final elements is intended to be drawn from each PSU (such as 10 students at each campus), then if PSUs are drawn with equal probability, students at small institutions will have a higher probability of being in the sample and the final sample will over-represent students at small campuses. Therefore, a sampling design to result in approximately equal probabilities of selection of lowest level units involves selecting PSUs directly proportionate to the size of the PSU, such that small PSUs have smaller chances to be in the sample and vice versa. This technique is further discussed in Desideratum 8.

Sampling designs can incorporate any one or all of the previously mentioned strategies. For example, a stratified, multistage systematic sample might be used by researchers wanting to survey freshmen at colleges and universities in the U.S. First, a list of 4-year institutions would be obtained and each might be identified by sector (public/private). Researchers might specifically select institutions using SRS from each sector to guarantee that the sample would include a specified number or percentage of the sample within each sector. Once institutions are selected, and assuming the registrars were not permitted to provide lists of all enrolled students, the sample of students could be selected using stratified systematic sampling by day and time. On specific selected days and times researchers would be on campus and approach every $i$th student who passes a certain location or locations.

Because the statistical analysis approach taken depends on the sampling strategy used with probability samples, authors should report the specific sampling strategy taken and at each level if a multistage sample is drawn. Authors also should report the selection rate either overall, with SRS, or at each level of the selection process for more complex designs. If a stratified sampling design was used, the strata should be defined.

Two additional sampling strategies, both non-probability designs, are prevalent in current literature. Convenience sampling and chain-referral sampling are both seen, with the latter being a more defensible strategy under some circumstances. *Convenience sampling* is a process by which respondents are identified in a manner that is convenient to the researcher: undergraduate students at a university, patients in a given clinic, people on the street. With this sampling approach, the sample is not drawn from the full population of interest to the researcher and therefore the generalizability of any of the researcher's findings is questionable. If researchers indicate that they want to generalize findings to all U.S. university students and then report on a sample of psychology students who participate in required subject pool research at University X, then there is a severe disconnect between the target population and the sample; coverage errors would be extremely likely. Authors who report studies based on convenience samples should defend why they believe that the findings can be generalized to the population of interest and also provide information regarding the similarity of the sample participants with target population characteristics. Even if such a similarity can be demonstrated (such as with commonly available data such as age and race/ethnicity), there are other ways that the convenience sample might not be similar to the target population. For example, subjects might differ on motivational or attitudinal characteristics not often measured. In general, convenience sampling is more acceptable with experimental designs, given that the desire usually is not to report on the finite population relations among measures but rather to examine the effect of a manipulated variable under the control of researchers.

The second non-probability sampling strategy, referred to as *chain-referrals* (including snowball sampling), has recently gained in use and opinion. With this sampling strategy, a representative sample of some given population is first found, and those individuals with the characteristic of interest are asked to refer the researcher to similar individuals. Such a strategy can be used to determine population estimates of "hidden" populations such as drug users or the homeless (Frank & Snijders, 1994). If chain-referrals are used, authors should describe the method used to obtain the initial sample and demonstrate that it obtained a fairly representative initial group and provide information about the number of waves or links in the chain referral collection.

## 5. Survey Administration

Because the method of survey administration has been found to influence both the survey response rate (and thus non-response error) and the responses themselves (and thus measurement error), authors should provide information about the administration process, including the mode of administration, the number and type of contacts, and whether anonymity or confidentiality was ensured.

The most typical survey modes include paper and pencil, face-to-face interview, phone interview, and web-based administration. There is no preferred survey administration mode, as its success depends on such considerations as the content area of the survey, the target population for which the survey is intended to be used, the anonymity, the length, and the time and labor resources available to administer the survey. Researchers should take care to determine whether the chosen mode might pose a problem in terms of response rate or response bias. For example, for sensitive topics, a face-to-face interview, in which an interviewer might gain rapport with the respondent, can yield more valid data, as can an anonymous web survey. A web survey also can be efficient and inexpensive, but is subject to a specific non-response bias for those lacking access and/or familiarity with the internet. A paper and pencil mail survey tends to yield poorer response rates but is a fairly inexpensive mode and can accommodate longer surveys. Phone surveys provide quick turnaround but can be limited to shorter surveys and require a sizable labor force to administer.

In addition to the mode of survey used, authors should provide information about the survey administration process. Dillman (2000) expressed many practical suggestions for obtaining high response rates and valid information from each item response. Particularly, a multi-contact system is viewed as necessary, which includes possible pre-notification that a survey is coming, the sending of the survey, and multiple non-response contacts with the final contact being of a different mode than the initial contact (for example, by phone if the initial survey was attempted to be administered by mail). An additional consideration in the survey administration process is the anonymity of the survey. If anonymous, responses have been found to have less bias under some contexts, however researchers lose the ability to track respondents and thus typically must send non-response contacts to all members of the sample or use a method whereby respondents send back a postcard indicating that they have responded. If survey response is desired to be tracked, researchers must put an identifying number or code on each survey; note that a name is not necessary.

For surveys that involve interviewer interaction with the respondent (such as with a face-to-face or telephone interview), authors should indicate the level of training that interviewers received or experience that they have and whether interviewer effects might have yielded bias in measurement (for example, having male interviewers ask female adolescents about depressive symptoms).

Survey administration also includes the process of transcribing data into a data set and this data-processing step can involve error. In the case of web surveys and computer-assisted telephone interviews, the data are already entered, although authors should verify that testing of the system has occurred prior to use. For other types of surveys, the data entry process should be outlined, including methods of quality control such as double entry and random quality checking.

## 6. Response Rate

Response rates inform researchers about the possibility for nonresponse bias as a potential component in survey error. Although the calculation of response rates may appear simple on the surface, it is actually fairly complex. In the late 1990s a group of survey research organizations agreed upon a set of standards in calculating survey outcome rates including contact rates, refusal rates, and eligibility rates. These definitional standards were published by the American Association for Public Opinion Research (2006). To illustrate the definitional complexities in reporting response rates, consider the following example. With a telephone survey, calls are made but when there is no answer, a message on an answer machine or voicemail is *not* left. Should that household or person be considered a non-respondent? Did he or she have the opportunity to respond to the survey? The standards are particularly helpful in determining how to report rates under complex sampling schemes. Researchers should use these reporting standards to inform their readers of the amount of non-response and thus the possibility of non-response error.

Although a common concern in survey administration is the level of response rate that is needed for appropriate inference to the population, such standards do not exist. In the late 1970s, the U.S. Office of Management and Budget (OMB) indicated that data collection needs a minimum response rate of 75% and proposed federal projects with anticipated response rates less than 50% should not be approved. The current OMB guidelines, however, have been altered to indicate that surveys need to yield "reliable results" and indicate that a high response rate is one source of reliability (Smith, 2002).

## 7. Sources of Non-Response

Non-response can be at an item or a unit (person) level. Issues influencing possible non-response should be delineated. Groves and Couper (1998) grouped the influences on non-response into four categories, two of which are under the control of the researcher and two that are not. The first two include survey protocols and interviewer training as addressed in Desiderata 4 and 5. The second two include the social climate for surveys in general (sometimes referred to as "survey fatigue") and the personal characteristics of the respondent.

As an example of the latter influence, suppose that a researcher wanted to report the average alcohol consumption of university students and finds that disproportionately more women than men responded to a survey. Such disproportionate non-response rates would likely result in non-response bias in parameter estimates. Assuming men consume more alcohol than women, the average in the obtained sample likely would be lower than the average in the population because women are over-represented in the sample due to their higher response rate.

Non-response does not necessarily indicate that estimates from the sample will be biased or not accurately reflect population parameters. A challenge to the researcher, then, is to determine whether non-response bias exists. Several approaches can be taken to evaluate possible non-response bias, including examining disproportional response rates using sampling frame information, analyzing external data to evaluate whether the sample is distributed similarly to the population, dissecting information from interviews about possible reasons why item and/or unit non-response exists, and examining whether the number of contacts needed to convert a sample element to a responder is related to sample characteristics. Additionally, one may conduct a subsequent survey of non-responders; although difficult to implement, it can provide valuable information about why some members of the selected sample did not participate in the survey. Authors should share results from these non-response analyses with the reader.

If a possible non-response bias exists, post-stratification weighting can be used to compensate for the disproportionate non-response rate. For the alcohol consumption survey example, sampling weights for responses from men would be adjusted higher to account for their lower response rate (see Desideratum 8 for a discussion of the creation and use of weights in an analysis).

## 8. Sampling Weights

The sampling weight is the inverse of the selection probability, $1/\pi_i$, where $\pi_i$ is the selection probability for the $i$th sample member, as introduced in Desideratum 4. The sampling weight can be thought of as the number of people in the population that this specific sample member is representing. If a SRS has been used in selecting the sample, then all sampling weights for observations will be equal and can be ignored in the analysis. However, if the sampling design includes stratification, multiple stages, disproportionate selection probabilities across strata, or adjustments for non-response, then inclusion of sampling weights in the analysis typically will be necessary to obtain unbiased estimates of population parameters. Care should be taken in understanding how the selected software utilizes the sampling weights. Some software (such as SPSS) uses weights as frequency weights and assumes that the sample size is equivalent to the sum of the weights and therefore the raw sampling weight will need to be scaled.

If not properly accounted for in the analysis, disproportionate selection of elements into the sample can adversely affect the resulting population estimates from an analysis. There are many reasons for the use of disproportionate sampling rates, and three situations that result in sampling weights that differ across sample members are discussed here: multistage sampling, over-sampling by stratum, and post-stratification adjustments. First, consider the example of a simple two-stage sampling design. Suppose we want to obtain a sample of 5,000 freshmen from 4-year colleges and universities in the U.S., and that there are 2,000,000 freshmen in about 2,000 institutions in the population. One way to obtain the desired sample would be to randomly sample 5,000 students out of the total pool of 2,000,000 students. To do this, we would use a selection probability of $\pi = \dfrac{5{,}000}{2{,}000{,}000} = .0025$ (or 25 out of every 10,000 students). We could, for example, take a randomly sorted list of the 2,000,000 students, select every 400th name on the list, and we would obtain a sample of 5,000 names; each of the names had a .0025 chance of being selected into the sample. The sampling interval of 400 was obtained by dividing the total population (2,000,000) by the number of desired sample elements (5,000). Note that this sampling interval is the reciprocal of the selection probability: $\dfrac{1}{.0025} = 400$, or the *sampling weight*. Each person in our hypothetical sample represents 400 people from the original population.

In actuality, we do not have a list of all freshmen in the U.S., so it is not feasible to simply randomly sample 5,000 of the 2,000,000 students. We might, instead, draw a multistage sample (as discussed in Desideratum 4) by selecting institutions and then sampling students within each selected institution. The difficulty with this approach, however, is that we now need to determine two selection probabilities, one for the institutions ($\pi_j$) and one for the freshmen within the selected institutions ($\pi_{i|j}$). In order to obtain a sample of 5,000 students, the product of these two probabilities must equal the overall desired selection probability of .0025. We can arbitrarily select a value for one of these probabilities and the other value will therefore be determined. For example, suppose we decide to sample 5% (or .05) of the students within each selected college. We would then need to select 5% of the colleges because $.05 \times .05 = .0025$. Sampling 5% of the colleges ($\pi_j = .05$) and sampling 5% of the freshmen at each selected college ($\pi_{i|j} = .05$), we would obtain a sample with an expected size of 5,000 and each element in the population would have an overall selection rate of .0025 ($\pi_{ij} = \pi_j \times \pi_{i|j} = .05 \times .05 = .0025$). But note that the number of freshmen typically varies across colleges and with this proposed process we might sample a relatively large number of students in very large colleges (for example, with a sampling rate of 5% and 5,000 freshmen in a college we would have a sample of 250 freshmen at that college) and we might sample only 1 freshman at another college (because there might be only 20 freshmen in total at the college). For the sake of efficiency, it is more typical to conduct surveys in a standardized manner across PSUs and with this sampling plan of using a fixed sampling rate within all institutions we could not guarantee a specific size of the sample at each college. An alternate plan might be to sample institutions at a fixed selection rate and then to sample a specified number of freshmen at each college, for example, 20 students. With this sampling design of taking a specified number of participants at each site, in a two-stage sample, the students in small schools have a very high probability of selection into the sample if their college is selected (given the example numbers above, the conditional selection rate would be $\pi_{i|j} = \dfrac{20}{20} = 1.00$). Conversely, students who are in very large colleges have a relatively small chance of being selected for the sample if their college is chosen (for example, $\pi_{i|j} = \dfrac{20}{5000} = .004$). If the colleges are sampled with equal probabilities, then students from these two different colleges would have very different overall rates of selection: $\pi_{ij} = \pi_j \times \pi_{i|j} = .05 \times 1.00 = .05$ for students in small colleges and $\pi_{ij} = \pi_j \times \pi_{i|j} = .05 \times .004 = .0002$ for students in large colleges. Therefore, freshmen in small schools would be over-represented in the sample, selected at a rate 250 times that of the students in the larger colleges. This overrepresentation can be handled in analyses by using sampling weights. Students in our example small college would have a weight of $w_s \frac{1}{.05} = 20$ and for students in our larger college, $w_l = \frac{1}{.0025} = 500$.

A way to avoid having unequal numbers of selected students per colleges or vastly unequal overall probabilities of selection for individual students across colleges is to sample colleges not with equal probability but rather to select them with a method that samples larger institutions at higher rates and smaller institutions at lower rates. This method, called *probability proportionate to size* (PPS) *sampling*, was introduced in Desideratum 4 and is commonly used in national data collection. With PPS sampling, the overall selection probabilities for sample members ($\pi_{ij}$) will tend to be similar, but their conditional probabilities ($\pi_{i|j}$) within their respective PSUs will differ, as will the selection probability for their PSU. For example, suppose with our previous example that we want to select about 20 freshmen at each college no matter the size. A college with 20 freshmen would need to have a .0025 chance of selection into the sample (that is, $\pi_j = .0025$), to result in an overall probability of inclusion for a student in that college of .0025 ($\pi_{ij} = \pi_j \times \pi_{i|j} = .0025 \times 1.0 = .0025$). Alternately, a college with 5,000 freshmen would need to have a chance of selection of .625 in the sample ($\pi_j = .625$), to result in an overall selection probability for the students in those colleges of .0025 ($\pi_{ij} = \pi_j \times \pi_{i|j} = .625 \times .004 = .0025$). The use of PPS is one source of obtaining sample members with approximately equal sampling weights but this equivalency is not guaranteed.

Disproportionate sampling rates across strata may also be used, and are employed to obtain sufficient numbers of elements to undertake subgroup reporting. When the desire is to report estimates by subgroup, sample designers might employ a higher rate of sampling in certain strata than the rate used in other strata. For example, continuing our example, special interest might lie in reporting estimates for African American freshmen and, therefore, instead of selecting 20 freshmen at random at each institution, researchers might select 5 African American freshmen and 15 freshmen of other race/ethnic groups. This method of selection will likely result in conditional sampling rates for African American students that are greater than that of other students within the same institution (for example, if there are 100 African American freshmen and 900 non African American freshmen at the college, then the conditional sampling rate for African American students is $\pi_{i|j} = \frac{5}{100} = .05$ and for other students in the same institution the sampling rate would be $\pi_{i|j} = \frac{15}{100} = .017$. These differing conditional probabilities will lead to different sampling weights.

Finally, as suggested in Desideratum 7, post-stratification weighting adjustments might be employed by the survey designers to adjust for non-response. Although equal selection probabilities might have been used for all elements in the initial sampling plan, some groups typically respond at lower rates than others. After survey data are collected, sampling weights for responses from under-represented groups (given their response rate) would be set higher to reflect their true proportion in the population (if it is known or can be approximated) and the sampling weights for proportionately over-represented groups would be adjusted lower. It is important to note that the use of post-stratification weighting for non-response can be somewhat controversial (Lohr, 1999) and several methods exist to adjust for the non-response including cell weighting, raking, regression estimation, and more complex modeling approaches.

Statistically, the use of sampling weights to address non-response reduces bias in parameter estimates but it should be noted that bias is reduced at the expense of precision. If the non-response in fact is not related to the variable of interest, the parameter estimate will not be biased and weights to adjust for disproportional non-response will not be needed. However the use of weights (either developed through post-stratification adjustment or due to initial disproportionate selection probabilities) in this situation will result in estimated standard errors that will be larger than if they had been estimated as if from an equal probability of selection sample. Therefore, the weighted analysis becomes less powerful. Authors should be explicit about the sampling weights available for their data and whether and how they were included in the analysis. Some authors may opt to analyze the data both weighted and unweighted to examine the effects of the inclusion of weights and provide both pieces of information in their manuscript.

## 9. Design Effects

A *design effect* refers to an inflation or deflation in the sampling variance (or square of the standard error) of a statistic due to the chosen sampling design. Under an assumption of simple random sampling (SRS), the sampling variance (*sv*) for the estimate of a population mean is typically defined in textbooks and in software packages as

$$sv(\hat{\mu}) = \frac{s_y^2}{n} \tag{1}$$

where

$$s_y^2 = \frac{\sum_{i=1}^{n}(y_i - \hat{\mu})^2}{(n-1)}. \tag{2}$$

Thus, as the sample size increases, the estimate of the sampling variance of $\hat{\mu}$ decreases and the estimate of $\hat{\mu}$ becomes more *efficient*. Technically, Equation (1) assumes that observations were sampled with replacement, which is not typically the case in practice. However, when sampling has been undertaken without replacement, the use of Equation (1) to obtain an estimate of the sampling variance can be acceptable if the sampling fraction is small. The procedure of determining the precision of a parameter estimate is referred to as *variance estimation*. It is more typical in applied research, however, for authors to refer to the estimate of the standard error (*se*) for the parameter estimates. This *se* is just the square root of the sampling variance:

$$se(\hat{\mu}) = \sqrt{sv(\hat{\mu})} = \frac{\sqrt{s_y^2}}{\sqrt{n}} = \frac{s_y}{\sqrt{n}}. \tag{3}$$

When analyzing data that have been collected through some sampling designs described in Desideratum 4, one of the main problems is that the estimates of sampling variances (and thus standard errors) will be biased when using the traditional formulas in Equations (1) and (3). Use of these formulas includes the important assumption that observations are independent. When data have been collected using complex sampling designs, the observations often are *not* independent and thus sampling variances calculated via Equation (1) will be biased. A measure of how biased a traditional sampling variance estimate will be is called the *design effect* (DEFF). DEFF is the ratio of the correct sampling variance of a statistic under the complex sampling design over the sampling variance that would have been obtained had SRS been used (Kish, 1965). The square root of DEFF (termed DEFT) is the estimate of the bias in the standard error estimate. If the complex sampling design has no effect on the sampling variance, the value of the DEFF would be 1.0. If the sample design improves the precision of the parameter estimate (such as with stratification), the DEFF will be less than 1.0, and if the design lessens the precision (as found with multistage sampling), the DEFF will be greater than 1.0. To estimate the DEFF for a sample mean, one must obtain an estimate of the appropriate sampling variance using software that can accommodate the sampling design (such as STATA, SUDAAN, WesVar, and select procedures in SAS) and calculate the "naïve" sampling variance assuming SRS with any typical software. The estimate of DEFF for the sample mean, then, is

$$\text{DEFF} = \frac{sv(\hat{\mu})_{complex}}{sv(\hat{\mu})_{SRS}}. \tag{4}$$

Authors can provide these DEFFs for key variables in analyses to allow the reader to determine whether statistical procedures to accommodate the sampling designs should be used or whether traditional statistical procedures will yield sufficiently robust standard error estimates.

If software that can accommodate survey design information is not available to the researcher, it is still possible to obtain an estimate of the appropriate sampling variance of the sample mean for select sampling designs. Under multistage sampling with no stratification, an estimate of the DEFF of the sample mean can be obtained based on the *intraclass correlation* (ICC). The ICC is a measure of the amount of variability in the response variable $y$ that can be explained by the fact that sample members were selected from within PSUs as opposed to selected randomly. The ICC typically ranges from 0 to 1 and a value close to 1 indicates that all of the elements in the PSU are nearly identical and therefore most variance is found between the PSU means as opposed to within PSUs. An ICC value near zero indicates that, within PSUs, the individuals differ on the response variable and the PSU means do not differ greatly. An estimate of the ICC for a given sample, $\hat{\rho}$, can be obtained using components from an analysis of variance on the variable of interest using the PSU identifier as the between subjects factor:

$$\hat{\rho} = \frac{MS_B - MS_W}{MS_B + (n. - 1)MS_W}. \tag{5}$$

Where $MS_B$ is the model mean square, $MS_W$ is the mean square error, and $n.$ is the sample size per group if balanced. The design effect of the mean is thus

$$\text{DEFF} = 1 + (n. - 1)\hat{\rho}. \tag{6}$$

Providing DEFF information for key variables in the analysis alerts the reader to information about possible violation (or lack of violation) of independence assumptions.

## 10. Variance Estimation

*Variance estimation* refers to the estimation of appropriate sampling variances (or standard errors) for a given sample statistic (such as a mean, regression coefficient, or factor loading). When data are not collected with SRS, typical formulas for sampling variance and standard error estimations are not appropriate.

The first decision that a researcher must make is whether the sampling strategy is to be part of the analytic model or whether the sampling strategy will be accounted for in the analysis. The former, referred to as a *model-based analysis*, presumes that the sampling information is helpful in explaining the hypothesized relations within the data. For example, if a two-stage survey of students within universities was conducted and students were asked about alcohol consumption and location of residence (on-campus housing, fraternity or sorority, off-campus with friends, off-campus with parents), a researcher might posit that the relation between location of residence and alcohol consumption actually depends on the type of university the student is attending, referred to as a cross-level moderation. In that case, the researcher might want to include university information (sampling information) into the analytic model. Techniques such as hierarchical linear modeling (see Chapter 10, this volume) allow the researcher to model the nested structure of the data resulting from the two-stage sampling design.

For those researchers who do not hypothesize that the sampling information is part of the analytic model, that information can be used to appropriately estimate the sampling variances (or standard errors) for the traditional single-level analysis. Such analyses are referred to as *design-based analyses*. The remainder of this section will refer to sampling variance estimation in design-based analyses.

When a sample has been collected via multistage sampling, the estimates of parameters (such as regression coefficients, means, or correlations) typically will be less precise than had a sample of the same size been collected through SRS. Therefore, if a researcher analyzes the data assuming independence of observations, the standard errors for estimates and thus the confidence intervals around the estimates will be underestimated or too narrow; the degree of this bias will depend on the homogeneity of the response variable(s) across clusters as can be measured by the design effect (see Desideratum 9).

If stratification is used as part of the sampling design and the response variable is homogeneous within strata, the estimates from the sample will be more precise than had a sample of the same size been obtained through SRS and thus DEFF is less than 1.0. Therefore, if a researcher analyzes the data ignoring the fact that stratification was used in the sampling design, standard errors associated with parameter estimates will be overestimated and the researcher will lose power and increase the likelihood of making Type II errors. If a sampling design includes both stratification and multistage sampling, the increase in precision of estimates resulting from stratification tends to be smaller than the decrease in precision found with multistage sampling.

Several methods are available to estimate sampling variances, standard errors, and test statistics when analyzing data collected through complex sampling designs. Design effect adjustments, linearization, and replication techniques are some of these methods, and while some of these methods can be estimated by hand, most will depend on having access to software that can accommodate the sampling information.

A simple method to adjust for a complex sampling design is to inflate the standard errors obtained from a conventional weighted analysis by the square root of the design effect (DEFT) of the mean of the dependent variable(s) in the analysis. Equivalently, a design effect adjusted sampling weight could be calculated and used in the analysis. The procedures of adjusting a standard error estimate by the DEFT or using an adjusted sampling weight result in accurate adjustments of the standard error for a simple statistic such as the mean. However, these procedures often result in conservative estimates of the sampling errors in more complex statistical procedures such as regression and structural equation modeling. Additionally, the researcher needs very good estimates of DEFF, which would require software that can accommodate the sampling design information. Therefore, it is suggested that researchers plan to use one of the more advanced techniques as described below.

Estimating sampling variances using a linearization method is a more appropriate approach. Because complex sample statistics are actually nonlinear functions, their sampling variances are often obtained by creating an approximate linear function, and then the variance of the new function is used as the sampling variance estimate. This approach to variance estimation is referred to by several additional terms in statistical analysis literature: the *delta method*, *Taylor Series approximation*, and *propagation of variance*. In the specific case of complex sample data derived from a stratified multistage sample, linearization results in a variance estimate that is a combination of the variation among PSUs within the same stratum. For example, for a stratified multi-stage sample with equal sample sizes within each PSU in a stratum, the standard error of the mean would be estimated in two steps. First, the sampling variance within each stratum would be estimated

$$s_h^2 = \frac{\sum\limits_{\alpha=1}^{a} (\bar{y}_{h\alpha} - \bar{y}_h)^2}{a-1} \tag{7}$$

where $\alpha$ represents the PSU, $a$ is the total number of PSUs within the stratum, $\bar{y}_{h\alpha}$ is the mean on the response variable in PSU $a$ within stratum $h$, and $\bar{y}_h$ is mean of the response variable within stratum $h$ across all PSUs. Then these estimates of sampling variance within each stratum are combined to obtain the overall standard error

$$se_{\hat{\mu}} = \sqrt{\sum_{h=1}^{H} \frac{s_h^2}{n}} \tag{8}$$

where $h$ represents the stratum, $H$ is the total number of strata, $n$ is the total sample size, and $s_h^2$ is the variance of PSU means within stratum $h$ as calculated in Equation (7).

There are many options to determine an approximate linear estimate and the choice of these depend on the complexity of the sampling design and the complexity of the parameter being estimated. Equations for linearized estimates for sampling variances for a range of different sampling schemes are available in Kalton (1983). Most researchers, however, use computer software that has been specially designed for complex sample data to provide these linearized estimates; information about software options is provided at the end of this section.

Another option for appropriate estimation of standard errors when using complex sample data is to use a replication method. The phrase "replication method" means that repeated samples are taken of the elements in the original sample to constitute new samples. For each of these new samples, the statistic of interest (a mean, a regression coefficient, etc.) is calculated. Then, the empirical distribution of those statistics is used to determine the estimated sampling distribution. With complex sample data, researchers most often use the following replication methods: *Jackknife Repeated Replication*, *Balanced Repeated Replication*, and *bootstrapping*. A very nice description of each of these methods is available in Rust and Rao (1996). The choice of any of these methods can depend on the statistical software available, whether replicate weights are already provided with the data, and the nature of the complex sample design.

Jackknife repeated replication (JRR) entails temporarily dropping one or more observations from the original dataset, obtaining the estimated statistic based on this subsample, and repeating this process until each observation has been dropped once. With multistage data, this process is usually accomplished by dropping all of the elements in one PSU at a time: the first-stage sampling unit, the PSU, is seen as the "dropped" observation. Sampling weights for elements from the other PSUs in the same strata are then adjusted to account for the dropped observation(s) yielding a sum of weights that is the same as the original sum of weights.

The standard error of the statistic is then calculated as a function of the variability of replicate estimates from the original estimate, although the calculation depends on whether a stratified sample was taken at the first-stage of selection. On some national and international data sets that are publicly available, the adjusted weights are already provided for the analyst. For example, a dataset might include 90 jackknife weights, called JACK1, JACK2, …, JACK90, indicating that the analyst might choose a replication method for standard error estimation, running the analysis 90 times, each time with a different weight. Those 90 estimates will then be used to determine the sampling variability of the original full sample estimates.

Balanced repeated replication (BRR), also referred to as *half-sample replicates*, is an approach where each replicate is created using half of the PSUs in the sample, one from each stratum. A second replicate, the complement replicate, can then be created out of the remaining PSUs. BRR can only be accomplished when the sampling design has been undertaken with the selection of two PSUs from each stratum. If the sample design did not include two PSUs from each stratum, similar strata and/or PSUs can be grouped to obtain such a design but such realignment must be done with caution. The term "Balanced" in the name BRR refers to the need to choose orthogonal replicates. There is a complication creating replicates using half of the PSUs (chosen at random) because dependent replicates can result, providing estimated statistics that are correlated across replicates. For example, if we have four strata in our sampling design (strata 1–4) and within each stratum sampled two universities (with IDs of 1 to 8 continuous across strata), we could obtain the following replicates with selected university IDs:

Replicate 1: 1, 3, 5, 7
Replicate 2: 1, 4, 5, 8
Replicate 3: 2, 4, 5, 8

Because replicates 2 and 3 share three of the same universities in the replicate sample, the estimated statistic in these two replicates will be very similar. For this reason, design matrices are used to select the appropriate PSUs for each BRR replicate sample; they are not chosen at random). The standard error of the statistic is then determined based on the variability of the estimate across replicates.

Bootstrapping is similar to JRR and BRR in that the observations from the original sample are used to form replicate samples. In bootstrapping, however, observations from the original sample are sampled with replacement to obtain a dataset that can be of the same size as the original dataset and no reweighting of the observations is typically undertaken. Bootstrapping is not an easy task with complex sample data but is a very flexible method. The process of creating bootstrapped replicates will depend on the complex sampling design. The standard error of the estimate will be a function of the variability of the estimate across the many bootstrap replicate samples. Bootstrap methods typically require many more replications than JRR or BRR. Additionally, while JRR and BRR estimation are available in many survey software packages already, bootstrapping must usually be programmed by the researcher.

There are several software options for the researcher who would like to use linearization or replication techniques, in conjunction with their sampling design information, to appropriately estimate sampling variances of statistics. Starting with version 8, SAS has included linearized sampling variance estimates for select procedures (such as means, regression coefficients, and frequency tables.) The WesVar software was developed specifically for analyzing complex sample data and relies on replication techniques to variance estimation; both BRR and JRR are accommodated in this software. SUDAAN supports both JRR and BRR replication methods, as well as linearized variance estimates. Stata, like SAS, is a full data base management and statistical package that also includes a complex sample modeling component, relying on linearization for variance estimation. Unfortunately, the base version of SPSS does not have complex sample variance estimation functions available. An add-on module, called COMPLEX, allows researchers access to advanced functions. Finally, specialized statistical software packages, such as LISREL and Mplus for structural equation modeling, are starting to include estimation techniques for data that arise from complex sampling designs.

If authors have used data from a complex sampling design, at a minimum they should alert readers to the violation of assumptions for traditional statistical analyses and the likely effects on standard errors and inference that these violations would present. With the many software programs that are available to appropriately analyze complex sample data, authors really should attempt to accommodate the sampling design using one of the sampling variance estimation techniques identified above.

## 11. Limitations in Survey Data Collection

Researchers are encouraged to remark about anything in the survey data collection and analysis that was not in line with the guidelines above. The context of the survey, however, is the most important consideration and decisions pertinent to this context can easily override the desiderata and other guidelines presented. Remarks about the likely effects of each of the problems or issues that arise for each desideratum would be appropriate, including coverage error in the discussion of the sample frame, measurement error in the discussion of questionnaire development and measures used, non-response error in the discussion of the procedures and response rate, editing and process errors, and, of course, sampling error in the statistical analysis portion.

## References

American Association for Public Opinion Research. (2006). *Standard definitions: Final dispositions of case codes and outcome rates for surveys.* http://www.aapor.org/uploads/standarddefs_4.pdf. Retrieved January 9, 2008.

American Association for Public Opinion Research. (2008). *Best practices for survey and public opinion research.* http://www.aapor.org/bestpractices. Retrieved January 10, 2008.

de Leeuw, E. D., Hox, J. J., & Dillman, D. A. (Eds.). (2008). *International handbook of survey methodology.* New York: Lawrence Erlbaum Associates.

Dillman, D. A. (2000). *Mail and internet surveys: The tailored design method.* New York: Wiley.

Forsyth, B. H., & Lessler, J. T. (1991). Cognitive laboratory methods: A taxonomy. In P. P. Biemer, R. M. Groves, L. E. Lyberg, N. A. Mathiowetz, & S. Sudman, S. (Eds.), *Measurement errors in surveys* (pp. 393–419). New York: Wiley.

Frank, O., & Snijders, T. (1994). Estimating the size of hidden populations using snowball sampling. *Journal of Official Statistics, 10,* 53–67.

Groves, R. M. (1989). *Survey errors and survey costs.* New York: Wiley.

Groves, R. M., & Couper, M. P. (1998). *Nonresponse in household interview surveys.* New York: Wiley.

Kalton, G. (1983). *Introduction to survey sampling.* Newbury Park, CA: Sage.

Kish, L. (1965). *Survey sampling.* New York: Wiley.

Lee, E. S., Forthofer, R. N., & Lorimor, R. J. (1989). *Analyzing complex survey data.* Newbury Park, CA: Sage.

Lohr, S. L. (1999). *Sampling: Design and analysis.* Pacific Grove, CA: Duxbury Press.

Lohr, S. L. (2008). Coverage and sampling. In E. D. de Leeuw, J. J. Hox, & D. A. Dillman (Eds.), *International handbook of survey methodology* (pp. 97–112). New York: Lawrence Erlbaum Associates.

Rust, K. F., & Rao, J. N. K. (1996). Variance estimation for complex surveys using replication techniques. *Statistical Methods in Medical Research, 5,* 283–310.

Smith, T. W. (2002) Developing nonresponse standards. In R. M. Groves, D. A. Dillman, J. L. Eltinge, & R. J. A. Little (Eds). *Survey nonresponse* (pp. 27–40). New York: Wiley.

Stapleton, L. M. (2008). Analysis of data from complex surveys. In E. D. de Leeuw, J. J. Hox, & D. A. Dillman (Eds.). *International handbook of survey methodology* (pp. 342–369). New York: Erlbaum.

Tourangeau, R., Rips, L. J., & Rasinski, K. (2000). *The psychology of survey response.* New York: Cambridge University Press.

# 31
# Survival Analysis

Paul D. Allison

Survival analysis is a collection of statistical methods that are used to describe, explain, or predict the occurrence and timing of events. The name *survival analysis* stems from the fact that these methods were originally developed by biostatisticians to analyze the occurrence of deaths. However, these same methods are perfectly appropriate for a vast array of social phenomena including births, marriages, divorces, job terminations, promotions, arrests, migrations, and revolutions. Other names for survival analysis include *event history analysis*, *failure time analysis*, *hazard analysis*, *transition analysis*, and *duration analysis*. Although some methods of survival analysis are purely descriptive (e.g., Kaplan-Meier estimation of survival functions), most applications involve estimation of regression models, which come in a wide variety of forms. These models are typically very similar to linear or logistic regression models, except that the dependent variable is a measure of the timing or rate of event occurrence. A key feature of all methods of survival analysis is the ability to handle *right censoring*, a phenomenon that is almost always present in longitudinal data. Right censoring occurs when some individuals do not experience any events, implying that an event time cannot be measured. Introductory treatments of survival analysis for social scientists can be found in Teachman (1983), Allison (1984, 1995), Tuma and Hannan (1984), Kiefer (1988), Blossfeld and Rohwer (2001), and Box-Steffensmeier and Jones (2004). For a biostatistical point of view, see Collett (2003), Hosmer and Lemeshow (2003), Kleinbaum and Klein (2005), or Klein and Moeschberger (2003). Specific desiderata for applied studies that use survival analysis are presented in Table 31.1 and later explained in detail.

## 1. Definition of the Event

The first step in any application of survival analysis is to define, operationally, the event that is to be modeled. Ideally, an event is a qualitative change that occurs at some specific, observed point in time. Classic examples include a death, a marriage, or a promotion. In such cases, where there is little ambiguity, there may be no need to explicitly define the event. Other applications may not be so clear cut, however. Some changes (e.g., menopause) take a while to "occur," so it is necessary to make decisions about criteria for determining the timing of the event. It is also possible to define events with respect to quantitative variables, especially if they undergo sharp, sudden changes. For instance, a "stock

Table 31.1 Desiderata for Survival Analysis

| Desideratum | Manuscript Section(s)* |
|---|---|
| 1. The event is defined in a clear and unambiguous way. | I |
| 2. The observation period is specified with careful consideration of origin time and possible late entry. | M |
| 3. Censoring is discussed, with indications of amount, type and reasons for censoring. | M |
| 4. An appropriate choice is made between a discrete versus a continuous time method. | M |
| 5. An appropriate choice is made between a parametric versus a semi-parametric method. | M |
| 6. Choice of covariates is discussed and justified. Possible omitted covariates are considered. | M, D |
| 7. Any time-varying covariates are appropriately defined, and a method for handling them is chosen. | M |
| 8. If there are multiple events per individual, an appropriate method is chosen to handle the possible dependence among those events. | M |
| 9. If there are competing risks, an appropriate method is chosen and appropriate tests are reported. | M |
| 10. Sampling method and sample size are explained and justified. | M |
| 11. The treatment of missing data is addressed. | M, R |
| 12. The name and version of the software package is reported. | M,R |
| 13. Summary statistics of measured variables are presented; information on how to gain access to the data is provided. | R |
| 14. Graphs of the survivor function(s) are presented. | R |
| 15. The proportional hazards (or equivalent) assumption is evaluated. | R |
| 16. For competing models, comparisons are made using statistical tests (for nested models) or information criteria (for non-nested models). | R |
| 17. Coefficients (or hazard ratios) are reported, together with standard errors, confidence intervals and $p$-values. | R |
| 18. Conditional survivor and/or hazard functions may be presented. | R |
| 19. Potential methodological limitations are discussed. | D |

\* *Note*: I = Introduction, M = Methods, R = Results, D = Discussion

market crash" could be said to occur if a particular market index falls more than 30% during a single week. Clearly, this definition involves some arbitrary choices that must be carefully considered and justified. A person could be said to "fall into poverty" if his income falls below some specified threshold. But this demands a rationale for choosing that threshold.

Another decision that must be made is whether to treat all events the same or to distinguish different types of events. If the event is an arrest, for example, one could either treat all arrests the same or distinguish between arrests for misdemeanors and arrests for felonies. All deaths could be treated alike, or one could distinguish between different kinds of deaths according to reported causes. Of course, such distinctions are only possible if data are available to differentiate the event types. Why do it? Usually, it is done because there are reasons to believe that predictor variables have different effects on different event types. In such cases, the prevailing strategy is to estimate competing risks models (see Desideratum 9). The downside of distinguishing different event types is that fewer events are available to estimate each set of parameters, which might substantially reduce statistical power.

Lastly, when events are repeatable for each individual, one must decide whether to focus on a single (usually the first) event for each individual, or to use a method that incorporates all the repeated events. If the average number of events per individual is small, say, less than two, it is usually better to restrict attention to the first event.

## 2. Observation Period

Survival analysis requires that each individual be observed over some defined interval of time; if events occurred during that interval, their times are recorded. If events are not repeatable, observation is often terminated at the occurrence of an event. Decisions about the starting and stopping times for the observation period should be reported and justified.

Most methods of survival analysis (e.g., Cox regression) require that the event time be measured with respect to some *origin time*. The choice of origin time is substantively important because it implies that the risk of the event varies as a function of time since that origin. In many cases, the choice of origin is obvious. If the event is a divorce, the natural origin time is the date of the marriage. In other cases, the choice is not so clear cut. If the event is a retirement, do you model age at retirement or time in the labor force?

Ideally, the origin time is the same as the time at which observation begins, and most software programs for survival analysis presume that this is the case. Frequently, however, observation does not begin until some time after the origin time. For example, although we may use date of marriage as the origin time in a study of divorce, couples may not be recruited into the study until years later. This is called *late entry* or *left truncation*. Because individuals are not at risk of an observed event until observation begins, special methods are necessary to take this into account. For more details, see Allison (1995, pp. 161–165)

## 3. Censoring

Censoring is endemic to survival analysis data, and any report of a survival analysis should discuss the types, causes, and treatment of censoring. By far the most common type of censoring is *right censoring*, which occurs when observation is terminated before an individual experiences an event. For example, in a study of divorce, couples that do not divorce during the observation period are right censored. All survival analysis software is designed to handle right censoring, and it is essential to include the right censored observations in the analysis.

Standard methods for dealing with right censoring presume that such censoring is *non-informative*. Roughly speaking, that means that the fact that an individual is censored at particular point in time does not tell us anything about that individual's risk of the event. That assumption is necessarily satisfied if the censoring time (or potential censoring time) is the same for everyone in the sample. However, the censoring could be informative if it occurs at varying times because individuals drop out of the study, which could lead to biased estimates of the parameters. Unfortunately, there is no test for the non-informative assumption and little that can be done to correct for bias due to violation of this assumption. But the lesson is that survival studies should be designed and executed so as to minimize censoring due to drop outs. In any case, the proportion of censoring cases due to drop outs should be reported.

A slightly less common type of censoring is *interval censoring*, which means that an individual is known to have an event between two points in time, but the exact time is unknown. For example, if a person reports being unmarried at wave 1 of a panel study but married at wave 2, then the marriage time is interval censored. If the censoring times are regularly spaced, interval censoring can often be handled by discrete-time methods (see the next section). However, most survival analysis software cannot handle irregular patterns of interval censoring.

The least common type of censoring is *left censoring*, which happens when an event is known to have occurred before some particular time, but the exact time is unknown. For example, in a study of first marriage, if a person is known only to have married before age 20, that person's marriage age is left censored. Note that the term left censoring is often used with a quite different meaning in the social science literature. In this alternative meaning, left censoring is said to occur when we begin observing an individual at some arbitrary point in time, but we do not know the origin time (i.e., how long it has been since the individual has been at risk of the event).

## 4. Discrete-Time vs. Continuous-Time Methods

If you know the exact times at which events occur, it is appropriate to use methods that treat time as continuous. If, on the other hand, you know only the month or the year of the event, you might be better off using discrete-time methods. One of the best indications of the need for discrete-time methods is the presence of large numbers of *ties*. A tie is said to occur if two individuals experience an event at the same recorded time. Occasionally, time is truly discrete in the sense that events can only occur at certain discrete points in time. For example, in most universities, faculty can only be promoted at the end of an academic year.

Most survival analysis software is designed for continuous-time data. If you want to go the discrete-time route, you must choose between a logit model and a complementary log-log model. Logit is more appropriate for event times that are truly discrete, while complementary log-log is more appropriate for events that can happen at any time but are only observed to occur in discrete intervals. In practice, the choice is usually not consequential.

Having chosen a model, you must then choose an estimation method. Some Cox regression programs (e.g., SAS, Stata, S-Plus) have options for estimating either of the two models using partial likelihood estimation. But partial likelihood can be very computationally intensive for large samples with lots of ties. The alternative is to do maximum likelihood using conventional binary regression software. The trick is to break up each individual's event history into a set of distinct records, one for each unit of time in which the individual is observed, with a dependent variable coded 1 if an event occurred in that time unit, otherwise 0. One can then estimate the logit model using standard logistic regression software (Allison 1982, 1995). Many packages also have options for estimating the complementary log-log model.

## 5. Parametric vs. Semi-parametric Methods

By far, the most popular method for regression analysis of survival data is Cox regression, which combines the proportional hazards model with the partial likelihood method of estimation. Cox regression is sometimes described as *semi-parametric* because, although it is based on a parametric regression model, it does not make specific assumptions about the probability distribution of event times. By contrast, parametric regression models assume particular families of probability distributions, such as exponential, Weibull, Gompertz, lognormal, log-logistic, or gamma.

Although Cox regression is probably the better default method, there are two goals that are easily accomplished with parametric methods but difficult or impossible with Cox regression. First, parametric methods are much better at handling left censoring or interval censoring (especially if the intervals differ across individuals). Second, it is easy to generate predicted times to events with parametric methods, but awkward (and sometimes impossible) to do so with Cox regression. Sometimes people choose parametric methods because they worry that their data do not satisfy the proportional hazards assumption (see Desideratum 15). However, parametric models typically make assumptions that are at least as restrictive as the proportional hazards assumption.

## 6. Covariates

Issues regarding covariates (also known as predictor variables, independent variables, regressors) are mostly the same in survival analysis as in linear regression and logistic regression (with the important exception described in Desideratum 7). Although it is desirable to provide a rationale for the inclusion of each covariate in the regression model, it is not essential. The consequences of including a variable that actually has no effect are minimal. The real danger, as with any regression analysis of observational data, comes from omitting variables that really have an effect on the outcome. This can lead to severe bias, especially if the omitted variable is moderately to strongly correlated with included variables. So any report of a survival regression should discuss the possibility of important variables that have not been included.

As with other kinds of regression, it is important to consider whether the covariates have nonlinear effects on the outcome and whether there are interactions among the covariates in their effects on the outcome. Strategies for testing and including such nonlinearities and interactions are basically the same as in linear regression, except that there are some special graphical diagnostics available for nonlinearities in Cox regression (Therneau & Grambsch, 2001). Multicollinearity is also a potential problem. Although survival analysis programs typically don't provide collinearity diagnostics, one can simply do a preliminary check with a linear regression program, while specifying the event time as the dependent variable. Because multicollinearity is all about linear relations among the covariates, it is not necessary to evaluate it within the context of a survival analysis.

## 7. Time-Dependent Covariates

One major difference between survival regression and conventional linear regression is the possibility of time-dependent (time-varying) covariates. These are predictor variables whose values may change over the course of observation. For example, suppose that over a five-year period, information is recorded on any changes in marital status. Then, marital status (updated on a daily basis) could be used as a time-varying predictor of some other event, such as an arrest.

Not all survival analysis methods and/or software can handle time-dependent covariates. For example, most programs for parametric survival models do not allow for time-dependent covariates (although that feature is available in recent releases of Stata). On the other hand, such variables are usually easy to incorporate into discrete-time methods based on logistic (or complementary log-log) regression. That is because each discrete time point is treated as a separate observation, so that any time-dependent covariates can be updated for each observation.

Cox regression is also well known for its ability to handle time-dependent covariates. However, there are two quite different approaches for implementing this capability in software packages. The "episode splitting" method requires that the data be configured so that there is a separate record for each interval of time during which all the covariates remain constant. The "programming statements" method expects one record per individual, with the time-varying covariates appearing as separate variables for each time at which the variables are measured. The time-dependent covariates are then defined in programming statements prior to model specification. Properly implemented, these two methods will give identical results.

One potential issue with time-dependent covariates is that the frequency with which they are measured may not correspond to the precision with which event times are measured. For example, we may know the exact day on which person died of a heart attack. Ideally, a time-dependent covariate, like smoking status, would also be measured on a daily basis. Instead, we may only have annual reports. Some form of imputation is necessary in such cases. The simplest and most common form of imputation is "last value carried forward," although other methods should be considered.

One should also keep in mind that there may be several plausible ways of representing a time-dependent covariate. For example, smoking status could be coded as "person smoked on this day," "number of days out of the last 30 in which the person smoked," or "number of years of smoking prior to the current day," and so forth. Decisions among the alternatives should be carefully considered, and may be based on empirical performance.

## 8. Repeated Events

If the data contain information on more than one event for each individual, special methods are needed to take advantage of this additional information and to deal with the problems that may arise. If repeated events are observed for an individual, the standard strategy is to reset the clock to 0 each time an event occurs and treat the intervals between events as distinct observations. Thus, if a person is observed to have three arrests over a five-year interval, four observations would be created. The last observation would be a right-censored interval, extending from the third arrest until the end of the observation period.

Repeated events provide more statistical power, and also make it possible to control or adjust for unobservable variables that are constant over time. However, whenever there are multiple observations per individual, there is also likely to be statistical dependence among those observations. Unless some correction is made for this dependence, standard errors and $p$-values will be too low. There are four widely available methods for repeated events that provide appropriate corrections for dependence.

1. Robust standard errors (also known as Huber-White or sandwich estimates) yield accurate standard errors and $p$-values, but leave coefficient estimates unchanged.
2. The method of generalized estimating equations (GEE) also gives corrected standard errors and $p$-values but, in addition, produces more efficient coefficient estimates.
3. Random effects (mixed) models provide the same benefits as GEE, but also correct the coefficients for "heterogeneity shrinkage." This is the tendency of coefficient estimates to be attenuated toward zero because of unobserved heterogeneity.
4. Fixed effects methods also correct for dependence and heterogeneity shrinkage. In addition, they actually control for all stable characteristics of the individual.

For more details, see Allison (1995).

Some of these methods may not be available for some survival regression models or software. For example, Stata will estimate random effects models for Cox regression but SAS will not. Also, note that while fixed effects methods seem to offer the most advantages, they also come with important disadvantages. First, one cannot estimate the effects of variables that are constant over time, like sex or race, although such variables are implicitly controlled. Second, standard errors may be substantially larger because the estimates are based only on variation within individuals.

## 9. Competing Risks

If a decision has been made to distinguish different kinds of events, an appropriate method must be chosen to handle the different event types. In the competing risks approach, a separate model is specified for the timing of each type of event. These could be any of the models already discussed. If one has continuous time data, each of these models can be estimated separately using standard software for single kinds of events. The trick is that events other than the focal event type are treated as though the individual is censored at that point in time. For example, suppose you want to estimate Cox regression models for job terminations, while distinguishing between quittings and firings. You would estimate one model for quittings, treating firings as censored observations. Then you would estimate a model for firings, treating the quittings as censored observations.

Test statistics are available for testing whether coefficients for a particular variable are the same across event types (Allison, 1995). There are also statistics for testing whether *all* variables have the same coefficients across event types. These statistics can be helpful in determining whether it is really necessary to distinguish the event types. As noted earlier, one disadvantage of distinguishing event types is that the number of events may be small for each event type, leading to a loss of statistical power.

If event times are discrete, maximum likelihood estimation requires that models for competing risks be estimated simultaneously rather than separately. An attractive model that can be estimated with conventional software is the multinomial logit model, also known as the generalized logit model. Unfortunately, there is no comparable multinomial model for the complementary log-log specification.

In some situations with multiple event types, a "conditional" approach may make more sense than competing risks (Allison, 1984). In this approach, the first step is to estimate a model for event timing without distinguishing the different event types. Then, restricting the sample to those individuals who experienced events, the second step is to estimate a binary or multinomial logit model predicting the type of event. This approach is attractive when the event types represent alternative means for achieving a single goal. For example, the event might be the purchase of a computer, and computers are distinguished by whether the operating system is Windows, Linux, or Macintosh.

## 10. Sampling Issues

There are three questions about sampling that should be addressed: What kind of sample is used? Are the analysis methods appropriate for the sampling method? Is the sample big enough? With regard to the first question, the ideal is a well-designed and executed probability sample. Nevertheless, many survival analyses are carried out on a complete population (e.g., the 50 states in the U.S.) or on convenience samples (e.g., students who volunteered to participate). Although others may disagree, I take the position that survival analysis—including the calculation of confidence intervals and hypothesis tests—is perfectly appropriate for analyzing a complete population. The statistical models that underlie such analyses are based on a hypothesis of inherent randomness in the phenomenon itself, and they do not require any randomization in the study design to justify the application of inferential techniques. The same argument could be made about convenience samples, although any conclusions might only apply to the sample at hand.

Regarding analysis, most survival analysis packages presume, by default, that the sample is a simple random sample. For many samples, however, there will be a need to adjust for clustering, stratification, and/or weighting. Although some packages are explicitly designed for survival analysis with complex samples (e.g., SUDAAN), conventional software can often do the job. Clustering can be accommodated by the methods described above for dependence with repeated events (although it might be difficult to adjust for both repeated events and cluster sampling). Stratification can usually be handled by including the stratification variables as covariates. Finally, most packages allow for differential weighting of observations. However, even if the sampling design involved disproportionate weights, it may not be necessary or desirable to incorporate those weights into the analysis (Winship & Radbill, 1994). This is most likely to be the case if the goal is to estimate an underlying causal model rather than some population regression function.

With regard to sample size, the most important thing to keep in mind is that censored observations contribute much less information than uncensored observations (events). Conventional wisdom has it that there should be at least five (some say 10) events for each parameter in the model, in order for maximum likelihood (or partial likelihood) estimates to have reasonably good properties. As for power considerations, there are numerous software packages and applets that will calculate power and sample size for a single dichotomous covariate. Vaeth and Skovlund (2004) showed how these programs can be easily extended to handle more complex regression problems. Some packages (e.g., Stata, PASS) have routines that will do power calculations for Cox regression analyses.

## 11. Missing Data

Reports of survival analysis should say something about the extent of missing data and the methods used to handle it. Of course, the default in virtually all survival packages is to do listwise deletion (complete case analysis). And if the proportion of cases lost to missing data is small (say, 10% or less), listwise deletion is probably the best choice. Other conventional methods, like (single) imputation or dummy variable adjustment, typically lead to biased parameter estimates, biased standard error estimates, or both.

For larger fractions of missing data, much better results can be obtained with multiple imputation (Allison, 2001). In this method, imputed values are random draws from the predictive distribution of the missing values given the observed values. Several data sets are created (typically five or more), each with slightly different imputed values. The analysis is performed on each data set using standard software. Then, using a few simple rules, the results are combined into a single set of parameter estimates, standard errors, and test statistics. Multiple imputation uses all the data to produce parameter estimates that are approximately unbiased and efficient. In calculating standard errors and test statistics, multiple imputation, unlike conventional imputation, also incorporates the inherent uncertainty about the values of the missing observations.

Although there are many stand-alone packages for doing multiple imputation, the process is much easier if the imputation is done within the same package used to do the analysis. Software for doing this is available for Stata, SAS, and S-Plus. These also happen to be great packages for survival analysis. Nearly all standard multiple imputation routines are based on the assumption that data are missing at random. This means, roughly, that the probability of missingness may depend on variables that are observed but does not depend on the values of the variables that are missing. Multiple imputation can be done under other assumptions, but the implementation is tricky and must be carefully tailored to each application.

For survival analysis, multiple imputation should only be done for missing values on the predictor variables. Cases that have missing values on the dependent variable should simply be deleted because conventional imputation software is not suited for missing data on event timing and censoring. In setting up the imputation model, however, it is generally a good idea to include both the (logged) event time and the censoring indicator variable so that the relations between these variables and the predictors are adequately reproduced for the imputed variables.

## 12. Software

Nearly all the major statistical packages have programs for doing Cox regression and Kaplan-Meier estimation of survivor functions. And all can do discrete-time maximum likelihood estimation via logistic regression. Not all can estimate parametric regression models, however, and those that do may vary widely in their capabilities. For example, SAS can estimate parametric models with left and interval censoring but cannot handle time-dependent covariates. With Stata, it is just the reverse. Cox regression programs may also vary widely in their features and capabilities. SPSS, for instance, can handle time-dependent covariates, but its programming functions for defining those covariates are rather limited compared with SAS. As of this writing, I would rate SAS, Stata, and S-Plus as the three best packages for doing survival analysis. Although they vary to some degree in their capabilities, all three have a wide array of programs, functions and options for survival analysis.

Some survival regression programs allow for the incorporation of unobserved heterogeneity into the model. In my judgment, this is a useful feature if individuals have repeated events because it allows for dependence among the multiple observations. However, I would caution against using this option in the more typical case of non-repeated events. In that situation, unobserved heterogeneity models are only weakly identified, and results may depend too critically on the particular specification.

### 13. Summary Statistics and Data Accessibility

As with other regression methods, it is good practice to report summary statistics for the predictor variables, usually their means and standard deviations. There is a potential complication, however, with time-dependent covariates. If you are using a method that requires multiple records per individual, like discrete-time maximum likelihood or Cox regression using the episode splitting method, you can simply calculate the means and standard deviations over the multiple records. On the other hand, if you are doing Cox regression with programming statements, the time-dependent covariates are created during the estimation process and are not available for calculating descriptive statistics. In that case, I would simply report such statistics for the baseline measurements of the variables.

### 14. Survivor and Hazard Functions

Although not essential, it is commonplace and informative to present a graph of the survivor function, usually estimated via the Kaplan-Meier method. Such graphs are helpful in giving the reader a sense of the rates of event occurrence and censoring, and how those change over time. In some fields, a cumulative failure graph is preferred over a survivor graph. The two graphs give the same information, however, because the failure probability is just 1 minus the survivor probability.

Even more informative than the survivor function is a graph of the estimated hazard function because it more directly quantifies the rate of event occurrence and how that rate changes over time. But the problem with the hazard function is that non-parametric estimates based on Kaplan-Meier require smoothing, and different smoothing algorithms can yield markedly different graphs. Therefore, if hazard graphs are to be presented, I recommend using the actuarial (life table) method. Although this requires an arbitrary choice of time intervals, results tend to be more stable than those produced by smoothing methods.

### 15. Proportional Hazards Assumption

Cox regression is based on the proportional hazards model. The proportional hazards assumption says, in essence, that the dependence of the hazard on time has the same basic shape for everyone, even as the magnitude of the hazard varies across individuals as a function of their predictor values. A crucial implication of this assumption is that predictor variables have the same effects at all points in time, that is, there are no interactions with time.

Although many researchers get very concerned about whether their data satisfy this assumption, I believe that those concerns are often unwarranted. If the assumption is violated for a particular predictor variable, it simply means that the coefficient for this variable represents a kind of "average" effect over the period of observation. For many applications, this may be sufficient. In some cases, however, the violations may be so severe that they lead to biases in the effects of other variables. In other cases, there may be direct interest in how the effect of a variable on the hazard changes over time.

A quick check of the proportional hazards assumption can be obtained by computing correlations between time (or some function of time) and "Schoenfeld residuals" which are calculated separately for each predictor. Non-zero correlations are evidence against the proportionality assumption. Several Cox regression software packages have an option to compute these statistics.

A more definitive check is to directly include interactions between predictors and time, which are specified as time-dependent covariates. Significant interactions indicate violation of the assumption. However, in this case the method of diagnosis is also the cure. By including the interactions, the Cox model is extended to allow for non-proportional hazards.

Another way to allow for non-proportional hazards is the method of stratification, which allows for different hazard functions for different categories of a categorical variable (like sex or marital status). This is a good method for controlling for a variable without imposing the proportional hazards assumption. But it does not yield any estimates of the effect of that variable, nor does it give a test of the proportional hazards assumption.

## 16. Model Comparisons

Researchers typically want to know how well their statistical models fit the data. Unfortunately, global or absolute measures of fit are generally not available for survival analysis models. Usually, the best we can do is to compare the relative fit of different models. If the models are nested (i.e., one model can be obtained from another by imposing restrictions on the parameters), likelihood ratio tests can be calculated by taking twice the positive difference in the log-likelihoods for the two models. Such tests can tell you whether the more complicated model is significantly "better" than the simpler model. These tests are especially useful when estimating parametric models because some of the better-known parametric distributions are nested within the generalized gamma distribution.

If two models are not nested, informal comparisons can be accomplished with Akaike's information criterion (AIC) or Schwarz's Bayesian Information Criterion (SBC or BIC). These statistics "penalize" the log-likelihood for the number of covariates in the model, enabling one to validly compare models with different sets of covariates. Many software packages report one or both of these statistics, both for parametric models and for Cox regression models. Preference is given to models with lower values of these statistics, although no $p$-values can be calculated.

## 17. Reports of Coefficients and Associated Statistics

Results for Cox regression may be reported as either beta ($\beta$) coefficients or hazard ratios (a few authors report both). Beta coefficients are more easily interpreted with respect to sign (positive, negative, or zero). However, their numerical magnitudes are difficult to interpret. Hazard ratios (which are always positive) may confuse some readers because a value of 1 means no effect. But the numerical magnitude has a more straightforward meaning: if HR denotes the hazard ratio, $100(\text{HR}-1)\%$ is the percentage change in the hazard for a one-unit increase in the predictor. In this respect, they behave just like odds ratios in logistic regression. In the biomedical sciences, there is a clear preference for reporting hazard ratios, and this preference seems to be spreading to other fields as well.

If you report $\beta$ coefficients, you should also report either standard errors or 95% confidence intervals. Because hazard ratios have asymmetric distributions, standard errors are not generally reported. Instead, the convention is to report 95% confidence intervals. It is optional but desirable to report $p$-values for testing the null hypothesis of no effect for each coefficient. Also desirable is a chi-square test for the null hypothesis that all coefficients are zero. Many authors ritualistically report the log-likelihood for each model, but this is usually not informative (unless it can be used to compare nested models).

## 18. Conditional Survivor or Hazard Functions

In Desideratum 14, I discussed the use of survivor or hazard functions as a descriptive device. After estimating a regression model, it is often desirable to illustrate its implications by displaying a model-based survivor function or hazard function. For example, if interest centers on the effect of some treatment, one could plot survivor functions for the treated vs. control groups in such a way that the plots embody any model assumptions (e.g., proportional hazards) and also control for any covariates in the model. If the variable of interest is quantitative, one can produce plots for several selected values of that variable, again while adjusting for any covariates.

## 19. Potential Methodological Limitations

Any application of statistical methods to real-world data is vulnerable to errors of one sort or another. Researchers need to be acutely aware of potential problems with their data and with the analytic methods they apply to those data. They also need to be upfront with their readers regarding any problems that they suspect could compromise their conclusions.

As noted in Desideratum 6, the most serious potential problem with survival analysis regression methods is the same as that for any other regression method applied to observational (non-experimental) data: the omission of variables (confounders) that affect the outcome and that are also correlated with the included variables. The omission of confounders can produce biases so severe that they lead to conclusions that are the exact opposite of the true state of affairs.

A problem peculiar to survival analysis is informative censoring (see Desideratum 3). Once the data are in hand, there is not much that can be done about this. But, if the number of randomly censored cases is substantial, research reports should discuss their potential impact. A sensitivity analysis can help to discern the potential direction of biases resulting from informative censoring.

Another potential danger comes from fitting an incorrect model. Some of the comparative statistics discussed in Desideratum 16 can be helpful in finding a good model. But it is also desirable to fit rather different models to the data and see if the results are consistent across models. For example, there is no good way to compare the fit of a Cox regression model with parametric gamma model. But it can be quite useful to fit both models to see if they lead to the same conclusions. If they do, well and good. If not, then your confidence in the results should be appropriately reduced.

## References

Allison, P. D. (1982). Discrete time methods for the analysis of event histories. In S. Leinhardt (Ed.), *Sociological methodology 1982* (pp. 61–98). San Francisco: Jossey-Bass.

Allison, P. D. (1984). *Event history analysis*. Thousand Oaks, CA: Sage.

Allison, P. D. (1995). *Survival analysis using SAS: A practical guide*. Cary, NC: SAS Institute.

Allison, P. D. (2001). *Missing data*. Thousand Oaks, CA: Sage.

Blossfeld, H.-P., & Rohwer, G. (2001) *Techniques of event history modeling: New approaches to causal analysis*. Mahwah, NJ: Erlbaum.

Box-Steffensmeier, J., & Jones, B. (2004). *Event history modeling: A guide for social scientists*. Cambridge, UK: Cambridge University Press.

Collett, D. (2003). *Modeling survival data in medical research*. New York: Chapman & Hall.

Hosmer, D. W., & Lemeshow, S. (2003). *Applied survival analysis: Regression modeling of time to event data*. New York: Wiley.

Kiefer, N. M. (1988). Economic duration data and hazard functions. *Journal of Economic Literature, 26*, 646–679.

Klein, J. P., & Moeschberger, M. L. (2003). *Survival analysis: Techniques for censored and truncated data*. New York: Springer-Verlag.

Kleinbaum, D. G., & Klein, M. (2005). *Survival analysis: A self-learning text*. New York: Springer-Verlag.

Teachman, J. D. (1983). Analyzing social processes: Life tables and proportional hazards models. *Social Science Research, 12*, 263–301.

Therneau, T. M., & Grambsch, P. M. (2001). *Modeling survival data: Extending the Cox model.* New York: Springer-Verlag.

Tuma, N. B., & Hannan, M. T. (1984). *Social dynamics: Models and methods.* Orlando, FL: Academic Press.

Væth, M., & Skovlund, E. (2004). A simple approach to power calculations in regression models. *Statistics in Medicine, 23,* 1781–1792.

Winship, C., & Radbill, L. (1994). Sampling weights and regression analysis. *Sociological Methods & Research, 23,* 230–257.

# Contributors

**Paul D. Allison** (allison@ssc.upenn.edu) is Professor of Sociology at the University of Pennsylvania, and President of Statistical Horizons LLC. His current research is focused on methods for handling missing data and on causal inference with longitudinal data. He has published eight books and more than 50 articles dealing with a wide variety of statistical methods. His most recent books include *Fixed Effects Regression Methods for Longitudinal Data Using SAS* (SAS Institute, 2005) and *Fixed Effects Regression Models* (Sage, 2009). In 2001 he received the Lazarsfeld award for distinguished contributions to sociological methodology.

**K. Rivet Amico** (rivetamico@comcast.net) is Assistant Research Professor in the University of Connecticut's Department of Psychology and the Center for Health Intervention and Prevention. Her area of research focuses primarily on the evaluation of risk and adherence behavior cross-sectionally and longitudinally. Her areas of interest reflect the unique challenges of working with count-based, percentage, and largely non-normal distributions inherent in modeling risk and risk-reduction, with a strong focus on logistic, ordinal, and Poisson-based approaches with generalized estimating equations and hierarchical linear modeling. Her work has appeared in *JAIDS, AIDS, AIDS and Behavior*, and *American Journal of Public Health*.

**Deborah L. Bandalos** (bandalos@uga.edu) is Professor and Coordinator of the program in Research, Evaluation, Measurement, and Statistics at the University of Georgia. Her research areas include structural equation modeling, exploratory factor analysis, scale development techniques, reliability and validity studies, and educational accountability and assessment systems. Her research has appeared in journals including *Structural Equation Modeling: A Multidisciplinary Journal*, *Multivariate Behavioral Research, Applied Measurement in Education*, and the *Journal of Educational Psychology*, as well as in several book chapters. Dr. Bandalos is the Associate Editor of *Structural Equation Modeling: A Multidisciplinary Journal* and serves on the editorial boards of *Psychological Methods* and *Applied Measurement in Education*.

**Christine H. Barthold** (cbarthol@udel.edu) is Assistant Professor in the Department of Human Development and Family Studies at the University of Delaware. A Board Certified Behavior Analyst-Doctoral (BCBA-D™), her work can be found in journals such as *Behavior Analyst Today* and the *Journal of Applied Behavior Analysis*. In addition, she serves on the editorial board of the *Journal of Speech-Language Pathology and Applied Behavior Analysis*. Her teaching and research interests include integrating technology for teaching students at the postsecondary level and the education of students with autism. She is the co-chair of the Teaching Behavior Analysis Special Interest Group for the Association for Behavior Analysis International.

**S. Natasha Beretvas** (tasha.beretvas@mail.utexas.edu) is Associate Professor and Chair of the Quantitative Methods Program, Department of Educational Psychology, the University of Texas at Austin, where she teaches statistics courses, including a graduate seminar on meta-analysis. Her research interests lie in the application and evaluation of statistical and psychometric modeling with a focus on methodological dilemmas in meta-analysis and multilevel modeling. Her research has appeared in methodological journals including *Multivariate Behavioral Research*, *Psychological Methods*, *Applied Psychological Measurement*, and *Journal of Educational Measurement*.

**Geoff Cumming** (g.cumming@latrobe.edu.ac) is Emeritus Professor, School of Psychological Science, La Trobe University, Melbourne, Australia. His main research area is statistical cognition, which is the study of how people understand, and misunderstand, statistical concepts, and how statistical communication can be improved. His research goal is to reform the way psychology does its statistics, especially by changing emphasis to effect sizes, confidence intervals, and meta-analytic thinking (ESCIMAT). He develops Exploratory Software for Confidence Intervals (ESCI; www.latrobe.edu.au/psy/esci).

**Sharon Anderson Dannels** (sdannels@gwu.edu) is Associate Professor of Educational Research and Coordinator of Research Methods in the Graduate School of Education and Human Development at The George Washington University. She teaches quantitative and qualitative research design and a range of data analysis courses. Her past research addressed subcortical arousal related to the biological basis of personality, cognitive effect of caffeine, effect of prophylactic drugs on cognitive performance, and the etiology of obesity. More recently, she was involved in explorations of leadership orientations of women in academic medicine and benefits and shortcomings of online advanced research methods courses.

**Mark L. Davison** (mld@umn.edu) is the American Guidance Service, Inc. and John P. Yackel Professor of Educational Assessment and Measurement in the Quantitative Methods in Education Program of the Department of Educational Psychology at the University of Minnesota. He co-directs the Minnesota Interdisciplinary Training in Educational Research (MITER) Program, a pre-doctoral training program funded by the Institute of Education Sciences in the U.S. Department of Education. Dr. Davison is Editor-in-Chief of the journal *Applied Psychological Measurement* and author of the book *Multidimensional scaling*. His publications include works in *Multivariate Behavioral Research*, *Psychological Assessment*, *Psychological Methods*, *Psychometrika*, and *Structural Equation Modeling: A Multidisciplinary Journal*.

**C. Mitchell Dayton** (cdayton@umd.edu) is Professor Emeritus, Department of Measurement, Statistics and Evaluation at the University of Maryland. His primary area of interest is latent class analysis which is a specialized field within the realm of discrete mixture models. In 1999, he published a Sage book dealing with latent class scaling models. His research has appeared in such journals as *Psychometrika*, *Journal of the American Statistical Association*, *American Statistician*, *Multivariate Behavioral Research*, *Applied Psychological Measurement*, *Journal of Educational and Behavioral Statistics*, *British Journal of Mathematical and Statistical Psychology*, *Psychological Methods*, and *Journal of Educational Measurement*.

**R. J. De Ayala** (rdeayala2@unl.edu) is Professor and Chair of the Department of Educational Psychology at the University of Nebraska, Lincoln. His research interests in psychometrics, item response theory, and computerized adaptive testing have appeared in *Applied Psychological Measurement, Applied Measurement in Education,* the *British Journal of Mathematical and Statistical Psychology, Educational and Psychological Measurement,* the *Journal of Applied Measurement,* and the *Journal of Educational Measurement.* He has authored *The Theory and Practice of Item Response Theory* and serves on the editorial boards of *Educational and Psychological Measurement* and *Applied Measurement in Education.* He is a Fellow of the APA and the AERA.

**Cody S. Ding** (dingc@umsl.edu) is Associate Professor of Educational Psychology, Research, and Evaluation at the University of Missouri-St. Louis. His research focuses on program evaluation, psychological and educational measurement, and growth modeling using both exploratory and confirmatory techniques, particularly with multidimensional scaling models. His writing appears in such journals as *International Journal of Behavioral Development, International Journal of Psychology, Educational and Psychological Measurement, Structural Equation Modeling: A Multidisciplinary Journal, Quality and Quantity,* and *Journal of Educational Measurement.* He has been a chief evaluator for several federally funded multi-million dollar programs.

**Andrew L. Egel** (aegel@umd.edu) is Professor in the Department of Special Education at the University of Maryland. His current research interests focus on the development and evaluation of innovative instructional methodologies for use with children with Autism Spectrum Disorders and other severe disabilities. Dr. Egel's research has been funded by the U.S. Office of Special Education and Rehabilitative Services and published in journals such as *Journal of Applied Behavior Analysis, Journal of Autism and Developmental Disorders,* and *Research in Developmental Disabilities.* He has also co-edited *Educating and Understanding Autistic Children* and *School Success for Kids with Autism.*

**Xitao Fan** (xfan@virginia.edu) is Curry Memorial Professor of Education, Curry School of Education, University of Virginia. His areas of interest include multivariate methods, structural equation modeling and model fit assessment, Monte Carlo simulation, and educational measurement issues. His work has appeared in journals such as *Structural Equation Modeling: A Multidisciplinary Journal, Multivariate Behavioral Research, Educational and Psychological Measurement,* and *Journal of Experimental Education.* He is currently the editor of *Educational and Psychological Measurement,* and serves on the editorial boards of several other journals (e.g., *American Education Research Journal* and *Effective Education*).

**Fiona Fidler** (f.fidler@latrobe.edu.au) is an Australian Research Council Postdoctoral Fellow in the School of Psychological Science at La Trobe University in Melbourne, Australia, where she works with Professor Geoff Cumming on various projects related to statistical cognition. Her Ph.D. in philosophy of science examined criticisms of Null Hypothesis Significance Testing and calls for statistical reform in Psychology, Medicine, and Ecology. Her dissertation is under contract to be published by Taylor and Francis in 2010.

**Sara J. Finney** (finneysj@jmu.edu) has a dual appointment at James Madison University as Associate Professor in the Department of Graduate Psychology and as Associate Assessment Specialist in the Center for Assessment and Research Studies. In addition to teaching statistics courses for the Assessment and Measurement Ph.D. program, she is Coordinator of the Quantitative Psychology Concentration within the Psychological Sciences M.A. program. Much of her research involves the application of structural equation modeling to better understand the functioning of self-report instruments, and has appeared in such journals as *Journal of Educational Psychology, Contemporary Educational Psychology,* and *Educational and Psychological Measurement.*

**Gregory R. Hancock** (ghancock@umd.edu) is Professor and Chair, Department of Measurement, Statistics and Evaluation at the University of Maryland, and Director of the Center for Integrated Latent Variable Research (CILVR). His area of interest includes structural equation modeling, with particular focus on latent means and latent growth models, and his work has appeared in such journals as *Psychometrika, Multivariate Behavioral Research, Structural Equation Modeling: A Multidisciplinary Journal, British Journal of Mathematical and Statistical Psychology,* and *Journal of Educational and Behavioral Statistics.* He co-edited the volumes *Structural Equation Modeling: A Second Course* and *Advances in Latent Variable Mixture Models.*

**Amy Hendrickson** (ahendrickson@collegeboard.org) is Associate Psychometrician, Research and Development, at the College Board. Previously, she was Assistant Professor in the Department of Measurement, Statistics and Evaluation at the University of Maryland. She is currently working on the review and redesign of the Advanced Placement examinations, including the design of generalizability studies for assessing rater reliability. Her other areas of interest include equating, large-scale assessment, test specifications, and item response theory, and her work has appeared in such journals as *Educational Measurement: Issues and Practice, Journal of Educational Measurement,* and *Applied Measurement in Education.*

**William T. Hoyt** (wthoyt@education.wisc.edu) is Professor and Training Director of the Ph.D. program, Department of Counseling Psychology, University of Wisconsin-Madison. His methodological interests include reliability and validity of observer ratings in research in the behavioral sciences, and his publications in this area have appeared in *Psychological Methods, Journal of Counseling Psychology, Journal of Clinical Child and Adolescent Psychology,* and *The Counseling Psychologist.* He is currently Associate Editor of the *Journal of Counseling Psychology.*

**Carl J Huberty** (chuberty@uga.edu) is Professor Emeritus at the University of Georgia, having retired in 2002 after being there since 1969. He taught all levels of statistical methods courses. His instructional focus was on multivariate methods. This focus is reflected in his 1994 textbook, *Applied Discriminant Analysis,* and its second edition (as senior author) in 2006. He has authored 15 book chapters and 58 journal articles. Journal writings have appeared in *Multivariate Behavioral Research, Journal of Educational Statistics, Educational and Psychological Measurement, Review of Educational Research, Psychological Bulletin,* and *Teaching Statistics.* He was very active in the American Educational Research Association, and the American Statistical Association.

**Ken Kelley** (kkelley@nd.edu) is Assistant Professor in the Department of Management at the University of Notre Dame. His research interests lie in quantitative methodology with a focus on sample size planning, effect size estimation, and confidence interval formation. His contributions include the development and implementation of sample size planning for statistical power and accuracy in parameter estimation for effect sizes commonly used in the managerial, behavioral, educational, and social sciences. He is the author of the MBESS R package. Currently, he serves on the editorial board of *Psychological Methods.*

**H. J. Keselman** (kesel@cc.umanitoba.ca) is Professor and Head of Psychology at the University of Manitoba, Winnipeg, Manitoba, Canada. He is a fellow of the American Educational Research Association, the American Psychological Association, and the American Psychological Society. He has published over 150 journal articles and book chapters related to his areas of interest: the analysis of repeated measurements, multiple comparison procedures, and robust estimation and testing. His publications have appeared in such journals as *British Journal of Mathematical and Statistical*

*Psychology, Educational and Psychological Measurement, Journal of Educational and Behavioral Statistics, Psychological Methods, Psychometrika, Psychophysiology,* and *Statistics in Medicine.*

**Se-Kang Kim** (sekim@fordham.edu) is Assistant Professor and Director of the Psychometrics Program in the Department of Psychology at Fordham University. His research and teaching interests include profile analysis utilizing multidimensional scaling, dual scaling/correspondence analysis, and factor analysis. He serves as Associate Editor for *Applied Psychological Measurement* and Consulting Editor for *Psychological Assessment.*

**Alan J. Klockars** (klockars@u.washington.edu) is Professor of Education at the University of Washington in the area of Measurement, Statistics, and Research Design. A major area of interest centers around the application of statistical methods to experimental designs with papers focusing on topics including multiple comparisons and the analysis of trait×treatment interactions. His work appears in such journals as *Psychological Bulletin, Psychological Methods, Journal of Educational and Behavioral Statistics, Journal of Modern Applied Statistical Methods, Journal of Educational Measurement,* and *Educational and Psychological Measurement.* He is Assistant Editor of the *Journal of Modern Applied Statistical Methods.*

**Thomas R. Knapp** (tknapp5@juno.com) is Professor Emeritus of Education and Nursing, University of Rochester and the Ohio State University. His specialty is the reliability of measuring instruments. He has published five books and about 100 articles in peer-reviewed journals such as *American Educational Research Journal, Educational and Psychological Measurement, Psychological Bulletin, Journal of Educational Statistics, Nursing Research,* and *Research in Nursing & Health.* He served for many years as a reviewer for research journals in education and nursing.

**Timothy R. Konold** (tk2e@virginia.edu) is Associate Professor and Director of the Research, Statistics, and Evaluation program at the University of Virginia. His research seeks to infuse contemporary quantitative methods into work that has direct implications for policy and the education of children and youth. Recent publications with applications of advanced contemporary multivariate procedures focus on large-scale test use as it pertains to construction, interpretation, classification and errors of measurement; with particular focus on errors of measurement associated with informant based assessment systems. His work has been published in *Educational and Psychological Measurement* and *Multivariate Behavioral Research.*

**Lisa M. Lix** (lisa.lix@usask.ca) is Associate Professor and Centennial Chair of the School of Public Health, and Associate Member of the Department of Mathematics and Statistics, at the University of Saskatchewan. Her research interests include analysis of repeated measures/longitudinal data, multivariate statistics, and multiple testing. She collaborates widely on projects about population health and the association between chronic disease and quality of life. Her research has appeared in such journals as *Psychological Methods, Quality of Life Research, British Journal of Mathematical and Statistical Psychology,* and *Computer Methods and Programs in Biomedicine.* She has served on the Board of the Statistical Society of Canada since 2005.

**Richard G. Lomax** (glubke@nd.edu) is Professor of Education and Human Ecology at the Ohio State University. He focuses on multivariate analysis, specifically structural equation modeling, resulting in three textbooks (e.g., *A Beginner's Guide to Structural Equation Modeling*) and articles in such diverse journals as *Reading Research Quarterly, Parenting: Science and Practice, Structural Equation Modeling: A Multidisciplinary Journal, Journal of Experimental Education, The School Community Journal,*

*Journal of Early Adolescence, Research Quarterly for Exercise and Sport, Journal of Educational Psychology*, and *Research in the Teaching of English*. He twice served as a Fulbright Scholar, teaching the first SEM course in Estonia.

**Gitta Lubke** (glubke@nd.edu) is Associate Professor of Quantitative Psychology at the University of Notre Dame. Her area of research is in the field of general latent variable modeling and cluster analysis. In addition to the analysis of complex phenotypes (e.g., attention problems), she is interested in the analysis of genetic data. A new topic is the development of multivariate methods for genome-wide association studies (GWAS). Other areas of expertise include twin models, measurement invariance, multi-group factor analysis, longitudinal analyses, and the analysis of categorical data. She serves on the editorial boards of *Psychological Methods* and *Structural Equation Modeling: A Multidisciplinary Journal*, and is Associate Editor of *Multivariate Behavioral Research*.

**Scott E. Maxwell** (smaxwell@nd.edu) is Fitzsimons Professor of Psychology at the University of Notre Dame, where he served as department chair from 1991 to 1994 and 1995 to 1998. His research interests are in the areas of experimental design and behavioral statistics, with special interests in statistical power analysis and longitudinal data analysis. He is an author (with Harold D. Delaney) of *Designing Experiments and Analyzing Data*. He served as Associate Editor of *Psychological Bulletin* from 1994 to 1996 and as Associate Editor of *Psychological Methods* from 2001 to 2007. He is currently serving as Editor of *Psychological Methods*.

**D. Betsy McCoach** (betsy.mccoach@uconn.edu) is Associate Professor of Educational Psychology in the Neag School of Education at the University of Connecticut, where she teaches courses in hierarchical linear modeling, structural equation modeling, and quantitative research methods. Her areas of interest include multilevel modeling, structural equation modeling, affective instrument design, and gifted education. She is the co-editor of *Multilevel Modeling of Educational Data*. In addition, she is co-editor of the *Journal of Advanced Academics*. She serves on the editorial boards of several journals, including *American Educational Research Journal, Journal of Educational Psychology, Journal of Educational Research*, and *Gifted Child Quarterly*.

**Ralph O. Mueller** (rmueller@hartford.edu) is Dean of the College of Education, Nursing and Health Professions, and Professor of Educational Leadership and of Psychology at the University of Hartford. Until 2009, he served as Professor of Educational Research, Public Policy and Public Administration and Chair of the Department of Educational Leadership at The George Washington University. His writings include textbooks, chapters, and articles on proper applications of multivariate statistical techniques, especially structural equation modeling. Among others, he serves on the editorial boards of *Educational and Psychological Measurement, Measurement and Evaluation in Counseling and Development*, and was a charter member of the editorial board of *Structural Equation Modeling: A Multidisciplinary Journal*.

**Kevin R. Murphy** (krmurphy@psu.edu) is Professor of Psychology and Information Sciences and Technology at the Pennsylvania State University. He served as Editor of the *Journal of Applied Psychology*, has served on four National Academy of Sciences committees, and is the recipient of the Society for Industrial and Organizational Psychology's 2004 Distinguished Scientific Contribution Award. He is the author of over 150 articles and book chapters, and author or editor of eleven books, in areas ranging from psychometrics and statistical analysis to individual differences and performance assessment honesty in the workplace.

**Ann A. O'Connell** (aoconnell@ehe.osu.edu) is Associate Professor in the School of Educational Policy and Leadership at the Ohio State University. Her interests are in appropriate application of multivariate and multilevel methodologies for education and health intervention research, particularly for analyzing discrete or non-normally distributed response variables. Her work has been published in *Evaluation and the Health Professions, Journal of Educational Psychology, Women and Health, Measurement and Research in Counseling and Development, Morbidity and Mortality Weekly Report,* and the *Journal of Modern Applied Statistical Methods.* She has written a book on ordinal regression models and co-edited the volume *Multilevel Modeling of Educational Data.*

**Stephen Olejnik** (olejnik@uga.edu) is Professor in the Department of Educational Psychology and Instructional Technology at the University of Georgia in Athens, Georgia. His areas of interest include applied univariate and multivariate statistical models with particular interest in effect size estimation, focused tests, and analysis of covariance. His research has appeared in such journals as *Psychological Methods, Journal of Educational and Behavioral Statistics, Multivariate Behavioral Research, Review of Educational Research,* and *British Journal of Mathematical and Statistical Psychology.* He co-authored the multivariate statistics textbook, *Applied MANOVA and Discriminant Analysis* (2nd ed.)

**Jason W. Osborne** (jason_osborne@ncsu.edu) is Associate Professor of Educational Psychology in the Department of Curriculum and Instruction at North Carolina State University and Senior Research Fellow for Evaluation at the William and Ida Friday Institute for Educational Innovation at the North Carolina State University. He writes on best practices in quantitative methods, and recently edited a handbook for Sage entitled *Best Practices in Quantitative Methods.* He researches issues relating to stereotype threat, self-concept, and motivation. He is the author of over 50 articles spanning the social and biomedical sciences.

**Dena A. Pastor** (pastorda@jmu.edu) has a dual appointment at James Madison University (JMU) as Associate Professor in the Department of Graduate Psychology and as Associate Assessment Specialist in the Center for Assessment and Research Studies. She received her doctoral degree in quantitative methods from the University of Texas in Austin in 2001 and has been teaching courses in measurement and statistics at JMU for over seven years. Her research, which has appeared in *Applied Psychological Measurement* and *Contemporary Educational Psychology,* typically involves the application of latent variable modeling to the study of college student learning and development.

**Kristopher J. Preacher** (preacher@ku.edu) is Assistant Professor of Quantitative Psychology, University of Kansas. His research concerns the use of structural equation modeling and multilevel modeling to analyze longitudinal and correlational data. Other interests include developing techniques to test mediation and moderation hypotheses, bridging the gap between theory and practice, and studying model evaluation and model selection in the application of multivariate methods to social science questions. He co-authored the book *Latent Growth Curve Modeling,* and serves on the editorial board of *Psychological Methods.*

**David Rindskopf** (drindskopf@gc.cuny.edu) is Distinguished Professor of Educational Psychology and Psychology at the City University of New York Graduate School. He is a Fellow of the American Statistical Association and the American Educational Research Association, Past President of the Society for Multivariate Experimental Psychology, and Editor of the *Journal of Educational and Behavioral Statistics.* His research interests are categorical data, latent variable models, and multilevel models. Current projects include (i) showing how people subconsciously use complex statistical methods to make decisions in everyday life, (ii) introducing floor and ceiling effects into logistic regression to model response probabilities constrained to a limited range, (iii) using multilevel models to analyze data from single case designs.

**Karen M. Samuelsen** (kmsamuelsen@yahoo.com) is Assistant Professor in the Research, Evaluation, Measurement, and Statistics program at the University of Georgia. She teaches graduate level courses in applied statistics, applied educational and psychological assessment methods, and measurement theory. Her research focuses on the validation of educational assessments and psychological measurement instruments, and on mixture models. Her research on validity has been the focus of two articles in *Educational Researcher*, a book chapter, another article, and several grant proposals. She has co-edited a volume titled *Advances in Latent Variable Mixture Models*.

**Michael A. Seaman** (mseaman@mailbox.sc.edu) is Associate Professor of Educational Research and Associate Dean for Administration and Research in the College of Education at the University of South Carolina. His research interests are nonparametric statistics and statistics applied to educational problems. His writings have appeared in *Psychological Bulletin*, *Psychological Methods*, *Communications in Statistics*, *Educational and Psychological Measurement*, *Journal of Educational Measurement*, and *The Journal of Experimental Education*.

**Ronald C. Serlin** (rcserlin@wisc.edu) is Professor and former Chair, Department of Educational Psychology at the University of Wisconsin-Madison. His areas of interest include nonparametric statistics and the philosophy of science and statistics, and his publications have appeared in such journals as *Psychological Methods*, *Journal of Modern Applied Statistical Methods*, *Journal of Consulting and Clinical Psychology*, *British Journal of Mathematical and Statistical Psychology*, *American Psychologist*, and *Journal of Educational and Behavioral Statistics*. He co-authored the text *Statistical Methods for the Social and Behavioral Sciences* and co-edited the volume *The SAGE Handbook for Research in Education: Engaging Ideas and Enriching Inquiry*.

**Laura M. Stapleton** (lstaplet@umbc.edu) is Associate Professor of quantitative psychology in the Department of Psychology at the University of Maryland, Baltimore County, and teaches graduate courses in intermediate statistics, structural equation modeling, and measurement. Her research interests include multilevel latent variable models and the analysis of survey data obtained under complex sampling designs. Her scholarship in these areas has appeared in such journals as *Structural Equation Modeling: A Multidisciplinary Journal*, *Multivariate Behavioral Research*, and *Educational and Psychological Measurement*, as well as numerous book chapters. She serves on the editorial board of the *Journal of Educational Psychology*.

**Keith F. Widaman** (kfwidaman@ucdavis.edu) is Distinguished Professor, Department of Psychology at the University of California at Davis. His areas of interest include common factor analysis, linear and nonlinear regression and latent variable models, topics in measurement, and economic and family influences on psychological development in normal and abnormal populations. His writings have appeared in journals such as *Psychological Methods*, *Multivariate Behavioral Research*, *Structural Equation Modeling: A Multidisciplinary Journal*, *American Journal on Mental Retardation*, and *Child Development*, and he serves on editorial boards of several journals, including *Psychological Methods*, *Multivariate Behavioral Research*, *Structural Equation Modeling: A Multidisciplinary Journal*, *Intelligence*, and *Applied Psychological Measurement*.

**Ping Yin** (pyin@questarai.com) is Senior Psychometrician in the Research and Development Department at Questar Assessment, Inc. Her areas of interest include generalizability theory, equating and scaling, and standard setting. Her work has appeared in such journals as *Educational and Psychological Measurement*, *Applied Psychological Measurement*, *Journal of Educational Measurement*, *International Journal of Testing*, and *Journal of Experimental Education*.